Lecture Notes in Computer Science

Lecture Notes in Artificial Intelligence 14647

Founding Editor

Jörg Siekmann

The series Lecture Notes in Artificial Intelligence (LNAI) was established in 1988 as a topical subseries of LNCS devoted to artificial intelligence.

The series publishes state-of-the-art research results at a high level. As with the LNCS mother series, the mission of the series is to serve the international R & D community by providing an invaluable service, mainly focused on the publication of conference and workshop proceedings and postproceedings.

De-Nian Yang · Xing Xie · Vincent S. Tseng ·
Jian Pei · Jen-Wei Huang · Jerry Chun-Wei Lin
Editors

Advances in Knowledge Discovery and Data Mining

28th Pacific-Asia Conference on Knowledge Discovery and Data Mining, PAKDD 2024
Taipei, Taiwan, May 7–10, 2024
Proceedings, Part III

Editors
De-Nian Yang
Academia Sinica
Taipei, Taiwan

Vincent S. Tseng
National Yang Ming Chiao Tung University
Hsinchu, Taiwan

Jen-Wei Huang
National Cheng Kung University
Tainan, Taiwan

Xing Xie
Microsoft Research Asia
Beijing, China

Jian Pei
Duke University
Durham, NC, USA

Jerry Chun-Wei Lin
Silesian University of Technology
Gliwice, Poland

ISSN 0302-9743 ISSN 1611-3349 (electronic)
Lecture Notes in Artificial Intelligence
ISBN 978-981-97-2261-7 ISBN 978-981-97-2259-4 (eBook)
https://doi.org/10.1007/978-981-97-2259-4

LNCS Sublibrary: SL7 – Artificial Intelligence

This Springer imprint is published by the registered company Springer Nature Singapore Pte Ltd.
The registered company address is: 152 Beach Road, #21-01/04 Gateway East, Singapore 189721, Singapore

Paper in this product is recyclable.

General Chairs' Preface

On behalf of the Organizing Committee, we were delighted to welcome attendees to the 28th Pacific-Asia Conference on Knowledge Discovery and Data Mining (PAKDD 2024). Since its inception in 1997, PAKDD has long established itself as one of the leading international conferences on data mining and knowledge discovery. PAKDD provides an international forum for researchers and industry practitioners to share their new ideas, original research results, and practical development experiences across all areas of Knowledge Discovery and Data Mining (KDD). This year, after its two previous editions in Taipei (2002) and Tainan (2014), PAKDD was held in Taiwan for the third time in the fascinating city of Taipei, during May 7–10, 2024. Moreover, PAKDD 2024 was held as a fully physical conference since the COVID-19 pandemic was contained.

We extend our sincere gratitude to the researchers who submitted their work to the PAKDD 2024 main conference, high-quality tutorials, and workshops on cutting-edge topics. The conference program was further enriched with seven high-quality tutorials and five workshops on cutting-edge topics. We would like to deliver our sincere thanks for their efforts in research, as well as in preparing high-quality presentations. We also express our appreciation to all the collaborators and sponsors for their trust and cooperation. We were honored to have three distinguished keynote speakers joining the conference: Ed H. Chi (Google DeepMind), Vipin Kumar (University of Minnesota), and Huan Liu (Arizona State University), each with high reputations in their respective areas. We enjoyed their participation and talks, which made the conference one of the best academic platforms for knowledge discovery and data mining. We would like to express our sincere gratitude for the contributions of the Steering Committee members, Organizing Committee members, Program Committee members, and anonymous reviewers, led by Program Committee Chairs De-Nian Yang and Xing Xie. It is through their untiring efforts that the conference had an excellent technical program. We are also thankful to the other Organizing Committee members: Workshop Chairs, Chuan-Kang Ting and Xiaoli Li; Tutorial Chairs, Jiun-Long Huang and Philippe Fournier-Viger; Publicity Chairs, Mi-Yen Yeh and Rage Uday Kiran; Industrial Chairs, Kun-Ta Chuang, Wei-Chao Chen and Richie Tsai; Proceedings Chairs, Jen-Wei Huang and Jerry Chun-Wei Lin; Registration Chairs, Chih-Ya Shen and Hong-Han Shuai; Web and Content Chairs, Cheng-Te Li and Shan-Hung Wu; Local Arrangement Chairs, Yi-Ling Chen, Kuan-Ting Lai, Yi-Ting Chen, and Ya-Wen Teng. We feel indebted to the PAKDD Steering Committee for their constant guidance and sponsorship of manuscripts. We are also grateful to the hosting organizations, National Yang Ming Chiao Tung University and Academia Sinica, and all our sponsors for continuously providing institutional and financial support to PAKDD 2024.

May 2024

Vincent S. Tseng
Jian Pei

General Chairs' Preface

PC Chairs' Preface

It is our great pleasure to present the 28th Pacific-Asia Conference on Knowledge Discovery and Data Mining (PAKDD 2024) as Program Committee Chairs. PAKDD is one of the longest-established and leading international conferences in the areas of data mining and knowledge discovery. It provides an international forum for researchers and industry practitioners to share their new ideas, original research results, and practical development experiences in all KDD-related areas, including data mining, data warehousing, machine learning, artificial intelligence, databases, statistics, knowledge engineering, big data technologies, and foundations.

This year, PAKDD received a record number of 720 submissions, among which 86 submissions were rejected at a preliminary stage due to policy violations. There were 595 Program Committee members and 101 Senior Program Committee members involved in the double-blind reviewing process. For submissions entering the double-blind review process, each one received at least three quality reviews from PC members. Furthermore, each valid submission received one meta-review from the assigned SPC member, who also led the discussion with the PC members. The PC Co-chairs then considered the recommendations and meta-reviews from SPC members and looked into each submission as well as its reviews and PC discussions to make the final decision.

As a result of the highly competitive selection process, 175 submissions were accepted and recommended to be published, with 133 oral-presentation papers and 42 poster-presentation papers. We would like to thank all SPC and PC members whose diligence produced a high-quality program for PAKDD 2024. The conference program also featured three keynote speeches from distinguished data mining researchers, eight invited industrial talks, five cutting-edge workshops, and seven comprehensive tutorials.

We wish to sincerely thank all SPC members, PC members, and external reviewers for their invaluable efforts in ensuring a timely, fair, and highly effective paper review and selection procedure. We hope that readers of the proceedings will find the PAKDD 2024 technical program both interesting and rewarding.

May 2024

De-Nian Yang
Xing Xie

PC Chairs' Preface

Organization

Organizing Committee

Honorary Chairs

Philip S. Yu	University of Illinois at Chicago, USA
Ming-Syan Chen	National Taiwan University, Taiwan

General Chairs

Vincent S. Tseng	National Yang Ming Chiao Tung University, Taiwan
Jian Pei	Duke University, USA

Program Committee Chairs

De-Nian Yang	Academia Sinica, Taiwan
Xing Xie	Microsoft Research Asia, China

Workshop Chairs

Chuan-Kang Ting	National Tsing Hua University, Taiwan
Xiaoli Li	A*STAR, Singapore

Tutorial Chairs

Jiun-Long Huang	National Yang Ming Chiao Tung University, Taiwan
Philippe Fournier-Viger	Shenzhen University, China

Publicity Chairs

Mi-Yen Yeh	Academia Sinica, Taiwan
Rage Uday Kiran	University of Aizu, Japan

Industrial Chairs

Kun-Ta Chuang	National Cheng Kung University, Taiwan
Wei-Chao Chen	Inventec Corp./Skywatch Innovation, Taiwan
Richie Tsai	Taiwan AI Academy, Taiwan

Proceedings Chairs

Jen-Wei Huang	National Cheng Kung University, Taiwan
Jerry Chun-Wei Lin	Silesian University of Technology, Poland

Registration Chairs

Chih-Ya Shen	National Tsing Hua University, Taiwan
Hong-Han Shuai	National Yang Ming Chiao Tung University, Taiwan

Web and Content Chairs

Shan-Hung Wu	National Tsing Hua University, Taiwan
Cheng-Te Li	National Cheng Kung University, Taiwan

Local Arrangement Chairs

Yi-Ling Chen	National Taiwan University of Science and Technology, Taiwan
Kuan-Ting Lai	National Taipei University of Technology, Taiwan
Yi-Ting Chen	National Yang Ming Chiao Tung University, Taiwan
Ya-Wen Teng	Academia Sinica, Taiwan

Steering Committee

Chair

Longbing Cao	Macquarie University, Australia

Vice Chair

Gill Dobbie	University of Auckland, New Zealand

Treasurer

Members

Life Members

Ning Zhong	Maebashi Institute of Technology, Japan
Zhi-Hua Zhou	Nanjing University, China

Past Members

Arbee L. P. Chen	Asia University, Taiwan
Hongjun Lu	Hong Kong University of Science and Technology, Hong Kong SAR
Takao Terano	Tokyo Institute of Technology, Japan

Senior Program Committee

Aijun An	York University, Canada
Aris Anagnostopoulos	Sapienza Università di Roma, Italy
Ting Bai	Beijing University of Posts and Telecommunications, China
Elisa Bertino	Purdue University, USA
Arnab Bhattacharya	IIT Kanpur, India
Albert Bifet	Université Paris-Saclay, France
Ludovico Boratto	Università degli Studi di Cagliari, Italy
Ricardo Campello	University of Southern Denmark, Denmark
Longbing Cao	University of Technology Sydney, Australia
Tru Cao	UTHealth, USA
Tanmoy Chakraborty	IIT Delhi, India
Jeffrey Chan	RMIT University, Australia
Pin-Yu Chen	IBM T. J. Watson Research Center, USA
Bin Cui	Peking University, China
Anirban Dasgupta	IIT Gandhinagar, India
Wei Ding	University of Massachusetts Boston, USA
Eibe Frank	University of Waikato, New Zealand
Chen Gong	Nanjing University of Science and Technology, China
Jingrui He	UIUC, USA
Tzung-Pei Hong	National University of Kaohsiung, Taiwan
Qinghua Hu	Tianjin University, China
Hong Huang	Huazhong University of Science and Technology, China
Jen-Wei Huang	National Cheng Kung University, Taiwan
Tsuyoshi Ide	IBM T. J. Watson Research Center, USA
Xiaowei Jia	University of Pittsburgh, USA
Zhe Jiang	University of Florida, USA

Toshihiro Kamishima	National Institute of Advanced Industrial Science and Technology, Japan
Murat Kantarcioglu	University of Texas at Dallas, USA
Hung-Yu Kao	National Cheng Kung University, Taiwan
Kamalakar Karlapalem	IIIT Hyderabad, India
Anuj Karpatne	Virginia Tech, USA
Hisashi Kashima	Kyoto University, Japan
Sang-Wook Kim	Hanyang University, Korea
Yun Sing Koh	University of Auckland, New Zealand
Hady Lauw	Singapore Management University, Singapore
Byung Suk Lee	University of Vermont, USA
Jae-Gil Lee	KAIST, Korea
Wang-Chien Lee	Pennsylvania State University, USA
Chaozhuo Li	Microsoft Research Asia, China
Gang Li	Deakin University, Australia
Jiuyong Li	University of South Australia, Australia
Jundong Li	University of Virginia, USA
Ming Li	Nanjing University, China
Sheng Li	University of Virginia, USA
Ying Li	AwanTunai, Singapore
Yu-Feng Li	Nanjing University, China
Hao Liao	Shenzhen University, China
Ee-peng Lim	Singapore Management University, Singapore
Jerry Chun-Wei Lin	Silesian University of Technology, Poland
Shou-De Lin	National Taiwan University, Taiwan
Hongyan Liu	Tsinghua University, China
Wei Liu	University of Technology Sydney, Australia
Chang-Tien Lu	Virginia Tech, USA
Yuan Luo	Northwestern University, USA
Wagner Meira Jr.	UFMG, Brazil
Alexandros Ntoulas	University of Athens, Greece
Satoshi Oyama	Nagoya City University, Japan
Guansong Pang	Singapore Management University, Singapore
Panagiotis Papapetrou	Stockholm University, Sweden
Wen-Chih Peng	National Yang Ming Chiao Tung University, Taiwan
Dzung Phan	IBM T. J. Watson Research Center, USA
Uday Rage	University of Aizu, Japan
Rajeev Raman	University of Leicester, UK
P. Krishna Reddy	IIIT Hyderabad, India
Thomas Seidl	LMU München, Germany
Neil Shah	Snap Inc., USA

Yingxia Shao	Beijing University of Posts and Telecommunications, China
Victor S. Sheng	Texas Tech University, USA
Kyuseok Shim	Seoul National University, Korea
Arlei Silva	Rice University, USA
Jaideep Srivastava	University of Minnesota, USA
Masashi Sugiyama	RIKEN/University of Tokyo, Japan
Ju Sun	University of Minnesota, USA
Jiliang Tang	Michigan State University, USA
Hanghang Tong	UIUC, USA
Ranga Raju Vatsavai	North Carolina State University, USA
Hao Wang	Nanyang Technological University, Singapore
Hao Wang	Xidian University, China
Jianyong Wang	Tsinghua University, China
Tim Weninger	University of Notre Dame, USA
Raymond Chi-Wing Wong	Hong Kong University of Science and Technology, Hong Kong SAR
Jia Wu	Macquarie University, Australia
Xindong Wu	Hefei University of Technology, China
Xintao Wu	University of Arkansas, USA
Yiqun Xie	University of Maryland, USA
Yue Xu	Queensland University of Technology, Australia
Lina Yao	University of New South Wales, Australia
Han-Jia Ye	Nanjing University, China
Mi-Yen Yeh	Academia Sinica, Taiwan
Hongzhi Yin	University of Queensland, Australia
Min-Ling Zhang	Southeast University, China
Ping Zhang	Ohio State University, USA
Zhao Zhang	Hefei University of Technology, China
Zhongfei Zhang	Binghamton University, USA
Xiangyu Zhao	City University of Hong Kong, Hong Kong SAR
Yanchang Zhao	CSIRO, Australia
Jiayu Zhou	Michigan State University, USA
Xiao Zhou	Renmin University of China, China
Xiaofang Zhou	Hong Kong University of Science and Technology, Hong Kong SAR
Feida Zhu	Singapore Management University, Singapore
Fuzhen Zhuang	Beihang University, China

Program Committee

Zubin Abraham	Robert Bosch, USA
Pedro Henriques Abreu	CISUC, Portugal
Muhammad Abulaish	South Asian University, India
Bijaya Adhikari	University of Iowa, USA
Karan Aggarwal	Amazon, USA
Chowdhury Farhan Ahmed	University of Dhaka, Bangladesh
Ulrich Aïvodji	ÉTS Montréal, Canada
Esra Akbas	Georgia State University, USA
Shafiq Alam	Massey University Auckland, New Zealand
Giuseppe Albi	Università degli Studi di Pavia, Italy
David Anastasiu	Santa Clara University, USA
Xiang Ao	Chinese Academy of Sciences, China
Elena-Simona Apostol	Uppsala University, Sweden
Sunil Aryal	Deakin University, Australia
Jees Augustine	Microsoft, USA
Konstantin Avrachenkov	Inria, France
Goonmeet Bajaj	Ohio State University, USA
Jean Paul Barddal	PUCPR, Brazil
Srikanta Bedathur	IIT Delhi, India
Sadok Ben Yahia	University of Southern Denmark, Denmark
Alessandro Berti	Università di Pisa, Italy
Siddhartha Bhattacharyya	University of Illinois at Chicago, USA
Ranran Bian	University of Sydney, Australia
Song Bian	Chinese University of Hong Kong, Hong Kong SAR
Giovanni Maria Biancofiore	Politecnico di Bari, Italy
Fernando Bobillo	University of Zaragoza, Spain
Adrian M. P. Brasoveanu	Modul Technology GmbH, Austria
Krisztian Buza	Budapest University of Technology and Economics, Hungary
Luca Cagliero	Politecnico di Torino, Italy
Jean-Paul Calbimonte	University of Applied Sciences and Arts Western Switzerland, Switzerland
K. Selçuk Candan	Arizona State University, USA
Fuyuan Cao	Shanxi University, China
Huiping Cao	New Mexico State University, USA
Jian Cao	Shanghai Jiao Tong University, China
Yan Cao	University of Texas at Dallas, USA
Yang Cao	Hokkaido University, Japan
Yuanjiang Cao	Macquarie University, Australia

Sharma Chakravarthy	University of Texas at Arlington, USA
Harry Kai-Ho Chan	University of Sheffield, UK
Zhangming Chan	Alibaba Group, China
Snigdhansu Chatterjee	University of Minnesota, USA
Mandar Chaudhary	eBay, USA
Chen Chen	University of Virginia, USA
Chun-Hao Chen	National Kaohsiung University of Science and Technology, Taiwan
Enhong Chen	University of Science and Technology of China, China
Fanglan Chen	Virginia Tech, USA
Feng Chen	University of Texas at Dallas, USA
Hongyang Chen	Zhejiang Lab, China
Jia Chen	University of California Riverside, USA
Jinjun Chen	Swinburne University of Technology, Australia
Lingwei Chen	Wright State University, USA
Ping Chen	University of Massachusetts Boston, USA
Shang-Tse Chen	National Taiwan University, Taiwan
Shengyu Chen	University of Pittsburgh, USA
Songcan Chen	Nanjing University of Aeronautics and Astronautics, China
Tao Chen	China University of Geosciences, China
Tianwen Chen	Hong Kong University of Science and Technology, Hong Kong SAR
Tong Chen	University of Queensland, Australia
Weitong Chen	University of Adelaide, Australia
Yi-Hui Chen	Chang Gung University, Taiwan
Yile Chen	Nanyang Technological University, Singapore
Yi-Ling Chen	National Taiwan University of Science and Technology, Taiwan
Yi-Shin Chen	National Tsing Hua University, Taiwan
Yi-Ting Chen	National Yang Ming Chiao Tung University, Taiwan
Zheng Chen	Osaka University, Japan
Zhengzhang Chen	NEC Laboratories America, USA
Zhiyuan Chen	UMBC, USA
Zhong Chen	Southern Illinois University, USA
Peng Cheng	East China Normal University, China
Abdelghani Chibani	Université Paris-Est Créteil, France
Jingyuan Chou	University of Virginia, USA
Lingyang Chu	McMaster University, Canada
Kun-Ta Chuang	National Cheng Kung University, Taiwan

Robert Churchill	Georgetown University, USA
Chaoran Cui	Shandong University of Finance and Economics, China
Alfredo Cuzzocrea	Università della Calabria, Italy
Bi-Ru Dai	National Taiwan University of Science and Technology, Taiwan
Honghua Dai	Zhengzhou University, China
Claudia d'Amato	University of Bari, Italy
Chuangyin Dang	City University of Hong Kong, China
Mrinal Das	IIT Palakkad, India
Debanjan Datta	Virginia Tech, USA
Cyril de Runz	Université de Tours, France
Jeremiah Deng	University of Otago, New Zealand
Ke Deng	RMIT University, Australia
Zhaohong Deng	Jiangnan University, China
Anne Denton	North Dakota State University, USA
Shridhar Devamane	KLE Institute of Technology, India
Djellel Difallah	New York University, USA
Ling Ding	Tianjin University, China
Shifei Ding	China University of Mining and Technology, China
Yao-Xiang Ding	Zhejiang University, China
Yifan Ding	University of Notre Dame, USA
Ying Ding	University of Texas at Austin, USA
Lamine Diop	EPITA, France
Nemanja Djuric	Aurora Innovation, USA
Gillian Dobbie	University of Auckland, New Zealand
Josep Domingo-Ferrer	Universitat Rovira i Virgili, Spain
Bo Dong	Amazon, USA
Yushun Dong	University of Virginia, USA
Bo Du	Wuhan University, China
Silin Du	Tsinghua University, China
Jiuding Duan	Allianz Global Investors, Japan
Lei Duan	Sichuan University, China
Walid Durani	LMU München, Germany
Sourav Dutta	Huawei Research Centre, Ireland
Mohamad El-Hajj	MacEwan University, Canada
Ya Ju Fan	Lawrence Livermore National Laboratory, USA
Zipei Fan	Jilin University, China
Majid Farhadloo	University of Minnesota, USA
Fabio Fassetti	Università della Calabria, Italy
Zhiquan Feng	National Cheng Kung University, Taiwan

Len Feremans	Universiteit Antwerpen, Belgium
Edouard Fouché	Karlsruher Institut für Technologie, Germany
Dongqi Fu	UIUC, USA
Yanjie Fu	University of Central Florida, USA
Ken-ichi Fukui	Osaka University, Japan
Matjaž Gams	Jožef Stefan Institute, Slovenia
Amir Gandomi	University of Technology Sydney, Australia
Aryya Gangopadhyay	UMBC, USA
Dashan Gao	Hong Kong University of Science and Technology, China
Wei Gao	Nanjing University, China
Yifeng Gao	University of Texas Rio Grande Valley, USA
Yunjun Gao	Zhejiang University, China
Paolo Garza	Politecnico di Torino, Italy
Chang Ge	University of Minnesota, USA
Xin Geng	Southeast University, China
Flavio Giobergia	Politecnico di Torino, Italy
Rosalba Giugno	Università degli Studi di Verona, Italy
Aris Gkoulalas-Divanis	Merative, USA
Djordje Gligorijevic	Temple University, USA
Daniela Godoy	UNICEN, Argentina
Heitor Gomes	Victoria University of Wellington, New Zealand
Maciej Grzenda	Warsaw University of Technology, Poland
Lei Gu	Nanjing University of Posts and Telecommunications, China
Yong Guan	Iowa State University, USA
Riccardo Guidotti	Università di Pisa, Italy
Ekta Gujral	University of California Riverside, USA
Guimu Guo	Rowan University, USA
Ting Guo	University of Technology Sydney, Australia
Xingzhi Guo	Stony Brook University, USA
Ch. Md. Rakin Haider	Purdue University, USA
Benjamin Halstead	University of Auckland, New Zealand
Jinkun Han	Georgia State University, USA
Lu Han	Nanjing University, China
Yufei Han	Inria, France
Daisuke Hatano	RIKEN, Japan
Kohei Hatano	Kyushu University/RIKEN AIP, Japan
Shogo Hayashi	BizReach, Japan
Erhu He	University of Pittsburgh, USA
Guoliang He	Wuhan University, China
Pengfei He	Michigan State University, USA

Yi He	Old Dominion University, USA
Shen-Shyang Ho	Rowan University, USA
William Hsu	Kansas State University, USA
Haoji Hu	University of Minnesota, USA
Hongsheng Hu	CSIRO, Australia
Liang Hu	Tongji University, China
Shizhe Hu	Zhengzhou University, China
Wei Hu	Nanjing University, China
Mengdi Huai	Iowa State University, USA
Chao Huang	University of Hong Kong, Hong Kong SAR
Congrui Huang	Microsoft, China
Guangyan Huang	Deakin University, Australia
Jimmy Huang	York University, Canada
Jinbin Huang	Hong Kong Baptist University, Hong Kong SAR
Kai Huang	Hong Kong University of Science and Technology, China
Ling Huang	South China Agricultural University, China
Ting-Ji Huang	Nanjing University, China
Xin Huang	Hong Kong Baptist University, Hong Kong SAR
Zhenya Huang	University of Science and Technology of China, China
Chih-Chieh Hung	National Chung Hsing University, Taiwan
Hui-Ju Hung	Pennsylvania State University, USA
Nam Huynh	JAIST, Japan
Akihiro Inokuchi	Kwansei Gakuin University, Japan
Atsushi Inoue	Eastern Washington University, USA
Nevo Itzhak	Ben-Gurion University, Israel
Tomoya Iwakura	Fujitsu Laboratories Ltd., Japan
Divyesh Jadav	IBM T. J. Watson Research Center, USA
Shubham Jain	Visa Research, USA
Bijay Prasad Jaysawal	National Cheng Kung University, Taiwan
Kishlay Jha	University of Iowa, USA
Taoran Ji	Texas A&M University - Corpus Christi, USA
Songlei Jian	NUDT, China
Gaoxia Jiang	Shanxi University, China
Hansi Jiang	SAS Institute Inc., USA
Jiaxin Jiang	National University of Singapore, Singapore
Min Jiang	Xiamen University, China
Renhe Jiang	University of Tokyo, Japan
Yuli Jiang	Chinese University of Hong Kong, Hong Kong SAR
Bo Jin	Dalian University of Technology, China

Ming Jin	Monash University, Australia
Ruoming Jin	Kent State University, USA
Wei Jin	University of North Texas, USA
Mingxuan Ju	University of Notre Dame, USA
Wei Ju	Peking University, China
Vana Kalogeraki	Athens University of Economics and Business, Greece
Bo Kang	Ghent University, Belgium
Jian Kang	University of Rochester, USA
Ashwin Viswanathan Kannan	Amazon, USA
Tomi Kauppinen	Aalto University School of Science, Finland
Jungeun Kim	Kongju National University, Korea
Kyoung-Sook Kim	National Institute of Advanced Industrial Science and Technology, Japan
Primož Kocbek	University of Maribor, Slovenia
Aritra Konar	Katholieke Universiteit Leuven, Belgium
Youyong Kong	Southeast University, China
Olivera Kotevska	Oak Ridge National Laboratory, USA
P. Radha Krishna	NIT Warangal, India
Adit Krishnan	UIUC, USA
Gokul Krishnan	IIT Madras, India
Peer Kröger	CAU, Germany
Marzena Kryszkiewicz	Warsaw University of Technology, Poland
Chuan-Wei Kuo	National Yang Ming Chiao Tung University, Taiwan
Kuan-Ting Lai	National Taipei University of Technology, Taiwan
Long Lan	NUDT, China
Duc-Trong Le	Vietnam National University, Vietnam
Tuan Le	New Mexico State University, USA
Chul-Ho Lee	Texas State University, USA
Ickjai Lee	James Cook University, Australia
Ki Yong Lee	Sookmyung Women's University, Korea
Ki-Hoon Lee	Kwangwoon University, Korea
Roy Ka-Wei Lee	Singapore University of Technology and Design, Singapore
Yue-Shi Lee	Ming Chuan University, Taiwan
Dino Lenco	INRAE, France
Carson Leung	University of Manitoba, Canada
Boyu Li	University of Technology Sydney, Australia
Chaojie Li	University of New South Wales, Australia
Cheng-Te Li	National Cheng Kung University, Taiwan
Chongshou Li	Southwest Jiaotong University, China

Fengxin Li	Renmin University of China, China
Guozhong Li	King Abdullah University of Science and Technology, Saudi Arabia
Huaxiong Li	Nanjing University, China
Jianxin Li	Beihang University, China
Lei Li	Hong Kong University of Science and Technology (Guangzhou), China
Peipei Li	Hefei University of Technology, China
Qian Li	Curtin University, Australia
Rong-Hua Li	Beijing Institute of Technology, China
Shao-Yuan Li	Nanjing University of Aeronautics and Astronautics, China
Shuai Li	Cambridge University, UK
Shuang Li	Beijing Institute of Technology, China
Tianrui Li	Southwest Jiaotong University, China
Wengen Li	Tongji University, China
Wentao Li	Hong Kong University of Science and Technology (Guangzhou), China
Xin-Ye Li	Bytedance, China
Xiucheng Li	Harbin Institute of Technology, China
Xuelong Li	Northwestern Polytechnical University, China
Yidong Li	Beijing Jiaotong University, China
Yinxiao Li	Meta Platforms, USA
Yuefeng Li	Queensland University of Technology, Australia
Yun Li	Nanjing University of Posts and Telecommunications, China
Panagiotis Liakos	University of Athens, Greece
Xiang Lian	Kent State University, USA
Shen Liang	Université Paris Cité, France
Qing Liao	Harbin Institute of Technology (Shenzhen), China
Sungsu Lim	Chungnam National University, Korea
Dandan Lin	Shenzhen Institute of Computing Sciences, China
Yijun Lin	University of Minnesota, USA
Ying-Jia Lin	National Cheng Kung University, Taiwan
Baodi Liu	China University of Petroleum (East China), China
Chien-Liang Liu	National Yang Ming Chiao Tung University, Taiwan
Guiquan Liu	University of Science and Technology of China, China
Jin Liu	Shanghai Maritime University, China
Jinfei Liu	Emory University, USA
Kunpeng Liu	Portland State University, USA

Ning Liu	Shandong University, China
Qi Liu	University of Science and Technology of China, China
Qing Liu	Zhejiang University, China
Qun Liu	Louisiana State University, USA
Shenghua Liu	Chinese Academy of Sciences, China
Weifeng Liu	China University of Petroleum (East China), China
Yang Liu	Wilfrid Laurier University, Canada
Yao Liu	University of New South Wales, Australia
Yixin Liu	Monash University, Australia
Zheng Liu	Nanjing University of Posts and Telecommunications, China
Cheng Long	Nanyang Technological University, Singapore
Haibing Lu	Santa Clara University, USA
Wenpeng Lu	Qilu University of Technology, China
Simone Ludwig	North Dakota State University, USA
Dongsheng Luo	Florida International University, USA
Ping Luo	Chinese Academy of Sciences, China
Wei Luo	Deakin University, Australia
Xiao Luo	UCLA, USA
Xin Luo	Shandong University, China
Yong Luo	Wuhan University, China
Fenglong Ma	Pennsylvania State University, USA
Huifang Ma	Northwest Normal University, China
Jing Ma	Hong Kong Baptist University, Hong Kong SAR
Qianli Ma	South China University of Technology, China
Yi-Fan Ma	Nanjing University, China
Rich Maclin	University of Minnesota, USA
Son Mai	Queen's University Belfast, UK
Arun Maiya	Institute for Defense Analyses, USA
Bradley Malin	Vanderbilt University Medical Center, USA
Giuseppe Manco	Consiglio Nazionale delle Ricerche, Italy
Naresh Manwani	IIIT Hyderabad, India
Francesco Marcelloni	Università di Pisa, Italy
Leandro Marinho	UFCG, Brazil
Koji Maruhashi	Fujitsu Laboratories Ltd., Japan
Florent Masseglia	Inria, France
Mohammad Masud	United Arab Emirates University, United Arab Emirates
Sarah Masud	IIIT Delhi, India
Costas Mavromatis	University of Minnesota, USA

Mario Prado-Romero	Gran Sasso Science Institute, Italy
Bardh Prenkaj	Sapienza Università di Roma, Italy
Jianzhong Qi	University of Melbourne, Australia
Buyue Qian	Xi'an Jiaotong University, China
Huajie Qian	Columbia University, USA
Hezhe Qiao	Singapore Management University, Singapore
Biao Qin	Renmin University of China, China
Zengchang Qin	Beihang University, China
Tho Quan	Ho Chi Minh City University of Technology, Vietnam
Miloš Radovanović	University of Novi Sad, Serbia
Thilina Ranbaduge	Australian National University, Australia
Chotirat Ratanamahatana	Chulalongkorn University, Thailand
Chandra Reddy	IBM T. J. Watson Research Center, USA
Ryan Rossi	Adobe Research, USA
Morteza Saberi	University of Technology Sydney, Australia
Akira Sakai	Fujitsu Laboratories Ltd., Japan
David Sánchez	Universitat Rovira i Virgili, Spain
Maria Luisa Sapino	Università degli Studi di Torino, Italy
Hernan Sarmiento	UChile & IMFD, Chile
Badrul Sarwar	CloudAEye, USA
Nader Shakibay Senobari	University of California Riverside, USA
Nasrin Shabani	Macquarie University, Australia
Ankit Sharma	University of Minnesota, USA
Chandra N. Shekar	RGUKT RK Valley, India
Chih-Ya Shen	National Tsing Hua University, Taiwan
Wei Shen	Nankai University, China
Yu Shen	Peking University, China
Zhi-Yu Shen	Nanjing University, China
Chuan Shi	Beijing University of Posts and Telecommunications, China
Yue Shi	Meta Platforms, USA
Zhenwei Shi	Beihang University, China
Motoki Shiga	Tohoku University, Japan
Kijung Shin	KAIST, Korea
Kai Shu	Illinois Institute of Technology, USA
Hong-Han Shuai	National Yang Ming Chiao Tung University, Taiwan
Zeren Shui	University of Minnesota, USA
Satyaki Sikdar	Indiana University, USA
Dan Simovici	University of Massachusetts Boston, USA
Apoorva Singh	IIT Patna, India

Bikash Chandra Singh	Islamic University, Bangladesh
Stavros Sintos	University of Illinois at Chicago, USA
Krishnamoorthy Sivakumar	Washington State University, USA
Andrzej Skowron	University of Warsaw, Poland
Andy Song	RMIT University, Australia
Dongjin Song	University of Connecticut, USA
Arnaud Soulet	Université de Tours, France
Ja-Hwung Su	National University of Kaohsiung, Taiwan
Victor Suciu	University of Wisconsin, USA
Liang Sun	Alibaba Group, USA
Xin Sun	Technische Universität München, Germany
Yuqing Sun	Shandong University, China
Hirofumi Suzuki	Fujitsu Laboratories Ltd., Japan
Anika Tabassum	Oak Ridge National Laboratory, USA
Yasuo Tabei	RIKEN, Japan
Chih-Hua Tai	National Taipei University, Taiwan
Hiroshi Takahashi	NTT, Japan
Atsuhiro Takasu	National Institute of Informatics, Japan
Yanchao Tan	Fuzhou University, China
Chang Tang	China University of Geosciences, China
Lu-An Tang	NEC Laboratories America, USA
Qiang Tang	Luxembourg Institute of Science and Technology, Luxembourg
Yiming Tang	Hefei University of Technology, China
Ying-Peng Tang	Nanjing University of Aeronautics and Astronautics, China
Xiaohui (Daniel) Tao	University of Southern Queensland, Australia
Vahid Taslimitehrani	PhysioSigns Inc., USA
Maguelonne Teisseire	INRAE, France
Ya-Wen Teng	Academia Sinica, Taiwan
Masahiro Terabe	Chugai Pharmaceutical Co. Ltd., Japan
Kia Teymourian	University of Texas at Austin, USA
Qing Tian	Nanjing University of Information Science and Technology, China
Yijun Tian	University of Notre Dame, USA
Maksim Tkachenko	Singapore Management University, Singapore
Yongxin Tong	Beihang University, China
Vicenç Torra	University of Umeå, Sweden
Nhu-Thuat Tran	Singapore Management University, Singapore
Yash Travadi	University of Minnesota, USA
Quoc-Tuan Truong	Amazon, USA

Yi-Ju Tseng	National Yang Ming Chiao Tung University, Taiwan
Turki Turki	King Abdulaziz University, Saudi Arabia
Ruo-Chun Tzeng	KTH Royal Institute of Technology, Sweden
Leong Hou U	University of Macau, Macau SAR
Jeffrey Ullman	Stanford University, USA
Rohini Uppuluri	Glassdoor, USA
Satya Valluri	Databricks, USA
Dinusha Vatsalan	Macquarie University, Australia
Bruno Veloso	FEP - University of Porto and INESC TEC, Portugal
Anushka Vidanage	Australian National University, Australia
Herna Viktor	University of Ottawa, Canada
Michalis Vlachos	University of Lausanne, Switzerland
Sheng Wan	Nanjing University of Science and Technology, China
Beilun Wang	Southeast University, China
Changdong Wang	Sun Yat-sen University, China
Chih-Hang Wang	Academia Sinica, Taiwan
Chuan-Ju Wang	Academia Sinica, Taiwan
Guoyin Wang	Chongqing University of Posts and Telecommunications, China
Hongjun Wang	Southwest Jiaotong University, China
Hongtao Wang	North China Electric Power University, China
Jianwu Wang	UMBC, USA
Jie Wang	Southwest Jiaotong University, China
Jin Wang	Megagon Labs, USA
Jingyuan Wang	Beihang University, China
Jun Wang	Shandong University, China
Lizhen Wang	Yunnan University, China
Peng Wang	Southeast University, China
Pengyang Wang	University of Macau, Macau SAR
Sen Wang	University of Queensland, Australia
Senzhang Wang	Central South University, China
Shoujin Wang	Macquarie University, Australia
Sibo Wang	Chinese University of Hong Kong, Hong Kong SAR
Suhang Wang	Pennsylvania State University, USA
Wei Wang	Fudan University, China
Wei Wang	Hong Kong University of Science and Technology (Guangzhou), China
Weicheng Wang	Hong Kong University of Science and Technology, Hong Kong SAR

Wei-Yao Wang	National Yang Ming Chiao Tung University, Taiwan
Wendy Hui Wang	Stevens Institute of Technology, USA
Xiao Wang	Beihang University, China
Xiaoyang Wang	University of New South Wales, Australia
Xin Wang	University of Calgary, Canada
Xinyuan Wang	George Mason University, USA
Yanhao Wang	East China Normal University, China
Yuanlong Wang	Ohio State University, USA
Yuping Wang	Xidian University, China
Yuxiang Wang	Hangzhou Dianzi University, China
Hua Wei	Arizona State University, USA
Zhewei Wei	Renmin University of China, China
Yimin Wen	Guilin University of Electronic Technology, China
Brendon Woodford	University of Otago, New Zealand
Cheng-Wei Wu	National Ilan University, Taiwan
Fan Wu	Central South University, China
Fangzhao Wu	Microsoft Research Asia, China
Jiansheng Wu	Nanjing University of Posts and Telecommunications, China
Jin-Hui Wu	Nanjing University, China
Jun Wu	UIUC, USA
Ou Wu	Tianjin University, China
Shan-Hung Wu	National Tsing Hua University, Taiwan
Shu Wu	Chinese Academy of Sciences, China
Wensheng Wu	University of Southern California, USA
Yun-Ang Wu	National Taiwan University, Taiwan
Wenjie Xi	George Mason University, USA
Lingyun Xiang	Changsha University of Science and Technology, China
Ruliang Xiao	Fujian Normal University, China
Yanghua Xiao	Fudan University, China
Sihong Xie	Lehigh University, USA
Zheng Xie	Nanjing University, China
Bo Xiong	Universität Stuttgart, Germany
Haoyi Xiong	Baidu, Inc., China
Bo Xu	Donghua University, China
Bo Xu	Dalian University of Technology, China
Guandong Xu	University of Technology Sydney, Australia
Hongzuo Xu	NUDT, China
Ji Xu	Guizhou University, China

Tong Xu	University of Science and Technology of China, China
Yuanbo Xu	Jilin University, China
Hui Xue	Southeast University, China
Qiao Xue	Nanjing University of Aeronautics and Astronautics, China
Akihiro Yamaguchi	Toshiba Corporation, Japan
Bo Yang	Jilin University, China
Liangwei Yang	University of Illinois at Chicago, USA
Liu Yang	Tianjin University, China
Shaofu Yang	Southeast University, China
Shiyu Yang	Guangzhou University, China
Wanqi Yang	Nanjing Normal University, China
Xiaoling Yang	Southwest Jiaotong University, China
Xiaowei Yang	South China University of Technology, China
Yan Yang	Southwest Jiaotong University, China
Yiyang Yang	Guangdong University of Technology, China
Yu Yang	City University of Hong Kong, Hong Kong SAR
Yu-Bin Yang	Nanjing University, China
Junjie Yao	East China Normal University, China
Wei Ye	Tongji University, China
Yanfang Ye	University of Notre Dame, USA
Kalidas Yeturu	IIT Tirupati, India
Ilkay Yildiz Potter	BioSensics LLC, USA
Minghao Yin	Northeast Normal University, China
Ziqi Yin	Nanyang Technological University, Singapore
Jia-Ching Ying	National Chung Hsing University, Taiwan
Tetsuya Yoshida	Nara Women's University, Japan
Hang Yu	Shanghai University, China
Jifan Yu	Tsinghua University, China
Yanwei Yu	Ocean University of China, China
Yongsheng Yu	Macquarie University, Australia
Long Yuan	Nanjing University of Science and Technology, China
Lin Yue	University of Newcastle, Australia
Xiaodong Yue	Shanghai University, China
Nayyar Zaidi	Monash University, Australia
Chengxi Zang	Cornell University, USA
Alexey Zaytsev	Skoltech, Russia
Yifeng Zeng	Northumbria University, UK
Petros Zerfos	IBM T. J. Watson Research Center, USA
De-Chuan Zhan	Nanjing University, China

Huixin Zhan	Texas Tech University, USA
Daokun Zhang	Monash University, Australia
Dongxiang Zhang	Zhejiang University, China
Guoxi Zhang	Beijing Institute of General Artificial Intelligence, China
Hao Zhang	Chinese University of Hong Kong, Hong Kong SAR
Huaxiang Zhang	Shandong Normal University, China
Ji Zhang	University of Southern Queensland, Australia
Jianfei Zhang	Université de Sherbrooke, Canada
Lei Zhang	Anhui University, China
Li Zhang	University of Texas Rio Grande Valley, USA
Lin Zhang	IDEA Education, China
Mengjie Zhang	Victoria University of Wellington, New Zealand
Nan Zhang	Wenzhou University, China
Quangui Zhang	Liaoning Technical University, China
Shichao Zhang	Central South University, China
Tianlin Zhang	University of Manchester, UK
Wei Emma Zhang	University of Adelaide, Australia
Wenbin Zhang	Florida International University, USA
Wentao Zhang	Mila, Canada
Xiaobo Zhang	Southwest Jiaotong University, China
Xuyun Zhang	Macquarie University, Australia
Yaqian Zhang	University of Waikato, New Zealand
Yikai Zhang	Guangzhou University, China
Yiqun Zhang	Guangdong University of Technology, China
Yudong Zhang	Nanjing Normal University, China
Zhiwei Zhang	Beijing Institute of Technology, China
Zike Zhang	Hangzhou Normal University, China
Zili Zhang	Southwest University, China
Chen Zhao	Baylor University, USA
Jiaqi Zhao	China University of Mining and Technology, China
Kaiqi Zhao	University of Auckland, New Zealand
Pengfei Zhao	BNU-HKBU United International College, China
Pengpeng Zhao	Soochow University, China
Ying Zhao	Tsinghua University, China
Zhongying Zhao	Shandong University of Science and Technology, China
Guanjie Zheng	Shanghai Jiao Tong University, China
Lecheng Zheng	UIUC, USA
Weiguo Zheng	Fudan University, China

Aoying Zhou	East China Normal University, China
Bing Zhou	Sam Houston State University, USA
Nianjun Zhou	IBM T. J. Watson Research Center, USA
Qinghai Zhou	UIUC, USA
Xiangmin Zhou	RMIT University, Australia
Xiaoping Zhou	Beijing University of Civil Engineering and Architecture, China
Xun Zhou	University of Iowa, USA
Jonathan Zhu	Wheaton College, USA
Ronghang Zhu	University of Georgia, China
Xingquan Zhu	Florida Atlantic University, USA
Ye Zhu	Deakin University, Australia
Yihang Zhu	University of Leicester, UK
Yuanyuan Zhu	Wuhan University, China
Ziwei Zhu	George Mason University, USA

External Reviewers

Zihan Li	University of Massachusetts Boston, USA
Ting Yu	Zhejiang Lab, China

Sponsoring Organizations

Accton

ACSI

Appier

Chunghwa Telecom Co., Ltd

DOIT, Taipei

ISCOM

Metaage

NSTC

Pegatron

Quanta Computer

TWS

Wavenet Co., Ltd

Contents – Part III

Semi-supervised and Unsupervised Learning

Big Data

Interpretability and Explainability

Neural Additive and Basis Models with Feature Selection and Interactions

Yasutoshi Kishimoto, Kota Yamanishi, Takuya Matsuda, and Shinichi Shirakawa(✉)

Graduate School of Environment and Information Sciences, Yokohama National University, Kanagawa, Japan
shirakawa-shinichi-bg@ynu.ac.jp

Abstract. Deep neural networks (DNNs) exhibit attractive performance in various fields but often suffer from low interpretability. The neural additive model (NAM) and its variant called the neural basis model (NBM) use neural networks (NNs) as nonlinear shape functions in generalized additive models (GAMs). Both models are highly interpretable and exhibit good performance and flexibility for NN training. NAM and NBM can provide and visualize the contribution of each feature to the prediction owing to GAM-based architectures. However, when using two-input NNs to consider feature interactions or when applying them to high-dimensional datasets, training NAM and NBM becomes intractable due to the increase in the computational resources required. This paper proposes incorporating the feature selection mechanism into NAM and NBM to resolve computational bottlenecks. We introduce the feature selection layer in both models and update the selection weights during training. Our method is simple and can reduce computational costs and model sizes compared to vanilla NAM and NBM. In addition, it enables us to use two-input NNs even in high-dimensional datasets and capture feature interactions. We demonstrate that the proposed models are computationally efficient compared to vanilla NAM and NBM, and they exhibit better or comparable performance with state-of-the-art GAMs.

Keywords: Neural additive models · Feature selection · Interpretable model

1 Introduction

Deep neural networks (DNNs) have become a standard tool in artificial intelligence owing to their high representation ability and flexible model training. DNNs have shown remarkable performance, such as in computer vision and natural language processing, and have become a promising model even for tabular datasets [3,8,20]. A well-known drawback of DNNs is the low interpretability and explainability of the prediction mechanism due to multi-layered nonlinear computations, making them difficult to use in applications that require high

D.-N. Yang et al. (Eds.): PAKDD 2024, LNAI 14647, pp. 3–16, 2024.
https://doi.org/10.1007/978-981-97-2259-4_1

reliability, such as medical and legal applications. Numerous methods have been examined to interpret trained DNNs, including locally approximating the decision boundary by an interpretable model [22] and visualizing important input image regions using gradient information [24]. However, these methods do not explain the global behavior of the models, and their explanations are often not faithful [23]. The alternative choice is to adopt interpretable models, such as shallow decision trees and linear models, which often suffer from low prediction accuracy.

The neural additive model (NAM) [1] is a type of generalized additive model (GAM) that possesses high interpretability and good prediction accuracy. GAMs are formed as the sum of nonlinear transformations for each input feature, referred to as shape functions. NAM uses one-input and one-output neural networks (NNs) as shape functions in GAMs and jointly trains all NNs through backpropagation. Because interactions between features do not exist in the shape functions, the global behavior of NAMs can be easily interpreted by visualizing shape functions. The neural basis model (NBM) [21] is another type of GAM, which uses a single shared NN for the basis of shape functions. The number of parameters in NBM can be significantly smaller than that in NAM because of the shared NN. Additionally, the implementation of NBM is simpler than that of NAM, making it easy to realize efficient implementation. Thanks to the NN shape functions, NAM and NBM can represent complex shape functions, which contributes to good prediction accuracy. They can also exploit the flexibility of DNN training, for example, extending it to multi-task or multi-label learning [1].

To enhance the prediction performance of GAMs without losing interpretability, adding pairwise interactions (two input shape functions) is promising. Such models are called GAMs plus interactions (GA^2Ms) [14]. Because the global behavior of two-input shape functions, such as a two-input NN, can be visualized using heat maps, GA^2Ms maintain high interpretability and are expected to improve performance. The naive extension method of NAM and NBM to consider pairwise feature interactions involves preparing the shape functions for all possible feature pairs. However, such a naive extension for high-dimensional datasets is intractable in NAM because the number of NNs and trainable parameters increases, leading to failure to run. Although the increase in the number of parameters of NBM is not as significant as that of NAM, the increase in the calculation cost becomes intractable. Even if only one-input NNs are used, the computations of NAM and NBM are still slow for high-dimensional ($> 1,000$) datasets. Additionally, increasing the shape functions may also compromise the interpretability of NAMs because considering and visualizing a large number of shape functions becomes unmanageable.

We incorporate a simple feature selection mechanism into NAM and NBM to reduce the number of shape functions and resolve their computational bottlenecks. The proposed method determines the features for one- and two-input NNs during model training. The feature selection allows us to reduce the computational complexity of NAM and NBM, making it possible to train NAM and NBM on high-dimensional datasets and easily incorporate pairwise interactions. We also discuss the model complexities of our models. We then compare the prediction performance of our models, termed NAM-FS and NBM-FS, with that of existing GAMs, other interpretable models, and non-interpretable but accurate models using high-dimensional classification datasets. The experimental results demonstrate that our models perform better or are comparable to other existing GAMs and GA^{2}Ms. We further compare the prediction performance of our models to vanilla NAM and NBM with pre-selected features by mutual information-based feature selection. Our models perform better than NAM and NBM with pre-selected features, demonstrating the effectiveness of the feature selection during model training.

Our contributions are as follows. (i) We incorporate a simple feature selection method into NAM and NBM, which enables us to apply them to high-dimensional datasets and consider pairwise interactions. (ii) We show that our proposed models achieve better throughput than vanilla NAM and NBM on high dimensions and better or comparable performance to other GAMs.

2 Generalized Additive Models (GAMs)

This section describes GAMs [13], NAM [1], and NBM [21]. GAMs are highly transparent due to their structure, which prepares one nonlinear function per input feature and uses the sum of the outputs of the nonlinear functions as the model's prediction. Given a D-dimensional input vector $\boldsymbol{x} \in \mathbb{R}^D$, the form of GAMs with a single output is given by $y(\boldsymbol{x}) = \sum_{i=1}^{D} f_i(x_i) + b$, where f_i is called the shape function of the i-th feature, which is an univariate nonlinear function, x_i is the i-th element of $\boldsymbol{x}$, and b is the bias term. Because each shape function value is decided by a single feature and the shape function values are added to make the model's output, the shape function value can be considered as its contribution to the model's output. Owing to this structure, we can interpret how each feature contributes to the model's prediction by showing the output of each nonlinear model. However, GAM cannot make predictions considering the interactions between features, which may lead to a degradation in prediction accuracy compared to complicated models.

Generalized additive models plus interactions (GA^{2}Ms) [14] are advanced versions of GAMs that consider interactions between features by adding bivariate shape functions to GAMs. The form of GA^{2}M is expressed as

$$y(\boldsymbol{x}) = \sum_{i=1}^{D} f_i(x_i) + \sum_{i=1}^{D} \sum_{j>i}^{D} f_{ij}(x_i, x_j) + b \ , \tag{1}$$

where f_{ij} is the bivariate nonlinear shape function of the (i, j)-feature pair. For two-input shape functions, we can visualize the input-output relationship using a heat map to interpret the prediction mechanism. For all interactions, there are $D(D-1)/2$ possible pairs in the D features; thus, the computational complexity increases with the number of features.

When applying GAMs to tasks with multiple outputs, such as multiclass classification, we can linearly combine shape functions to obtain multiple outputs as in [21]. The prediction function y_l for the l-th output is given by

$$y_l(\boldsymbol{x}) = \sum_{i=1}^{D} f_i(x_i) w_{li} + b_l \ , \tag{2}$$

where w_{li} and b_l indicate the coefficient of i-the shape function and the bias term for the l-th output, respectively. For GA^2M, we also introduce the coefficient for each bivariate shape function for the $(i,\ j)$-feature pair to support multiple outputs.

Several versions of GAMs and GA^2Ms have been proposed. Explainable boosting machine (EBM) [16] is a tree-based GAM and GA^2M. NODE-GAM [4] is based on neural oblivious decision ensembles (NODE) [20] that are ensembles of oblivious decision trees with gradient-based training. In the following, we describe the neural network versions of GAMs as our baseline models.

2.1 Neural Additive Model (NAM)

NAM [1], a variant of GAMs, uses NNs as nonlinear shape functions f_i. For the GAM form, a one-input and one-output NN f_i with trainable parameters is used for the i-th shape function. The training of NAM is performed by stochastic gradient descent (SGD), as in standard NNs, and all NNs composing the NAM are trained simultaneously. In NAM, the feature dropout and the output penalty are used for regularization, in addition to the standard dropout [25] and the weight decay for individual NN. In feature dropout, the dropout technique is applied to the NN outputs, i.e., the shape function outputs. The output penalty is the mean of the squares of the NN outputs. Due to the end-to-end learning manner, NAMs inherit the flexibility of NNs. NAM can be extended to consider pairwise interactions by adding two-input NNs for all possible feature pairs as bivariate shape functions f_{ij}. We denote NAM with pairwise interactions as NA^2M. However, the number of parameters and the computational cost increase by $O(D^2)$ as the number of features increases in NA^2M.

2.2 Neural Basis Model (NBM)

NBM [21] uses multiple basis functions shared across all features to make predictions. NBM decomposes each shape function f_i into a linear combination of a small set of basis functions that are shared among all shape functions. NBM uses only a single one-input and B-output NN to represent the shared basis functions. The training of NBM is performed by SGD. Due to the shared basis functions, the number of parameters in NBM can be reduced compared to NAM. The shape function in NBM is given by

$$f_i(x_i) = \sum_{k=1}^{B} h_k(x_i) a_{ik} \ , \tag{3}$$

where $\{h_1, h_2, \ldots, h_B\}$ denotes a set of B basis functions represented by a one-input and B-output NN, and a_{ik} are the coefficients of k-th basis function. When extending NBM to the form of GA^{2}M, the two-input shape function f_{ij} in (1) is given by

$$f_{ij}(x_i, x_j) = \sum_{k=1}^{B} u_k(x_i, x_j) c_{ijk} \ , \tag{4}$$

where $\{u_1, u_2, \ldots, u_B\}$ denotes a set of B bivariate basis functions represented by a two-input and B-output NN, and c_{ijk} are the coefficients of k-th output of NN. We denote NBM with pairwise interactions as NB^{2}M.

3 NAM and NBM with Feature Selection

3.1 Motivation

Although NAM and NBM possess promising interpretability, prediction performance, and flexible training schemes, there is a problem when applying them to high-dimensional datasets.[1] In NAM, preparing NNs for all features brings a disadvantageous increase in the number of parameters and computational cost, which is notable in NA^{2}M. Although the increase in the number of parameters in NBM can be relaxed due to the shared basis functions, NBM still results in an increase in the calculation cost in (dense) high-dimensional datasets. In addition, the large number of shape functions to be managed are intractable for visualizing and interpreting the models' behavior. Reducing the number of single features and pairs of features in NAM and NBM will be useful in applying them to high-dimensional datasets. Therefore, we incorporate feature selection layers using entmax [19] into NAM and NBM to select the predefined numbers of single features and pairs of features. The feature selection is performed through model training by optimizing the feature selection weights. That is, our method inherits the end-to-end learning properties of NAM and NBM. Our method is simple, but it is valuable to develop scalable NAM and NBM for high-dimensional datasets. It should be noted that our method does not compromise the interpretability of NAM and NBM.

3.2 Model Architecture

We prepare K_1 one-input shape functions $\{f_1, \ldots, f_{K_1}\}$ and K_2 two-input shape functions $\{g_1, \ldots, g_{K_2}\}$, where the total number of shape functions is $K = K_1 + K_2$. Let us consider an additive model with interaction terms, $\tilde{y} : \mathbb{R}^{K_1} \times \mathbb{R}^{K_2} \times \mathbb{R}^{K_2} \to \mathbb{R}$, which takes $K_1 + 2K_2$ input variables as

$$\tilde{y}(\boldsymbol{z}_1, \boldsymbol{z}_2, \boldsymbol{z}_3) = \sum_{i=1}^{K_1} f_i\left(z_{1,i}\right) + \sum_{i=1}^{K_2} g_i\left(z_{2,i},\ z_{3,i}\right) + b \ , \tag{5}$$

[1] Although high-dimensional datasets with sparse features can be handled by NBM with the specialized implementation, as shown in [21], it cannot be applied to dense features. In our experiments, NA^{2}M and NB^{2}M could not run on more than hundred features, and training NAM and NBM slowed down on more than thousand features in dense feature datasets.

where $\boldsymbol{z}_1 \in \mathbb{R}^{K_1}$ and $\boldsymbol{z}_2, \boldsymbol{z}_3 \in \mathbb{R}^{K_2}$ are the input variables for one- and two-input shape functions, respectively, $z_{j,i}$ indicates the i-th element of $\boldsymbol{z}_j$, and b is a bias term. The form of (5) can reduce the number of shape functions to be managed compared to (1) by setting $K_1 < D$ for univariate functions and $K_2 < D(D-1)/2$ for bivariate functions. Setting $K_2 = 0$ means that the model only considers univariate effects, whereas $K_2 > 0$ can consider interactions of K_2 feature pairs.[2]

Let us consider selecting $K_1 + 2K_2$ features from D to be fed into (5). To maintain end-to-end learning, we consider differentiable feature selection that enables us to select features through SGD-based training. Given a one-hot vector $\bar{\boldsymbol{v}} \in \{0,1\}^D$, the selection of a feature from a D-dimensional feature vector $\boldsymbol{x}$ can be represented by the dot product of $\bar{\boldsymbol{v}} \cdot \boldsymbol{x}$. As $\bar{\boldsymbol{v}}$ is one-hot and discrete, optimizing it directly using a gradient method is not feasible. Therefore, we replace it with continuous variables given by entmax [19] to make feature selection differentiable. Entmax is the generalized version of the softmax and sparsemax functions, and it has a hyperparameter α used to tune the sparsity of the output probability. Entmax is equivalent to softmax and sparsemax if $\alpha = 1$ and $\alpha = 2$, respectively. We set $\alpha = 1.5$ in our experiments. We also introduce the temperature parameter τ and anneal it in model training to make the entmax values approach one-hot, similar to [4]. Note that other continuous relaxation approaches can be used in our method, such as the Gumbel-Softmax [11,15] function.

Denoting the value of entmax with trainable logit parameters $\boldsymbol{\pi} \in \mathbb{R}^D$ as $\boldsymbol{v} = \mathrm{entmax}_\alpha(\boldsymbol{\pi}/\tau) \in \Delta^{D-1}$, where Δ^{D-1} is the standard simplex in $\mathbb{R}^D$, we get the relaxed feature selection process as $\boldsymbol{v} \cdot \boldsymbol{x}$. With a low-temperature parameter τ, the vector $\boldsymbol{v}$ is expected to approach one-hot, and we can pick up one feature from $\boldsymbol{x}$. By using this feature selection relaxation, we can get our additive model with feature selection as follows:

$$y(\boldsymbol{x}) = \sum_{i=1}^{K_1} f_i\left(\boldsymbol{v}_{1,i} \cdot \boldsymbol{x}\right) + \sum_{i=1}^{K_2} g_j\left(\boldsymbol{v}_{2,i} \cdot \boldsymbol{x},\ \boldsymbol{v}_{3,i} \cdot \boldsymbol{x}\right) + b\ , \tag{6}$$

where $\{\boldsymbol{v}_{1,i}\}_{i=1}^{K_1}$, $\{\boldsymbol{v}_{2,i}\}_{i=1}^{K_2}$, and $\{\boldsymbol{v}_{3,i}\}_{i=1}^{K_2}$ indicate the entmax values from different trainable logit parameters.

The form of (6) can be extended to multiple outputs, e.g. multiclass classification, as in (2). By introducing the coefficients for each shape function, $\{w_{li}^{(1)}\}_{i=1}^{K_1}$ and $\{w_{li}^{(2)}\}_{i=1}^{K_2}$, we obtain the l-th output of the model as

$$y_l(\boldsymbol{x}) = \sum_{i=1}^{K_1} f_i\left(\boldsymbol{v}_{1,i} \cdot \boldsymbol{x}\right) w_{li}^{(1)} + \sum_{i=1}^{K_2} g_j\left(\boldsymbol{v}_{2,i} \cdot \boldsymbol{x},\ \boldsymbol{v}_{3,i} \cdot \boldsymbol{x}\right) w_{li}^{(2)} + b_l\ . \tag{7}$$

In model training, all trainable parameters in the NN shape functions and entmax functions are optimized using SGD as the usual NN training. We gradually decrease the temperature parameter τ during model training. When the

[2] Our method can be extended to three or more input shape functions to capture high-order feature interactions while it compromises the interpretability.

temperature parameter τ becomes sufficiently small, the input features of shape functions are fixed based on the logit parameters $\boldsymbol{\pi}$ in the entmax function, meaning that the calculation of $\boldsymbol{v} \cdot \boldsymbol{x}$ can be ignored. We select the feature that corresponds to the largest logit parameter in these phases, that is, an input feature z_i becomes x_k, where $k = \mathrm{argmax}_j \pi_{i,j}$. In the inference phase, we also select the features in the same way.

NAM with Feature Selection (NAM-FS). We call the additive model of (6) or (7) using NNs as shape functions NAM-FS. The model is denoted as NAM-FS$_{(K_1)}$ when $K_2 = 0$, that is, no feature interactions, and as NA^{2}M-FS$_{(K_1,K_2)}$ when considering feature interactions ($K_2 > 0$).

NBM with Feature Selection (NBM-FS). Similar to NAM-FS, we call the additive model of (6) or (7) using the shape functions given by (3) and (4) NBM-FS. The model is denoted as NBM-FS$_{(K_1)}$ when $K_2 = 0$ and NB^{2}M-FS$_{(K_1,K_2)}$ when $K_2 > 0$.

3.3 Implementation Remark

The actual computational cost significantly depends on the implementation and library used. Radenovic et al. [21] reimplemented NAM and optimized the computation using grouped 1-D convolution in PyTorch [17] We also used the PyTorch library (version 1.13.1) and further improved the implementation of NAM. We found that using the `torch.bmm` function (batch matrix-matrix product) to implement MLP layers in NAM is more efficient than using the 1-D convolution function. Our implementation of the NAM layer speeds up by approximately $\times 2$–$\times 6$ compared to when using the 1-D convolution function on an NVIDIA A100 GPU machine.

4 Discussion of Model Complexities

While NAM requires as many NNs as the number of features D, NAM-FS requires only K_1 NNs and $K_1 D$ parameters for feature selection. The number of parameters in NAM-FS is significantly small when setting $K_1 \ll D$. In the case of NA^{2}M, $D + D(D-1)/2$ NNs are required, which results in a significant increase in the number of parameters and computational cost as D increases. However, the increase in the number of parameters in our NA^{2}M-FS can be reduced because the number of NNs in NA^{2}M-FS is $K_1 + K_2$. In the NAM family, the total multiply-accumulate operations (MACs) in the forward calculation, which affect the computational cost, are equivalent to the number of parameters. In NBMs, the number of parameters does not increase significantly compared to NAMs as D increases, while MACs increase. Due to the increase in MACs, NBM cannot run in the case of a large feature dimension and when considering pairwise interactions. Our NBM-FS and NB^{2}M-FS can save MACs because the calculations are required only for K_1 and $K_1 + K_2$ shape functions, respectively.

We then measured the throughput rate, defined as the number of data instances processed per second (denoting $\boldsymbol{x}$/sec), in order to check the actual computational cost. The throughput directly affects the training time. The NN architectures in NAM and NBM are the same as those used in Sect. 5. The batch size was set to 1,024, and the numbers of one- and two-input shape functions were set to $K_1 \in \{50, 500\}$ for NAM-FS and NBM-FS and to $(K_1, K_2) \in \{(50, 50), (500, 500)\}$ for NA^{2}M-FS and NB^{2}M-FS. We ran each model on an NVIDIA A100 GPU (80GB memory) machine. Figures 1 and 2 show the throughput ($\boldsymbol{x}$/sec) of the NAM and NBM families for forward calculation in training mode, respectively. We observe that the throughputs of NA^{2}M and NB^{2}M rapidly worsen as D increases, and they cannot run on more than 100 features due to the huge memory consumption. Even in NAM and NBM, the throughput slows down for more than 1,000 features. In contrast, our models are not affected by the increase in the number of features. Figure 1 shows that NAM-FS$_{(50)}$, NAM-FS$_{(500)}$, and NA^{2}M-FS$_{(50,50)}$ exhibit good throughput for any D (up to 5×10^4). NA^{2}M-FS$_{(500,500)}$ was slower than NA^{2}M-FS$_{(50,50)}$ due to the large settings of K_1 and K_2. However, it can run on more than a hundred features while considering feature interactions of 500 pairs. Figure 2 shows a similar trend to that in Fig. 1. In summary, NAM-FS and NBM-FS are advantageous in terms of throughput with more than 1,000 features, and NA^{2}M-FS and NB^{2}M-FS enable us to exploit feature interactions in datasets with more than 100 features that vanilla NA^{2}M and NB^{2}M cannot handle.

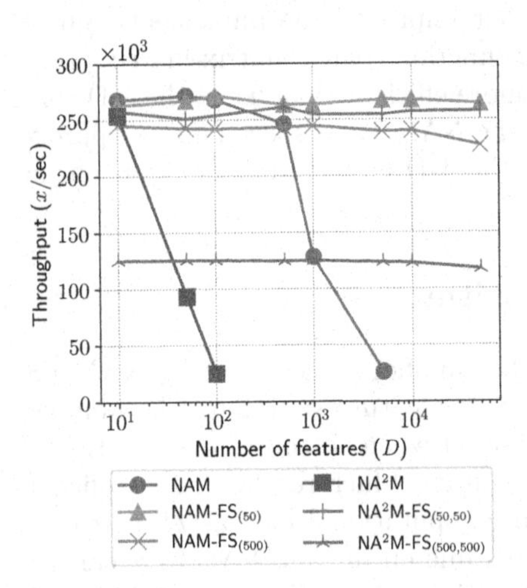

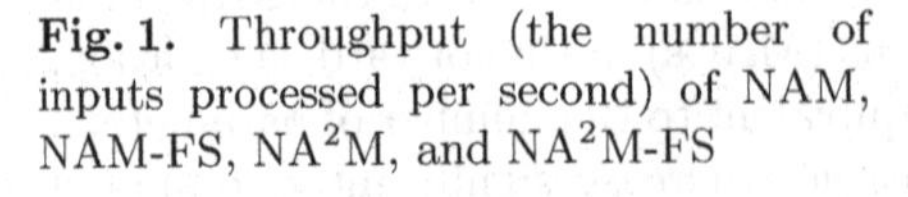

Fig. 1. Throughput (the number of inputs processed per second) of NAM, NAM-FS, NA^{2}M, and NA^{2}M-FS

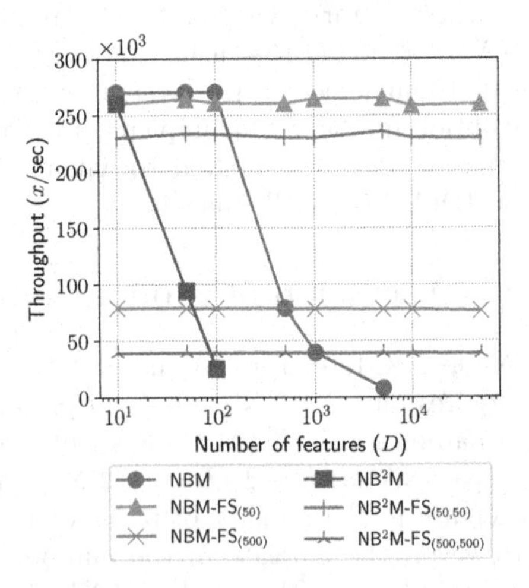

Fig. 2. Throughput (the number of inputs processed per second) of NBM, NBM-FS, NB^{2}M, and NB^{2}M-FS

5 Experiments

We compared the performance of our methods with that of other GAMs and well-known models on high-dimensional datasets ($D = 500$ to 5000) that cannot run NA^{2}M and NB^{2}M due to high computational costs. We further evaluated our feature-selection mechanism by comparing its performance with those of vanilla NAM and NBM using pre-selected features. We used the six classification datasets listed in Table 1. For F-MNIST and Epsilon, we used the test data provided in the datasets. We randomly split the data into 80% and 20% for training and testing, respectively, for the remaining datasets. Each data was standardized using the training dataset.

5.1 Experimental Settings

The NN architectures used in the NAM and NBM families were determined by referring to original papers [1,21]. We used multi-layer perceptrons (MLPs) containing three hidden layers with (64, 64, 32) units and those with (256, 128, 128) for the NAM and NBM families, respectively. Batch normalization and ReLU activation were applied to each layer. The number of bases was set to $B = 100$ for NBM and NBM-FS, and $B = 200$ for NB^{2}M-FS by following the default setting in [21]. We varied the number of shape functions, K_1 and K_2, in our models as $(K_1, K_2) \in \{(50, 0), (500, 0), (50, 50), (500, 500)\}$ to investigate their impact on performance. Our models with $K_2 > 0$ can capture pairwise feature interactions.

We used cross-entropy loss and Adam optimizer [12] with an initial learning rate of 0.01. For the training procedure, we early stopped the training if the validation accuracy did not improve for 11,000 iterations, and we decayed the learning rate to 1/5 if the validation accuracy did not improve for 5,000 iterations. This training procedure was based on [4] and was applied to all variants of NAM and NBM. The mini-batch size was set to 1,024. For the annealing of the temperature parameter in the proposed method, we linearly decreased it from 1 to 0.01 during the first 4,000 iterations to make the feature selection weights one-hot. After 4,000 iterations, we fixed the features to be fed into NNs and continued to train the model.

Other hyperparameters, dropout rate, weight decay coefficient, output penalty, feature dropout in the NAM family, basis dropout in the NBM family, were tuned by grid search. We divided the training dataset into 90% and 10% for the model training and validation data, respectively. The validation data were used for hyperparameter tuning and early stopping. By using tuned hyperparameters, we report the average test accuracy and standard deviation over 10 runs with different random seeds. Each NAM- and NBM-based model was run on a single NVIDIA A100 GPU (40GB or 80GB of memory).

Table 1. Summary of datasets. #Feat, #Train, #Test, and #Class indicate the numbers of features, training data, test data, and classes.

Dataset	#Feat	#Train	#Test	#Class
HAR [2]	561	8,239	2,060	6
ISOLET [7]	617	6,237	1,560	26
F-MNIST [26]	784	60,000	10,000	10
Epsilon [6]	2,000	400,000	100,000	2
guillermo [9]	4,296	16,000	4,000	2
Gisette [10]	5,000	5,600	1,400	2

5.2 Baselines

In addition to the NAM and NBM families, we evaluated six other models: EBM [16], NODE-GAM [4], logistic regression (LR), decision tree (DT), MLP, and XGBoost [5]. EBM and NODE-GAM are state-of-the-art GAMs and support to consider pairwise feature interactions. We denote EBM and NODE-GAM with feature interactions as EB^{2}M and NODE-GA^{2}M, respectively. It should be noted that EB^{2}M supports only regression and binary classification tasks. MLP and XGBboost are non-interpretable models but will be accurate. The hyperparameters of each model were tuned by grid search in the same manner as in the proposed model.

5.3 Results

Comparison with Baselines. Table 2 lists the test accuracies of each model. The models in the first group belong to GAMs and only use univariate shape functions. Among GAMs, NBM and NAM-FS$_{(500)}$ exhibit the best performance for two datasets. For high-dimensional datasets, guillermo ($D = 4,296$) and Gisette ($D = 5,000$), our NAM-FS$_{(500)}$ and NBM-FS$_{(500)}$ yield the highest accuracy among GAMs, respectively. In such high-dimensional datasets, NAM-FS$_{(500)}$ and NBM-FS$_{(500)}$ are also advantageous in terms of computational efficiency compared to vanilla NAM and NBM, as discussed in Sect. 4. However, the performance of NAM-FS$_{(50)}$ and NBM-FS$_{(50)}$ is worse than that of other GAMs. This is because the number of features to be selected ($K_1 = 50$) is insufficient to obtain a better accuracy, while such a setting can speed up the training process and reduce the number of shape functions to be managed in high-dimensional datasets.

Focusing on the models belonging to GA^{2}Ms (second group in Table 2), either the proposed NA^{2}M-FS$_{(500,500)}$ or NB^{2}M-FS$_{(500,500)}$ exhibits the highest accuracy among GA^{2}Ms and outperforms state-of-the-art GA^{2}Ms. Even NA^{2}M-FS$_{(50,50)}$ and NB^{2}M-FS$_{(50,50)}$ that have only 50 shape functions for feature interactions are better than EB^{2}M and NODE-GA^{2}M on ISOLET and guillermo, which implies the potential of our feature selection. The user parameters K_1

Table 2. Comparison of test accuracy with baselines. Models in the first and second groups belong to GAMs and GA^{2}M, respectively. The **bold** font indicates the model with the highest accuracy among GAMs or GA^{2}Ms. The underline means the best-performing model in considered models.

Model	HAR	ISOLET	F-MNIST	Epsilon	guillermo	Gisette
NAM	0.9850 ±0.0021	0.9534 ±0.0039	0.8634 ±0.0018	**0.8950** ±0.0008	0.7714 ±0.0071	0.9715 ±0.0053
NAM-FS$_{(50)}$	0.9764 ±0.0035	0.9257 ±0.0108	0.8203 ±0.0047	0.8443 ±0.0054	0.7545 ±0.0052	0.9438 ±0.0064
NAM-FS$_{(500)}$	0.9877 ±0.0024	**0.9580** ±0.0027	0.8601 ±0.0018	0.8927 ±0.0007	**0.7802** ±0.0050	0.9684 ±0.0048
NBM	**0.9879** ±0.0024	0.9494 ±0.0029	**0.8668** ±0.0014	0.8943 ±0.0027	0.7280 ±0.0183	0.9711 ±0.0033
NBM-FS$_{(50)}$	0.9739 ±0.0051	0.9297 ±0.0033	0.8212 ±0.0027	0.8499 ±0.0038	0.7564 ±0.0104	0.9509 ±0.0076
NBM-FS$_{(500)}$	0.9858 ±0.0025	0.9547 ±0.0051	0.8618 ±0.0021	0.8948 ±0.0005	0.7739 ±0.0114	**0.9740** ±0.0024
EBM	0.9835 ±0.0007	0.9493 ±0.0009	0.8639 ±0.0007	0.8788 ±0.0002	0.7440 ±0.0039	0.9575 ±0.0013
NODE-GAM	0.9845 ±0.0028	0.9438 ±0.0047	0.8621 ±0.0022	0.8945 ±0.0004	0.7710 ±0.0068	0.9706 ±0.0031
NA^{2}M-FS$_{(50,50)}$	0.9859 ±0.0024	0.9471 ±0.0036	0.8634 ±0.0022	0.8772 ±0.0015	0.7908 ±0.0094	0.9665 ±0.0027
NA^{2}M-FS$_{(500,500)}$	**0.9900** ±0.0016	**0.9645** ±0.0031	**0.8949** ±0.0017	0.8931 ±0.0003	**0.8012** ±0.0026	0.9611 ±0.0356
NB^{2}M-FS$_{(50,50)}$	0.9816 ±0.0034	0.9482 ±0.0064	0.8606 ±0.0038	0.8826 ±0.0019	0.7892 ±0.0074	0.9597 ±0.0048
NB^{2}M-FS$_{(500,500)}$	0.9863 ±0.0022	0.9597 ±0.0033	0.8925 ±0.0025	**0.8946** ±0.0007	0.7702 ±0.0059	**0.9759** ±0.0019
EB^{2}M	N/A	N/A	N/A	0.8794 ±0.0003	0.7834 ±0.0024	0.9671 ±0.0011
NODE-GA^{2}M	0.9867 ±0.0031	0.8876 ±0.0541	0.8817 ±0.0016	0.8941 ±0.0003	0.7861 ±0.0050	0.9721 ±0.0039
LR	0.9820 ±0.0000	0.9583 ±0.0000	0.8470 ±0.0000	0.8990 ±0.0000	0.7203 ±0.0000	0.9786 ±0.0000
Decision Tree	0.9388 ±0.0023	0.8158 ±0.0027	0.8067 ±0.0008	0.6826 ±0.0000	0.7588 ±0.0006	0.9433 ±0.0027
MLP	0.9852 ±0.0024	0.9663 ±0.0026	0.9099 ±0.0024	0.8970 ±0.0005	0.7738 ±0.0043	0.9732 ±0.0025
XGBoost	0.9907 ±0.0013	0.9567 ±0.0029	0.8979 ±0.0023	0.8836 ±0.0006	0.8195 ±0.0040	0.9711 ±0.0021

and K_2 balance the trade-off between performance and computational cost. Setting these values as large as possible within the available computational resource will be a possible choice when pursuing performance.

Finally, checking the performance of other well-known models, logistic regression shows the best performance on Epsilon and Gisette, whereas it does not work well on F-MNIST and guillermo. Both MLP and XGBoost archive the best performance on two datasets out of six, which may be attributed to the complicated model architectures. Among the interpretable models (GAMs, GA^{2}Ms, logistic regression, and decision tree), the proposed NA^{2}M-FS$_{(500,500)}$ and NB^{2}M-FS$_{(500,500)}$ consistently achieve good performances, which demonstrates the effectiveness of our models.

Table 3. Test accuracy of NAM and NBM with pre-selected features. The underline means that its model is better than the corresponding NAM-FS$_{(K_1)}$ or NBM-FS$_{(K_1)}$.

Model	HAR	ISOLET	F-MNIST	Epsilon	guillermo	Gisette
NAM$_{(50)}$	0.9281 ±0.0055	0.7474 ±0.0065	0.5665 ±0.0117	0.8007 ±0.0005	0.7678 ±0.0031	0.9053 ±0.0037
NAM$_{(500)}$	0.9814 ±0.0020	0.9587 ±0.0034	0.8613 ±0.0014	0.8839 ±0.0005	0.7256 ±0.0587	0.9709 ±0.0028
NBM$_{(50)}$	0.9246 ±0.0022	0.7535 ±0.0068	0.5901 ±0.0023	0.7995 ±0.0004	0.7927 ±0.0035	0.9046 ±0.0046
NBM$_{(500)}$	0.9863 ±0.0011	0.9490 ±0.0043	0.8634 ±0.0017	0.8848 ±0.0006	0.7735 ±0.0068	0.9736 ±0.0022

Evaluation of Feature Selection Mechanism. We validated whether the feature-selection mechanism used in the proposed models works well. A naive approach for reducing the number of features to be considered is to pre-select the features before the training of NAM and NBM. We used mutual information (MI)-based feature selection implemented in the `scikit-learn` library [18] and selected the top K_1 features according to the MI scores. We then trained NAM and NBM using only the selected K_1 features. We performed this two-stage method with $K_1 = 50, 500$, denoted by $\text{NAM}_{(K_1)}$ and $\text{NBM}_{(K_1)}$.

Table 3 shows the test accuracies of NAM and NBM with pre-selected features. In the case of $K_1 = 50$, we observe that our $\text{NAM-FS}_{(50)}$ and $\text{NBM-FS}_{(50)}$ are better than NAM and NBM with pre-selected features, except for guillermo. In addition, the performances of NAM and NBM with pre-selected features are greatly worse for the datasets other than guillermo. For $K_1 = 500$, $\text{NAM-FS}_{(500)}$ and $\text{NBM-FS}_{(500)}$ tend to be better than $\text{NAM}_{(500)}$ and $\text{NBM}_{(500)}$ for a large number of features ($D \leq 2,000$). This result demonstrates that our models can select appropriate features for NAM and NBM through end-to-end learning.

6 Conclusion

In this paper, we have proposed extensions of NAM and NBM by incorporating the feature selection mechanism in an end-to-end learning manner. We demonstrated the computational efficiency of the proposed models. Although the proposed models are simple, they allow us to incorporate pairwise feature interactions efficiently and scale for high-dimensional datasets. The experimental results demonstrate that our proposed models are better than or comparable to state-of-the-art GAMs.

Although the proposed NAM-FS and NBM-FS enable us to reduce the NN shape functions in NAM and NBM and show promising performance among GAMs in high-dimensional datasets, the advantages of our models fade with fewer than 100 features. Another limitation is that we attempted to capture only pairwise interactions. It is possible to extend our NAM-FS and NBM-FS to consider more than two feature interactions, although this will compromise interpretability. An efficient implementation of NBM for sparse features was proposed in [21]. Introducing such sparse feature support is a possible future work. Moreover, applying our models to real-world datasets, such as biological and medical datasets, is also an important direction.

Acknowledgements. This work was partially supported by JSPS KAKENHI (JP20H04240, JP20H04254, JP22H03590, JP23H00491, JP23H03466), JST PRESTO (JPMJPR2133), NEDO (JPNP18002, JPNP20006), and a grant from the Kanagawa Prefectural Government of Japan.

References

1. Agarwal, R., et al.: Neural additive models: interpretable machine learning with neural nets. In: Advances in Neural Information Processing Systems, vol. 34 (2021)
2. Anguita, D., Ghio, A., Oneto, L., Parra, X., Reyes-Ortiz, J.L.: A public domain dataset for human activity recognition using smartphones. In: 21th European Symposium on Artificial Neural Networks, Computational Intelligence and Machine Learning (ESANN) (2013)
3. Arik, S.Ö., Pfister, T.: TabNet: attentive interpretable tabular learning. In: AAAI Conference on Artificial Intelligence, vol. 35, no. 8, pp. 6679–6687 (2021). https://doi.org/10.1609/aaai.v35i8.16826
4. Chang, C., Caruana, R., Goldenberg, A.: NODE-GAM: neural generalized additive model for interpretable deep learning. In: International Conference on Learning Representations (ICLR) (2022)
5. Chen, T., Guestrin, C.: XGBoost: a scalable tree boosting system. In: 22nd ACM SIGKDD International Conference on Knowledge Discovery and Data Mining, pp. 785–794 (2016). https://doi.org/10.1145/2939672.2939785
6. Epsilon: Large scale learning challenge (2008). https://k4all.org/project/large-scale-learning-challenge/
7. Fanty, M., Cole, R.: Spoken letter recognition. In: Advances in Neural Information Processing Systems, vol. 3 (1990)
8. Gorishniy, Y., Rubachev, I., Khrulkov, V., Babenko, A.: Revisiting deep learning models for tabular data. In: Advances in Neural Information Processing Systems, vol. 34 (2021)
9. Guillermo: ChaLearn AutoML challenge (2000). http://automl.chalearn.org/data/
10. Guyon, I., Gunn, S., Ben-Hur, A., Dror, G.: Result analysis of the NIPS 2003 feature selection challenge. In: Advances in Neural Information Processing Systems, vol. 17 (2004)
11. Jang, E., Gu, S., Poole, B.: Categorical reparameterization with Gumbel-softmax. In: International Conference on Learning Representations (ICLR) (2017)
12. Kingma, D.P., Ba, J.: Adam: a method for stochastic optimization. In: International Conference on Learning Representations (ICLR) (2015). https://doi.org/10.48550/arXiv.1412.6980
13. Lou, Y., Caruana, R., Gehrke, J.: Intelligible models for classification and regression. In: 18th ACM SIGKDD International Conference on Knowledge Discovery and Data Mining, pp. 150–158 (2012). https://doi.org/10.1145/2339530.2339556
14. Lou, Y., Caruana, R., Gehrke, J., Hooker, G.: Accurate intelligible models with pairwise interactions. In: 19th ACM SIGKDD International Conference on Knowledge Discovery and Data Mining, pp. 623–631 (2013). https://doi.org/10.1145/2487575.2487579
15. Maddison, C.J., Mnih, A., Teh, Y.W.: The concrete distribution: a continuous relaxation of discrete random variables. In: International Conference on Learning Representations (ICLR) (2017)
16. Nori, H., Jenkins, S., Koch, P., Caruana, R.: InterpretML: a unified framework for machine learning interpretability (2019). https://doi.org/10.48550/arXiv.1909.09223
17. Paszke, A., et al.: PyTorch: an imperative style, high-performance deep learning library. In: Advances in Neural Information Processing Systems, vol. 32 (2019)
18. Pedregosa, F., et al.: Scikit-learn: machine learning in python. J. Mach. Learn. Res. **12**(85), 2825–2830 (2011)

19. Peters, B., Niculae, V., Martins, A.F.T.: Sparse sequence-to-sequence models. In: 57th Annual Meeting of the Association for Computational Linguistics (ACL). pp. 1504–1519 (2019). https://doi.org/10.18653/v1/P19-1146
20. Popov, S., Morozov, S., Babenko, A.: Neural oblivious decision ensembles for deep learning on tabular data. In: International Conference on Learning Representations (ICLR) (2020)
21. Radenovic, F., Dubey, A., Mahajan, D.: Neural basis models for interpretability. In: Advances in Neural Information Processing Systems, vol. 35 (2022)
22. Ribeiro, M.T., Singh, S., Guestrin, C.: "Why Should I Trust You?": explaining the predictions of any classifier. In: 22nd ACM SIGKDD International Conference on Knowledge Discovery and Data Mining, pp. 1135–1144 (2016). https://doi.org/10.1145/2939672.2939778
23. Rudin, C.: Stop explaining black box machine learning models for high stakes decisions and use interpretable models instead. Natur. Mach. Intell. **1**(5), 206–215 (2019). https://doi.org/10.1038/s42256-019-0048-x
24. Selvaraju, R.R., Cogswell, M., Das, A., Vedantam, R., Parikh, D., Batra, D.: Grad-CAM: visual explanations from deep networks via gradient-based localization. In: International Conference on Computer Vision (ICCV), pp. 618–626 (2017). https://doi.org/10.1109/ICCV.2017.74
25. Srivastava, N., Hinton, G., Krizhevsky, A., Sutskever, I., Salakhutdinov, R.: Dropout: a simple way to prevent neural networks from overfitting. J. Mach. Learn. Res. **15**(56), 1929–1958 (2014)
26. Xiao, H., Rasul, K., Vollgraf, R.: Fashion-MNIST: a novel image dataset for benchmarking machine learning algorithms (2017). https://doi.org/10.48550/arXiv.1708.07747

Random Mask Perturbation Based Explainable Method of Graph Neural Networks

Xinyue Yang, Hai Huang(✉), and Xingquan Zuo

Beijing University of Posts and Telecommunications, Beijing 100876, China
{yang_xy,hhuang,zuoxq}@bupt.edu.cn

Abstract. Graph Neural Networks (GNNs) have garnered considerable attention due to their potential applications across multiple domains. However, enhancing their interpretability is a significant challenge for the crucial application. This paper proposes an innovative node perturbation-based method to explicate GNNs and unveil their decision-making processes. Categorized as a black-box method, it generates explanations of node importance solely through the input-output analysis of the model, obviating the necessity for internal access. The method employs fidelity as a metric for calculating the significance of perturbation masks and utilizes a sparsity threshold to filter the computation results. Furthermore, recognizing the impact of different node combinations on model prediction outcomes, we treat the mask as a random variable. By randomly sampling various masks, we compute perturbed node importance, facilitating the generation of user-friendly explanations. Comparative experiments and ablation studies conducted on both real and synthetic datasets substantiate the efficacy of our approach in interpreting GNNs. Additionally, through a case study, we visually demonstrate the method's compelling interpretative evidence regarding model prediction outcomes.

Keywords: Interpretability · Perturbation-based Method · Graph Neural Network

1 Introduction

Graph Neural Networks (GNNs) have been extensively employed in diverse domains, including recommendation systems, molecular structure prediction, and remote sensing analysis. Nevertheless, as a deep learning model, GNNs encounter the challenge of interpretability [1], wherein humans struggle to comprehend the rationale behind the specific decisions made by the model. This challenge implies that biases or prejudices in the model may elude human detection, eroding user trust. Consequently, this obstacle prevents the application of GNNs in critical fields with a low tolerance for errors, such as autonomous driving and medical diagnosis.

D.-N. Yang et al. (Eds.): PAKDD 2024, LNAI 14647, pp. 17–29, 2024.
https://doi.org/10.1007/978-981-97-2259-4_2

Recently, there has been a growing focus among researchers on elucidating GNNs, categorizing explanatory methods into four types [24]. Notably, gradient-based and decomposition-based methods employ backpropagation to calculate the importance of nodes or edges. Both fall under the category of white-box methods, necessitating an understanding of the GNN's structure, and may encounter challenges such as gradient saturation [18]. Surrogate models employ an interpretable model to replicate the behavior of the GNN. Such approaches necessitate consideration of the consistency between the surrogate model and the original model [6]. Perturbation-based methods ascertain the significance of nodes or edges by perturbing the input and observing the output. In comparison to surrogate models, they are simpler and more feasible. Most existing works on perturbation-based methods utilize soft masks to enable optimization through gradient descent, potentially reducing the interpretability of the results. Additionally, some works utilize hard masks [5,20]. These methods employ a greedy algorithm to choose significant nodes. Due to the interdependence among nodes in the graph, the optimal selection of nodes does not adhere to the property of memorylessness. This results in suboptimal performance for the greedy algorithm.

In this paper, we propose a perturbation-based black-box explainable method for graph neural networks. Primarily designed for graph classification tasks, this method interprets the results by identifying a set of highly important nodes in the graph. First, we perturb the nodes in the graph using masks, with Fidelity serving as the metric for mask importance. Diverging from traditional white-box methods and soft mask methods, we treat the mask as a random variable with values of 1 or 0, rather than a learnable parameter, thus avoiding reliance on the specific structure of the GNN. Subsequently, due to the large scale of node masks, we employ Monte Carlo sampling to randomly sample hard masks for node importance calculation. In this process, we use Sparsity thresholds to filter the results of random sampling, facilitating the generation of more user-friendly explanations. Finally, by calculating the importance of the sampled masks, we weight the importance of perturbed graph nodes, thereby producing a set of nodes that significantly impact the decision outcome as the interpretative result. The major contributions of this paper are as follows:

- We propose a Random Mask Perturbation based Explainable Method (RMPEM) for GNNs. In the process of calculating node importance, this method jointly considers the impact of mask fidelity and sparsity, aiming to generate a node importance map to explain model's predicted outcomes. Furthermore, the method treats the mask as a random variable and calculates node importance through random sampling, operating independently of the specific internal structure of GNNs.
- We conduct extensive experiments using both real-world and synthetic datasets. The experiments demonstrate that the proposed method excels in identifying the most crucial nodes, providing high discriminability for different predicted classes within the same instance, and achieving a commendable balance between accuracy and sparsity. Moreover, ablation studies confirm

the efficacy of the Fidelity measure and the Sparsity threshold constraint in our approach.
- Through a case study on a synthetic dataset, we visually demonstrate that our approach outperforms in identifying effective interpretative evidence for the model's predictive outcomes.

2 Related Work

Based on the scope of explanation, methods for explaining GNN can be divided into two categories: local and global methods. Among the global methods, XGNN [23] generates graph patterns through reinforcement learning to maximize the prediction scores for a certain category within the GNN model; Xuanyuan et al. [21] attempt to match semantic concepts to neurons in GNN. Global methods seek to abstract the overall predictive logic of the GNN through simple concepts. However, such explanations may remain elusive to average users.

Local methods hunt the important features of single instances, which can be divided into four categories [24]: gradient-based method, decomposition-based method, surrogate method, and perturbation-based method. Gradient-based methods, such as SA [2] and Grad-CAM [13], leverage gradients to determine the node importance through backpropagation. These methods are mostly derived from the methods for image domain. Moreover, they are categorized as white-box methods, requiring the knowledge of the specific structure of GNN to generate explanations. Decomposition-based methods, such as DEGREE [4] and GNN-LRP [14], regard the model's predictions as scores, decomposing and backpropagating them layer by layer to the input features, then determine the node or edge importance by the decomposed scores. Due to the necessity of accessing the GNN's parameters, these methods are also classified as white-box methods. Surrogate methods approximate the original GNN model's predictions through training an inherently interpretable model. PGM-Explainer [19] offers explanations by locally fitting an interpretable Bayesian network to the specific instance; Huang et al. [8] apply the LIME method to GNNs, but it can only elucidate the importance of node features, and falls short of interpreting the graph's structure. Perturbation-based methods can be further subdivided into three categories [3]. The first category consider the mask as a learnable parameter, optimized through gradient descent. Examples include GNN-Explainer [22] and CF-Explainer [10]. These methods transition the mask from discrete to continuous values to apply optimization algorithms, which could potentially affect interpretability. The second category identifies the optimal mask using search algorithms, such as ZORRO [5], Causal Screening [20], and GstarX [25]. These methods select the most important nodes or edges to be included in the explanation step by step based on a greedy algorithm. Due to the unique structure of graph data, where nodes are interrelated and lack the property of memorylessness, greedy algorithms exhibit suboptimal performance. Both of the aforementioned categories are considered black-box methods. The third category, represented by PGExplainer [11], adopts reparameterization techniques and trains a multilayer perceptron that predicts the mask corresponding to different instances.

This method, requiring not only the original model's input and output but also its intermediate results within the model, is considered a white-box method.

3 Problem Statement

The input of a GNN can be represented as $G = (A, X)$, where $X = \{x_1, x_2, ..., x_N\}$, $x_i \in \mathbb{R}^d$ denotes the feature matrix, $A \in \{0,1\}^{N \times N}$ denotes the adjacency matrix, N is the number of nodes, and d is the dimension of the node feature vector. It can also be represented as $G = \{V, \varepsilon\}$, where $V = \{v_1, v_2, ..., v_N\}$ denotes the set of nodes, $\varepsilon = \{e_1, e_2, ..., e_{Ne}\}$ denotes the set of edges and N_e is the number of edges.

For graph classification task, the objective of this task is to classify graph G into a specific category, i.e.$G \rightarrow \{1, ..., C\}$. GNN model can be represented as $y = \Phi(G(A, X); W)$, where W represents the parameters of the GNN model and the output $y \in \mathbb{R}^C$ is a confidence scores vector of graph G belonging to each category. The optimization objective of the GNN model is to maximize the confidence score corresponding to the actual category of the graph.

The interpretability problem in graph classification can be described as follows:

Given: A GNN model $y = \Phi(G\{V, \varepsilon\})$, with an input $G = \{V, \varepsilon\}$, where the prediction result for this input is category c.

Select: A node subset $V_c' \in V$, which includes nodes that have the most significant impact on the prediction result c.

4 Explainable Method

4.1 Node Importance Based on Fidelity

It is necessary to compute node importance to address the interpretability problem in graph classification. Node importance should elucidate which nodes' variations may affect the decision outcomes of the GNN model. We employ a node perturbation-based approach, utilizing fidelity as the metric for measuring node importance. Fidelity is defined as the magnitude of the impact on decision outcomes caused by perturbing nodes. If perturbing a subset of nodes results in a significant change in the predicted outcome, the perturbed nodes are evidently crucial to the original prediction.

To assess the impact of different combinations of perturbed nodes on model decisions, inspired by the RISE [12] method for image tasks, we introduce a node random mask $M = (m_1, m_2, ..., m_N)$, where $m_i \in \{0, 1\}$, and with a probability p of taking the value 0. When $m_i = 0$, it signifies that node v_i is masked. Utilizing a random mask assists in understanding the model's overall behavior for different types of inputs. The graph perturbed by the mask can be denoted as $\overline{G} = \{V \odot M, \varepsilon\}$, where $\odot$ denotes element-wise multiplication.

The importance of node mask can be calculated using fidelity, as illustrated by the following formula:

$$fidelity(M) = \phi_c(G\{V, \varepsilon\}) - \phi_c(G\{V \odot M, \varepsilon\}) \tag{1}$$

The importance of node v_i can be expressed as follows:

$$I(v_i) = \mathbb{E}_M[fidelity(M)|m_i = 0] \tag{2}$$

The expected form of Eq. 2 can be rewritten in terms of a probability distribution, namely:

$$\begin{aligned} I(v_i) &= \sum_M fidelity(\mathcal{M}) \cdot P[M = \mathcal{M}|m_i = 0] \\ &= \frac{1}{P[m_i = 0]} \sum_M fidelity(\mathcal{M}) \cdot P[M = \mathcal{M}, m_i = 0] \end{aligned} \tag{3}$$

The joint probability expression in Eq. 3 can be substituted as follows:

$$P[M = \mathcal{M}, m_i = 0] = \begin{cases} 0, m_i = 1 \\ P[M = \mathcal{M}], m_i = 0 \end{cases} = (1 - m_i) \cdot P(M = \mathcal{M}) \tag{4}$$

Ultimately, we can derive the calculation formula for the importance of node v_i as follows:

$$\begin{aligned} I(v_i) &= \frac{1}{P[m_i = 0]} \sum_M fidelity(\mathcal{M}) \cdot (1 - m_i) \cdot P(M = \mathcal{M}) \\ &= \frac{1}{P[m_i = 0]} \mathbb{E}_M[fidelity(M) \cdot (1 - m_i)] \end{aligned} \tag{5}$$

The mask M is a high-dimensional random variable, making direct solutions challenging. Thus, we can use Monte Carlo approximation to compute the expectation. Based on the definition of the Monte Carlo method, we can approximate Eq. 5 as follows:

$$I(v_i) \approx \frac{1}{P[m_i = 0] \cdot S} \sum_M fidelity(\mathcal{M}) \cdot (1 - m_i) \tag{6}$$

where S denotes the number of randomly sampled masks, each m_i is drawn with a probability p, meaning $P[m_i = 0] = p$.

Using Eq. 6, we can compute the importance indicators for all nodes. If we set the importance threshold as τ, the node importance interpretation for the GNN's classification result on a particular instance can be expressed as $V' = \{v_i | I(v_i) > \tau, v_i \in V\}$.

4.2 Explanation Sparsity

For the node importance explanation V', it should be as sparse as possible, containing only the most important nodes [3]. When generating explanations, we sample a large number of masks to enhance the precision of the Monte Carlo method. To optimize the explanation' sparsity, we set a threshold β to filter out certain masks. Specifically, in Eq. 6, if a mask M meets the condition $fidelity(M) < \beta$, the $fidelity(M)$ for that mask is set to 0. The implementation process of the algorithm is shown in Algorithm 1.

Algorithm 1: The algorithm for RMPEM

Input: GNN model $\Phi(\cdot)$, input graph G, probability p of the node mask, sampling size S, sparsity threshold β, class to be explained c
Output: Node importance explanation V_c'
for $i = 1$ *to* S **do**
 generate node mask $M_i = (m_{i1}, m_{i2}, ..., m_{ij}, m_{iN})$: **for** $m_{ij} \in M_i$ **do**
  m_{ij} is 0 with probability p, otherwise 1
 set the features of node v_j to zero where $m_{ij} = 0$ to obtain $G\{V \odot M, \varepsilon\}$
 calculate the $fidelity(M_i) = \Phi_c(G\{V, \varepsilon\}) - \Phi_c(G\{V \odot M, \varepsilon\})$
 if $fidelity(M_i) < \beta$ **then**
  $fidelity(M_i) = 0$
calculate $I(v_j) = \frac{1}{p.S} \sum_{i=1}^{S} fidelity(M_i) \cdot (1 - m_{ij})$
Return: $V_c' = \{v_i | I(v_j) > \tau, v_i \in V\}$

5 Experiments

Six baselines are compared with the method we proposed, and a quantitative analysis is conducted based on four evaluation metrics with results shown in Table 2. Further, the effectiveness of the fidelity value and sparsity threshold is explored through ablation studies. Finally, visualization of explanations generated by different methods based on the BA-2motif dataset is provided for qualitative analysis.

5.1 Experimental Setup

We conduct the experiments on five datasets and a classic GNN model, GCN, as the black-box model for evaluating these methods. We utilized Pytorch version 1.5 for our experiments, with the GCN model being built upon the PyG framework. The hyperparameter settings used in our method include: mask sampling probability $p = 0.5$, sparsity threshold $\beta = 0$, and significance threshold $\tau = 0.5$. The corresponding code is available at https://github.com/plumerxy/RMPEM.

Dataset: As detailed in Table 1, we use five binary classification datasets for graph classification tasks. The training and testing sets are split in an 8:2 ratio. BA-2motifs [11] is a synthetic dataset; half its graphs exhibit a "house" motif while the other half present a five-node "cycle" pattern. The other four datasets are from bioinformatics benchmarks [16]. MUTAG: Captures the molecular structures of aromatic nitro compounds. The classification within this dataset signifies if these aromatic entities possess mutagenic effects on bacteria. Tox21_AR: Its classification demarcates whether a compound is toxic or not. NCI-H23: Describe the active characteristics of human non-small cell lung cancer cells. PTC_FR: Describe the carcinogenic properties of compounds on female rats.

Table 1. Details of Datasets

Dataset	Classes	Graphs	Features	Nodes(avg.)	Edges(avg.)
BA-2motifs	2	1000	7	25	50
MUTAG	2	4337	14	30	60
Tox21_AR	2	12000	42	18	19
NCI-H23	2	2500	18	27	29
PTC_FR	2	350	27	15	15

Black-Box: We employ the GNN model structure as used in the GNNExplainer [22]. This model contains three GCN layers [9] followed by a linear layer, and it employs mean pooling to obtain a representation of the entire graph. The model's parameters are set according to the details in the GNNExplainer method.

Baselines: We employ six baseline methods. Two of these methods are gradient-based, as described in reference [16]. Firstly, CG [17] calculates the node importance directly from the input gradient, ignoring the impact of negative gradients. Secondly, Grad-CAM [15] assesses node importance by computing gradients at the final convolutional layer. Additionally, we explore ZORRO [5], a perturbation-based black-box explanation approach that utilizes a greedy algorithm to incrementally select the most significant nodes for node importance explanations. The remaining two baselines are also perturbation-based methods, focusing on edge importance, implemented following the guidelines in [7]. GNNExplainer [22] generates explanations by perturbing the edges to minimize the output error with the original prediction. PG-Explainer [11] operates similarly to GNNExplainer, offering enhanced generalization capabilities. Lastly, CF-GNNExplainer [10] is another perturbation-based method, yet it differs in its optimization objective, aiming to maximize the output error through perturbations to generate counterfactual explanations.

It should be noted that the latter three methods provide edge importance, whereas our explanation format pertains to node importance, needs a conversion when computing evaluation metrics. To address this issue, Funke et al. [5]

distributed the edge importance equally between the two nodes connected by that edge, but it would make the degree of the node a key factor affecting its importance, failing to reflect the original distribution of importance. Therefore, we use the highest edge importance among all edges connected to a node as that node's importance.

Evaluation Metrics: We select four objective metrics for evaluation, with the calculation methods referencing those used by Shun et al. [16].

- Accuracy: Calculate the ratio of nodes included in the explanation within those included in the ground truth.
- Fidelity: Calculate the area under the curve (AUC) for the GNN classifier on the test set before and after masking important nodes. The fidelity value is the difference between these two values.
- Contrastivity: When GNN classifier makes a prediction for an input graph, node importance explanations can be generated separately for the positive and negative classes. These explanations can be treated as binary strings, assigning '1' to important nodes and '0' to others. Calculate the Hamming distance between the two strings to indicate their difference.
- Sparsity: Calculate the proportion of the number of nodes relative to the total number of nodes.

All the above metrics are positive metrics, meaning that the larger the value, the better the performance of the method. It should be noted that since only the BA-2motif dataset is a synthetic dataset with the ground truth, the Accuracy metric is used exclusively for it. For the remaining four datasets, the Fidelity metric is employed to measure the performance of the explanations. Among the four metrics mentioned, the first two are prerequisites for generating meaningful explanations: the explanation must accurately capture the ground truth. Only when these requirements are satisfied can the explanation be considered meaningful. Building on this, the contrastivity metric demands significant differences between explanations for different classes, and the sparsity metric requires for conciseness in explanations. These latter two metrics enhance human comprehensibility of the explanations.

5.2 Quantitative Experiments

Accuracy/Fidelity: For the BA-2motif dataset, we utilize the Accuracy metric to precisely evaluate the capability to identify ground truth. As can be seen in Table 2, the RMPEM method performs the best in locating ground truth. For other datasets, fidelity metric is used for evaluation. RMPEM continues to excel, indicating a strong alignment between the generated explanations and the GNN model's behavior. Within gradient-based methods, CG shows better fidelity performance, while Grad-CAM lags across several datasets. Among the perturbation-based methods, PGExplainer and GNNExplainer demonstrate moderate performance; CF-Explainer excels in BA-2motif but poorly on

Tox21_AR, showing a less stable ability to locate important nodes. ZORRO performs well on some datasets but falls short on BA-2motif. Considering ZORRO's reliance on random perturbations using original node features, such perturbations are less effective for these particular datasets, impeding the generation of effective explanations.

Contrasitivity: The RMPEM method stands out with top performance in four different datasets, whereas the Grad-CAM method performed the best on another dataset. Both ZORRO and CG face challenges in identifying crucial nodes across various datasets. Other baseline methods have average performance, neither excelling nor falling behind significantly.

Sparisity: According to Table 2, the RMPEM method performs best on the PTC-FR and Tox21_AR datasets but is less dominant on others. CG method consistently shows good sparsity, while perturbation-based methods tend to have

Table 2. Acc. stands for the Accuracy metric, Fid. stands for the Fidelity metric, Con. stands for the Contrasitivity metric, and Spa. stands for the Sparsity metric. The bolded values indicate the best values in methods other than RMPEM(NF/NS).

	BA-2motif			MUTAG			NCI-H23		
	Acc.	Con.	Spa.	Fid.	Con.	Spa.	Fid.	Con.	Spa.
CG	0.896	0.014	**0.830**	0.316	0.014	0.722	0.167	0.023	**0.733**
Grad-CAM	0.024	**0.423**	0.766	0.129	0.308	0.086	0.038	0.379	0.478
GNNExplainer	0.284	0.320	0.477	0.175	0.329	0.425	0.154	0.345	0.470
PGExplainer	0.412	0.354	0.406	0.245	0.361	0.632	0.121	0.165	0.511
CF-Explainer	0.742	0.214	0.746	0.294	0.250	**0.738**	0.129	0.276	0.674
ZORRO	0.049	0.124	0.353	0.438	0.180	0.549	0.180	0.264	0.481
RMPEM	**0.950**	0.386	0.734	**0.833**	**0.463**	0.732	**0.759**	**0.456**	0.642
RMPEM(NF)	0.738	0.437	0.587	0.719	0.436	0.531	0.704	0.413	0.509
RMPEM(NS)	0.950	0.480	0.720	0.833	0.460	0.631	0.744	0.465	0.589
	PTC_FR			**Tox21_AR**					
	Fid.	Con.	Spa.	Fid.	Con.	Spa.			
CG	0.320	0.013	0.666	0.313	0.02	0.641			
Grad-CAM	0.136	0.391	0.340	0.064	0.371	0.445			
GNNExplainer	0.047	0.315	0.557	0.365	0.337	0.467			
PGExplainer	0.050	0.256	0.384	0.243	0.113	0.537			
CF-Explainer	0.111	0.252	0.374	0.132	0.306	0.646			
ZORRO	0.098	0.040	0.070	0.201	0.002	0.020			
RMPEM	**0.587**	**0.393**	**0.694**	**0.749**	**0.414**	**0.702**			
RMPEM(NF)	0.493	0.322	0.584	0.661	0.358	0.508			
RMPEM(NS)	0.555	0.433	0.629	0.724	0.445	0.651			

moderate sparsity performance, except for ZORRO, whose overall performance may be affected by its perturbation technique.

5.3 Ablation Study

Fidelity: To explore the effectiveness of using the fidelity to measure node importance, we modify Eq. 2 to the Eq. 7, which simply uses the output of the GNN model after applying the mask to measure the node importance:

$$I'(v_i) = \mathbb{E}_M[\Phi_c(G\{V \odot M, \varepsilon\})|m_i = 1] \tag{7}$$

The modified method is referred to as RMPEM(NF). The results show that RMPEM(NF) underperforms RMPEM across three different evaluation metrics, confirming the significance of our adopted node importance measure in enhancing interpretability.

Sparsity: Similar to the previous experiment, we remove the sparsity constraint and referred to the modified method as RMPEM(NS). As shown in Table 2, the experiment tries to find the impacts of this change. With the sparsity constraint removed, RMPEM(NS) sees a decrease in performance on fidelity and a significant reduction in sparsity. However, there is a slight improvement in contrastivity compared to RMPEM, which may be due to the decreased likelihood of similarity when explanations become more complex. Considering that the sparsity constraint optimized not only sparsity but also enhanced the fidelity, this further proves the importance of incorporating this constraint.

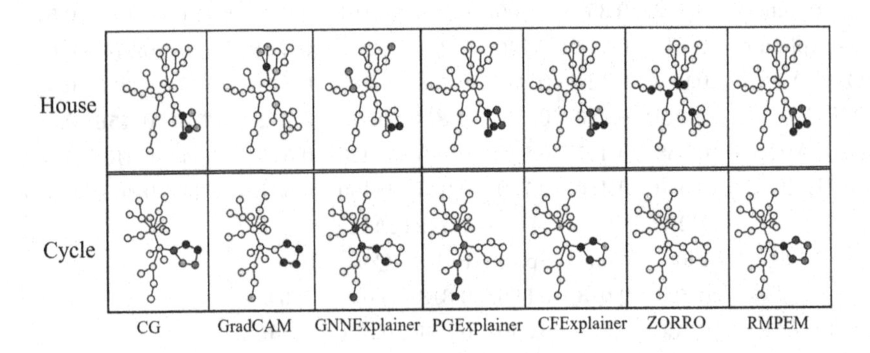

Fig. 1. Visualization of Explanations for an Instance in BA-2motif.

5.4 Use Case

We select two instances of different classes from the synthetic dataset BA-2motif, with the corresponding explanation ground truth being a house and a cycle shape, respectively. All seven methods are used to generate the corresponding node importance explanations. As shown in Fig. 1, our method RMPEM can capture these two shapes well. Besides, only the CG method and CF-Explainer have better capability in capturing the baseline shapes; the Grad-CAM method can capture the circular shape but is completely unable to handle the house shape; PGExplainer, on the other hand, easily captures the house shape but has difficulty in capturing the circular shape; the GNN-Explainer method is able to partially capture both shapes; finally, the ZORRO method does not perform well on the BA-2motif dataset.

6 Conclusion

We propose RMPEM, a method for explaining GNN models that can be used without understanding the GNN models' architecture. Masks are considered as random variables, indicative of the vast array of interaction scenarios among graph nodes. We introduce the use of Fidelity to describe node importance and optimize the sparsity of the explanations through a sparsity threshold. As the experiment results, RMPEM achieves the best performance in Accuracy/Fidelity across all datasets and also ensures good performance on the Contrastivity and Sparsity metrics. From the visualization results, RMPEM can also accurately identify the ground truth of either class in binary classification problems. Additionally, ablation studies confirm the effectiveness of our fidelity node importance measure and the sparsity threshold constraint in the RMPEM method.

The discussions above have been conducted with graph classification tasks. However, node classification tasks are similar and might benefit from employing the computational graph of GCN [22] for further optimization of the number of nodes that require sampling. Additionally, since the Monte Carlo method requires a sufficient number of samples to approximate those under the probability distribution, the current approach may perform less effectively on large-scale graph data. Therefore, there's a need to consider more suitable sampling techniques for optimization.

References

1. Adadi, A., Berrada, M.: Peeking inside the black-box: a survey on explainable artificial intelligence (XAI). IEEE Access **6**, 52138–52160 (2018)
2. Baldassarre, F., Azizpour, H.:. Explainability techniques for graph convolutional networks. arXiv preprint arXiv:1905.13686 (2019)
3. Dai, E., et al.: A comprehensive survey on trustworthy graph neural networks: privacy, robustness, fairness, and explainability. arXiv preprint arXiv:2204.08570 (2022)
4. Feng, Q., Liu, N., Yang, F., Tang, R., Du, M., Hu, X.: Degree: decomposition based explanation for graph neural networks. arXiv preprint arXiv:2305.12895 (2023)
5. Funke, T., Khosla, M., Rathee, M., Anand, A.: ZORRO: valid, sparse, and stable explanations in graph neural networks. IEEE Trans. Knowl. Data Eng. **35**, 8687–8698 (2022)

6. Guidotti, R., Monreale, A., Ruggieri, S., Turini, F., Giannotti, F., Pedreschi, D.: A survey of methods for explaining black box models. ACM Comput. Surv. (CSUR) **51**(5), 1–42 (2018)
7. Holdijk, L., Boon, M., Henckens, S., de Jong, L.: [re] parameterized explainer for graph neural network. In: ML Reproducibility Challenge 2020 (2021)
8. Huang, Q., Yamada, M., Tian, Y., Singh, D., Chang, Y.: GraphLIME: local interpretable model explanations for graph neural networks. IEEE Trans. Knowl. Data Eng. **35**, 6968–6972 (2022)
9. Kipf, T.N., Welling, M.: Semi-supervised classification with graph convolutional networks. arXiv preprint arXiv:1609.02907 (2016)
10. Lucic, A., Ter Hoeve, M.A., Tolomei, G., De Rijke, M., Silvestri, F.: CF-GNNexplainer: counterfactual explanations for graph neural networks. In: International Conference on Artificial Intelligence and Statistics, pp. 4499–4511. PMLR (2022)
11. Luo, D., et al.: Parameterized explainer for graph neural network. In: Advance Neural Information Processing System, vol. 33, pp. 19620–19631 (2020)
12. Petsiuk, V., Das, A., Saenko, K.: Rise: randomized input sampling for explanation of black-box models. arXiv preprint arXiv:1806.07421 (2018)
13. Pope, P.E., Kolouri, S., Rostami, M., Martin, C.E., Hoffmann, H.: Explainability methods for graph convolutional neural networks. In: Proceedings of the IEEE/CVF Conference on Computer Vision and Pattern Recognition, pp. 10772–10781 (2019)
14. Schnake, T., et al.: Higher-order explanations of graph neural networks via relevant walks. IEEE Trans. Pattern Anal. Mach. Intell. **44**(11), 7581–7596 (2021)
15. Selvaraju, R.R., Cogswell, M., Das, A., Vedantam, R., Parikh, D., Batra, D.: Grad-CAM: visual explanations from deep networks via gradient-based localization. In: Proceedings of the IEEE International Conference on Computer Vision, pp. 618–626 (2017)
16. Shun, K.T.T., Limanta, E.E., Khan, A.: An evaluation of backpropagation interpretability for graph classification with deep learning. In: 2020 IEEE International Conference on Big Data (Big Data), pp. 561–570. IEEE (2020)
17. Simonyan, K., Vedaldi, A., Zisserman, A.: Deep inside convolutional networks: visualising image classification models and saliency maps. arXiv preprint arXiv:1312.6034 (2013)
18. Sundararajan, M., Taly, A., Yan, Q.: Axiomatic attribution for deep networks. In: International Conference on Machine Learning, pp. 3319–3328. PMLR (2017)
19. Vu, M., Thai, M.T.: PGM-Explainer: probabilistic graphical model explanations for graph neural networks. In: Advances in Neural Information Processing Systems, vol. 33, pp. 12225–12235 (2020)
20. Wang, X., Wu, Y., Zhang, A., He, X., Chua, T.S.: Causal screening to interpret graph neural networks (2020)
21. Xuanyuan, H., Barbiero, P., Georgiev, D., Magister, L.C., Liò, P.: Global concept based interpretability for graph neural networks via neuron analysis. In: Proceedings of the AAAI Conference on Artificial Intelligence, vol. 37, pp. 10675–10683 (2023)
22. Ying, Z., Bourgeois, D., You, J., Zitnik, M., Leskovec, J.: GNNExplainer: generating explanations for graph neural networks. In: Advances in Neural Information Processing Systems, vol. 32 (2019)
23. Yuan, H., Tang, J., Hu, X., Ji, S.: XGNN: towards model-level explanations of graph neural networks. In: Proceedings of the 26th ACM SIGKDD International Conference on Knowledge Discovery & Data Mining, pp. 430–438 (2020)

24. Yuan, H., Haiyang, Yu., Gui, S., Ji, S.: Explainability in graph neural networks: a taxonomic survey. IEEE Trans. Pattern Anal. Mach. Intell. **45**(5), 5782–5799 (2022)
25. Zhang, S., Liu, Y., Shah, N., Sun, Y.: GStarX: explaining graph neural networks with structure-aware cooperative games. In: Advance Neural Information Processing System, vol. 35, pp. 19810–19823 (2022)

RouteExplainer: An Explanation Framework for Vehicle Routing Problem

Daisuke Kikuta(✉), Hiroki Ikeuchi, Kengo Tajiri, and Yuusuke Nakano

NTT Corporation, Tokyo, Japan
daisuke.kikuta@ntt.com

Abstract. The Vehicle Routing Problem (VRP) is a widely studied combinatorial optimization problem and has been applied to various practical problems. While the explainability for VRP is significant for improving the reliability and interactivity of practical VRP applications, it remains unexplored. In this paper, we propose RouteExplainer, a post-hoc explanation framework that explains the influence of each edge in a generated route. Our framework realizes this by rethinking a route as the sequence of actions and extending counterfactual explanations based on the action influence model to VRP. To enhance the explanation, we additionally propose an edge classifier that infers the intentions of each edge, a loss function to train the edge classifier, and explanation-text generation by Large Language Models (LLMs). We quantitatively evaluate our edge classifier on four different VRPs. The results demonstrate its rapid computation while maintaining reasonable accuracy, thereby highlighting its potential for deployment in practical applications. Moreover, on the subject of a tourist route, we qualitatively evaluate explanations generated by our framework. This evaluation not only validates our framework but also shows the synergy between explanation frameworks and LLMs. See https://ntt-dkiku.github.io/xai-vrp for code, appendices, and demo.

Keywords: Vehicle Routing Problem · Explainability · Structural Causal Model · Counterfactual Explanation · Large Language Models

1 Introduction

The Vehicle Routing Problem (VRP) is a combinatorial optimization problem that aims to find the optimal routes for a fleet of vehicles to serve customers. Since Dantzig and Ramser [5] introduced its concept, various VRPs that model real-world problems have been proposed, imposing constraints such as time windows [6], vehicle capacity [5], and minimum prize [16]. Concurrently, various solvers have been proposed, ranging from exact solvers [20] to heuristics [9], Neural Network (NN) solvers [3,11,23], and combinations of them [13,15,24].

While existing works have successfully developed various VRPs and their solvers, explainability for a generated route still remains unexplored. In this paper, we argue that explainability is essential for practical applications such as

D.-N. Yang et al. (Eds.): PAKDD 2024, LNAI 14647, pp. 30–42, 2024.
https://doi.org/10.1007/978-981-97-2259-4_3

responsible or interactive route generation. Furthermore, we argue that explaining how each edge influences the subsequent route is one of the most effective ways to explainability for VRP. For example, in a route for emergency power supply, when asked why the vehicle went to the location instead of other locations at a certain time, a responsible person can justify the decision with the subsequent influence of the movement (i.e., edge). In interactive tourist route generation, the influence of an edge provides hints to tourists who try to modify, based on their preferences, an edge in an automatically generated route.

By rethinking that a route is created by a chain of cause-and-effects (i.e., actions/movements), we can evaluate the subsequent influence of each edge through causal analysis. The Structural Causal Model (SCM) [8] is one of the most popular models to analyze causality. In SCM, causal dependencies among variables are represented by a Directed Acyclic Graph (DAG). Recently, to adapt causal analysis to reinforcement learning models, Madumal et al. [18] introduced the Action Influence Model (AIM), where variables and causal edges are replaced with environment states and actions, respectively. They furthermore proposed a counterfactual explanation based on AIM, which answers *why* and *why-not* questions: why was the action selected instead of other actions at a certain step?

Inspired by this, we propose RouteExplainer, a post-hoc (solver-agnostic) explanation framework that explains the influence of each edge in a generated route. We modify AIM for VRP (we name it Edge Influence Model (EIM)) considering a route as the sequence of actions (i.e., movements/edges). Based on EIM, our framework generates counterfactual explanations for VRP, which answer why the edge was selected instead of other edges at a specific step. To enhance the explanation, we additionally incorporate the intentions of each edge as a metric in the counterfactual explanation, and Large Language Models (LLMs) in explanation-text generation. We here propose a many-to-many sequential edge classifier to infer the intentions of each edge, a modified class-balanced loss function, and in-context learning for LLMs to take in our framework.

In experiments, we evaluate our edge classifier on four different VRP datasets. Our edge classifier outperforms baselines by a large margin in terms of calculation time while maintaining reasonable accuracy (e.g., 85–90% in macro-F1 score) on most of the datasets. The results demonstrate its capability for handling a huge number of requests in practical applications. Lastly, we qualitatively evaluate counterfactual explanations generated by our framework on a practical tourist route, demonstrating the validity of our framework and the effectiveness of the additional components: intentions of each edge and LLM-powered text generation.

The main contributions of this paper are organized as follows:

1) We are the first to argue for the importance of explainability in VRP and propose a novel explanation framework for VRP: its pipeline and EIM.
2) We propose a many-to-many sequential edge classifier that infers the intentions of each edge in a route, which enhances the explanation.
3) To train the edge classifier, we propose a modified class-balanced loss function for step-wise imbalanced classes emerging in our datasets.

4) We leverage LLMs to generate the explanation text in our framework, showing the promise of combining explanation frameworks and LLMs.

2 Related Work

Vehicle Routing Problem. The applications of VRP range widely from truck dispatching [5] and school bus routing [6] to tourist routing [2]. In each application, we need to select an appropriate VRP solver according to its requirements, including problem size, time limit, accuracy, etc. Today, a variety of solvers exist, ranging from exact solvers to heuristics, neural network (NN) solvers, and their combinations. Exact solvers such as Branch-Cut-and-Price [20] may provide the optimal solution, but its calculation cost is expensive when the problem size is large. Heuristics such as LKH [9] provide a near-optimal solution within a reasonable calculation time. In contrast to these conventional ones, NN solvers with supervised learning [12,23] or reinforcement learning [3,14] have been recently proposed. NN solvers realize faster computation and automatic design of (data-driven) heuristics without domain knowledge for a specific VRP. More recently, to take advantage of both NNs and heuristics, combinations of the two have emerged [13,15,24]. To address this diversification of VRP applications and their solvers, this paper proposes a post-hoc explanation framework that can be applied to any VRP solvers.

Explainability. One of the definitions of explainability is the ability to explain the outputs of a model in a way understandable by humans. In the context of eXplainable Artificial Intelligence (XAI), various explainability methods have been proposed [21]. In particular, post-hoc explainability methods such as feature importance-based methods [1,7,17] and causal explanations [8,18] are promising to address the explainability for VRP in a solver-agnostic manner. The former explains the input-output relationships through the future-importance analysis. The latter, on the other hand, structuralizes the dependencies between features (variables) and explains their importance based on the causal analysis. Considering that the former is limited to NN solvers and the latter provides intermediate cause-and-effect results, the latter is more suitable for VRP. Recently, Madumal et al. [18] have adapted causal explanations to actions in reinforcement learning, where variables and directed edges in a DAG are replaced with environment states and actions, respectively. They approximate structural equations with a simple regression model to construct CF examples and define a pipeline to generate counterfactual explanations.

3 Proposed Framework: RouteExplainer

In this section, we introduce the EIM and RouteExplainer. We here discuss them in terms of the Traveling Salesman Problem with Time Windows (TSPTW). Given sets of nodes, their positions, and time windows, TSPTW aims to find the shortest route that visits each node exactly once and returns to the original node, where each node must be visited within its time window.

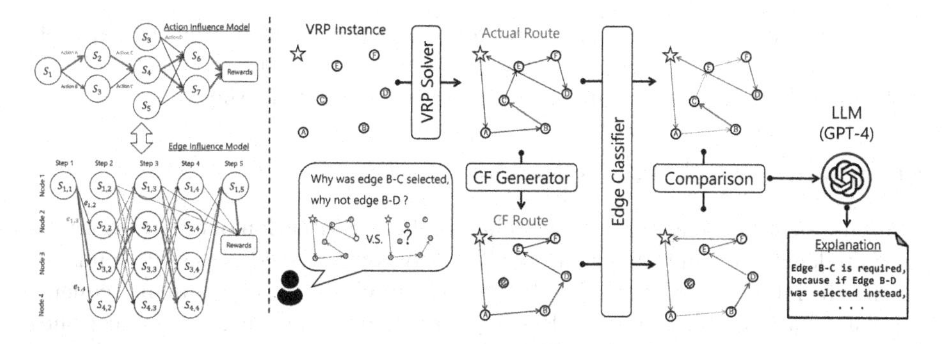

Fig. 1. Left: Action Influence Model [18] v.s. Edge Influence Model (ours). Right: the pipeline of RouteExplainer; it first takes a why and why-not question for VRP and simulates the CF route. The edge classifier then identifies the intentions of each edge in the actual and CF routes, and an LLM (e.g., GPT-4 [19]) generates a counterfactual explanation by comparing the influences of the actual and CF edges. (Color figure online)

The l.h.s. of Fig. 1 shows the comparison between AIM and EIM. In AIM, vertices[1] are environment states $S_{1,\ldots,7}$, and their causal relationships are linked by actions. By contrast, in EIM, vertices are nodes with a *global state* $S_{(i,t)}$ (for TSPTW, intermediate route length/travel time), and their causal relationships are linked by edges of the VRP instance. The bold red path represents a route, which is created by the chain of passed edges. We can intuitively interpret causal relationships in EIM such that the previously visited node affects the global state at the next visited node, e.g., $e_{3,2}$ and $e_{4,2}$ provide different intermediate route lengths at node 2. The structural equation is $f_{\pi_t \to \pi_{t+1}} = S_{(\pi_{t+1},t+1)} = S_{(\pi_t,t)} + \mathrm{d}(\boldsymbol{x}_{v_{\pi_t}}, \boldsymbol{x}_{v_{\pi_{t+1}}})$, where $\mathrm{d}(\cdot,\cdot)$ is a function that computes the distance between two nodes. Counterfactual explanations are generated based on EIM.

The r.h.s of Fig. 1 shows the pipeline of RouteExplainer. Given an automatically generated route (we name it *actual route*), we consider a situation where a user asks why edge B-C was selected at the step and why not edge B-D, based on the user's thought or preference. Our framework answers this question by comparing the influence of the two edges. It first takes this question and simulates a counterfactual (CF) route, which is an alternative route where edge B-D is instead selected at the step. Then, the edge classifier infers the intentions of edges in both the actual and counterfactual routes. Finally, an LLM generates counterfactual explanations for comparing the influence of edges B-C and B-D.

Hereafter, we describe the edge classifier and explanation generation in detail.

Notation. Let $\mathcal{G} = (\mathcal{V}, \mathcal{E}, X)$ be an undirected graph of a VRP instance, where nodes $v_i \in \mathcal{V}; i \in \{1, ..., N\}$ are destinations, $\boldsymbol{x}_{v_i} \in X$ is the node feature, edges $e_{ij} \in \mathcal{E}$ are feasible movements between v_i and v_j; $i \neq j$, and $N(= |\mathcal{V}|)$ is the number of nodes. In this paper, we consider only complete graphs, i.e., all movements between any two nodes are feasible. Given the visiting order of

[1] We call "vertices" for nodes in a DAG and "nodes" for nodes in a VRP instance.

nodes in a route $\boldsymbol{\pi} = (\pi_1, ..., \pi_T)$, the route is represented by the sequence of edges $\boldsymbol{e} = (e_1, ..., e_{T-1}) = (e_{\pi_1\pi_2}, ..., e_{\pi_{T-1}\pi_T})$, where $\pi_t \in \{1, ..., N\}$ is the index of the node visited at step t and T is the sequence length of the route.

3.1 Many-to-Many Edge Classifier

Assuming that each edge in the routes has a different intention of either prioritizing route length or time constraint in a TSPTW route, the edge classifier aims to classify the intentions of each edge. We formulate this as a many-to-many sequential classification, i.e., the edge classifier takes a route (sequence of edges) and sequentially classifies each edge. We here propose a Transformer-based edge classifier, which consists of a node encoder, edge encoder, and classification decoder. In the following, we describe the details of each component and how to train the classifier in supervised learning.

Node Encoder. The node encoder aims to convert input node features into higher-dimensional representations that consider the dependencies with other nodes. As in [14], we here employ the Transformer encoder by Vaswani et al. [22], but without the positional encoding since nodes are permutation invariant. First, the i-th node's input features $\boldsymbol{x}_{v_i} \in \{\mathbb{R}^{D'}, \mathbb{R}^{D}\}$ is projected to node embeddings $\boldsymbol{h}_{v_i} \in \mathbb{R}^H$ with different linear layers for the depot and other nodes.

$$\boldsymbol{h}_{v_i}^{(0)} = \begin{cases} W_{\text{depot}}\boldsymbol{x}_{v_i} + \boldsymbol{b}_{\text{depot}} & i = 1, \\ W\boldsymbol{x}_{v_i} + \boldsymbol{b} & i = 2, \ldots, N, \end{cases} \tag{1}$$

where $W_{\text{depot}} \in \mathbb{R}^{H \times D'}, W \in \mathbb{R}^{H \times D}$, $\boldsymbol{b}_{\text{depot}}, \boldsymbol{b} \in \mathbb{R}^H$ are projection matrices and biases for the depot and other nodes, respectively. See Appendix B for details of input features. Then we obtain the final node embeddings by stacking L Transformer encoder layers XFMR, as follows:

$$\boldsymbol{h}_{v_i}^{(l)} = \text{XFMR}_{v_i}^{(l)}(\boldsymbol{h}_{v_1}^{(l-1)}, \cdots, \boldsymbol{h}_{v_N}^{(l-1)}), \tag{2}$$

where l indicates l-th layer.

Edge Encoder. The edge encoder generates edge embeddings for the edges in the input route. For the embedding of the edge at step t, it simply concatenates the final node embeddings of both ends of the edge and the global state at that step.

$$\boldsymbol{h}_{e_t}^{(0)} = W_{\text{edge}}\left[\boldsymbol{h}_{v_{\pi_t}}^{(L)} || \boldsymbol{h}_{v_{\pi_{t+1}}}^{(L)} || \boldsymbol{s}\right], \tag{3}$$

where $\boldsymbol{s} \in \mathbb{R}^{D_{\text{st}}}$ is the environment state value, $||$ indicates concatenation w.r.t. feature dimensions, and $W_{\text{edge}} \in \mathbb{R}^{H \times (2H + D_{\text{st}})}$ is the projection matrix.

Classification Decoder. The decoder takes the sequence of the edge embeddings and outputs probabilities of each edge being classified into classes. Similar to the node encoder, we here employ the Transformer encoder, but with causal masking, i.e., it considers only the first t edges when computing the embedding of the edge at step t. This causality ensures the consistency of the predicted labels of common edges in the actual and CF routes. In addition, we do not use any positional encodings as positional information is already included in the environment state value $\boldsymbol{s}$. We obtain the new edge embeddings by stacking L' Transformer encoder layers, as follows:

$$\boldsymbol{h}_{e_t}^{(l)} = \text{XFMR}_{e_t}^{(l)}(\boldsymbol{h}_{e_1}^{(l-1)}, \ldots, \boldsymbol{h}_{e_t}^{(l-1)}). \tag{4}$$

Finally, we obtain the probabilities of the edge at step t being classified into the classes of intentions with a linear projection and the softmax function.

$$\boldsymbol{p}_{e_t} = \text{Softmax}\left(W_{\text{dec}}\boldsymbol{h}_{e_t}^{(L')} + \boldsymbol{b}_{\text{dec}}\right), \tag{5}$$

where $W_{\text{dec}} \in \mathbb{R}^{C \times H}, \boldsymbol{b}_{\text{dec}} \in \mathbb{R}^{C}$ are the projection and bias that adjust the output dimension to the number of classes C. In inference, the predicted class is determined by the argmax function, i.e., $\hat{y}_{e_t} = \text{argmax}_c(\boldsymbol{p}_{e_t})$.

Training. For supervised learning, we need labels of edges in routes. In this paper, we employ a rule-based edge annotation for simplicity and to remove human biases (see Appendix B for the details of annotation). Importantly, the advantage of machine learning models is that they can also accommodate manually annotated data. We train the edge classifier with the generated labels. A challenge in our problem setting is that the class ratio changes over each step. In TSPTW, we observe a transitional tendency such that time window priority is the majority class in the early steps, whereas route length priority becomes the majority as the steps progress. Usually, a weighted loss function and class-balanced loss function [4] are used for class-imbalanced data. However, they only consider the class ratio in the entire training batch, which fails to capture step-wise class imbalances as seen in TSPTW datasets. Therefore, we propose a cross-entropy loss that considers step-wise class imbalance. We adjust the class-balanced loss to many-to-many sequential classification by calculating the weights separately at each step, as follows,

$$J = -\sum_{T \in \mathcal{T}} \sum_{b=1}^{B_T} \sum_{t=1}^{T} \sum_{c=0}^{C-1} \frac{1-\beta}{1-\beta^{\sum_b \mathrm{I}\left(y_{e_t^b}=c\right)}} y_{e_t^b} \log\left((\boldsymbol{p}_{e_t^b})_c\right), \tag{6}$$

where $y_{e_t^b}$ is the true label of the edge at t in the route of the instance b, β is a hyperparameter that corresponds to the ratio of independent samples, $\mathrm{I}(\cdot)$ is the Boolean indicator function, $\mathcal{T}$ is the set of sequence lengths included in the training batch, B_T is the number of samples whose sequence length is T, and $(\boldsymbol{p}_{e_t^b})_c$ indicates the c-th element of $\boldsymbol{p}_{e_t^b}$.

3.2 Counterfactual Explanation for VRP

In this section, we describe the explanation generation by defining why and why-not questions in VRP, CF routes, and counterfactual explanations.

Why and Why-Not Questions in VRP. In the context of VRP, a why and why-not question asks *why edge B-C was selected at the step and why not edge B-D*. Formally, we define the why and why-not question in VRP as a set:

$$\mathcal{Q} = (\boldsymbol{e}^{\text{fact}}, t_{\text{ex}}, e^{\text{fact}}_{t_{\text{ex}}}, e^{\text{cf}}_{t_{\text{ex}}}), \tag{7}$$

where $\boldsymbol{e}^{\text{fact}}$ is the actual route, $t_{\text{ex}} \in \{1, \ldots, T-1\}$ is the explained step, $e^{\text{fact}}_{t_{\text{ex}}}$ is the actual edge, and $e^{\text{cf}}_{t_{\text{ex}}} (\neq e^{\text{fact}}_{t_{\text{ex}}})$ is the CF edge, which is an edge that was not selected at t_{ex} in the actual route but could have been (i.e., *edge B-D* in the above example).

Counterfactual Routes. Given a why and why-not question $\mathcal{Q}$, our framework generates its CF route. The CF route $\boldsymbol{e}^{\mathcal{Q}}$ is the optimal route that includes both the sequence of edges before the explained step in the actual route and the CF edge:

$$\boldsymbol{e}^{\mathcal{Q}} = \text{SOLVER}\left(\text{VRP}, \mathcal{G}, \boldsymbol{e}_{\text{fixed}}\right), \tag{8}$$

$$\boldsymbol{e}_{\text{fixed}} = (e^{\text{fact}}_1, \ldots, e^{\text{fact}}_{(t_{\text{ex}}-1)}, e^{\text{cf}}_{t_{\text{ex}}}), \tag{9}$$

where $\boldsymbol{e}^{\mathcal{Q}}[: t_{\text{ex}}] = \boldsymbol{e}^{\text{fact}}[: t_{\text{ex}}]$, $\boldsymbol{e}[: t'] := (e_1, \ldots, e_{t'-1})$, $e^{\mathcal{Q}}_{t_{\text{ex}}} = e^{\text{cf}}_{t_{\text{ex}}}$, VRP is the VRP where the actual route was solved, and SOLVER is the VRP solver. Intuitively, the CF route simulates the best-effort route for the remaining subsequent steps when a CF edge is selected instead of the actual edge at the explained step.

Explanation Generation. To evaluate the influence of an edge, we leverage the global state of visited nodes in EIM, the intentions of each edge, and other metrics such as feasibility ratio. Given a route $\boldsymbol{\pi} = (\pi_1, \ldots, \pi_T)$ and the intentions of edges in the route $\hat{\boldsymbol{y}} = (\hat{y}_1, \ldots, \hat{y}_{T-1})$, we generally define the influence of the edge e_t as a tuple:

$$\boldsymbol{I}_{e_t} = \left(S_{(\pi_{t+1},t+1)}, \hat{y}_{t+1}, S_{(\pi_{t+2},t+2)}, \ldots, \hat{y}_{T-1}, S_{(\pi_T,T)}\right). \tag{10}$$

Furthermore, we shall construct a *minimally complete explanation* [18] so that one can effectively understand the influence of an edge. We here use representative values calculated from Eq. (10), which include objective values, the ratio of each class, the feasibility ratio, and so on:

$$\mathcal{R}_{e_t} = \mathcal{F}_{\text{rep}}(\boldsymbol{I}_{e_t}), \tag{11}$$

where $\mathcal{R}_{e_t}$ is the set of representative values and $\mathcal{F}_{\text{rep}}$ is the set of representative value functions. For TSPTW, $\mathcal{F}_{\text{rep}} = (f^{\text{s}}_{\text{obj}}, f^{\text{l}}_{\text{obj}}, f_{\text{class}}, f_{\text{feas}})$, where the short-term objective value $f^{\text{s}}_{\text{obj}}(\boldsymbol{I}_{e_t}) = S_{(\pi_{t+1},t+1)}$, the long-term objective value $f^{\text{l}}_{\text{obj}}(\boldsymbol{I}_{e_t}) = S_{(\pi_T,T)}$, the class ratio $f_{\text{class}}(\boldsymbol{I}_{e_t}) = \frac{1}{T-t-1}\sum_{t'=t+1}^{T-1} \mathrm{I}(\hat{y}_{t'} = 0)$, and the feasibility ratio $f_{\text{feas}}(\boldsymbol{I}_{e_t}) = \frac{|\{v_{\pi_{t+1}}, \ldots, v_{\pi_T}\}|}{|\mathcal{V} \setminus \{v_{\pi_1}, \ldots, v_{\pi_t}\}|}$.

The counterfactual explanation is generated by comparing the representative values of actual and CF edges, as follows,

$$\mathcal{X}_{\mathcal{Q}} = \mathcal{F}_{\text{compare}}\left(\mathcal{R}_{e^{\text{fact}}_{t_{\text{ex}}}}, \mathcal{R}_{e^{\mathcal{Q}}_{t_{\text{ex}}}}\right), \tag{12}$$

where $\mathcal{X}_{\mathcal{Q}}$ is the counterfactual explanation for $\mathcal{Q}$, $\mathcal{R}_{e^{\text{fact}}_{t_{\text{ex}}}}, \mathcal{R}_{e^{\mathcal{Q}}_{t_{\text{ex}}}}$ are the sets of representative values of actual and CF edges, and $\mathcal{F}_{\text{compare}}$ is the function that compares the two sets of representative values (we here use element-wise difference operation). Finally, we generate an explanation text with the compared representative values using GPT-4 [19], an LLM. NLP templates require complicated conditional branches to generate user-friendly explanations, whereas LLMs realize this only with simple natural language instructions. We incorporate our framework into GPT-4 by writing the description of our framework and an example of explanation texts in the input context (i.e., In-context learning). See Appendix E for details of GPT-4 configurations, including the system architecture and system prompts. In practice, we also leverage visual information, which is demonstrated in the experiments.

4 Experiments

Datasets. We evaluate our edge classifier on ACTUAL-ROUTE-DATASETS and CF-ROUTE-DATASETS. Each includes routes and their edge labels in TSPTW, Prize Collecting TSP (PCTSP), PCTSP with Time Windows (PCTSPTW), and Capacitated VRP (CVRP) with $N = 20, 50$. The labels are annotated by the rule-based annotation in the previous section. ACTUAL-ROUTE-DATASETS are split into training, validation, and test splits. Figure 2 shows the statistics of the training split in each VRP of ACTUAL-ROUTE-DATASETS. See Appendix B for the details of VRP datasets.[2]

Baselines. As baselines against our edge classifier, we use the annotation strategy with different VRP solvers, including LKH [9], Concorde,[3] and Google OR Tools[4] (one of them corresponds to the ground truth). Note that the baselines are limited to classification labeled by the rule-based annotation, whereas our edge classifier is capable of handling classification labeled by other strategies such as manual annotation. See Appendix C for the details of the baselines.

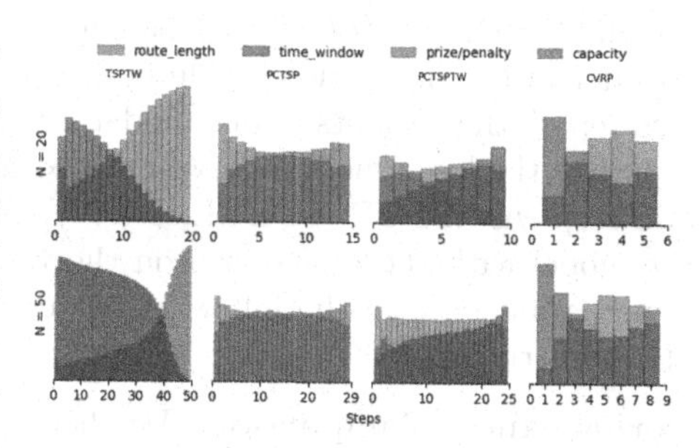

Fig. 2. The class ratio of edges w.r.t. steps, on each training split. The class ratios are for samples in which the number of visited nodes is the mode.

4.1 Quantitative Evaluation of the Edge Classifier

Table 1 shows the quantitative results in classification on ACTUAL-ROUTE-DATASETS and CF-ROUTE-DATASETS: the macro f1 score (MF1) and total

[2] Our code is publicly available at https://github.com/ntt-dkiku/route-explainer.
[3] https://www.math.uwaterloo.ca/tsp/concorde/.
[4] https://github.com/google/or-tools.

Table 1. Macro-F1 (MF1) and calculation time (Time) on ACTUAL-ROUTE-DATASETS and CF-ROUTE-DATASETS (10K evaluation samples). The ground truth is grayed out.

	ACTUAL-ROUTE-DATASETS															
	TSPTW				PCTSP				PCTSPTW				CVRP			
	$N=20$		$N=50$		$N=20$		$N=50$		$N=20$		$N=50$		$N=20$		$N=50$	
	MF1	Time	MF1	Time	MF1	Time	MF1	Time	MF1	Time	MF1	Time	MF1	Time	MF1	Time
LKH	100	(2 min)	100	(4 min)	100	(1 min)	100	(2 min)	–	–	–	–	100	(2 min)	100	(5 min)
Concorde	**100**	(2 min)	**99.9**	(8 min)	**99.9**	(2 min)	**99.6**	(3 min)	–	–	–	–	90.5	(3 min)	**96.3**	(12 min)
ORTools	96.5	(16 s)	94.3	(4 min)	93.5	(14 s)	84.9	(3 min)	100	(25 s)	100	(7 min)	**93.5**	(24 s)	89.1	(6 min)
EC^{-enc}_{scbce}	**93.1**	(1 s)	**92.0**	(1 s)	78.4	(1 s)	75.5	(1 s)	66.8	(1 s)	60.8	(1 s)	83.5	(2 s)	82.8	(2 s)
EC^{-dec}_{cbce}	90.9	(1 s)	90.8	(1 s)	**84.4**	(1 s)	**79.2**	(1 s)	**76.1**	(1 s)	**65.6**	(1 s)	**88.6**	(2 s)	**87.1**	(2 s)
EC_{ce}	**94.5**	(1 s)	**93.1**	(1 s)	**90.2**	(1 s)	**84.8**	(1 s)	78.4	(1 s)	**69.2**	(1 s)	**91.1**	(2 s)	**88.9**	(3 s)
EC_{cbce}	**94.5**	(1 s)	**93.1**	(1 s)	89.9	(1 s)	84.4	(1 s)	**78.7**	(1 s)	69.0	(1 s)	**91.1**	(2 s)	**88.9**	(3 s)
EC_{scbce}	94.2	(1 s)	92.5	(1 s)	89.9	(1 s)	84.7	(1 s)	76.6	(1 s)	67.4	(1 s)	90.5	(2 s)	88.2	(3 s)
	CF-ROUTE-DATASETS															
	TSPTW				PCTSP				PCTSPTW				CVRP			
	$N=20$		$N=50$		$N=20$		$N=50$		$N=20$		$N=50$		$N=20$		$N=50$	
	MF1	Time	MF1	Time	MF1	Time	MF1	Time	MF1	Time	MF1	Time	MF1	Time	MF1	Time
LKH	100	(2 min)	100	(4 min)	100	(1 min)	100	(3 min)	-	-	-	-	100	(2 min)	100	(5 min)
Concorde	**99.7**	(2 min)	**99.9**	(7 min)	**100**	(2 min)	**99.6**	(6 min)	-	-	-	-	90.8	(3 min)	**96.2**	(10 min)
ORTools	96.8	(15 s)	94.6	(4 min)	**93.9**	(14 s)	85.7	(4 min)	100	(24 s)	100	(6 min)	94.0	(25 s)	89.5	(6 min)
EC^{-enc}_{scbce}	**91.3**	(1 s)	90.2	(1 s)	78.2	(1 s)	76.0	(1 s)	66.1	(1 s)	62.0	(1 s)	80.4	(2 s)	81.5	(3 s)
EC^{-dec}_{cbce}	88.8	(1 s)	**90.6**	(1 s)	**83.9**	(1 s)	**80.0**	(1 s)	**73.9**	(1 s)	**66.7**	(1 s)	**86.4**	(2 s)	**86.3**	(2 s)
EC_{ce}	**93.2**	(1 s)	**92.2**	(1 s)	**88.2**	(1 s)	84.1	(1 s)	76.3	(1 s)	**69.6**	(1 s)	89.1	(2 s)	88.5	(3 s)
EC_{cbce}	93.1	(1 s)	91.9	(1 s)	88.1	(1 s)	**84.4**	(1 s)	**76.9**	(1 s)	**69.6**	(1 s)	**89.3**	(2 s)	**88.6**	(3 s)
EC_{scbce}	92.7	(1 s)	92.0	(1 s)	88.0	(1 s)	84.2	(1 s)	75.2	(1 s)	67.9	(1 s)	**89.3**	(2 s)	88.0	(3 s)

inference time (Time). Here, we report the results of baselines and ablations of our model: $EC^{-enc}_{scbce}/EC^{-dec}_{cbce}$ is the edge classifier that replaces the Transformer encoder in the node encoder/decoder with Multi-Layer Perceptron (MLP). We also report the variants of our model: EC_{ce}, EC_{cbce}, and EC_{scbce}. The subscript indicates the loss function: cross-entropy loss (CE), class-balanced CE (CBCE), and step-wise CBCE (SCBCE, Eq. (6)). All the results of our model are from the model with the best epoch on the validation split. See Appendix D for the hyperparameters. In the following, we discuss the performance comparison and ablation studies.

Performance Comparison. We here focus on the variants of our model: EC_{ce}, EC_{cbce}, and EC_{scbce}. EC^{-enc}_{scbce} and EC^{-dec}_{cbce} are discussed in ablation study section. For ACTUAL-ROUTE-DATASETS, our models significantly improve the inference time while retaining 85–95% of MF1 on most VRPs. In PCTSPTW, MF1 remains around 69–78%, but it is the only three-class classification and not so low compared to random classification (i.e., 33.3%). Real-time response is essential for practical applications. The baselines require more than 3 min for 10K samples, whereas our models require less than 3 s, thereby demonstrating their potential for rapidly handling a huge number of requests in practical applications. For CF-ROUTE-DATASETS, the results are similar to those in ACTUAL-ROUTE-DATASETS. Comparing MF1 in both datasets, our models reduce MF1 in CF-ROUTE-DATASETS by only 1–2%. Rather, MF1 increases slightly in some

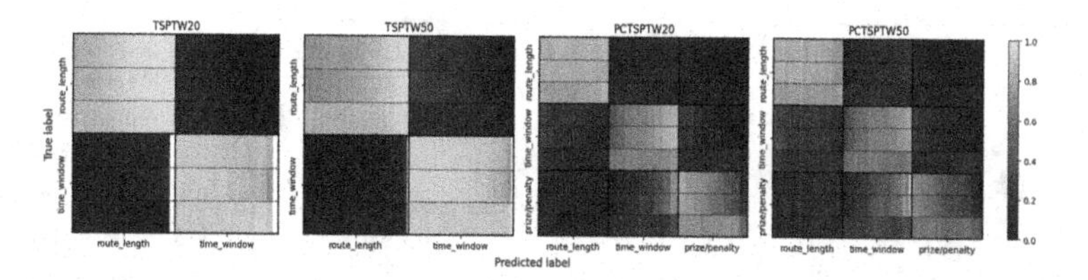

Fig. 3. Sequential confusion matrices that consist of $C \times C$ grids, where each grid visualizes the normalized confusion matrix value at each step as a heatmap in the order of steps from left to right. The top, middle, and bottom in each grid correspond to $\mathrm{EC}_{\mathrm{ce}}$, $\mathrm{EC}_{\mathrm{cbce}}$, and $\mathrm{EC}_{\mathrm{scbce}}$, respectively.

VRPs. This generalization ability for CF routes demonstrates that our model can be applied to both actual and CF routes in our framework.

Ablation Study. In terms of ablation studies of model architecture, we compare $\mathrm{EC}^{-\mathrm{enc}}_{\mathrm{scbce}}$ and $\mathrm{EC}^{-\mathrm{dec}}_{\mathrm{cbce}}$.[5] Table 1 shows that the tendency of the drop of MF1 by the ablation varies among VRPs. In TSPTW, $\mathrm{EC}^{-\mathrm{enc}}_{\mathrm{scbce}}$ drops 0.5–1.8% from $\mathrm{EC}_{\mathrm{scbce}}$, whereas $\mathrm{EC}^{-\mathrm{dec}}_{\mathrm{cbce}}$ drops 1.4–3.3% from $\mathrm{EC}_{\mathrm{cbce}}$. In the other VRPs, $\mathrm{EC}^{-\mathrm{enc}}_{\mathrm{scbce}}$ drops 5.4–11.5% from $\mathrm{EC}_{\mathrm{scbce}}$, whereas $\mathrm{EC}^{-\mathrm{dec}}_{\mathrm{cbce}}$ drops 1.8–5.5% from $\mathrm{EC}_{\mathrm{cbce}}$. These results show that the significance of the encoder and decoder increases in the other VRPs, in which the number of visited nodes per route varies (Fig. 2). Furthermore, the significance of the decoder outweighs that of the encoder in TSPTW, while the tendency is the opposite in the other VRPs.

For ablation studies of loss functions, we compare $\mathrm{EC}_{\mathrm{ce}}$, $\mathrm{EC}_{\mathrm{cbce}}$, and $\mathrm{EC}_{\mathrm{scbce}}$. Table 1 shows that MF1 of $\mathrm{EC}_{\mathrm{scbce}}$ slightly drops compared to the other two models on most datasets. However, it is insufficient to evaluate models based solely on a central metric (i.e., MF1), particularly in the presence of step-wise class imbalances. Figure 3 shows the sequential confusion matrix [10] of the three models on TSPTW and PCTSPTW with $N = 20, 50$. We observe that the error tendency of $\mathrm{EC}_{\mathrm{ce}}$ and $\mathrm{EC}_{\mathrm{cbce}}$ strongly depends on the step-wise class imbalance shown in Fig. 2, i.e., they tend to fail to classify the minority class at each step. On the other hand, $\mathrm{EC}_{\mathrm{scbce}}$ successfully reduces the misclassification of the minority class while retaining reasonable accuracy for the other classes. Therefore, it is advisable to employ $\mathrm{EC}_{\mathrm{scbce}}$ in the presence of step-wise class imbalances, while $\mathrm{EC}_{\mathrm{ce}}$ is more suitable when such imbalances are absent.

4.2 Qualitative Evaluation of Generated Explanations

In this section, we qualitatively evaluate explanations generated by our framework. As a case study in TSPTW, we consider a tourist route that visits historical buildings in Kyoto within their time windows. Each destination includes information about a user-defined time window, duration of stay, and remarks (e.g., take lunch). The travel time between two destinations is calculated using Google

[5] $\mathrm{EC}^{-\mathrm{dec}}_{\mathrm{cbce}}$ employs CBCE instead of SCBCE since MLP does not consider sequence.

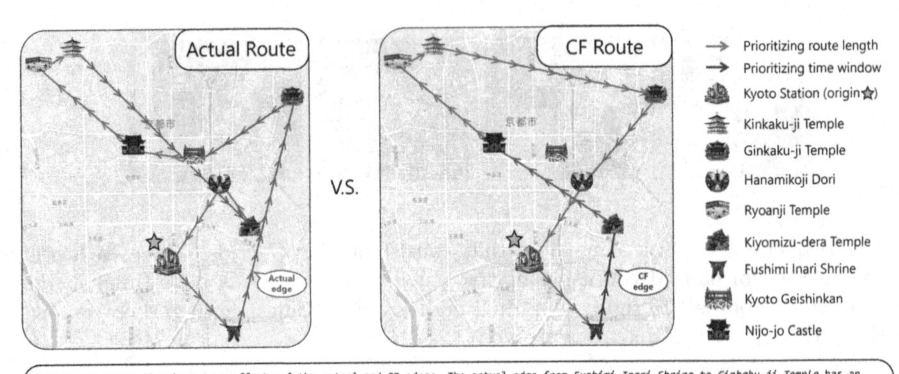

Firstly, we assess the short-term effects of the actual and CF edges. The actual edge from *Fushimi Inari Shrine* to *Ginkaku-ji Temple* has an immediate travel time of 26 minutes, while the CF edge to *Kiyomizu-dera Temple* reduces this time to 16 minutes, saving 10 minutes initially. Secondly, we consider the long-term effects. The actual route has a total travel time of 132.0 minutes, ensuring all nodes are visited. The CF route, however, has a total travel time of 115.0 minutes, which is 17.0 minutes shorter. This reduction comes at the cost of missing the *Kyoto Geishinkan*, where tourists can attend an English-guided tour and have lunch. Thirdly, the influence on subsequent routes is notable. The actual route includes a balance of time window and route length intentions, while the CF route, after the user preference edge, follows a pattern of prioritizing route length, potentially compromising adherence to time windows. In summary, the actual edge is selected to ensure that all destinations, including the culturally enriching experience at *Kyoto Geishinkan*, are visited within their time windows, despite a slightly longer travel time. The CF edge, while reducing travel time, leads to missing an important node and shifts the subsequent route intentions towards route length, potentially affecting the overall experience.

Fig. 4. An explanation generated by our framework, in the case study of TSPTW.

Maps. Figure 4 shows the generated explanation text with the visualization of the actual and CF routes. Here, a user tries to understand the influence of the actual edge from *Fushimi Inari Shrine* to *Ginkaku-ji Temple* so that the user considers whether the actual edge could be replaced with the user's preference (i.e., the CF edge towards *Kiyomizu-dera Temple*). The generated explanation successfully describes the importance of the actual edge while mentioning each component in Eq. (12). The visualization of the intentions of each edge helps the user understand the immediate effect of each edge. Thanks to LLMs, the explanation succeeds in detailing the losses from not visiting the *Kyoto Geishinkan*, incorporating the remarks. Overall, the generated explanation assists the user in making decisions to edit the edges of the automatically generated route.

5 Conclusion and Future Work

In this paper, we proposed a post-hoc explanation framework that explains the influence of each edge in a VRP route. We introduced EIM, a novel causal model, and the pipeline of generating counterfactual explanations based on EIM. Furthermore, we enhanced the explanation with the classifier inferring the intentions of each edge and LLM-powered explanation generation. Through quantitative evaluation of the classifier and qualitative evaluation of generated explanations, we confirmed the validity of our framework and the effectiveness of LLMs in explanation generation. We hope this paper sheds light on both explainability in VRP and the combination of explanation frameworks and LLMs.

In future work, we will address two current problems: inadequate classification performance for some VRPs and the limitation of the rule-based edge annotation for more complicated VRPs. We will address them by increasing training datasets and parameters of our model and leveraging LLM-powered annotation.

References

1. Bach, S., Binder, A., Montavon, G., Klauschen, F., Müller, K.R., Samek, W.: On pixel-wise explanations for non-linear classifier decisions by layer-wise relevance propagation. PLOS ONE **10**(7), 1–46 (2015)
2. Balas, E.: The prize collecting traveling salesman problem and its applications. In: Gutin, G., Punnen, A.P. (eds.) The Traveling Salesman Problem and Its Variations. Combinatorial Optimization, vol. 12, pp. 663–695. Springer, Boston (2007). https://doi.org/10.1007/0-306-48213-4_14
3. Bello, I., Pham, H., Le, Q.V., Norouzi, M., Bengio, S.: Neural combinatorial optimization with reinforcement learning (2017)
4. Cui, Y., Jia, M., Lin, T.Y., Song, Y., Belongie, S.: Class-balanced loss based on effective number of samples. In: IEEE/CVF Conference on Computer Vision and Pattern Recognition, pp. 9260–9269 (2019)
5. Dantzig, G.B., Ramser, J.H.: The truck dispatching problem. Manage. Sci. **6**(1), 80–91 (1959)
6. Dumas, Y., Desrosiers, J., Gelinas, E., Solomon, M.M.: An optimal algorithm for the traveling salesman problem with time windows. Oper. Res. **43**(2), 367–371 (1995)
7. Fong, R.C., Vedaldi, A.: Interpretable explanations of black boxes by meaningful perturbation. In: 2017 IEEE International Conference on Computer Vision, pp. 3449–3457 (2017)
8. Halpern, J.Y., Pearl, J.: Causes and explanations: a structural-model approach. Part I: causes. Br. J. Philos. Sci. **56**(4), 843–887 (2005)
9. Helsgaun, K.: An extension of the Lin-Kernighan-Helsgaun TSP solver for constrained traveling salesman and vehicle routing problems: Technical report (2017)
10. Hinterreiter, A., et al.: ConfusionFlow: a model-agnostic visualization for temporal analysis of classifier confusion. IEEE Trans. Vis. Comput. Graph. **28**(2), 1222–1236 (2022)
11. Hopfield, J., Tank, D.: Neural computation of decisions in optimization problems. Biol. Cybern. **52**, 141–152 (1985)
12. Joshi, C.K., Laurent, T., Bresson, X.: An efficient graph convolutional network technique for the travelling salesman problem (2019)
13. Kool, W., van Hoof, H., Gromicho, J., Welling, M.: Deep policy dynamic programming for vehicle routing problems (2021)
14. Kool, W., van Hoof, H., Welling, M.: Attention, learn to solve routing problems! In: International Conference on Learning Representations (2019)
15. Li, S., Yan, Z., Wu, C.: Learning to delegate for large-scale vehicle routing. In: Thirty-Fifth Conference on Neural Information Processing Systems (2021)
16. Lopez, L., Carter, M.W., Gendreau, M.: The hot strip mill production scheduling problem: a Tabu search approach. Eur. J. Oper. Res. **106**(2–3), 317–335 (1998)
17. Lundberg, S.M., Lee, S.I.: A unified approach to interpreting model predictions. In: Advances in Neural Information Processing Systems, vol. 30 (2017)
18. Madumal, P., Miller, T., Sonenberg, L., Vetere, F.: Explainable reinforcement learning through a causal lens. In: AAAI Conference on Artificial Intelligence (2019)
19. OpenAI: GPT-4 technical report (2023)
20. Pessoa, A.A., Sadykov, R., Uchoa, E., Vanderbeck, F.: A generic exact solver for vehicle routing and related problems. Math. Program. **183**, 483–523 (2020)

21. Saeed, W., Omlin, C.: Explainable AI (XAI): a systematic meta-survey of current challenges and future opportunities. Knowl. Based Syst. **263**, 110273 (2023). https://www.sciencedirect.com/science/article/pii/S0950705123000230
22. Vaswani, A., et al.: Attention is all you need. In: Advances in Neural Information Processing Systems, vol. 30 (2017)
23. Vinyals, O., Fortunato, M., Jaitly, N.: Pointer networks. In: Advances in Neural Information Processing Systems, vol. 28 (2015)
24. Xin, L., Song, W., Cao, Z., Zhang, J.: NeuroLKH: combining deep learning model with Lin-Kernighan-Helsgaun heuristic for solving the traveling salesman problem. In: Advances in Neural Information Processing Systems, vol. 34 (2021)

On the Efficient Explanation of Outlier Detection Ensembles Through Shapley Values

Simon Klüttermann(✉), Chiara Balestra, and Emmanuel Müller

TU Dortmund University, Dortmund, Germany
{simon.kluettermann,chiara.balestra,emmanuel.mueller}@cs.tu-dortmund.de

Abstract. Feature bagging models have revealed their practical usability in various contexts, among them in outlier detection, where they build ensembles to reliably assign outlier scores to data samples. However, the interpretability of so-obtained outlier detection methods is far from achieved. Among the standard black-box models interpretability approaches, we find Shapley values that clarify the roles of single inputs. However, Shapley values are characterized by high computational runtimes that make them useful in pretty low-dimensional applications. We propose *bagged Shapley values*, a method to achieve interpretability of feature bagging ensembles, especially for outlier detection. The method not only assigns local importance scores to each feature of the initial space, helping to increase the interpretability but also solves the computational issue; specifically, the *bagged Shapley values* can be exactly computed in polynomial time.

Keywords: Explainable Machine Learning · Polynomial Shapley values · Outlier Detection

1 Introduction

Detecting anomalous samples is a crucial task in various domains, ranging from fraud detection in financial systems [12] to identifying defective components in manufacturing processes [10]. Outlier detection can generally be categorized into two paradigms: *supervised* and *unsupervised*. Supervised methods rely on labeled data, explicitly defining and identifying anomalies during model training [11]. In contrast, unsupervised techniques operate without labeled anomalies, making them particularly valuable when labeled data is scarce or costly. Unsupervised outlier detection encompasses a myriad of approaches, each with

This work was supported by the Lamarr-Institute for ML and AI, the research training group *Dataninja*, the Research Center Trustworthy Data Science and Security, the Federal Ministry of Education and Research of Germany and the German federal state of NRW. The Linux HPC cluster at TU Dortmund University, a project of the German Research Foundation, provided the computing power.

D.-N. Yang et al. (Eds.): PAKDD 2024, LNAI 14647, pp. 43–55, 2024.
https://doi.org/10.1007/978-981-97-2259-4_4

unique strengths and limitations. One notable trend for unlabeled data involves ensemble methods; Ensemble techniques [31] leverage the diversity outputs of multiple potentially different base models to produce more robust and accurate predictions, thus directly enhancing the performance and reliability of outlier detection algorithms [1]. The nature of the base models divides the ensembles among heterogeneous or homogeneous [14]; Some algorithms, e.g., DEAN [5] and IForest [16], use homogeneity to profit from a higher number of submodels.

The importance of outlier detection underscores the need for accurate and interpretable methods. Shapley values [25] have emerged as a promising technique for interpreting the contributions of individual features in black-box models. They offer mathematical guarantees of fairness that make them an attractive choice for outlier detection as well. However, their practical application poses a significant challenge due to the requirement of training an exponentially large number of models. While significant progress has been made in anomaly detection interpretability [15], challenges persist. The trade-off between interpretability and model complexity transferred to the computational complexity of the feature importance scores, thus remains an interesting topic of investigation.

In response to this challenge, we propose an innovative approach that leverages modern ensemble methods to approximate Shapley values efficiently and makes outlier detection methods based on feature bagging interpretable. First, we delve into the details, defining the *bagged Shapley values* and presenting a theoretical proof of our approach. The experimental results demonstrate our method's effectiveness in achieving efficient interpretability in outlier detection tasks with complex, high-dimensional data. The code is available on Github[1].

2 Related Work

Ensemble Methods for Outlier Detection. Ensemble methods emerged as a powerful paradigm for improving outlier detection algorithms w.r.t. reliability and performance [14]. Ensemble methods comprehend bagging [3], boosting [24], and stacking [23]. Bagging involves training multiple base models (e.g., k-nearest neighbors, Support Vector Machines, neural networks, among others) on possibly bootstrapped data samples and aggregating their predictions. Adapted to outlier detection, the ensemble's collective decision provides more robust results [1]. Homogeneity among the base models' types characterizes *homogeneous* outlier ensembles: Submodels usually differ only by a different initialization. DEAN [5] and Isolation Forest (IForest) [16] are prime examples of outlier detection methods employing such homogeneous ensembles; DEAN is based on multiple neural networks, while IForest relies on a collection of isolation trees.

Shapley Values, for Intepretability and Beyond. Shapley values [25] originate from Cooperative Game Theory. Since their first applications, they gained

[1] https://github.com/KDD-OpenSource/ensemble_shapley.

prominence as a powerful tool for increasing the interpretability of machine learning black-box models [18,21,26]. Shapley values offer a theoretically sound framework for quantifying the impact of each feature or factor in a model's prediction; the scores, being the average marginal contribution across all possible feature combinations, are robust and interpretable. Attributing the contributions of the individual features revealed helpful for outlier detection [28,29], where Shapley values provide valuable insights into the importance of features in identifying anomalies. However, their practical use is contrasted by one significant challenge, namely their computational complexity. The exact computation of Shapley values requires evaluating a *value function* for every possible subset of players. Thus, the consequent exponential blow-up in computational cost soon renders their use for high-dimensional contexts infeasible. Approximation techniques have been implemented to make Shapley values more accessible; These include Monte Carlo sampling, stratification of players, and kernel approximations [4,6,7,18,27]. Each method addresses the efficient computation of Shapley values differently, with potential accuracy and computational cost trade-offs.

Interpretability for Anomaly Detection Methods Interpreting anomaly detection methods is essential for understanding *why* single data points are considered anomalous, e.g., in safety-critical applications. Feature importance analysis plays an essential role [15]: Techniques such as feature attribution [9] are employed to highlight which features have the most significant impact on the detection. Additionally, we find rule-based models [19], decision trees [20], and model-agnostic techniques like LIME [21] and SHAP [18] to shed light on the decision-making process of anomaly detection models. Furthermore, visualizations are essential for enhancing trust in complex scenarios. Examples are heatmaps, scatter plots, and time-series representations [2,13].

3 Outlier Detection Ensembles

In our context, a set $X \subseteq \mathbb{R}^N$ of data points can be parted into two subsets: the set of *normal observation* indicated with X_{nor}, and the set of abnormal observations, indicated with X_{abn}. In unlabeled data, distinguishing normal from anomalous data is not always straightforward. We consider a model for outlier detection f, that aims at classifying each data point $x \in X$ as either *normal* or *anomalous*. Among the various anomaly detection methods, we focus on methods that provide to each data point a score measuring its *outlierness*.

Definition 1. *Given a set of data points* X*, we call* model *a function* $a : X \mapsto \mathbb{R}$ *where* $a(x)$ *represents the outlier score assigned by* a *to the sample* x*.*

The higher the value $a(x)$, the more likely x is considered to be an anomaly compared to the set X. On the same set X, various outlier detection models can be constructed. We indicate with $\mathcal{M}_X$ the set of models constructed on X.

Definition 2 (Ensemble). *Given a set of (sub)models $\mathcal{M}_X$, an ensemble is a function $A_{\mathcal{M}_X} : X \mapsto \mathbb{R}$ that assigns to each $x \in X$ its average outlier score, i.e.,*

$$A_{\mathcal{M}_X}(x) = \frac{1}{\|\mathcal{M}_X\|} \sum_{a \in \mathcal{M}_X} a(x). \tag{1}$$

The ensemble prediction is the average submodel prediction in the set $\mathcal{M}_X$.

Using the trick of projected data points in lower dimensional spaces, we reach the definition of bagging. We indicate with $\mathcal{N}$ the set of coordinates of X and with X_I the set of data points in X projected only on the $I \subseteq \mathcal{N}$ coordinates (or *features*), i.e., given $x \in X$ the corresponding point $x_I = (x_i)_{i \in I}$ and $I \subset \mathcal{N}$. Now we can define a subset $\mathcal{M}_{X_I} \subseteq \mathcal{M}_X$ as the set of submodels that belongs to $\mathcal{M}_X$ trained only on X_I.

The bagging procedure is meant to randomly cover the information in X, considering only the projection of X in smaller-sized subsets. We refer to the size of the data points in the projection as *bag*. Having $X \subseteq \mathbb{R}^N$ fixed and bag $\leq N$, we can get $\binom{N}{\text{bag}}$ different subsets of size bag from the N features.

Definition 3 (Bagging). *After fixing the* bag, *the* bagging procedure *consists in defining the model $b_{S,a} \in \mathcal{M}_S$ such that $b_{S,a}(x)$ is the result of a model a when trained on the data set X_S and S is a subset of N whose size is $|S| = bag$.*

The bagging procedure does not fix either the model a from $\mathcal{M}_X$ or the set $S \subseteq \mathcal{N}$, thus potentially covering, using sufficiently many random seeds, all the information contained in X. We write $b_{S,a|\text{seed}}$ for the specific *bagging submodel* resulting after we fixed the seed for the random sampling of S and the model a. Finally, we can construct the so-called *feature bagging* ensemble based on the bagging technique.

Definition 4 (Feature Bagging). *Given a dataset X and a set of models $\mathcal{M}_X$, we define the function $f_{\mathcal{M}_X} : X \mapsto \mathbb{R}$ such that it assign to each $x \in X$ the score defined as*

$$f_{\mathcal{M}_X}(x) = \lim_{n \to \infty} \frac{1}{n} \sum_{j=0}^{n} b_{S,a|seed[j]}(x). \tag{2}$$

where seed is an eventually infinite vector of randomly drawn seeds.

A similar definition could also be made for non-outlier detection ensembles as long as the output is a linear combination of the submodel predictions. Still, feature bagging is most commonly used in outlier detection.

4 The *bagged Shapley Values*

A cooperative game is a pair $(\mathcal{N}, v)$ where $\mathcal{N}$ represents the set of *players*, and v is a function over the subsets of $\mathcal{N}$. v assigns to each *coalition* of players a worth, i.e., a positive real number representing the score obtained by the players as a team. Usually, the monotonicity of the value function is assumed, i.e., $v(\mathcal{A}) \leq v(\mathcal{B})$ if $\mathcal{A} \subseteq \mathcal{B} \subseteq \mathcal{N}$.

The Shapley values are a *fair* assignment of weights to the single players that consider the role of the single players in any single coalition. Given a game $(\mathcal{N}, v)$, $\phi_v(i)$ represents the Shapley value of player i:

Definition 5 (Shapley Value). *Given a game $(\mathcal{N}, v)$, the Shapley value $\phi_v(i)$ of player i is defined as*

$$\phi_v(i) = \sum_{\mathcal{S} \subseteq \mathcal{N}, i \notin \mathcal{S}} \frac{|\mathcal{S}|! \cdot (|\mathcal{N}| - |\mathcal{S}| - 1)!)}{|\mathcal{N}|!} [v(\mathcal{S} \cup \{i\}) - v(\mathcal{S})] \tag{3}$$

We refer to $v(\mathcal{S} \cup \{i\}) - v(\mathcal{S})$ as the *marginal contribution of i to $\mathcal{S}$*. Shapley values have a flexible and straightforward definition, depending only on v and the number of players; this made them the object of study in various circumstances and applications. However, their computation results in an NP-hard problem that approximation approaches can only partly solve. We show that the exact computation of Shapley value-similar scores for feature bagging ensembles can be easily reduced to a polynomial time.

We introduce the *bagged Shapley values*; their definition perfectly aligns with the impossibility of training an ensemble method with less than *bag* features. We rewrite the definition of Shapley values from Eq. (3) $\phi_{f_{\mathcal{M}_X}(x)}(i)$ for feature bagging ensembles, where $x \in X \subseteq \mathbb{R}^N$ is a data point, $f_{\mathcal{M}_X}$ is the feature bagging model and we are interested in assigning to the coordinate i of X an importance score in predicting the overall outlier score $f_{\mathcal{M}_X}(x)$. We define the bagged Shapley values:

Definition 6 (bagged Shapley Value). *Given a set of data points $X \subseteq \mathbb{R}^N$, a set of (sub)models $\mathcal{M}_X$ and a feature bagging model $f_{\mathcal{M}_X}$ defined over $\mathcal{M}_X$, the* bagged Shapley values *are the values*

$$\tilde{\phi}_{f_{\mathcal{M}_X}(x)}(i) = \sum_{S \subseteq \mathcal{N}, i \notin S, s \geq bag} \frac{N}{N - bag} \frac{s! \cdot (N - s - 1)!)}{N!} \left[f_{M_{X_{S \cup \{i\}}}}(x) - f_{M_{X_S}}(x) \right] \tag{4}$$

This equation removes terms with magnitude $\propto \frac{\text{bag}}{N}$, a necessary step, as defining an ensemble model with less than *bag* features is not possible. Notice that the higher the dimension of the data points in X is, the smaller the difference between $\tilde{\phi}_{f_{\mathcal{M}_X}(x)}(i)$ and $\phi_{f_{\mathcal{M}_X}(x)}(i)$. To somewhat correct for this difference, we add a factor $\frac{N}{N - \text{bag}}$ to compensate that we are summing over fewer subsets of $\mathcal{N}$.

5 Theoretical Guarantees for the Approximation

The main result of our studies regards the chance to express Shapley values with a limited number of selected bagging submodels, thus avoiding the exponential computational costs of Shapley values.

Theorem 1. *The* bagged Shapley values *can be expressed using a selection of submodels involved in the feature bagging ensemble $f_{\mathcal{M}_X}$. In particular, it holds*

$$\tilde{\phi}_{f_{\mathcal{M}_X}(x)}(i) \propto f_{\mathcal{M}_X}(x) - f_{\mathcal{M}_{X_{\mathcal{N}\setminus i}}}(x).$$

Proof. To increase readability, we use the notation

$$k(S, N) = \frac{N}{N - \text{bag}} \frac{s!(N - s - 1)!}{N!}$$

where $s = |S|$ and $N = |\mathcal{N}|$. For abuse of notation and readability, we write S instead of X_S throughout the whole proof.

Now, we can rewrite the *bagged Shapley values* in the following way $b_{S,a|\text{seed}}$ and substitute it with $b_{|\text{seed}} \in \mathcal{M}_S$

$$\begin{aligned}
\tilde{\phi}_{f_{\mathcal{M}_X}(x)}(i) &= \sum_{S \subseteq \mathcal{N}, i \notin S, s \geq \text{bag}} k(S, N) \left[f_{S \cup \{i\}}(x) - f_S(x) \right] \\
&= \lim_{n \to \infty} \sum_{S \subseteq \mathcal{N}, i \notin S, s \geq \text{bag}} k(S, N) \\
&\cdot \left(\frac{\sum_{j=0,\ldots,n, b \in \mathcal{M}_{S \cup \{i\}}} b_{|\text{seed}}(x)}{\|\mathcal{M}_{S \cup \{i\}}\|} - \frac{\sum_{j=0,\ldots,n, b \in \mathcal{M}_S} b_{|\text{seed}}(x)}{\|\mathcal{M}_S\|} \right)
\end{aligned}$$

where $\mathcal{M}_K = \{a \in \mathcal{M}_X \mid a \text{ restricted to features in } K\}$ is the subset of models that contain only features included in K.

From the previous equation, we see that $\tilde{\phi}_{f_{\mathcal{M}_X}(x)}(i)$ is a sum over the same bagging models multiple times, as they are part of various subsets. We can simplify the writing to evaluate each model only once but weight them.

$$\tilde{\phi}_{f_{\mathcal{M}_X}(x)}(i) = \lim_{n \to \infty} \frac{1}{\|\mathcal{M}_X\|} \sum_{b \in \mathcal{M}_X} \alpha_b \cdot b_{|\text{seed}}(x) - \frac{1}{\|\mathcal{M}_{\mathcal{N}\setminus i}\|} \sum_{b \in \mathcal{M}_{\mathcal{N}\setminus i}} \beta_b \cdot b_{|\text{seed}}(x) \quad (5)$$

Noting that we can shuffle our feature labels without changing Eq. 5, $\alpha_b = \alpha$ and $\beta_b = \beta$ have to be independent on the specific model $b_{|\text{seed}}$. By the same argument, α and β can not depend on the model outputs $b_{|\text{seed}}(x)$. This allows us to choose any model $b(x)$ to compute them; we pick here

$$b(x) = \begin{cases} 1 \text{ if model } b \text{ considers feature i} \\ 0 \text{ else} \end{cases} . \quad (6)$$

Using the proposed $b(x)$, the β term disappears, thus we can write α as:

$$\begin{aligned}
\alpha &= \lim_{n \to \infty} \frac{\sum_{S \subseteq \mathcal{N}, i \notin S, |S| \geq \text{bag}} k(S, N) \frac{\|\mathcal{M}_X\|}{\|\mathcal{M}_{X_{S \cup \{i\}}}\|} \sum_{b \in \mathcal{M}_{X_{S \cup \{i\}}}} b(x)}{\sum_{b \in \mathcal{M}_X} b(x)} \\
&= \lim_{n \to \infty} \frac{\sum_{S \subseteq \mathcal{N}, i \notin S, |S| \geq \text{bag}} k(S, N) \cdot \frac{\text{count}(\mathcal{M}_{X_{S \cup \{i\}}})}{\|\mathcal{M}_{X_{S \cup \{i\}}}\|}}{\frac{\text{count}(\mathcal{M}_X)}{\|\mathcal{M}_X\|}}
\end{aligned}$$

where $\text{count}(\mathcal{M}_{X_K})$ is the number of models in $\mathcal{M}_{X_K}$ that contain one specific feature in K. We can use $\lim_{n\to\infty} \frac{\text{count}(\mathcal{M}_{X_K})}{\|\mathcal{M}_{X_K}\|} = \frac{\binom{|K|-1}{\text{bag}-1}}{\binom{|K|}{\text{bag}}} = \frac{\text{bag}}{|K|}$ thus getting

$$\begin{aligned}
\alpha =& N\frac{N}{N-\text{bag}}\sum_{s=\text{bag}}^{N-1}\binom{N}{s}\cdot\frac{s!(N-s-1)!}{N!}\cdot\frac{1}{s+1}\\
=&\frac{N}{N-\text{bag}}\sum_{s=\text{bag}}^{N-1}\cdot\frac{1}{s+1}\\
=&\frac{N}{N-\text{bag}}\cdot(\psi^0(N+1)-\psi^0(\text{bag}+1))
\end{aligned}$$

with the digamma function ψ^0.

When instead of choosing $b(x)$ to be independent of i, we find that $\tilde{\phi}_{f_{\mathcal{M}_X}(x)}(i) \propto (\alpha-\beta)$. But since the feature is designed not to have any effect, we know that $\tilde{\phi}_{f_{\mathcal{M}_X}(x)}(i) = 0$ and thus $\alpha = \beta$. This concludes the proof. □

The results not only show that the bagged Shapley value is proportional to the difference of two feature bagging, respectively defined on $\mathcal{M}_X$ and $\mathcal{M}_{X_{\mathcal{N}\setminus i}}$, but also that when using bagging models, we can estimate the bagged Shapley values in polynomial time. This is because for deterministic submodels, instead of using ∞ of them, we only need to train $\binom{N}{bag} < N^{bag}$ submodels.

6 Experiments

We evaluate our approach on various freely available real-world datasets with varying numbers of features [8,17,30]. We conduct experiments on the correctness of the approximation (Sect. 6.1), the effectiveness (Sect. 6.2), and the scalability (Sect. 6.3) of our approach.

6.1 Quality of the Approximation

To fairly investigate the approximation accuracy of the bagged Shapley values, we use a low-dimensional dataset, i.e., the five-dimensional phoneme dataset [30], thus requiring the training of feature bagging ensemble models only 2^5 times. The low-dimensionality of the dataset allows us to compute the non-approximated version of Shapley values without incurring extremely long runtimes. We train isolation trees from [16] with a bagging size of 2 and simplify the anomaly score from [16] to fit our methodology by using the negative average path length over all trees as an indicator of anomalies. We train one million submodels and average the obtained results to guarantee consistent and robust results. The total training takes about 70min of CPU time[2]. The ROC-AUC score is 0.733.

[2] All experiments were performed on Intel Xeon E5 CPUs. In the paper, we stick to CPUs over GPUs also when we use neural network submodels; the choice is justified by the higher amount of parallelization they allow.

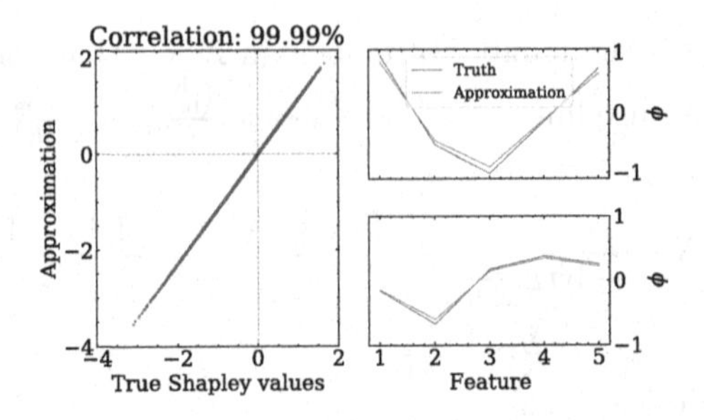

Fig. 1. Left: Plot of the bagged Shapley values against the exact Shapley values for each data sample in the phoneme dataset. Right: Shapley values and their approximation for two example samples. The color-coding of the features is represented in the legend.

We separate the trained models into ensembles for each subset of them and compute the exact Shapley values and the bagged Shapley values. We combine the values obtained into Fig. 1. As the mapping lies on the diagonal line, we conclude that the approximation works well on all data points.

6.2 Effectiveness

We can compute the bagged Shapley values for datasets whose dimensions are too high for an exact computation. We focus on the MNIST dataset [8], a collection of images of hand-written digits usually used to train image-recognition models. Following the approach of [22], we consider normal all images representing a handwritten 7, and anomalous the images representing other digits. Each image has a resolution of $28 \cdot 28$, i.e., we handle 784 features in each image. Computing the exact Shapley values for the single pixels requires $2^{784} \approx 10^{237}$ evaluations, a number significantly larger than the computational power available.

For the bagged Shapley values, we use the bagging size bag $= 32$. We train two models: we use DEAN, a deep learning model-based ensemble, and a shallow isolation forest [16]. We choose DEAN [5] because of its inherited feature bagging and relatively low training time per submodel. The training time is significantly longer than using IForest[3]. Note that we do not only train a model on each possible subset, as the number of subsets is still $\binom{768}{32} \approx 4 \cdot 10^{32}$. Instead, we train on random subsets until the result converges. This also helps deal with the random nature of our algorithms.

Figure 2 represents the plots of the Shapley values for five representative samples in the form of heatmaps; bright colors represent high score, i.e., features highly increasing the outlier score. Each heatmap, both for DEAN and IFOR, highlights the changes to the original input that would make it closer to a normal observation by highlighting the erroneous regions. From the left to the right

[3] The isolation forest takes about 220min of CPU time. DEAN requires about 113days; However, the independent ensembles are easy to parallelize, and less accurate results can already be achieved with ten thousand submodels (27hours).

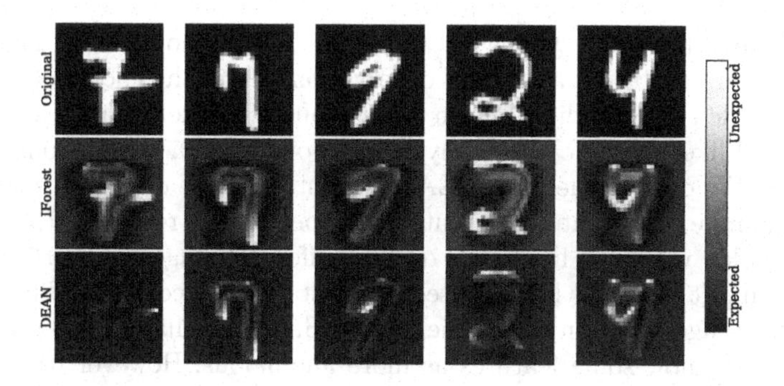

Fig. 2. MNIST dataset. The original images are in the top row. The bottom rows contain the derived *bagged Shapley values* heatmap for ISOR and DEAN. We rescaled the color legend to the upper and lower bound of the Shapley values in each plot.

side, the first two input images are labeled as normal; however, they still contain features that are not expected, e.g., the middle horizontal line in the first image. These unexpected features are highlighted in bright red/yellow. Similarly, the other three images obtain high outlier scores, although they contain typical features for normal input images. These features are also unexpected by the model and thus result in high Shapley values. Examples are represented by the nine and the four; removing the lower line from the circle would make the nine more similar to a normal observation, while adding a horizontal line to the top would make the four more similar to a seven.

Comparing DEAN and IForest, we see how the understanding of the normal concept, i.e., the digit "seven", of the isolation forest is too simple to explain the predictions entirely. In the second column of Fig. 2, we see that the isolation forest expects the tail of the seven to bend instead of going straight down. On the other hand, DEAN, based on a deep learning method, has less difficulty in learning a broader concept of seven. This is also reflected in the outlier detection performance: While DEAN reaches a ROC-AUC of 0.9698 on the dataset, the isolation forest only reaches a lower 0.9118 score. We strongly believe that the bagged Shapley images provide useful insights into what the model understood and learned from the training data, additionally to better performance measured by the ROC-AUC metric.

6.3 Scalability

We select the celebA dataset [17] to study how the approach scales to larger datasets. celebA contains images with $218 \cdot 178 = 38804$ pixels, which we convert to grayscale to simplify the plotting later. In the previous section, we showed how complex patterns can overwhelm outlier detection ensembles that struggle to learn a proper schema for normal and abnormal data points. Thus, we aim to maximize the separation between normal and abnormal classes in order to simplify the learning task. We divide the dataset into normal and anomalous instances, where we characterize a normal observation being labeled with the attributes "female", "young", "attractive", and "not bald". The inverse attributes characterize an abnormal observation. Here, the choice of attributes was only guided by the distribution of attributes in the dataset, and similar results would likely have followed any other choices for the anomalous and normal classes. We only trained the DEAN ensemble on the dataset, as the model proved to handle complicated attributed data better. We represent the obtained bagged Shapley values as heatmaps on five different images in Fig. 3. The first row is the input image, while the second contains the corresponding Shapley values. The images resulting from the bagged Shapley values plotting have high resolution and show some features as more anomalous; However, the designed features do not match the designed separation in normal and abnormal images. This can also be seen in the ROC-AUC score of 0.6184. The most anomalous features seem to be (from left to right) the bindi, the partially covered forehead, the shirt collar, the laugh lines, and the skin paint transition. These are rare features in the images of young women in celebA, thus considered anomalous by the model. Still, the complexity of the separation is likely too big for the available samples (≈ 72000), and thus, the learning, as shown by the ROC-AUC, is inaccurate. Although the features outlined are not the expected ones from our understanding of the separation between the two classes, it is worth noticing how the bagged Shapley value maps can be used to understand and improve the outlier detection models. The runtime of the training procedure for one million DEAN submodels is ≈ 468 days; training 500 submodels at the same time requires about 4 days of CPU time. We use 4 millions of submodels in our training setup, under parallelization assumption, and set up the bagging size to be bag $= 32$. A different bagging size might have achieved more accurate results, but we did not optimize it since, in most contexts, the outlier detection task sets the bagging size.

Fig. 3. Analysis on the celebA dataset. Heatmaps show the bagged Shapley value; Brighter colors indicate higher values.

We finally want to characterize the minimum number of submodels needed for our methodology to perform well. For this, we calculate the bagged Shapley values maps so that each feature is used 10, 100, and 1000 times. The corresponding maps for the center image from Fig. 3 are shown in Fig. 4. While some features are already visible at about 12000 submodels, the noise level being still very high, facial features are undetectable; with about 10, those become visible while extensively the number of submodels to about 100 times more, they have become clear. As a rule of thumb, we suggest training $10 \cdot N$ features to visualize the basic features and to train $10 \cdot N^{\frac{3}{2}}$ for clear images.

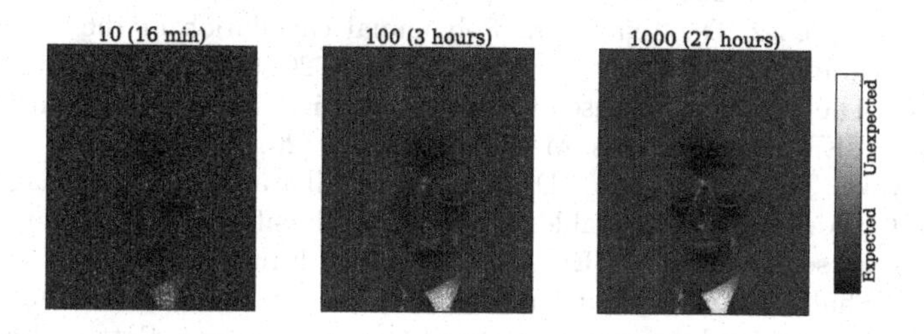

Fig. 4. Shapley value maps for different numbers of submodels. Here we use 12127, 121263 and 1212625 submodels, so that each feature is approximately sampled 10, 100 and 1000 times. The times stated assume a parallelization with 500 CPUs.

7 Conclusions

Detecting and explaining outlier can be highly complicated. Shapley values by their side offer a flexible definition, easily applicable to this context; However, their high computational costs represents an often insormontable downside that makes their exact computation often unfeasible.

We combine Shapley values with ensemble techniques, specifically focusing on feature-bagging ensembles for outlier detection. The *bagged Shapley values* offer an advantageous reduction of the computational costs, giving the chance to compute importance scores for settings with tens of thousands of features. Furthermore, we showed the value of highlighting anomalous features in images to obtain insights into the features learned by the outlier detection method.

We believe that combining Shapley values with ensemble methods can boost the use of Shapley values in the Machine Learning community, showing advantages from a computational and interpretability point of view, as well as lead to better, more reliable, outlier detection models.

References

1. Ali, K.M., Pazzani, M.J.: Error reduction through learning multiple descriptions. Mach. Learn. **24**, 173–202 (1996)
2. Balestra, C., Li, B., Müller, E.: slidshaps - sliding shapley values for correlation-based change detection in time series. In: DSAA (2023)
3. Breiman, L.: Bagging predictors. Mach. Learn. **24**, 123–140 (1996)

4. Burgess, M.A., Chapman, A.C.: Approximating the shapley value using stratified empirical bernstein sampling. In: IJCAI (2021)
5. Böing, B., Klüttermann, S., Müller, E.: Post-robustifying deep anomaly detection ensembles by model selection. In: ICDM (2022)
6. van Campen, T., Hamers, H., Husslage, B., Lindelauf, R.: A new approximation method for the shapley value applied to the WTC 9/11 terrorist attack. Soc. Netw. Anal. Min. **8**, 1–12 (2018)
7. Castro, J., Gómez, D., Tejada, J.: Polynomial calculation of the shapley value based on sampling. Comput. Oper. Res. **36**(5), 1726–1730 (2009)
8. Deng, L.: The MNIST database of handwritten digit images for machine learning research. IEEE Signal Process. Mag. **29**, 141–142 (2012)
9. Dissanayake, T., Fernando, T., Denman, S., Sridharan, S., Ghaemmaghami, H., Fookes, C.: A robust interpretable deep learning classifier for heart anomaly detection without segmentation. IEEE J. Biomed. Health Inform. **25**, 2162–2171 (2021)
10. Dong, L., Shulin, L., Zhang, H.: A method of anomaly detection and fault diagnosis with online adaptive learning under small training samples. Pattern Recogn. **64**, 374–385 (2017)
11. Han, S., Hu, X., Huang, H., Jiang, M., Zhao, Y.: Adbench: anomaly detection benchmark. In: NeurIPS (2022)
12. Hilal, W., Gadsden, S.A., Yawney, J.: Financial fraud: a review of anomaly detection techniques and recent advances. Expert Syst. Appl. **193**, 116429 (2022)
13. Kadir, T., Brady, M.: Saliency, scale and image description. Int. J. Comput. Vision **45**(2), 83–105 (2001)
14. Klüttermann, S., Müller, E.: Evaluating and comparing heterogeneous ensemble methods for unsupervised anomaly detection. In: IJCNN (2023)
15. Li, Z., Zhu, Y., Van Leeuwen, M.: A survey on explainable anomaly detection. ACM Trans. Knowl. Discovery Data **18**, 1–54 (2023)
16. Liu, F.T., Ting, K.M., Zhou, Z.H.: Isolation forest. In: ICDM (2008)
17. Liu, Z., Luo, P., Wang, X., Tang, X.: Deep learning face attributes in the wild. In: ICCV (2015)
18. Lundberg, S.M., Lee, S.I.: A unified approach to interpreting model predictions. In: Advances in Neural Information Processing Systems, vol. 30 (2017)
19. Müller, E., Keller, F., Blanc, S., Böhm, K.: Outrules: a framework for outlier descriptions in multiple context spaces. In: ECML PKDD (2012)
20. Park, C.H., Kim, J.: An explainable outlier detection method using region-partition trees. J. Supercomput. **77**, 3062–3076 (2021)
21. Ribeiro, M.T., Singh, S., Guestrin, C.: "why should i trust you?" explaining the predictions of any classifier. In: KDD (2016)
22. Ruff, L., et al.: Deep one-class classification. In: ICML (2018)
23. Sandim, M.O.: Using Stacked Generalization for Anomaly Detection. Ph.D. thesis
24. Schapire, R.E., et al.: A brief introduction to boosting. In: IJCAI (1999)
25. Shapley, L.S.: A value for n-person games. Contributions to the Theory of Games (1953)
26. Strumbelj, E., Kononenko, I.: An efficient explanation of individual classifications using game theory. J. Mach. Learn. Res. **11**, 1–18 (2010)
27. Štrumbelj, E., Kononenko, I.: Explaining prediction models and individual predictions with feature contributions. Knowl. Inf. Syst. **41**(3), 647–665 (2014)
28. Takahashi, T., Ishiyama, R.: FIBAR: fingerprint imaging by binary angular reflection for individual identification of metal parts. In: EST (2014)

29. Tallón-Ballesteros, A., Chen, C.: Explainable AI: using shapley value to explain complex anomaly detection ml-based systems. Mach. Learn. Artif. Intell. **332**, 152 (2020)
30. Triguero, I., et al.: Keel 3.0: An open source software for multi-stage analysis in data mining. Int. J. Comput. Intell. Syst. **10**, 1238–1249 (2017)
31. Zimek, A., Campello, R.J., Sander, J.: Ensembles for unsupervised outlier detection: challenges and research questions a position paper. SIGKDD Expl. Newslet. **15**, 11–22 (2014)

Interpreting Pretrained Language Models via Concept Bottlenecks

Zhen Tan[1(✉)], Lu Cheng[2], Song Wang[3], Bo Yuan[4], Jundong Li[3], and Huan Liu[1]

[1] Arizona State University, Tempe, AZ, USA
{ztan36,huanliu}@asu.edu

[2] University of Illinois Chicago, Chicago, IL, USA
lucheng@uic.edu

[3] University of Virginia, Charlottesville, VA, USA
{sw3wv,jundong}@virginia.edu

[4] Zhejiang University, Zhejiang, China
byuan@zju.edu.cn

Abstract. Pretrained language models (PLMs) have made significant strides in various natural language processing tasks. However, the lack of interpretability due to their "black-box" nature poses challenges for responsible implementation. Although previous studies have attempted to improve interpretability by using, e.g., attention weights in self-attention layers, these weights often lack clarity, readability, and intuitiveness. In this research, we propose a novel approach to interpreting PLMs by employing high-level, meaningful concepts that are easily understandable for humans. For example, we learn the concept of "Food" and investigate how it influences the prediction of a model's sentiment towards a restaurant review. We introduce C^{3}M, which combines human-annotated and machine-generated concepts to extract hidden neurons designed to encapsulate semantically meaningful and task-specific concepts. Through empirical evaluations on real-world datasets, we show that our approach offers valuable insights to interpret PLM behavior, helps diagnose model failures, and enhances model robustness amidst noisy concept labels.

Keywords: Language Models · Interpretability · Conceptual Learning

1 Introduction

Although Pretrained Language Models (PLMs) like BERT [6] have achieved remarkable success in various NLP tasks [35], they are frequently regarded as black boxes, posing significant obstacles to their responsible deployment in real-world scenarios, particularly in critical domains such as healthcare [14]. To date, many existing works [19] leverage attention weights extracted from the self-attention layers to provide token-level or phrase-level importance. These low-level explanations are found unfaithful [31] and lack readability and intuitiveness [17], leading to unstable or even unreasonable explanations. To address

D.-N. Yang et al. (Eds.): PAKDD 2024, LNAI 14647, pp. 56–74, 2024.
https://doi.org/10.1007/978-981-97-2259-4_5

these limitations, we seek to explain via human-comprehensible *concepts* that use more abstract features (e.g., general notions) as opposed to raw input features at the token level [32]. The foundation of this work is the Concept Bottleneck Models (CBMs) [14] that interprets deep models (e.g., ResNet [10]) for image classification tasks using high-level concepts (e.g., shape). For NLP tasks such as sentiment analysis, concepts can be Food, Ambiance, and Service as shown in Fig. 1, where each concept corresponds to a neuron in the concept bottleneck layer. The final decision layer is then a linear function of these concepts. Using concepts greatly improves the readability and intuitiveness of the explanations compared to low-level features such as "lobster".

We propose to study *Concept-Bottleneck-Enabled Pretrained Language Models* (CBE-PLMs). There are two key challenges: ❶ First, existing CBMs [14,32] require human-annotated concepts. This can be challenging for natural language since the annotator may need to read through the entire text to understand the context and label one concept [21]. This limits the practical usage and scalability of CBE-PLMs. ❷ Second, many studies have identified the tradeoff between interpretability and task accuracy using CBMs since the predetermined concepts may leave out important information for target task prediction [32]. Therefore, it is crucial to improve both interpretability and task performance to achieve optimal interpretability-utility tradeoff.

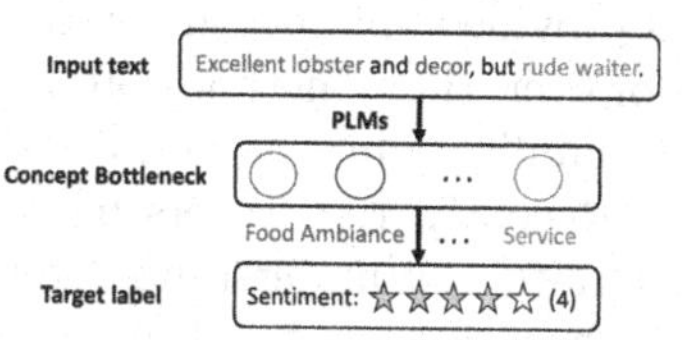

Fig. 1. The illustration of CBE-PLMs. Via PLMs, the input text x is mapped into an intermediate layer consisting of a set of human-comprehensible concepts c, which are then used to predict the target label y.

❶ To tackle the first challenge, we propose leveraging Large Language Models (LLMs) trained on extensive human-generated corpora and feedbacks, such as ChatGPT [23], to identify novel concepts in text and generate pseudo-labels (via prompting) for unlabeled concepts. Recent studies [23] exhibit that these LLMs encapsulate significant amounts of human common sense knowledge. By augmenting the small set of human-specified concepts with machine-generated concepts, we increase concept diversity and useful information for prediction. In addition, generated pseudo-labels offer us a large set of instances with noisy concept labels, complementing the smaller set of instances with clean labels. ❷ To further improve interpretability-utility tradeoff (second challenge), we propose to learn from noisy concept labels and incorporate a concept-level MixUp mechanism [33] that allows CBE-PLMs to cooperatively learn from both noisy and clean concept sets. We name our framework for training CBE-PLMs as *ChatGPT-guided Concept augmentation with Concept-level Mixup* (C^{3}M). In summary, our contributions include:

- We provide the first investigation of utilizing CBMs for interpreting PLMs.
- We propose C^{3}M, which leverages LLMs and MixUp to help PLMs learn from human-annotated and machine-generated concepts. C^{3}M liberates CBMs from predefined concepts and enhances the interpretability-utility tradeoff.
- We demonstrate the effectiveness and robustness of test-time concept intervention for the learned CBE-PLMs for common text classification tasks.

2 Related Work

2.1 Interpreting Pretrained Language Models

PLMs such as Word2Vec [20], BERT [6], and the more recent GPT series [23] have demonstrated impressive performance in various NLP tasks. However, their opaque nature poses a challenge in comprehending how PLMs work internally [7]. In order to improve the interpretability and transparency of PLMs, researchers have explored different approaches, such as visualizing attention weights [9], probing feature representations [3], and using counterfactuals [26], among others, to provide explanations at the local token-level, instance-level, or neuron-level. However, these methods often lack faithfulness and intuitiveness, and are of poor readability, which undermines their trustworthiness [19].

Recently, researchers have turned to global concept-level explanations that are naturally understandable to humans. Although this level of interpretability has been less explored in NLP compared to computer vision [12], it has gained attention. For instance, a study [28] investigates gender classification bias by examining the association of occupation words such as "nurse" with gender. In addition, the CBMs [14] have emerged as novel frameworks for achieving concept-level interpretability in lightweight image classification systems. CBMs typically involve a layer preceding the final fully connected classifier, where each neuron corresponds to a concept that can be interpreted by humans. CBMs also show advantages in improving accuracy through human intervention during testing. Yet, the application of CBMs to larger-scale PLMs interpretation is under-explored. Implementing CBMs necessitates human involvement in defining the concept set and annotating the concept labels. Such requirements are challenging for natural language as humans may need to read through the entire text to understand the context and label one concept [21].

2.2 Learning from Noisy Labels

Addressing inaccurately labeled or misclassified data in real-world scenarios is the goal of learning from noisy labels, with techniques including noise transition matrix estimation [15], robust risk minimization [8], and more. Recently, the resilience of semi-supervised learning methods like MixMatch [2] and FixMatch [27] to label noise has been discovered by using pseudo-labels for unlabeled data. Inspired by them, we propose to utilize an LLM (ChatGPT) as a fixed-label guesser, generating noisy intermediate concept labels to potentially predict task labels.

Notably, CBMs specialize in the interpretation and interactability of deep models for general classification tasks. While *Multi-Aspect Sentiment Analysis* [34] (MASA) shares similar goals when using aspects as concepts, it differs as concepts are not confined to fine-grained aspectual features and can be abstract ideas or broader notions throughout entire contexts. Aspect labels in MASA, primarily used for prediction accuracy, are not always mandatory. To summarize, this study pioneers the comprehensive exploration of utilizing concepts for interpreting large-scale PLMs, and provides a robust framework for harnessing the noisy signals from LLMs to achieve interpretable outcomes from lighter-weight PLMs, which can be easily understood by users.

3 Enable Concept Bottlenecks for PLMs

3.1 Problem Setup

We focus on interpreting the predictions of fine-tuned PLMs for both classification and regression tasks. Given data $\mathcal{D} = \{(x^{(i)}, y^{(i)}, c^{(i)})_{i=1}^{n}\}$, where $x \in \mathbb{R}^d$ is the original text input, $y \in \mathbb{R}$ is the target label, and $c \in \mathbb{R}^k$ is a vector of k concepts from the concept set $\mathcal{C}$ with $|\mathcal{C}| = k$. We consider a PLM f_θ encoder that embeds an input text $x \in \mathbb{R}^d$ into its latent representation $z \in \mathbb{R}^e$. Vanilla fine-tuning strategy, concretely defined in Supplimentary A, can be abstracted as $x \rightarrow z \rightarrow y$.

Concept-Bottleneck-Enabled Pretrained Language Models. The original concept bottlenecks in CBMs [14] come from resizing one of the layers in the CNN encoder to match the number of concepts. However, since PLM encoders typically provide text representations with much higher dimensions than the number of concepts, directly reducing the neurons in the layer would significantly impact the quality of learned text representation. To address this issue, we instead add a linear layer with the sigmoid activation, denoted as p_ψ, that projects the learned latent representation $z \in \mathbb{R}^e$ into the concept space $c \in \mathbb{R}^k$. This process can be represented as $x \rightarrow z \rightarrow c \rightarrow y$. Note that, unlike the previous works for image classification, each concept here does not need to be binary (i.e., present or not). We allow multi-class concepts, e.g., the concept "Food" in a restaurant review can be positive, negative, or unknown. We refer to the PLM and the projector (f_θ, p_ψ) together as the *concept encoder* and the complete model $(f_\theta, p_\psi, g_\phi)$ as *Concept-Bottleneck-Enabled Pretrained Language Models* (CBE-PLMs). During training, CBE-PLMs seek to achieve two goals: (1) align concept prediction $\hat{c} = p_\psi(f_\theta(x)))$ to x's ground-truth concept labels c and (2) align label prediction $\hat{y} = g_\phi(p_\psi(f_\theta(x)))$ to ground-truth task labels y. We accordingly adapt the three conventional strategies, *independent* training, *sequential* training, and *joint* training, proposed in [14] to learn the CBE-PLM. Their detailed formulations are given in Supplimentary A.

While initial findings from applying vanilla CBM [14] for interpreting PLMs appear encouraging, they require human-annotated concepts during training. This proves to be impractical in real-world situations due to the vast number of potential concepts and the time-intensive annotation process [21]. Often, only a limited number of texts come with manually labeled concepts. Moreover, as humans continuously acquire new concepts, it is desirable for the training framework to discover and incorporate new concepts automatically. Thus, we aim to design a general framework for training CBE-PLMs.

4 C³M: A General Framework for Learning CBE-PLMs

We define the following data potions according to the real-world scenarios. We refer to a dataset with human-annotated concepts as the *source concept dataset*, denoted as $\mathcal{D}_s = \{(x^{(i)}, y^{(i)}, c_s^{(i)})_{i=1}^{n_s}\}$, where n_s denotes the size and $c_s \in \mathbb{R}^{k_s}$ is a vector of k_s concepts from the pre-defined source concept set $\mathcal{C}_s$. We also consider another dataset without concept labels, referred to as the *unlabeled concept dataset*, denoted as $\mathcal{D}_u = \{(x^{(i)}, y^{(i)})_{i=1}^{n_u}\}$. The complete dataset is then the combination of these two datasets: $\mathcal{D} = \{\mathcal{D}_s, \mathcal{D}_u\}$. n_s and k_s are typically small, limiting the effectiveness of CBE-PLMs. Specifically, small n_s leads to sparse concept labels in $\mathcal{D}$, and vanilla CBM cannot be trained on datasets with unlabeled concepts $\mathcal{D}_u$. Additionally, small k_s indicates that we may not have sufficient information for model prediction. To address these limitations, we propose *ChatGPT-guided Concept augmentation with Concept-level Mixup* (C³M), a novel framework for training CBE-PLMs effectively. As illustrated in Fig. 2, at the high level, we augment the concept set $\mathcal{C}_s$ and annotate pseudo concept labels for the unlabeled concept dataset using ChatGPT. Since these pseudo labels are noisy, we propose a novel concept-level MixUp to train the CBE-PLMs effectively on the augmented dataset with noisy concept labels.

4.1 ChatGPT-Guided Concept Augmentation

In this section, we detail how to leverage ChatGPT (GPT4) to automatically (1) augment the concept set, and (2) annotate missing concept labels.

Concept Set Augmentation. The goal of concept set augmentation is to automatically generate high-quality concepts using human-specified concepts $\mathcal{C}_s$ as references. These generated concepts should be semantic meaningful and useful for target task prediction. Inspired by LF-CBM [22], we query ChatGPT with appropriate prompts to generate additional concepts. Our prompts are designed using "in-context learning" [4], and include examples from human annotations. Below is an example of a ChatGPT prompt designed for a sentiment classification task using the IMDB dataset [18]:

> Besides {*Acting*, *Storyline*, *Emotional Arousal*, *Cinematography*}, what are the additional important features to judge if a {*movie*} is good or not?

Parentheses represent fields that can be customized for different tasks. The concepts *Acting*, *Storyline*, *Emotional Arousal*, and *Cinematography* are from the source concept set $\mathcal{C}_s$ with labels manually annotated following procedures in Supplimentary B. Different from LF-CBM which generates concepts merely relying on GPT3 [4], we further include a small set of human-specified concepts in the prompt to improve the quality of generated concepts. This additional information can help effectively filter out undesired output without additional operations (e.g., deletions). Rarely seen concepts are discarded using a predefined threshold and the remaining generated concepts are referred to as *augmented concepts set* C_a with size k_a. Results are given in Table 4 in Supplimentary F.

Noisy Concept Label Annotation.

a. According to the review "{$text_1$}", the "{$concept_1$}" of the movie is "positive".
b. According to the review "{$text_2$}", the "{$concept_2$}" of the movie is "negative".
c. According to the review "{$text_3$}", the "{$concept_3$}" of the movie is "unknown".
d. According to the review "{$text_i$}", how is the "{$concept_i$}" of the movie? Please answer with one option in "positive, negative, or unknown".

Following a similar "in-context learning" strategy described in Sect. 4.1, Prompts a-c are three human-annotated examples randomly selected to represent positive, negative, and unknown concept labels, respectively. Prompt d is the query instance. The goal is to obtain noisy labels for any given $\{text_i\}$ and $\{concept_i\}$. There are three types of noisy concept annotations:

1. Noisy labels for human-specified concepts in $\mathcal{D}_s$. The resulting dataset $\tilde{\mathcal{D}}_s$ can be used to validate the quality of labels generated by ChatGPT only. Note that we annotate all the augmented concept manually for better validation (See Table 4 in Supplimentary F).
2. Noisy labels for ChatGPT-generated concepts in $\mathcal{D}_s$. The augmented concept set is denoted as $c_{sa} = (c_s || c_a) \in \mathbb{R}^{k_s+k_a}$, where $||$ refers to the concatenation operator and $c_a \in \mathbb{R}^{k_a}$ stands for the generated concepts. For example, we identify new important concepts such as *Soundtrack* using ChatGPT for the IMDB movie reviews.
3. Noisy labels for both human-specified and ChatGPT-generated concepts in unlabeled concept datasets $\mathcal{D}_u$. The augmented concept set is denoted as $\tilde{c_{sa}} = (\tilde{c_s} || \tilde{c_a}) \in \mathbb{R}^{k_s+k_a}$ and $\tilde{c_s} \in \mathbb{R}^{k_s}, \tilde{c_a} \in \mathbb{R}^{k_a}$ stand for the generated concept labels for human-specified and ChatGPT-generated concepts, respectively.

In summary, we transform the original dataset with sparse concept labels into an augmented dataset with new concepts and noisy labels: $\mathcal{D} = \{\mathcal{D}_s, \mathcal{D}_u\} \rightarrow \tilde{\mathcal{D}} = \{\tilde{\mathcal{D}}_{sa}, \tilde{\mathcal{D}}_u\}$. Examples of these queries are illustrated in Supplimentary I.

4.2 Learning from Noisy Concept Labels

While directly training CBE-PLMs on the transformed dataset $\tilde{\mathcal{D}}$ is straightforward, this method's drawback is its equal treatment of human annotations and ChatGPT-generated noisy labels, potentially leading to prediction and interpretation inaccuracies. To improve interpretability and accuracy, we introduce a novel *Concept-level MixUp* (CM) approach. It advocates for a convex behavior of PLMs between human-annotated and ChatGPT-generated concepts, thereby enhancing its robustness against noisy concept labels.

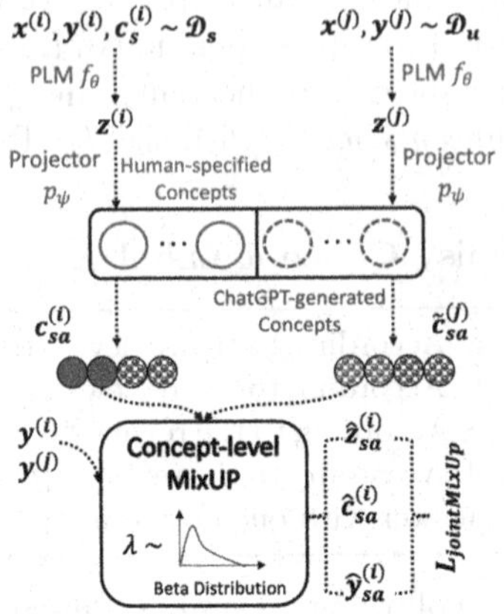

Fig. 2. The illustration of CBE-PLMs. Via PLMs, the input text x is mapped into an intermediate layer consisting of a set of human-comprehensible concepts c, which are then used to predict the target label y.

Concept-Level MixUp. To better utilize the noisy concept labels, CM linearly interpolates the texts and concept labels between human-annotated concepts ($\tilde{\mathcal{D}}_{sa}$) and ChatGPT-generated concepts ($\tilde{\mathcal{D}}_u$). Specifically, we interpolate any two text-concept-label ternaries $(x^{(i)}, c^{(i)}, y^{(i)})$, $(x^{(j)}, c^{(j)}, y^{(j)})$ for both their latent representation $(z^{(i)}, z^{(j)})$, concepts $(c^{(i)}, c^{(j)})$ and the task labels $(y^{(i)}, y^{(j)})$ via the MixUp $(\cdot)$ defined below:

$$
\begin{aligned}
\lambda \sim \text{Beta}(\alpha, \alpha);\;\; \hat{\lambda} &= \max(\lambda, 1-\lambda);\\
z^{(i)} = f_\theta(x^i);\;\; z^{(j)} &= f_\theta(x^j);\\
\hat{z}^{(i,j)} &= \hat{\lambda} z^{(i)} + (1-\hat{\lambda}) z^{(j)};\\
\hat{c}^{(i,j)} &= \hat{\lambda} c^{(i)} + (1-\hat{\lambda}) c^{(j)};\\
\hat{y}^{(i,j)} &= \hat{\lambda} y^{(i)} + (1-\hat{\lambda}) y^{(j)},
\end{aligned}
\tag{1}
$$

where α is a hyperparameter for the Beta distribution. Notably, $\hat{\lambda} \geq 0.5$ preserves the order of human-annotated concepts and ChatGPT-generated concepts for computing individual loss components in Eq. (4) appropriately. Then, we combine and shuffle human-annotated and ChatGPT-annotated data in the transformed dataset $\tilde{\mathcal{D}} = \{\tilde{\mathcal{D}}_{sa}, \tilde{\mathcal{D}}_u\}$:

$$
\mathcal{W} = \text{Shuffle}(\tilde{\mathcal{D}}) = \text{Shuffle}(\tilde{\mathcal{D}}_{sa} || \tilde{\mathcal{D}}_u), \tag{2}
$$

where $||$ indicates the concatenation of two potions of datasets. Next, we perform MixUp $(\cdot)$ for the ith instance as follows:

$$
\begin{aligned}
(\hat{z}_{sa}^{(i)}, \hat{c}_{sa}^{(i)}, \hat{y}_{sa}^{(i)}) &= \text{MixUp}(\tilde{\mathcal{D}}_{sa}^{(i)}, \mathcal{W}^{(i)}),\\
(\hat{z}_{u}^{(i)}; \hat{c}_{u}^{(i)}, \hat{y}_{u}^{(i)}) &= \text{MixUp}(\tilde{\mathcal{D}}_{u}^{(i)}, \mathcal{W}^{(i)}).
\end{aligned}
\tag{3}
$$

Through these steps, we can generate a "mixed version" for each instance in $\tilde{D}_{sa}$ and $\tilde{D}_u$, while preserving a larger portion of the original instance.

Loss Function The loss function $L_{jointMixUp}$ for training CBE-PLMs with the MixUped dataset is defined below:

$$\begin{aligned} L_{sa} &= L_{joint}(\hat{z}_{sa}^{(i)}, \hat{c}_{sa}^{(i)}, \hat{y}_{sa}^{(i)}); \\ [-4pt] L_u &= L_{joint}(\hat{z}_{u}^{(i)}, \hat{c}_{u}^{(i)}, \hat{y}_{u}^{(i)}); \\ [-4pt] L_{jointMixUp} &= L_{sa} + \tau L_u, \end{aligned} \tag{4}$$

where τ is a hyperparameter and L_{joint} is the joint training loss used in vanilla CBM. In this way, We backpropagate gradients of the mixed noisy concept labels and gold concept labels to update all the parameters.

5 Experiments

Datasets. In this section, we give detailed descriptions of the experimented datasets. Each of the datasets has two components: source concept dataset and unlabeled concept dataset ($\mathcal{D} = \{\mathcal{D}_s, \mathcal{D}_u\}$). Existing datasets with human-annotated concept labels are very limited. One source concept dataset is `CEBaB` [1], a common sentiment classification dataset for restaurant reviews. Its corresponding $\mathcal{D}_u$ is the restaurant reviews from the Yelp Dataset[1]. We also curate another dataset for movie reviews. Specifically, we randomly sample two portions of reviews from the IMDB datasets [18] to represent $\mathcal{D}_s$ and $\mathcal{D}_u$, respectively. Following a previous NLP work [5], we manually annotate the concept labels for $\mathcal{D}_s$ in the movie reviews. More annotation details are included in Supplimentary B. For convenience, we still refer to these two new datasets as `CEBaB` and `IMDB-C`. Each concept contains three values, i.e., Negative, Positive, and Unknown. As described in Sect. 4.1, each dataset $\mathcal{D}$ is then transformed into $\tilde{\mathcal{D}} = \{\tilde{\mathcal{D}_{sa}}, \tilde{\mathcal{D}_u}\}$. The basic statistics of the transformed datasets and their human-annotated concepts are given in Table 3 in Supplimentary E and Table 4 in Supplimentary F, respectively. Note that the last column in Table 4 indicates the accuracy of ChatGPT-labeled concepts. Both the human-annotated and ChatGPT-generated data, alone with the framework implementation are released[2].

PLM Backbones. We experiment with the same PLM backbones as in the CEBaB paper [1]: GPT2 [25], BERT [6], RoBERTa [16], and BiLSTM [11] with CBOW [20]. For better performance, we obtain the representations of the input texts by pooling the embedding of all tokens. Reported scores are the averages of six independent runs, each taking 5 to 40 min. More implementation details and parameter values are included in Supplimentary C and Table 2 in Supplimentary D.

[1] https://www.kaggle.com/datasets/omkarsabnis/yelp-reviews-dataset.
[2] https://github.com/Zhen-Tan-dmml/CBM_NLP.git.

Table 1. Comparisons of task accuracy and interpretability using CEBaB and IMDB-C datasets. Metrics for both task and concept labels are written as **Accuracy/Macro F1**. Scores are reported in %. Scores in **bold** indicate that the CBE-PLM under the current setting outperforms its standard PLM counterpart. CM denotes Concept-level MixUp.

Dataset		CEBaB				IMDB-C			
Model		$\mathcal{D}$		$\tilde{\mathcal{D}}$		$\mathcal{D}$		$\tilde{\mathcal{D}}$	
		Task	Concept	Task	Concept	Task	Concept	Task	Concept
PLMs	LSTM	40.57/60.67	-	43.34/64.47	-	68.25/53.37	-	90.5/90.46	-
	GPT2	66.69/77.25	-	67.26/78.81	-	71.67/67.53	-	97.64/97.55	-
	BERT	68.75/78.71	-	71.81/82.58	-	80.5/78.4	-	98.89/98.68	-
	RoBERTa	71.36/80.17	-	73.12/82.64	-	84.1/82.5	-	99.13/99.12	-
CBE-PLMs	LSTM	**56.47/67.82**	86.46/85.24	**54.54/65.84**	83.46/84.74	**68.5/55.4**	72.5/77.5	**93.02/91.53**	76.92/75.41
	GPT2	64.04/77.75	92.14/92.05	63.57/74.71	90.17/90.13	70.05/69.53	80.6/82.5	96.85/96.81	86.14/88.06
	BERT	67.27/79.24	93.65/92.75	68.23/78.13	89.64/90.45	77.42/74.57	80.2/83.7	97.62/97.58	92.57/92.05
	RoBERTa	70.98/79.89	96.12/95.34	69.85/79.29	91.45/92.23	82.33/80.13	86.7/85.3	98.45/98.12	93.99/94.28
CBE-PLMs-CM	LSTM	-	-	**59.67/70.53**	88.75/86.67	-	-	**94.35/92.32**	83.83/84.52
	GPT2	-	-	65.54/77.87	93.58/92.32	-	-	**97.89/97.88**	89.64/88.25
	BERT	-	-	70.58/80.07	94.43/93.26	-	-	98.18/98.06	94.87/94.32
	RoBERTa	-	-	72.88/81.91	96.3/98.5	-	-	**99.69/99.66**	96.35/96.36

Task Accuracy vs Interpretability. Table 1 presents the results for the two original datasets ($\mathcal{D}$) and their transformed versions ($\tilde{\mathcal{D}}$). We have the following observations:

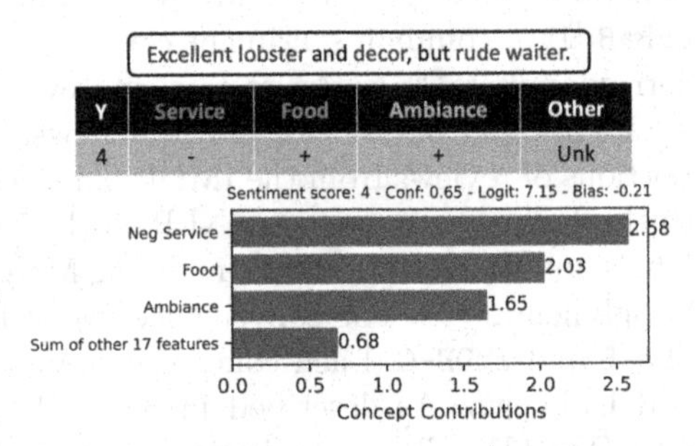

Fig. 3. Illustration of the explainable prediction for a toy example in restaurant review sentiment analysis.

▷ **CBE-PLMs offer interpretability and competitive task prediction performance.** Compared to standard PLMs (trained solely with task labels), CBE-PLMs provide concept-level interpretability with only a minor decrease in task prediction. Interestingly, a smaller PLM, i.e., LSTM with CBOW embeddings, achieves improved task accuracy when learning from concept labels. This suggests that the accuracy-interpretability tradeoff in concept learning is not necessary, as opposed to the prevailing view. Concepts can help guide PLMs trained on smaller corpora with fewer parameters.

▷ **Noisy concept labels can facilitate the training of CBE-PLMs on small datasets.** The extremely limited size of the IMDB-C source concept dataset (deliberately set to 100) yields unsurprisingly low test scores. Transforming $\mathcal{D}$ into $\tilde{\mathcal{D}}$ using ChatGPT for noisy labeled concept instances leads to significant improvements in both concept and task predictions for CBE-PLMs-CM.

▷ **Uncritical learning from noisy concept labels can impair performance.** Results for CEBaB in Table 1 demonstrate that, learning from the transformed dataset $\tilde{\mathcal{D}}$ directly leads to inferior performance for CBE-PLMs.

Unlike `IMDB-C`, the source concept dataset in `CEBaB` contains sufficient training instances, therefore, enforcing CBE-PLMs to learn from noisy concept labels will undesirably mislead the model, exacerbating both the concept and task prediction performance.

▷ **CBE-PLMs-CM trained via the proposed C^{3}M framework consistently deliver superior interpretability-utility trade-offs.** By encouraging the CBE-PLMs to linearly interpolate between examples with gold-labeled concepts and those with ChatGPT-generated concepts, the model gains semantic insight and noise resilience. The result is promising: We achieve the best concept-level prediction (interpretability measure) without sacrificing the task prediction performance, and in some cases, CBE-PLMs trained through C^{3}M can outperform their standard PLM counterparts.

Explainable Predictions. A unique advantage of CBMs is that its decision rules can be interpreted as a linear combination of comprehensible variables [14]. Inheriting this strength, our proposed CBE-PLMs can deliver intuitive concept-level explanations for predictions by assessing the activations of each concept. We measure concept contribution using the product of activation and the corresponding weight in the linear label predictor g_ϕ [22]. Concepts with negative activation are designated as "Neg Concept". We highlight the concepts contributing the most in our visualizations. Visualization results are demonstrated in Fig. 3 for a toy example, while real-world `CEBaB` and `IMDB-C` case studies can be found in Supplimentary G. These visualizations provide new intriguing insights into real-world applications. For instance, negative concepts (e.g., Service) contribute more to the final prediction of positive sentiment in Fig. 3, making the predicted sentiment second highest ($Y = 4$) rather than the highest ($Y = 5$). Moreover, results such as Fig. 5 in Supplimentary G imply that concepts such as "Food" and "Ambiance" weigh more heavily for restaurant evaluations compared to "Noise" and "Menu Variety".

Test-Time Intervention. Another strength of CBE-PLMs is that they allow test-time concept intervention (inherited from CBMs), facilitating deeper, user-friendly interactions. To assess this strength, we follow [14] to intervene in the predicted concepts and investigate the impact of such interventions on test-time prediction accuracy. Concept mispredictions arise from ChatGPT's incorrect labels or inaccurate concept activation. Recall that the input of the task label predictor is the predicted concept activations $\hat{a} = p_\phi(f_\theta(x))$ rather than the predicted ternary concepts $\hat{c}$. In a concept-level intervention I, the activation $\hat{a}_j$ of the jth concept with a target concept c_j is set to the 5th, 95th, or 50th percentile of $\hat{a}_j$ over the training distribution for Negative, Positive, or Unknown c_j respectively. Multiple concepts can be intervened upon by replacing all related predicted concept activations and updating the prediction. Experiments were conducted on the transformed version $\tilde{\mathcal{D}}$ of the `CEBaB` dataset. Figure 4 exhibits results for CBE-PLMs using BERT and GPT2 as the PLM backbones (with similar observations for LSTM and RoBERTa). A case study is further illustrated in

Supplimentary H. The results reveal that task accuracy improves substantially when more concepts are corrected by the oracle. Additionally, while the performance of CBE-PLMs declines as more concepts are intervened upon incorrectly (randomly), the proposed concept-level MixUp effectively mitigates this impact. Notably, the decline in performance is marginal when only two concepts are erroneously intervened upon. These findings underscore the pronounced advantages of test-time intervention for CBE-PLMs trained through C^3M. First, domain experts can interact with the model to rectify any inaccurately predicted concept values. Second, in reality, even experts might inadvertently implement incorrect interventions. Yet, despite this susceptibility, our proposed concept-level MixUp strategy effectively curbs performance degradation, particularly when inaccuracies affect only a small subset of the intervention. This attests to the robustness of the proposed framework.

6 Conclusion

Our analysis began with an exhaustive examination of three training strategies, identifying joint training as the most efficacious. Further, we proposed the C^3M framework, designed to streamline the training process of CBE-PLMs in the presence of incomplete concept labels. Moreover, we showcased the interpretability of our models in their decision-making process and elucidated how this comprehensibility can be harnessed to boost test accuracy via concept intervention.

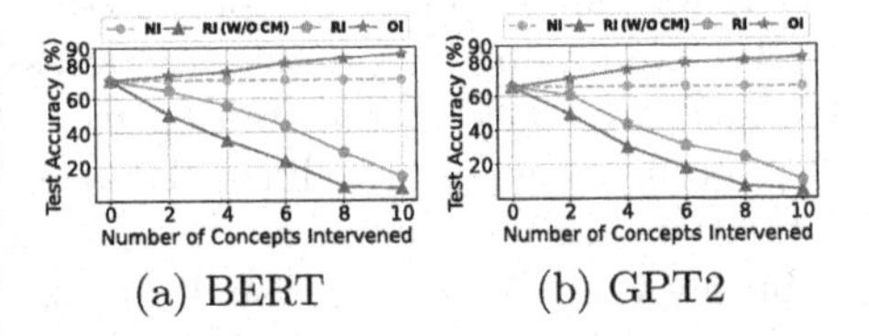

(a) BERT (b) GPT2

Fig. 4. The results of Test-time Intervention. "NI" denotes "no intervention", "RI (W/O CM)" denotes "random intervention on CBE-PLMs without the concept-level MixUp", "RI" denotes "random intervention on CBE-PLMs", and "OI" denotes "oracle intervention".

Acknowledgements. This work is supported by the National Science Foundation (NSF) under grants IIS-2229461.

A Definitions of Training Strategies

Given a text input $x \in \mathbb{R}^d$, concepts $c \in \mathbb{R}^k$ and its label y, the strategies for fine-tuning the text encoder f_θ, the projector p_ψ and the label predictor g_ϕ are defined as follows:

i) Vanilla fine-tuning a PLM: The concept labels are ignored, and then the text encoder f_θ and the label predictor g_ϕ are fine-tuned either as follows:

$$\theta, \phi = \text{argmin}_{\theta,\phi} L_{CE}(g_\phi(f_\theta(x), y),$$

or as follows (frozen text encoder f_θ):

$$\phi = \text{argmin}_{\phi} L_{CE}(g_\phi(f_\theta(x), y),$$

where L_{CE} indicates the cross-entropy loss. In this work we only consider the former option for its significant better performance.

ii) Independently training PLM with the concept and task labels: The text encoder f_θ, the projector p_ψ and the label predictor g_ϕ are trained separately with ground truth concepts labels and task labels as follows:

$$\theta, \psi = \text{argmin}_{\theta,\psi} L_{CE}(p_\psi(f_\theta(x)), c),$$
$$\phi = \text{argmin}_{\phi} L_{CE}(g_\phi(c), y).$$

During inference, the label predictor will use the output from the projector rather than the ground-truth concepts.

iii) Sequentially training PLM with the concept and task labels: We first learn the concept encoder as the independent training strategy above, and then use its output to train the label predictor:

$$\phi = \text{argmin}_{\phi} L_{CE}(g_\phi(p_\psi(f_\theta(x)), y).$$

iv) Jointly training PLM with the concept and task labels: Learn the concept encoder and label predictor via a weighted sum L_{joint} of the two objectives described above:

$$\begin{aligned}\theta, \psi, \phi &= \text{argmin}_{\theta,\psi,\phi} L_{joint}(x, c, y) \\ &= \text{argmin}_{\theta,\psi,\phi}[L_{CE}(g_\phi(p_\psi(f_\theta(x)), y) \\ &+ \gamma L_{CE}(p_\psi(f_\theta(x)), c)].\end{aligned}$$

It's worth noting that the CBE-PLMs trained jointly are sensitive to the loss weight γ. We report the most effective results here, tested value for γ are given in Table 2 in Appendix D.

B Details of the Manual Concept Annotation for the IMDB Dataset

Our annotation policy is following a previous work [5] for NLP datasets annotating. For the `IMDB-C` dataset, we annotate the four concepts (Acting, Stroyline, Emotional Arousal, Cinematography) manually. Even though the concepts are naturally understandable by humans, two Master students familiar with sentiment analysis are selected as annotators for independent annotation with the annotation tool introduced by [30]. The strict quadruple matching F1 score between two annotators is 85.74%, which indicates a consistent agreement between the two annotators [13]. In case of disagreement, a third expert will be asked to make the final decision.

C Implementation Detail

In this section, we provide more details on the implementation settings of our experiments. Specifically, we implement our framework with PyTorch [24] and

HuggingFace [29] and train our framework on a single 80 GB Nvidia A100 GPU. We follow a prior work [1] for backbone implementation. All backbone models have a maximum token number of 512 and a batch size of 8. We use the Adam optimizer to update the backbone, projector, and label predictor according to Sect. 3.1. The values of other hyperparameters (Table 2 in Appendix D) for each specific PLM type are determined through grid search. We run all the experiments on an Nvidia A100 GPU with 80 GB RAM.

D Parameters and Notations

In this section, we provide used notations in this paper along with their descriptions for comprehensive understanding. We also list their experimented values and optimal ones, as shown in Table 2.

Table 2. Key parameters in this paper with their annotations and evaluated values. Note that **bold** values indicate the optimal ones.

Notations	Specification	Definitions or Descriptions	Values
max_len	-	maximum token number of input	128/256/**512**
batch_size	-	batch size	8
plm_epoch	-	maximum training epochs for PLM and Projector	20
clf_epoch	-	maximum training epochs for the linear classifier	20
hidden_dim	-	hidden dimension size	128
emb_dim	LSTM	embedding dimension for LSTM	300
lr	LSTM	learning rate when the backbone is LSTM	1e−1e−**1e−2**/5e−2/1e−3/1e−4
	GPT2	learning rate when the backbone is GPT2	1e−3/5e−3/ **1e−4**/5e−4/ 1e−5
	BERT	learning rate when the backbone is BERT	1e−4/5e−4/ **1e−5**/3e−5/ 5e−5
	RoBERTa	learning rate when the backbone is RoBERTa	1e−4/5e−4/ **1e−5**/3e−5/ 5e−5
\gamma	-	loss weight in the joint loss L_{joint}	0.1/0.3/**0.5**/0.7/1.0
\tau	-	loss weight in the joint-MixUp loss $L_{jointMixUp}$	0.1/0.5/**1.0**/1.5/2.0

E Statistics of Data Splits

The Statistics and split policies of the experimented datasets, including the source concept dataset $\mathcal{D}_s$, the unlabeled concept dataset $\mathcal{D}_u$, and their augmented versions. The specific details are presented in Table 3.

Table 3. Statistics of experimented datasets. k denotes the number of concepts.

Dataset	$\mathcal{D}_s$		$\mathcal{D}_u$		$\tilde{\mathcal{D}}_{sa}$		$\tilde{\mathcal{D}}_u$		Task
	Train/Dev/Test	k	Train/Dev/Test	k	Train/Dev/Test	k	Train/Dev/Test	k	
CEBaB	1755/1673/1685	4	2000/500/500	0	1755/1673/1685	10	2000/500/500	10	5-way classification
IMDB-C	100/50/50	4	1000/1000/1000	0	100/50/50	8	1000/1000/1000	8	2-way classification

F Statistics of Concepts in Transformed Datasets

The Statistics and split policies of the transformed datasets of experimented datasets are presented in Table 4.

Table 4. Statistics of concepts in transformed datasets ($\tilde{\mathcal{D}}$). Human-specified concepts are underlined. Concepts shown in gray are not used in experiments as the portion of the "Unknown" label is too large.

Dataset	Concept	Negative	Positive	Unknown	Total	ChatGPT Acc.
CEBaB	Food	2043(25.2%)	4382(54.0%)	1688(20.8%)	8113	78.5%
	Ambiance	868(10.7%)	1659(20.4%)	5586(68.9%)	8113	76.4%
	Service	1543(19.0%)	2481(30.6%)	4089(50.4%)	8113	86.4%
	Noise	668(8.2%)	477(5.9%)	6968(85.9%)	8113	89.7%
	Cleanliness	55(0.7%)	610(7.5%)	7448(91.8%)	8113	76.9%
	Price	714(8.8%)	527(6.5%)	6872(84.7%)	8113	76.4%
	Location	303(3.7%)	2598(32.0%)	5212(64.2%)	8113	87.5%
	Menu Variety	238(2.9%)	2501(30.8%)	5374(66.2%)	8113	89.4%
	Waiting Time	572(7.1%)	608(7.5%)	6933(85.5%)	8113	90.4%
	Waiting Area	267(3.3%)	1136(14.0%)	6710(82.7%)	8113	92.8%
	Parking	53(0.7%)	107(1.3%)	7953(98.0%)	8113	-
	Wi-Fi	9(0.1%)	39(0.5%)	8065(99.4%)	8113	-
	Kids-Friendly	15(0.2%)	536(6.6%)	7562(93.2%)	8113	-
IMDB-C	Acting	663(20.7%)	1200(37.5%)	1337(41.8%)	3200	80.1%
	Storyline	1287(40.2%)	1223(38.2%)	690(21.6%)	3200	77.9%
	Emotional Arousal	1109(34.7%)	1136(35.5%)	955(29.8%)	3200	74.5%
	Cinematography	165(5.2%)	481(15.0%)	2554(79.8%)	3200	92.1%
	Soundtrack	107(3.3%)	316(9.9%)	2777(86.8%)	3200	93.8%
	Directing	537(16.8%)	850(26.6%)	1813(56.7%)	3200	74.3%
	Background Setting	288(9.0%)	581(18.2%)	2331(72.8%)	3200	72.7%
	Editing	304(9.5%)	240(7.5%)	2656(83.0%)	3200	-

G More Results on Explainable Predictions

Case studies on explainable predictions for both CEBaB and IMDB-C datasets are given in Fig. 5 and Fig. 6 respectively.

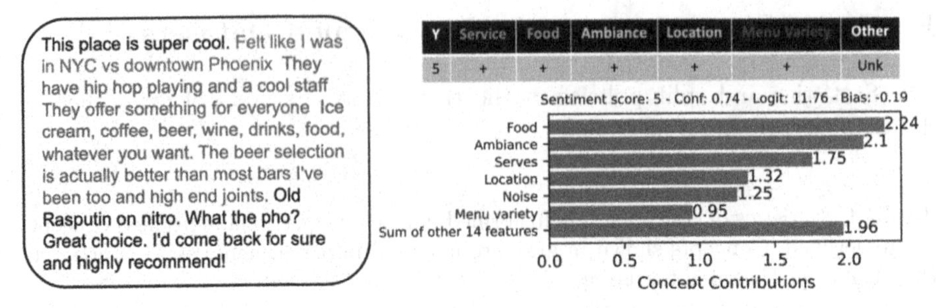

Fig. 5. Illustration of the explanable prediction for an example from the CEBaB dataset.

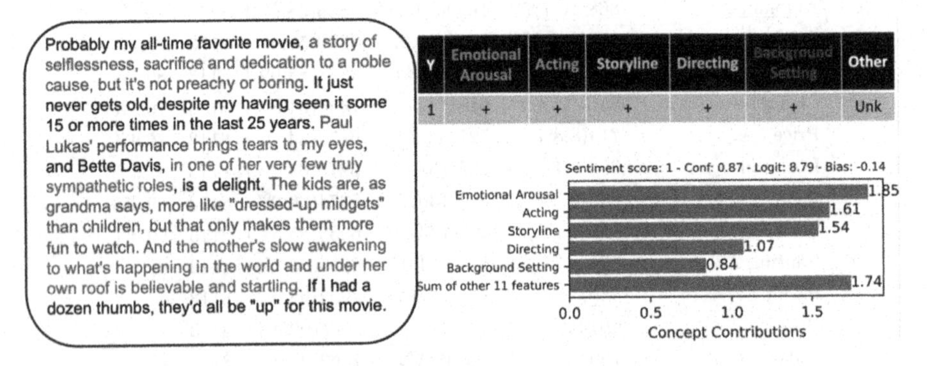

Fig. 6. Illustration of the explanable prediction for an example from the IMDB-C dataset.

H A Case Study on Test-Time Intervention

We present a case study of Test-time Intervention using an example from the transformed unlabeled concept data $\tilde{\mathcal{D}}_u$ of the CEBaB dataset, as shown in Fig. 7. The first row displays the target concept labels generated by ChatGPT. The second row shows the predictions from the trained CBE-PLM model, which mispredicts two concepts ("Waiting time" and "Waiting area"). The third row demonstrates test-time intervention using ChatGPT as the oracle, which corrects the predicted task labels. Finally, the fourth row implements test-time intervention with a human oracle, rectifying the concept that ChatGPT originally mislabeled.

Finally 85c bakery in Las Vegas! :D This place was one of my favorite places to go when I lived in Cali, so many asian style bread and sea salt cream drinks. Bread and drinks here are above average taste (at least to me) and also price is reasonable. This time I bought 4 bread and a fresh strawberry milk tea and it was like 12 bucks, which I think it's reasonable in the price wise. The store is also clean, spacious, and modernly decorated! Be aware that there might be a huge line when you come around lunch time in weekends.

$Y/\hat{Y}$	Clean-liness	Food	Ambiance	Location	Menu Variety	Price	Waiting time	Waiting area	Other
4	+	+	+	+	+	+	-	+	Unk
5	+	+	+	+	+	+	Unk	Unk	Unk
4	+	+	+	+	+	+	-	+	Unk
4	+	+	+	+	+	+	-	Unk	Unk

Fig. 7. Illustration of the explanable prediction for an example from the transformed unlabeled concept data $\tilde{\mathcal{D}}_u$ of the CEBaB dataset. The brown box with dash lines indicates the test-time intervention on corresponding concepts. (Color figure online)

I Examples of Querying ChatGPT

In this paper, we query ChatGPT for 1) augmenting the concept set, and 2) annotate missing concept labels. Note that in practice, we query ChatGPT (GPT4) via OpenAI API. Here we demonstrate examples from the ChatGPT (GPT4) GUI for better illustration. The illustrations are given in Fig. 8 and Fig. 9.

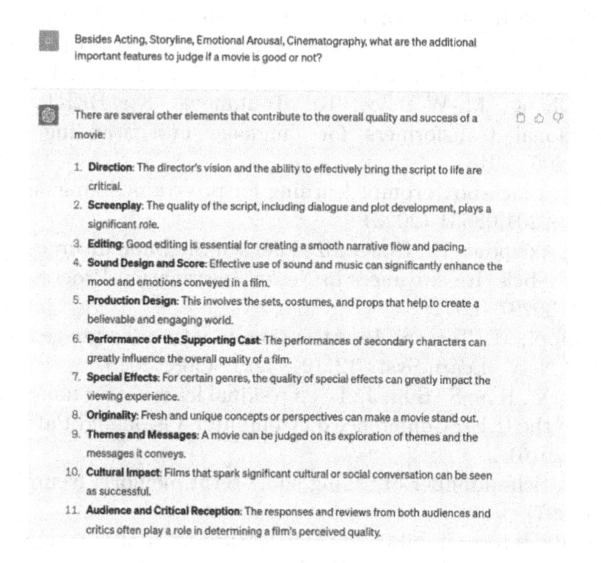

Fig. 8. The illustration of querying ChatGPT for additional concepts for the IMDB-C dataset.

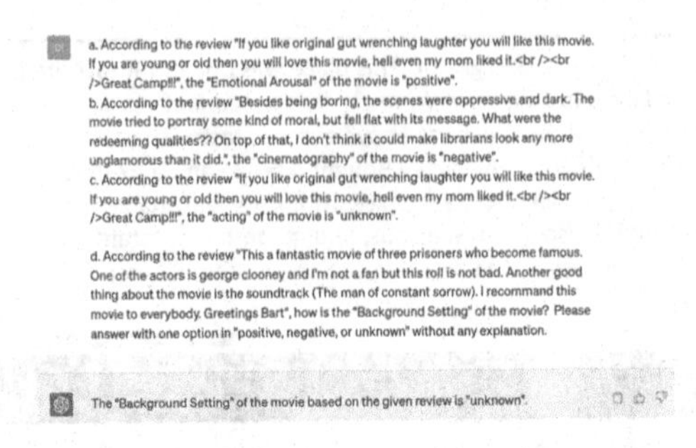

Fig. 9. The illustration of querying ChatGPT for annotating a missing concept label for the IMDB-C dataset.

References

1. Abraham, E.D., et al.: Cebab: estimating the causal effects of real-world concepts on NLP model behavior. In: Advances in Neural Information Processing Systems, vol. 35, pp. 17582–17596 (2022)
2. Berthelot, D., Carlini, N., Goodfellow, I., Papernot, N., Oliver, A., Raffel, C.A.: Mixmatch: a holistic approach to semi-supervised learning. In: Advances in Neural Information Processing Systems, vol. 32 (2019)
3. Bills, S., et al.: Language models can explain neurons in language models (2023). https://openaipublic.blob.core.windows.net/neuron-explainer/paper/index.html
4. Brown, T., et al.: Language models are few-shot learners. In: Advances in Neural Information Processing Systems, vol. 33, pp. 1877–1901 (2020)
5. Cai, H., Xia, R., Yu, J.: Aspect-category-opinion-sentiment quadruple extraction with implicit aspects and opinions. In: Proceedings of the 59th Annual Meeting of the Association for Computational Linguistics and the 11th International Joint Conference on Natural Language Processing (Volume 1: Long Papers) (2021)
6. Devlin, J., Chang, M.-W., Lee, K., Toutanova, K.: BERT: pre-training of deep bidirectional transformers for language understanding. arXiv preprint arXiv:1810.04805 (2018)
7. Diao, S., et al.: Black-box prompt learning for pre-trained language models. arXiv preprint arXiv:2201.08531 (2022)
8. Englesson, E., Azizpour, H.: Generalized Jensen-Shannon divergence loss for learning with noisy labels. In: Advances in Neural Information Processing Systems, vol. 34, pp. 30284–30297 (2021)
9. Galassi, A., Lippi, M., Torroni, P.: Attention in natural language processing. IEEE Trans. Neural Netw. Learn.Syst. **32**(10), 4291–4308 (2020)
10. He, K., Zhang, X., Ren, S., Sun, J.: Deep residual learning for image recognition. In: Proceedings of the IEEE Conference on Computer Vision and Pattern Recognition, pp. 770–778 (2016)
11. Hochreiter, S., Schmidhuber, J.: Long short-term memory. Neural Comput. **9**(8), 1735–1780 (1997)
12. Kim, B., et al.: Interpretability beyond feature attribution: quantitative testing with concept activation vectors (TCAV). In: International Conference on Machine Learning, pp. 2668–2677. PMLR (2018)

13. Kim, E., Klinger, R.: Who feels what and why? Annotation of a literature corpus with semantic roles of emotions. In: Proceedings of the 27th International Conference on Computational Linguistics, pp. 1345–1359 (2018)
14. Koh, P.W., et al.: Concept bottleneck models. In: International Conference on Machine Learning, pp. 5338–5348. PMLR (2020)
15. Liu, Y., Cheng, H., Zhang, K.: Identifiability of label noise transition matrix. arXiv preprint arXiv:2202.02016 (2022)
16. Liu, Y., et al.: RoBERTa: a robustly optimized BERT pretraining approach. arXiv preprint arXiv:1907.11692 (2019)
17. Losch, M., Fritz, M., Schiele, B.: Interpretability beyond classification output: semantic bottleneck networks. arXiv preprint arXiv:1907.10882 (2019)
18. Maas, A., Daly, R.E., Pham, P.T., Huang, D., Ng, A.Y., Potts, C.: Learning word vectors for sentiment analysis. In: Proceedings of the 49th Annual Meeting of the Association for Computational Linguistics: Human Language Technologies, pp. 142–150 (2011)
19. Madsen, A., Reddy, S., Chandar, S.: Post-hoc interpretability for neural NLP: a survey. ACM Comput. Surv. **55**(8), 1–42 (2022)
20. Mikolov, T., Chen, K., Corrado, G., Dean, J.: Efficient estimation of word representations in vector space. arXiv preprint arXiv:1301.3781 (2013)
21. Németh, R., Sik, D., Máté, F.: Machine learning of concepts hard even for humans: the case of online depression forums. Int. J. Qual. Methods **19**, 1609406920949338 (2020)
22. Oikarinen, T., Das, S., Nguyen, L.M., Weng, T.-W.: Label-free concept bottleneck models. In: The Eleventh International Conference on Learning Representations (2023)
23. OpenAI. Gpt-4 Technical report (2023)
24. Paszke, A., et al.: Automatic differentiation in PyTorch. In: NeurIPS (2017)
25. Radford, A., et al.: Language models are unsupervised multitask learners. OpenAI blog **1**(8), 9 (2019)
26. Ross, A., Marasović, A., Peters, M.E.: Explaining NLP models via minimal contrastive editing (mice). In: Findings of the Association for Computational Linguistics: ACL-IJCNLP 2021, pp. 3840–3852 (2021)
27. Sohn, K., et al.: Fixmatch: simplifying semi-supervised learning with consistency and confidence. In: Advances in Neural Information Processing Systems, vol. 33, pp. 596–608 (2020)
28. Vig, J., et al.: Investigating gender bias in language models using causal mediation analysis. In: Advances in Neural Information Processing Systems, vol. 33, pp. 12388–12401 (2020)
29. Wolf, T., et al.: Huggingface's transformers: state-of-the-art natural language processing (2020)
30. Yang, J., Zhang, Y., Li, L., Li, X.: Yedda: a lightweight collaborative text span annotation tool. arXiv preprint arXiv:1711.03759 (2017)
31. Yin, K., Neubig, G.: Interpreting language models with contrastive explanations. arXiv preprint arXiv:2202.10419 (2022)
32. Zarlenga, M.E., et al.: Concept embedding models. In: NeurIPS 2022 - 36th Conference on Neural Information Processing Systems (2022)

33. Zhang, H., Cisse, M., Dauphin, Y.N., Lopez-Paz, D.: Mixup: beyond empirical risk minimization. arXiv preprint arXiv:1710.09412 (2017)
34. Zhang, W., Li, X., Deng, Y., Bing, L., Lam, W.: A survey on aspect-based sentiment analysis: tasks, methods, and challenges. IEEE Trans. Knowl. Data Eng. (2022)
35. Zhu, J., et al.: Incorporating BERT into neural machine translation. In: International Conference on Learning Representations (2020)

Unmasking Dementia Detection by Masking Input Gradients: A JSM Approach to Model Interpretability and Precision

Yasmine Mustafa and Tie Luo(✉)

Computer Science Department, Missouri University of Science and Technology, Rolla, MO 65409, USA
{yam64,tluo}@mst.edu

Abstract. The evolution of deep learning and artificial intelligence has significantly reshaped technological landscapes. However, their effective application in crucial sectors such as medicine demands more than just superior performance, but trustworthiness as well. While interpretability plays a pivotal role, existing explainable AI (XAI) approaches often do not reveal *Clever Hans* behavior where a model makes (ungeneralizable) correct predictions using spurious correlations or biases in data. Likewise, current post-hoc XAI methods are susceptible to generating unjustified counterfactual examples. In this paper, we approach XAI with an innovative *model debugging* methodology realized through Jacobian Saliency Map (JSM). To cast the problem into a concrete context, we employ Alzheimer's disease (AD) diagnosis as the use case, motivated by its significant impact on human lives and the formidable challenge in its early detection, stemming from the intricate nature of its progression. We introduce an interpretable, multimodal model for AD classification over its multi-stage progression, incorporating JSM as a modality-agnostic tool that provides insights into volumetric changes indicative of brain abnormalities. Our extensive evaluation including ablation study manifests the efficacy of using JSM for model debugging and interpretation, while significantly enhancing model accuracy as well.

Keywords: Trustworthy AI · Interpretability · Explainability · Reliability · Jacobian saliency map · Alzheimer's disease

1 Introduction

Despite the remarkable successes of deep learning and artificial intelligence across various technological domains, achieving a high-performing system does not automatically guarantee its practical deployment and use, particularly in the field of medicine. Given the profound implications of medical decisions on human lives, doctors and patients often approach AI diagnoses with skepticism, notwithstanding claims of high precision, due to concerns surrounding trustworthiness.

In this study, we delve into the concept of trustworthy medical AI, focusing on two essential aspects. Firstly, *explainability* is crucial; without a clear explanation

D.-N. Yang et al. (Eds.): PAKDD 2024, LNAI 14647, pp. 75–90, 2024.
https://doi.org/10.1007/978-981-97-2259-4_6

of the rationale behind a diagnosis, patients would be much less receptive to AI-driven decisions. Secondly, *reliability* is paramount, ensuring that AI models make predictions based on pertaining patterns rather than exhibiting what is known as "Clever-Hans behavior." This phenomenon occurs when a machine learning model seemingly performs well but makes decisions based on irrelevant factors such as biases or coincidental correlations in data.

These two aspects are interrelated: a thorough model explanation not only provides the foundation for decision-making but also reveals whether predictions are influenced by Clever-Hans behavior. By addressing both explainability and reliability, we aim to enhance trust in medical AI systems and pave the way for their responsible and effective integration into healthcare practices.

A wide variety of explainable AI (XAI) tools have arisen to explain the predictions of trained *black-box* models, referred to as *post-hoc methods*. Although these methods have shown some promising prospects, they are vulnerable to the risk of generating unjustified counterfactual examples [12], hence may not be reliable. It is worth noting the nuances between *explainability* and *interpretability* in this context, although they are often loosely used interchangeably. *Explainability* refers to the ability to explain model decisions after training (post-hoc), while the original model is not interpretable by itself. On the other hand, *interpretability* is an inherent property of a model and means how easily and intuitively one can understand and make sense of the model's decision-making process. Examples of highly interpretable models include decision trees and linear regression, which provide easily traceable logic for what roles the features played in decision-making. However, such models are *shallow models* and typically under-perform deep neural networks (DNNs). As post-hoc explanation is often inadequate to unravel the full complexity of model behavior due to its after-training nature, we focus on designing *interpretable* models in this paper, taking a *during-modeling* approach.

In this paper, we introduce a novel approach that guides the decision-making process of neural networks during the training phase by not only directing the model toward correct predictions but also penalizing any (including correct) predictions based on *wrong cues*. Cues refer to patterns, relationships, or features within the input data that are deemed relevant to the task at hand by the model to make predictions or decisions. Misinterpreted or misidentified cues erode the trustworthiness and reliability of a model even if its predictions are correct since the performance would not generalize to future unseen data.

For a concrete problem context, we take Alzheimer's disease (AD) as the specific medical condition in this study, but highlight that our approach can be extended to similar problems without change of principle. AD is the predominant form of dementia and a major contributor to mortality. It impacts brain areas that are responsible for thought, memory, and language and is hard to cure. Although changes in the brain can manifest long before AD symptoms appear, it is challenging for medical professionals to detect manually until AD reaches late and severe stages, which however are no longer reversible. Therefore, AI-based diagnosis of AD has been researched actively. However, to date, newly developed models as such have rarely been adopted in clinical decision support systems (CDSS) for primary care, because such new models are almost exclusively based on DNNs which lack explainability.

Another crucial aspect of AD is that its clinical representation often involves multiple modalities such as computed tomography (CT) and magnetic resonance imaging (MRI) images. While combining such data modalities could lead to better precision [28], it poses a further challenge in designing interpretable models to interpret decisions involving all modalities.

In this paper, we propose an interpretable, multimodal model for classifying AD across its multi-stage progression, including early detection. We introduce a *Jacobian-Augmented Loss* function (JAL) that incorporates Jacobian Saliency Maps (JSM) as a model *self-debugger*. This approach is *modality-agnostic* as it is not limited to any specific imaging modality but can work seamlessly with various integrated modalities, making it versatile and adaptable to different data sources. In this study, we focus on images, given that other modalities such as Mini-Mental State Examination (MMSE) can be diagnosed effectively by medical professionals, and image-based early detection of AD proves challenging. Our methodology aligns with the broader concept of *model debugging* that involves troubleshooting unwanted behaviors and examining models' predictions. Our goal is to ensure that a model not only learns to make accurate predictions but also avoids wrong cues as in the *Clever Hans* phenomenon, thus enhancing model reliability. We achieve this using JSM, which is computed during image preprocessing. Besides debugging, JSM also allows us to enhance model precision by highlighting deformations in body issues (e.g., brain as in the context of AD).

In summary, this paper makes the following contributions:

- We design a novel loss function JAL which incorporates Jacobian saliency maps (JSM) to enable machine learning models to self-debug its decision-making process automatically. This approach ensures the model predictions to be based on genuine patterns and cues, and renders the model decision-making process to be more interpretable through a during-modeling methodology.
- We include multimodal data fusion into the process through two distinct fusion techniques, not only shedding light on their differences but also showcasing the adaptability of JAL to different modalities and fusion levels.
- Our approach enables models to provide fine-grained classification in terms of 4 classes including cognitively normal (CN) and three main AD stages: mild cognitive impairment (MCI), mild AD, and moderate to severe AD. On the contrary, existing approaches only provide binary classification or combine two or more stages into one class. Furthermore, although coarse-grained approaches are *easier* to attain higher accuracy due to less classes, our fine-grained diagnosis achieves higher accuracy than them.
- Our comprehensive evaluation including ablation study proves the efficacy of using JSM as a model self-debugger for producing both reliable predictions and trustworthy interpretations.

2 Related Work

El-sappagh et al. [6]'s primary goal in explainability is to provide post-hoc explanations for the decisions made by Random Forest (RF) on classifying AD based on multi-modal data. Hence, the method does not try to explain the internal workings of RF but aims to provide explanations to decisions for a better understanding by physicians. The framework consists of two RF models. The first RF performs multi-classification to categorize individuals as normal, MCI, or AD. The second RF model only comes into play in the case of MCI, to classify whether the MCI is stable (sMCI) or progressive (pMCI). The authors used FreeSurfer software to automatically label areas of the structural MRI scans in order to provide natural language explanations using the Fuzzy Unordered Rule Induction Algorithm (FURIA) proposed by [8]. As a result, it produces a compact set of If-THEN statements that are understandable by physicians. SHapley Additive exPlanations (SHAP) [14] was employed to show local/global feature contributions to final decisions of the RF models. Finally, they used a visualization tool called RF explainer on the tree decisions to provide explanation about individual modalities.

Khare et al. [10] used electroencephalogram (EEG) which is lower-cost and less prone to radiation, as the only modality to perform binary classification of AD. After channel and feature analysis to extract the most important channels and features, the authors used the explainable boosting machine (XBM) model with three model-agnostic explainers: SHAP [14], Local Interpretable Model-agnostic Explanations (LIME) [21], and Morris Sensitivity (MS) [17]. The study presented topographic maps to elucidate feature importance in terms of EEG channels. Although XBM has shown promising performance, the used dataset is small insofar as the results are not affirmative enough to support EEG signals as a standalone diagnostic tool for AD. In fact, analyzing EEG is a challenging task [19], and it may not provide comprehensive information on a large scale. Nevertheless, we note that integrating EEG into a multimodal setup could be a valuable complementary aid.

Zhang et al. [30] proposed a 3D explainable residual attention network (3D ResAttNet) which is a deep convolutional neural network (CNN) with the addition of self-attention residual blocks and Gradient-weighted Class Activation Mapping (Grad-CAM) [25]. The residual mechanism alleviates vanishing gradients in deep networks, while self-attention learns long-range dependencies. Grad-CAM is used to pinpoint relevant areas (regions associated with disease presence) in each brain scan, by calculating the gradient of the probabilities of those areas with respect to the activation of a particular unit located at a certain position in the last convolutional layer of the network. This gradient represents how sensitive the predicted probabilities are to changes in the activation of that unit, thus highlighting the contribution of each brain area to the model's decision. The authors used structural MRI (sMRI) as the single modality for two separate *binary* classifications: 1) AD vs CN and 2) pMCI vs sMCI.

Similarly, Yu et al. [29] used only sMRI too to perform binary diagnosis of AD (CN vs AD). Their objective was to create higher-resolution brain heatmaps to capture fine-grained details. To that end, they developed a network named MAXNet that consists of a Dual Attention Module (DAM) and a Multi-resolution Fusion Module (MFM), which learn representations that contain information at the voxel level. Additionally, they introduced High-resolution Activation Mapping (HAM) as a visualization method to enhance the quality of the heatmap. Although the algorithm can identify precise small regions in terms of voxels, validation cannot be provided as to whether the algorithm's predictions are actually *correct for the correct cues.*

Mulyadi et al. [18] tackled this issue by developing a method called eXplainable AD Likelihood Map Estimation (XADLiME), based on clinically-guided prototype learning. They measured the similarity between those prototypes and the latent features of clinical information, a clinical label, MMSE score, and age, and thereby created a *pseudo* likelihood map representing the likelihood of AD across different stages. The AD likelihood map was estimated from sMRI using a feature extractor network and a reference map for comparison was obtained by a neural network with a sigmoid activation function. The estimated likelihood map can be viewed from both clinical and morphological perspectives to interpret predictions as a diagnostic tool, to help understand the likelihood of AD progression based on sMRI imaging.

While most studies, including [10,18,29,30], utilize a single modality, our approach leverages multiple modalities for interpretable AD diagnosis and achieves enhanced performance. A perhaps more important differentiator of our work lies in our interpretation approach, which is rooted in our JSM framework, leading to more trustworthy medical diagnoses.

3 Methods

3.1 Jacobian Saliency Map (JSM)

Jacobian Saliency Maps (JSM) emerged recently [1] as a highly effective tool for deciphering the decision-making mechanisms of a deep learning model. It accomplishes this goal by defining specific zones within an input image and measuring their volumetric changes, thereby providing a precise understanding of feature attribution. Feature attribution, which ascribes significance and influence to individual features, enables us to identify the particular aspects of an input that exert significant influence on the model's output. Thus, JSM transforms data in a way that aligns with human intuition and enhances the model's interpretability before the actual deep learning pipeline, making it a promising choice for a diagnosis model debugger.

Our approach provides interpretation by computing the gradients of input with respect to *weighted* elements of the input image and optimizing them toward matching the patterns of deformations highlighted by the JSM. On manipulating the input gradients, we can explore two directions: enhance their significance in relevant brain areas or reduce their significance in irrelevant brain areas. However, because the appropriate magnitudes of gradients in relevant areas are typically unknown a priori, we choose to dampen the gradients in irrelevant areas as the normal regions are a reliable reference.

We are inspired by the work done by Ross et al. [23] who introduced a method to regularize a model's gradients with respect to input features based on a binary mask annotation matrix. However, the annotation term added to the model in [23] is a simple binary mask, which fails to capture correlations between different regions of a brain scan. This poses a significant limitation to the effectiveness because such correlations carry crucial diagnostic information. Additionally, the datasets used in their validation were small and synthetic, thus leaving considerable doubt on whether the method would perform well in real-world medical applications.

Moreover, we do not use the annotation matrix but a special weighted saliency map instead. The rationale is that we prefer a matrix to not only represent the varying degrees of sensitivity of each feature to the diagnosis accurately but also highlight brain regions that exhibit deformations; in the meantime, the normal regions have to be preserved for contrast purposes. These properties cannot be achieved by the binary annotation matrix. In addition, by using the special saliency map to characterize the deformations in the brain, it also allows us to align the constraints on gradients (a penalty we formulate later in (2), (4)) with the domain-specific knowledge related to the medical problem. We perform both registration and Jacobian using Advanced Normalization Tools (ANTs) [3].

Intuition Behind JSM. Ideally, the presence of a disease can be identified by comparing an individual's many scans over a long period of time. Since this is usually not feasible in practice, we can compare an individual's scan to a standardized healthy brain template to assess its local changes using a deformation map. JSM is derived from non-linear image registration, which is a process designed to both minimize variations among individual subjects and align all different images with a standardized template. JSM then examines the deformations that transform the anatomical structures of individuals onto a common standard space, through which we can deduce the relative volume differences between each individual and the template. This analysis aids in pinpointing statistically significant anatomical variations across diverse populations, such as distinguishing between AD patients and healthy elderly individuals. In our investigation, we utilized the Montreal Neurological Institute's 152 brain template (MNI152) as the reference. This process of image registration and JSM computation can be attributed to the realm of computational anatomy, specifically recognized as tensor-based morphometry [22], and has not been adequately explored in the context of dementia.

Formulation of JSM. In medical image processing, computing deformation involves first aligning a source image M with a target image F. This is achieved by using a transformation ϕ that maps points in M to their corresponding points in F. Then, the displacement between these corresponding points is represented as a Deformation Vector Field:

$$v(x, y, z) = \phi(x, y, z) - (x, y, z). \tag{1}$$

To transform a point (x, y, z) to $\phi(x, y, z)$, it is necessary to impose a regularization constraint in order to ensure that the deformation is seamless, one-to-one, and differentiable. This is framed as an optimization problem that minimizes the following cost function:

$$L(\phi, M, F) = -L_{sim}(\phi(M), F) + \alpha L_{Reg}(\phi) \tag{2}$$

where L_{sim} is a similarity measure between two images, the transformed image $\phi(M)$ and F, and L_{Reg} is a regularization term that enforces the desired properties on the deformation. We use Mattes Mutual Information (MI) [16] as our similarity measure, i.e.,

$$L_{sim} = MI(M', F) = \sum_{m'} \sum_{f} P(m', f) \log \left(\frac{P(m', f)}{Q_1(m')Q_2(f)} \right) \tag{3}$$

where M' denotes $\phi(M)$ for simplicity. $P(m', f)$ is the joint probability distribution of the intensity of voxel m' in image M' and that of voxel f in image F, and $Q_1(m')$ and $Q_2(f)$ represent the marginal probability distributions of m' and f, respectively. For the regularizer, we use the B-spline regularization from [27]:

$$L_{Reg} = L_{\text{B-spline}} = \int \left|\nabla^2 \phi(x, y, z)\right|^2 dV \tag{4}$$

where ∇^2 denotes the second-order derivative, dV represent seach voxel (x, y, z), and we integrate over the entire spatial domain.

After solving for the transformation ϕ, we compute a Jacobian matrix J from the deformation vector field $\boldsymbol{v}$, by calculating the first derivative of $\boldsymbol{v}$ at each voxel to encode local deformations including stretching, shearing, and rotation. That is,

$$J(v) = \begin{bmatrix} \frac{\partial v_x}{\partial x} & \frac{\partial v_x}{\partial y} & \frac{\partial v_x}{\partial z} \\ \frac{\partial v_y}{\partial x} & \frac{\partial v_y}{\partial y} & \frac{\partial v_y}{\partial z} \\ \frac{\partial v_z}{\partial x} & \frac{\partial v_z}{\partial y} & \frac{\partial v_z}{\partial z} \end{bmatrix} \tag{5}$$

Then, denoting the Jacobian determinant by $Det(J)$, we calculate it for every voxel $v(x, y, z)$ as

$$Det(J) = \frac{\partial v_x}{\partial x}\left(\frac{\partial v_y}{\partial y}\frac{\partial v_z}{\partial z} - \frac{\partial v_y}{\partial z}\frac{\partial v_z}{\partial y}\right) - \frac{\partial v_x}{\partial y}\left(\frac{\partial v_y}{\partial x}\frac{\partial v_z}{\partial z} - \frac{\partial v_y}{\partial z}\frac{\partial v_z}{\partial x}\right) + \frac{\partial v_x}{\partial z}\left(\frac{\partial v_y}{\partial x}\frac{\partial v_z}{\partial y} - \frac{\partial v_y}{\partial y}\frac{\partial v_z}{\partial x}\right) \tag{6}$$

which forms what we call a *Jacobian Saliency Map JSM* of the source image M:

$$JSM(M) = \begin{bmatrix} & \vdots & \\ \dots & Det(J(v(x, y, z)) & \dots \\ & \vdots & \end{bmatrix}_{\substack{x = 1...W \text{ (width)} \\ y = 1...H \text{ (height)} \\ z = 1...D \text{ (depth)}}}$$

$$\text{At each voxel: } \begin{cases} \text{volume expansion} & \text{if } Det(J) > 1 \\ \text{no change} & \text{if } Det(J) = 1 \\ \text{volume compression} & \text{if } Det(J) < 1 \end{cases} \tag{7}$$

Volumetric changes refer to the alteration in volume at the level of individual voxels within a medical image. This JSM helps us to identify the volumetric ratio of the brain image at the voxel level before and after transformation ϕ, which indicates the brain's volume change.

3.2 Jacobian-Augmented Loss Function (JAL)

By breaking down an input image into distinct regions and measuring how they are transformed, JSM provides precise insight into the complexities of feature attribution and thus model explainability. To this end, we take an innovative approach to explore the possibility of leveraging JSM as a powerful *model debugger*, for which we incorporate the JSM formulated above into the loss function $\mathcal{L}$ of a medical diagnostic model:

$$\mathcal{L}(\boldsymbol{x}, \boldsymbol{y}) = -\sum_{k=1}^{K} y_k \log(\hat{y}_k) + \lambda \sum_{d=1}^{D} \sum_{p=1}^{P} \left(w_{dp} \frac{\partial}{\partial x_{dp}} \log \left(\sum_{k=1}^{K} e^{\hat{y}_k} \right) \right)^2 \tag{8}$$

Given input data $\boldsymbol{x}$ and data label $\boldsymbol{y}$, the loss function $\mathcal{L}(\lambda, \boldsymbol{x}, \boldsymbol{y})$ consists of the training loss (first term) and a novel, JSM-based regularization term (second term). It introduces an emphasis on the importance of anatomical changes captured by the Jacobian map values. Specifically, JSM_{dp} represents the JSM values associated with the dth feature and p^{th} spatial dimension (width, height, and depth in 3D), $\log(\sum_{k=1}^{K} e^{\hat{y}_k})$ represents the natural logarithm of the sum of exponentials of the output logits of x belonging to class k, and $\frac{\partial}{\partial x_{dp}}$ reflects the partial derivative with respect to the dth feature of x in the pth spatial dimension. Thus, this regularization term aims to not only enhance interpretability but also rectify predictions by mitigating the influence of irrelevant cues.

With reference to Equation (7), we add a weight matrix W to give more importance to areas in the JSM that have volumetric changes (expansion or compression) and discourage the input gradients from being significant in areas with no volumetric changes (marked by 1). Hence each element of W is defined as follows:

$$w_{dp} = \begin{cases} \text{feature\_weight}, & \text{if } JSM_{dp} \neq 1 \\ \text{debug\_weight}, & \text{otherwise} \end{cases} \tag{9}$$

Both feature_weight and debug_weight are hyperparameters that indicate the level of importance in every region. We designate a debug weight (penalty) of 1. This choice is deliberate, allowing us to down-weight potentially misleading features while still retaining them as a reference for contrasting relevant features. On the other hand, since regions of volumetric changes are of higher importance, we assign them a feature weight of 0 (no penalty). Hence, we debug the model through weighting features according to its relevance. Note that we don't eliminate irrelevant areas to preserve correlations in the brain volume. More rigorous hyperparameter tuning of such weights can be incorporated in future work.

4 Experiments

4.1 Dataset

In our experiments, we utilized the recently published OASIS-3 dataset. This dataset includes multi-modality data such as MRI, PET, and CT scan images from a diverse group of 1377 participants. Among these participants, 755 were cognitively normal (CN) adults, while the remaining 622 individuals showed different levels of cognitive decline. The age range of the participants was extensive, spanning from 42 to 95 years. CT imaging was used to detect whether certain areas of the brain were shrinking, which can be an indication of AD. On the other hand, MRI provided detailed images of the body and a clear view of progressive cerebral atrophy, which is most visible through T1-weighted volumetric sequences. Consideration of PET data was deferred due to its temporal nature, making it more suitable for future spatiotemporal analyses.

During clinical assessments and diagnoses, the clinical dementia rating (CDR) scores of the participants were utilized. The scores range from 0 to 3, with 0 indicating no AD dementia and 3 indicating severe AD dementia. The very mild stage (rating 0.5) of dementia is similar to the Mild Cognitive Impairment (MCI) stage of AD. As mentioned before, we combined moderate and severe AD due to the very low number of subjects with severe AD. Ultimately, we created four classes based on CDR scores: normal, MCI, mild AD, and severe AD. Having the same subjects in multiple sets (train and test sets) can result in the model overfitting to those individuals, potentially leading to a subpar performance on new, unseen subjects. Hence, for patients who underwent multiple sessions, we preserved the initial MRI session and selected the CT session that was closest in date to that MRI.

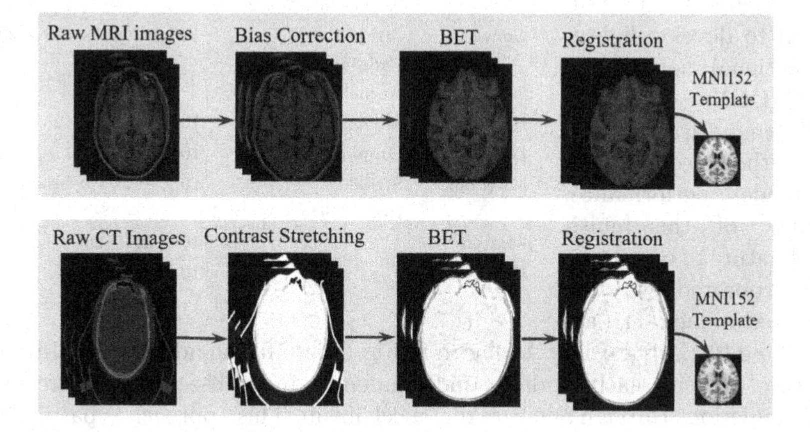

Fig. 1. Preprocessing pipelines for MRI and CT scans, involving bias field correction for MRI, contrast stretching for CT to enhance diagnostic values, BET for brain extraction, and registering CT and MRI to MNI152 brain template.

4.2 Preprocessing

Our preprocessing pipelines for MRI and CT scans are shown in Fig. 1. To minimize any spatially varying intensity bias that may result from factors such as magnetic field inhomogeneities and acquisition artifacts, we use the FMRIB's Linear Image Registration Tool (FLIRT) [9] for bias field correction on MRI images. Meanwhile, we utilize a technique called contrast stretching on CT images to enhance their diagnostic value and visual perception. This process involves adjusting the pixel intensities to fully utilize the display's dynamic range. For CT images, we followed the framework established by Kuijf et al. [11]. Also, we apply the Brain Extraction Tool (BET) [26] to eliminate non-brain portions from both MRI and CT images. Finally, we register both CT and MRI to the MNI152 brain template. Brain templates are typically created for MRI images, making it difficult to register a CT image to an MRI template. We overcame this by adhering to a method in [11] which involves identifying corresponding landmarks in the CT image and MRI template and then using these landmarks to align the images.

4.3 Multimodal Classification

Table 1. CNN Architecture for both branches

Layers	Parameters	Output Size
Input	Batch Size 10	$10 \times 1 \times 182 \times 256 \times 512$
Conv1	Stride 1 Padding 1 Kernel Size $3 \times 3 \times 3$	$10 \times 4 \times 182 \times 256 \times 512$
BatchNorm1	Momentum=0.9	$10 \times 4 \times 182 \times 256 \times 512$
Dropout1	Dropout rate 0.5	$10 \times 4 \times 182 \times 256 \times 512$
MaxPool1	Stride 2 Kernel Size $2 \times 2 \times 2$	$10 \times 4 \times 91 \times 128 \times 256$
Conv2	Stride 1 Padding 1 Kernel Size $3 \times 3 \times 3$	$10 \times 8 \times 91 \times 128 \times 256$
BatchNorm2	Momentum=0.9	$10 \times 8 \times 91 \times 128 \times 256$
Dropout2	Dropout rate 0.2	$10 \times 8 \times 91 \times 128 \times 256$
MaxPool2	Stride 2 Kernel Size $2 \times 2 \times 2$	$10 \times 8 \times 45 \times 64 \times 128$
Flatten		10×2949120
Full-Conn.		10×4

Medical images, especially those related to the brain, are typically 3D. This introduces a computational burden into complex deep neural networks. Our approach seeks to harness the guidance provided by the JSM, and by incorporating such insights from JSM, we aim to develop lighter convolutional neural networks (CNNs) that alleviate computational burdens without compromising model performance. In view of the multimodal nature of our study, we incorporated two data fusion techniques: 1) Late Fusion and 2) Early Fusion. As shown in Fig. 2, late fusion adopts a dual-branch structure, treating each modality independently and subsequently aggregating their predictions through an averaging mechanism. This approach is particularly advantageous for JAL, where debugging is performed separately for each modality through its own JSM. On the other hand, early fusion involves concatenating the input images as well as their corresponding JSM maps, which allows the model to glean correlations between the two modalities and concurrently debug predictions for the input holistically.

Our lightweight CNN model [20] contains two convolutional layers coupled with batch normalization, ReLU activation, dropout, and max-pooling operations, which capture spatial hierarchies and correlation patterns in the input data. The convolutional layers use a kernel size of $3 \times 3 \times 3$, stride of 1, and padding, with a dropout of rate 0.2 for regularization. The max-pooling operation reduces spatial dimensions to $2 \times 2 \times 2$. The 3D tensor is reshaped into a 1D tensor by a flattening layer, preparing it for the subsequent fully connected subnetwork, which performs the overall feature integration and classification. All the specifications are presented in Table 1. Finally, a softmax layer converts the predicted scores into probabilities. In the late fusion setup, the final prediction is computed by averaging the probabilities from both branches.

4.4 Performance Evaluation

We trained our model on a 40GB A100 GPU with batch size 10. Our model was able to quickly converge within only 20 epochs. To address the class imbalance problem in AD, we utilized the Adaptive Synthetic (ADASYN) [7] oversampling algorithm to generate synthetic samples for minority classes during training. Another challenge is that OASIS-3 included identical subjects from multiple sessions across training and test sets may contribute to overfitting, as the model can become excessively attuned to the characteristics of those particular subjects and sessions [2]. To address this, we take meticulous care to ensure that only subjects from distinct sessions are grouped within either the training set or the test set (but not both).

Table 2 summarizes the performance comparison in terms of accuracy, sensitivity, and specificity between our approach and the state-of-the-art. Please note that while *precision* and *recall* are more general terms used in machine learning, *sensitivity* and *specificity* are often preferred in medical and diagnostic fields due to their direct interpretation in the context of disease detection and diagnosis. Table 2 shows that our testing accuracy across four classes surpasses all the baselines that employ the same dataset. Massalimova et al. [15] achieved marginally higher sensitivity and specificity, but it is crucial to note that our model handles a larger number of classes, making it more challenging to achieve higher accuracy. In addition, [15] uses ResNet18 while our model is significantly lighter. In fact, our model is lighter than nearly all the baselines. Basheer et al. [4] used features like CDR, MMSE, age, gender, etc. along with MRI images, and found that age and gender had substantial positive impact on performance. In our case, we achieved superior performance solely with images as we aimed to test our model using spatial features.

Ablation Study. To provide an in-depth assessment of the impact of JAL on our model's performance, we conducted an ablation study on models with and without JAL. This was achieved by setting the JSM term in (8) to zero. The results are presented in Fig. 3, which provides histograms of model performance distribution across multiple mini-batches in the test set. The overlap represents the intersection between the *with JAL* and *without JAL* conditions, indicating

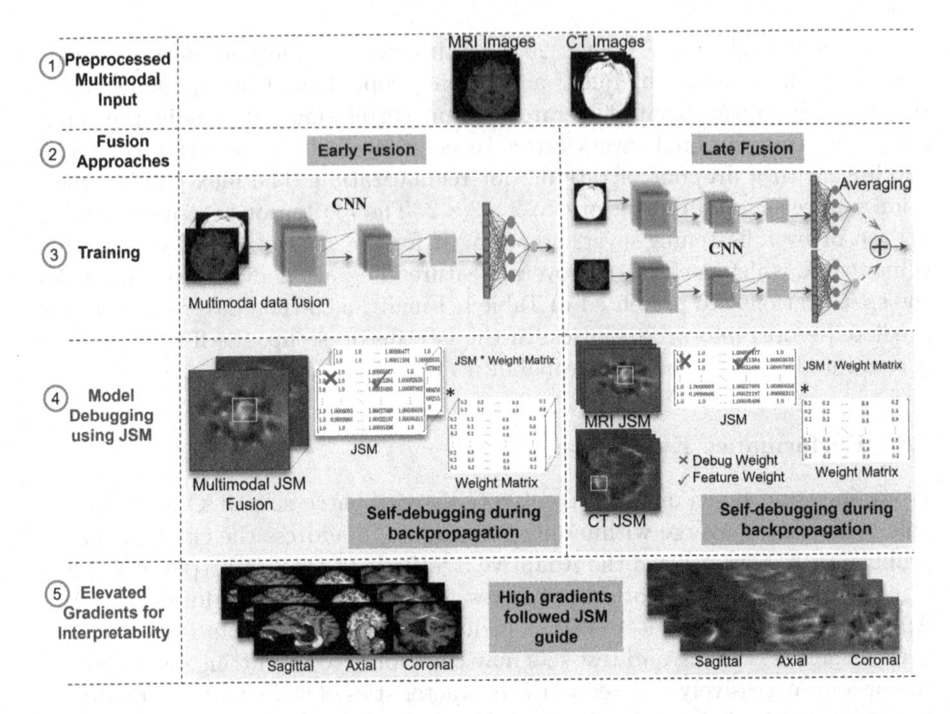

Fig. 2. The complete pipeline. Model debugging using JSM is integrated into training and takes effect during backpropagation, for each modality (late fusion) or both (early fusion). Final predictions are interpreted by plotting elevated gradients overlaid on input images.

the extent to which the model's performance remains consistent regardless of the presence or absence of JAL. The histogram demonstrates that the model performance significantly improves in terms of all the metrics (accuracy, sensitivity, and specificity) when JAL is incorporated, as evidenced by the rightward shift of the distributions in the blue bars in comparison with the yellow bars. This compellingly demonstrates the impact of incorporating JAL on the model's decisions. For a more comprehensive evaluation, Table 3 dissects the four classes and provides more detailed results. It shows substantial improvements over all the AD stages (CN, MCI, MLD, SEV), further affirming the efficacy of JAL. Table 3 also allows us to see that performance for CN and SEV (severe) are relatively higher than MCI and MLD, which is because MCI and MLD have more subtle differences in dementia patterns, making them more intricate to discern. Nevertheless, our model with JAL exhibits evenly promising outcomes for all stages. Scores in Table 2 are the macro average of the scores in Table 3.

Interpretability. We examined the similarity between the volumetric changes characterized by JSM and the decision-making process of the neural network. By plotting gradients overlaid on their corresponding input images, we observed that they closely aligned with the patterns highlighted by the JSM. This can

Table 2. Comparison with reported state-of-the-art using OASIS-3 dataset for AD classification

Model	Modalities	Classes	Sensitivity (%)	Specificity (%)	Accuracy (%)
Salami et al. [24]	MRI	AD, CN	86.01	85.04	87.75
Massalimova et al. [15]	MRI	CN, MCI, AD	96	96	96
Lazli et al. [13]	MRI, PET	AD, CN	92.00	91.78	91.46
Basheer et al. [4]	MRI, features	AD, CN	82.3	*NP	92.3
Castellano et al. [5]	PET	AD, CN	NP	NP	80
Our work	MRI, CT (Early)	CN, MCI, MOD, SEV	92.72	95	91.31
	MRI, CT (Late)	CN, MCI, MOD, SEV	93.5	93.5	95.37

*NP: Not Provided

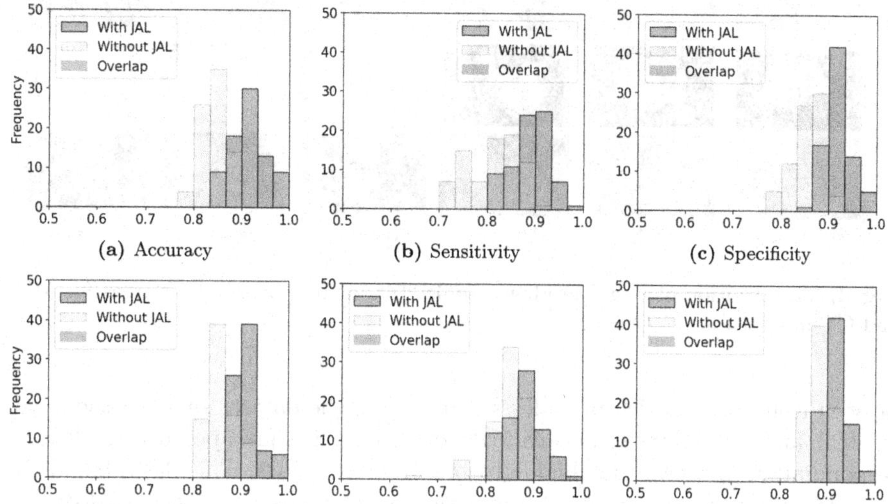

(a) Accuracy (b) Sensitivity (c) Specificity

(d) Accuracy (e) Sensitivity (f) Specificity

Fig. 3. Ablation study on JAL in terms of performance histograms. (a-c): Early fusion, (d-f): Late fusion.

Table 3. Ablation study comparing model performance with and without JAL in Late and Early Fusion setups.

Fusion	Loss Function	Sensitivity (%)				Specificity (%)				Accuracy (%)			
		CN	MCI	MLD	SEV	CN	MCI	MLD	SEV	CN	MCI	MLD	SEV
Early	**w/o JAL**	80.34	78.96	74.3	80.96	84.34	89.5	84.34	89	80	84.3	80.21	89
	w/ JAL	90.3	94.6	94	92	99.12	92.01	99	90	98.8	87.33	86.59	92.5
Late	**w/o JAL**	87.7	87.9	85.5	87.6	87.8	86.6	86.6	85.2	88.8	86.1	88.4	87.3
	w/ JAL	95.3	93.3	93.3	91.5	99.6	92	92	92.2	99.8	92	93	96.7

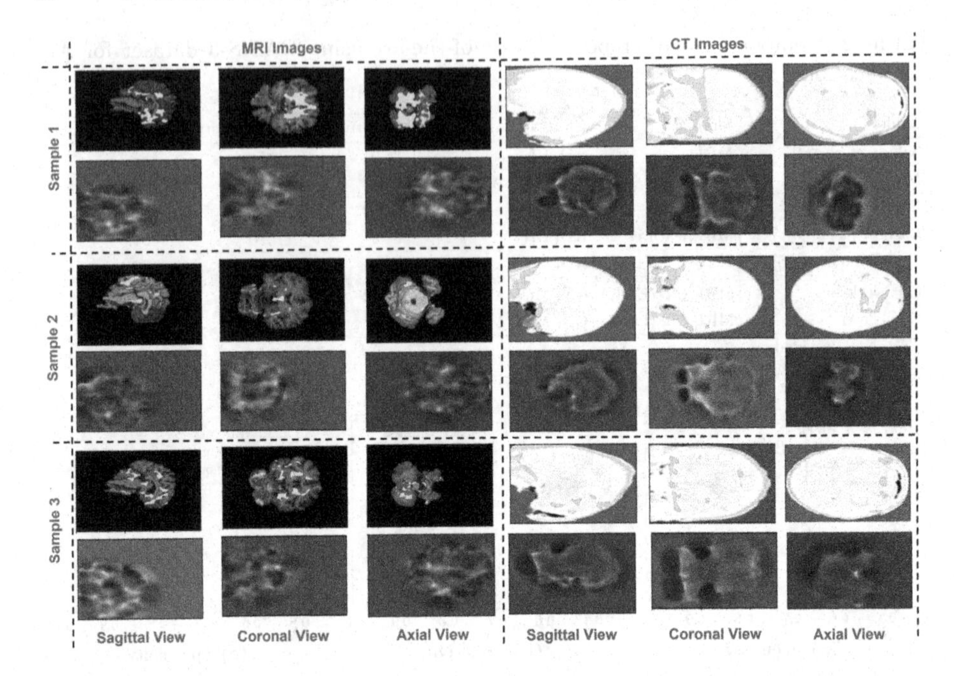

Fig. 4. Visualization of larger gradients in JSM-indicated deformation areas for MRI and CT modalities.

be seen from Fig. 4 which provides samples from the dataset showing the axial, sagittal, and coronal views of each MRI and CT images juxtaposed with the corresponding JSM views that highlight brain deformations. This desirable alignment is attributed to our JAL loss function which incorporates the JSM during the model debugging process, contributing to the model trustworthiness by promoting transparency.

Furthermore, incorporating JSM in JAL during model debugging also enables the model to learn and adapt to the highlighted deformations, and hence serves as a powerful tool for refining the model performance as well, fostering a more accurate and *informed* decision-making process.

5 Conclusion

This paper introduces a new approach to trustworthy medical diagnoses, by addressing two key challenges: model explainability and reliability. On explainability, we leverage Jacobian saliency maps (JSM) to provide informative and interpretable guide for feature learning, as well as capture subtle morphological changes associated with the disease. On reliability, we incorporate JSM into the loss function as a self-debugger to direct the model to critical (disease-relevant) regions during training, avoiding the *Clever-Hans* behavior. Our approach not only helps rectify erroneous predictions but also identifies regions in a post-hoc manner with elevated gradients for interpretability enhancement. (Note that post-hoc is for visualization only; our main approach of JAL/debugging is a during-modeling approach.) Our extensive evaluation underscores the success of JSM via a Jacobian-augmented loss function (JAL), leading to substantial accuracy improvement (by up to 10%) and greater model interpretability in identifying significant brain areas that lead to diagnostic predictions. Our XAI approach also works seamlessly with our multimodal data fusion methods and provides explanation in both early and late fusion setups.

References

1. Abbas, S.Q., et al.: Transformed domain convolutional neural network for Alzheimer's disease diagnosis using structural MRI. Pattern Recogn. **133**, 109031 (2023)
2. Altay, F., et al.: Preclinical stage Alzheimer's disease detection using magnetic resonance image scans. In: Proceedings of the AAAI Conference on Artificial Intelligence, vol. 35, pp. 15088–15097 (2021)
3. Avants, B.B., et al.: Advanced normalization tools (ants). Insight j **2**(365), 1–35 (2009)
4. Basheer, S., et al.: Computational modeling of dementia prediction using deep neural network: analysis on oasis dataset. IEEE Access **9**, 42449–42462 (2021)
5. Castellano, G., et al.: Detection of dementia through 3d convolutional neural networks based on amyloid pet. In: 2021 IEEE Symposium Series on Computational Intelligence (SSCI), pp. 1–6. IEEE (2021)
6. El-Sappagh, S., et al.: A multilayer multimodal detection and prediction model based on explainable artificial intelligence for Alzheimer's disease. Sci. Rep. **11**(1), 2660 (2021)
7. He, H., et al.: Adasyn: adaptive synthetic sampling approach for imbalanced learning. In: 2008 IEEE International Joint Conference on Neural Networks (IEEE World Congress on Computational Intelligence), pp. 1322–1328. IEEE (2008)
8. Hühn, J., Hüllermeier, E.: Furia: an algorithm for unordered fuzzy rule induction. Data Min. Knowl. Disc. **19**, 293–319 (2009)
9. Jenkinson, M., et al.: Improved optimization for the robust and accurate linear registration and motion correction of brain images. Neuroimage **17**(2), 825–841 (2002)
10. Khare, S.K., et al.: Adazd-net: automated adaptive and explainable Alzheimer's disease detection system using EEG signals. Knowl.-Based Syst. **278**, 110858 (2023)
11. Kuijf, H.J., et al.: Registration of brain CT images to an MRI template for the purpose of lesion-symptom mapping. In: Multimodal Brain Image Analysis: Third International Workshop, MBIA 2013, Held in Conjunction with MICCAI 2013, Japan, Proceedings 3, pp. 119–128. Springer (2013)
12. Laugel, T., et al.: The dangers of post-hoc interpretability: Unjustified counterfactual explanations. arXiv preprint arXiv:1907.09294 (2019)
13. Lazli, L., et al.: Computer-aided diagnosis system of Alzheimer's disease based on multimodal fusion: tissue quantification based on the hybrid fuzzy-genetic-possibilistic model and discriminative classification based on the SVDD model. Brain Sci. **9**(10), 289 (2019)

14. Lundberg, S.M., et al.: A unified approach to interpreting model predictions. Advances in neural information processing systems 30 (2017)
15. Massalimova, A., et al.: Input agnostic deep learning for Alzheimer's disease classification using multimodal MRI images. In: 2021 43rd Annual International Conference of the IEEE Engineering in Medicine & Biology Society (EMBC), pp. 2875–2878. IEEE (2021)
16. Mattes, D., et al.: Pet-ct image registration in the chest using free-form deformations **22**(1), 120–128 (2003)
17. Morris, M.D.: Factorial sampling plans for preliminary computational experiments. Technometrics **33**(2), 161–174 (1991)
18. Mulyadi, A.W., et al.: Estimating explainable Alzheimer's disease likelihood map via clinically-guided prototype learning. Neuroimage **273**, 120073 (2023)
19. Mustafa, Y., Elmahallawy, M., Luo, T., Eldawlatly, S.: A brain-computer interface augmented reality framework with auto-adaptive ssvep recognition. In: 2023 IEEE International Conference on Metrology for eXtended Reality, Artificial Intelligence and Neural Engineering (MetroXRAINE), pp. 799–804. IEEE (2023)
20. Mustafa, Y., Luo, T.: Diagnosing Alzheimer's disease using early-late multimodal data fusion with Jacobian maps. In: IEEE International Conference on E-health Networking, Application & Services (Healthcom) (2023)
21. Ribeiro, M.T., et al.: "Why should i trust you?" explaining the predictions of any classifier. In: Proceedings of the 22nd ACM SIGKDD International Conference on Knowledge Discovery and Data Mining, pp. 1135–1144 (2016)
22. Riyahi, S., et al.: Quantifying local tumor morphological changes with Jacobian map for prediction of pathologic tumor response to chemo-radiotherapy in locally advanced esophageal cancer. Phys. Med. Biol. **63**(14), 145020 (2018)
23. Ross, A.S., et al.: Right for the right reasons: training differentiable models by constraining their explanations. In: Proceedings of the 26th International Joint Conference on Artificial Intelligence (IJCAI), pp. 2662–2670 (2017)
24. Salami, F., et al.: Designing a clinical decision support system for Alzheimer's diagnosis on oasis-3 data set. Biomed. Signal Process. Control **74**, 103527 (2022)
25. Selvaraju, R.R., et al.: Grad-cam: visual explanations from deep networks via gradient-based localization. In: Proceedings of the IEEE International Conference on Computer Vision, pp. 618–626 (2017)
26. Smith, S.M.: Fast robust automated brain extraction. Hum. Brain Mapp. **17**(3), 143–155 (2002)
27. Tustison, N.J., et al.: Explicit b-spline regularization in diffeomorphic image registration. Front. Neuroinform. **7**, 39 (2013)
28. Venugopalan, J., et al.: Multimodal deep learning models for early detection of Alzheimer's disease stage. Sci. Rep. **11**(1), 3254 (2021)
29. Yu, L., Xiang, et al.: A novel explainable neural network for Alzheimer's disease diagnosis. Pattern Recogn. **131**, 108876 (2022)
30. Zhang, X., Han, et al.: An explainable 3d residual self-attention deep neural network for joint atrophy localization and Alzheimer's disease diagnosis using structural MRI. IEEE J. Biomed. Health Inform. **26**(11), 5289–5297 (2021)

Towards Nonparametric Topological Layers in Neural Networks

Gefei Shen and Dongfang Zhao(✉)

University of Washington, Seattle, USA
{gefeis3,dzhao}@uw.edu

Abstract. Various topological techniques and tools have been applied to neural networks in terms of network complexity, explainability, and performance. One fundamental assumption of this line of research is the existence of a global (Euclidean) coordinate system upon which the topological layer is constructed. Despite promising results, such a *topologization* method has yet to be widely adopted because the parametrization of a topologization layer takes a considerable amount of time and lacks a theoretical foundation, leading to suboptimal performance and lack of explainability. This paper proposes a learnable topological layer for neural networks without requiring an Euclidean space. Instead, the proposed construction relies on a general metric space, specifically a Hilbert space that defines an inner product. As a result, the parametrization for the proposed topological layer is free of user-specified hyperparameters, eliminating the costly parametrization stage and the corresponding possibility of suboptimal networks. Experimental results on three popular data sets demonstrate the effectiveness of the proposed approach.

Keywords: Applied topology · Nonparametric learning · Neural networks

1 Introduction

1.1 Background

Topological techniques and optimizations have been proposed for neural networks from various aspects, such as network complexity [16], explainability [19], and performance [23]. Contrary to the conventional geometrical standpoint for training a neural network, topological approaches focus on the *qualitative* properties of the underlying data. A geometrical approach typically imposes the *quantitative* metric between objects (e.g., Euclidean distance), while topologists are more interested in identifying the intrinsic properties of the targeted objects. Informally, an *intrinsic* property refers to a property that remains invariant under continuous deformation of the object's geometrical form.

The benefits of incorporating topological information into neural networks are evident, and we provide two illustrative examples:

D.-N. Yang et al. (Eds.): PAKDD 2024, LNAI 14647, pp. 91–102, 2024.
https://doi.org/10.1007/978-981-97-2259-4_7

One of the most widely used topological invariant is *connectedness*. In the conventional sense of graph theory (at least in $\mathbb{R}^2$), a graph is considered *connected* if there exists a path between any pair of vertices. For instance, a figure '8' (in $\mathbb{R}^2$) is connected, while an English letter 'i' is not. Furthermore, we can reasonably assume that handwritten '8' and 'i' would retain the connectedness property, allowing them to be distinguished based on this topological property.

Another less obvious (topological) invariant is the *homology group*. Roughly speaking, the homology group captures the cycles of the underlying object that are not boundaries of any region (i.e., subset) of the object. A useful geometrical intuition for these homology groups is to consider them as representing *holes* in the object. Importantly, the ranks of these homology groups (up to homomorphism) remain unchanged under continuous deformation. For example, '8' has two 1-dimensional holes, '0' has one 1-dimensional hole, and '1' has none.

Incorporating topological invariants has sparked considerable research interest, as illustrated in the aforementioned examples. The state-of-the-art topological extensions [3,9,13] of neural networks are built upon the concept of *persistent topology* [5], which essentially determines these topological invariants from a series of parametrizations over the underlying object.

1.2 Motivation and Challenges

Existing works on employing topology into neural networks commonly assume the existence of a global (Euclidean) coordinate system for the layer of persistent topology parametrization. However, the parameterization procedure in such works is ad-hoc, lacking a systematic approach for tuning the (hyper)parameters. State-of-the-art methods for constructing topological layers for neural networks often adopt polynomial [7] or exponential [8] expressions over the infinite $\mathbb{R}$ field, resulting in an *infinite* parameter space. Consequently, the parametrization process incurs high preprocessing costs, and there is no guarantee that the resulting model is optimal. Hence, a more efficient approach for integrating topology into neural networks is urgently needed.

To address this issue, this paper proposes a *nonparametric* construction of a learnable topological layer. The rationale behind our proposal lies in the challenge of parametrization: the Euclidean space is an *overly* rich topological (metric) space with every single point being globally identified by a sequence of real-valued coordinates (infinite-dimensional). If we can migrate the parametrization of these points (in terms of Euclidean coordinates) into intrinsic components implicitly characterized by the application, we can avoid dealing with the infinite parameter space. In *functional analysis*, which studies the relationships between functions rather than values, spaces that are more abstract than the Euclidean space, such as Banach spaces [2] and Hilbert spaces [26] (e.g., reproducing kernel Hilbert space, or RKHS), have drawn peculiar interest as they allow us to quantify the relationships among functions. This work is inspired by these ideas, where learnable topological layers can be embedded in a Hilbert space.

The technical challenges of employing nonparametric topological layers in a functional space are twofold. First, since the topological layer is constructed without a global Euclidean coordinate system, we need to ensure (i.e., prove) that the layer is *differentiable*, enabling well-defined metrics (e.g., gradients) in neural network applications. Second, it is unclear how to construct meaningful subsystems/criteria (e.g., loss functions) using a limited set of metrics (e.g., norms and inner products in a Hilbert space). To our knowledge, no prior work has investigated a nonparametric method for incorporating topological layers into neural networks.

1.3 Contributions

We present the construction of a learnable topological layer embedded in a Hilbert space. To the best of our knowledge, this is the first learnable topological layer constructed using only inner products, which eliminates the high cost of the topological layer's parametrization. This paper makes the following contributions:

1. We propose a nonparametric approach for integrating topological layers into neural networks, leveraging Hilbert spaces and inner products as the foundation for our construction.
2. We prove the continuity of the proposed metrics, ensuring stability and reliability in the topological layer's learning process.
3. We demonstrate that the derived loss function, used in a gradient-based learning procedure, is differentiable. This essential property allows for seamless incorporation of the topological layer into standard neural network training.

2 Preliminaries and Related Work

2.1 Basics of Topology

We briefly review concepts and facts (without proof) that will be used later. A *topology* of a set S is a collection of subsets of S, denoted $\mathcal{T}$. One example topology of S is then the power set of S, $\mathcal{P}(S)$, which consists of all the possible $2^{|S|}$ subsets of S. This is also called the *discrete topology* of S. The pair $(S, \mathcal{T})$ is called a *topological space*; in practice, we usually adopt more information to characterize the topological space with additional properties. Each of the subsets U from $\mathcal{T}$ is called an *open set*, and the complement set $S \setminus U$ is a *closed set* by definition. A function g from space X to Y is called *continuous* if: $\forall v$ is an open set in Y, then $g^{-1}(v)$ is an open set in X. The composition of two continuous functions is also continuous. If both g and g^{-1} are continuous and bijective, we call g a *homeomorphism*. Since a homeomorphism is defined on open and closed sets, two topological spaces are *equivalent up to homeomorphism* if the latter exists. A *metric space* is a set S of points where a function d is defined for every pair of points, $d : S \times S \rightarrow \mathbb{R}^*$, where $\mathbb{R}^*$ denotes the set of non-negative real numbers and d satisfies the following properties (axioms): $\forall x, y, z \in S$, $d(x, x) = 0$, $d(x, y) = d(y, x)$, and $d(x, y) \leq d(x, z) + d(z, y)$. A *normed space* is a metric space where a norm function, usually denoted $\|\cdot\|$, is well defined and satisfies the following properties: $\forall x, y \in S, \forall \lambda \in \mathbb{R}$,[1] $\|x + y\| \leq \|x\| + \|y\|$, $\|\lambda x\| = |\lambda| \|x\|$, and $\|x\| = 0 \Leftrightarrow x = 0$. A complete, normed space is also called a *Banach space*.

2.2 Topological Neural Network

Adopting topological information in neural networks (esp. machine learning applications) has drawn a lot of research interest. Much of this line of work focuses on constructing a kernel translating the bar codes into a metric, such as [12,22]. In addition, some works [1,2] show that the kernel can be incorporated with inner products (e.g., in a Banach space). It should be noted that the above works take the topological characteristics as a *static* feature of the input data. The first *learnable* layer of topological information for neural networks appeared in [8]. The *differentiable* property exhibited in the learned layer then sparkled a new series of works such as leveraging topological information for regularization [4]. Notably, it has been shown that a learnable topological layer can further promote or discount a specific topological feature, as reported in [7]. We remark that all of the above works require the users to parametrize the layer, e.g., through polynomial or exponential functions. The nonparametric topological layer proposed in this paper is implemented atop the code base of [7].

2.3 Functional Spaces for Machine Learning

Embedding functional spaces of topological invariant to statistical-learning-based systems has been an active research area for a while. In [25], a distance based on the Gaussian kernel was proposed in the reproducing kernel Hilbert space (RKHS). Later in [17], a *support vector machine* (SVM) in the RKHS was proposed for nonparametric inference. Extending from the Gaussian kernel, a *weighted Gaussian kernel* for persistent homology was proposed in [11]. In a more general setup, an RKHS embedding was proposed for quantum graphical modelling [24], a continuous-time reinforcement-learning framework with RKHS was proposed in [20], a nonparametric optimization technique in RKHS was proposed for regression applications [21], and a new theory was developed for self-distilling in deep learning in a Hilbert space [15]. To the best of our knowledge, however, there exists no prior work that systematically constructs a nonparametric learnable topological layer embedded in a Hilbert space.

3 Methodology

We initialize the topological layer with the norms of input data and update the node weights in a Hilbert space. Similar to TopoSig [8], the nonparametric

[1] Technically, λ could be an element in any other *field* $\mathbb{F}$. We restrict our discussion to the real numbers $\mathbb{R}$ (which is also a field) in the context of neural network applications.

topological layer (NTL) is prepended to the neural network. The topological information used to feed into the NTL is also called the input's (topological) *signature*, which is usually communicated through multisets (or, bag) with possibly duplicate values. In the following, we will be focusing on the construction of a nonparametric loss function in the Hilbert space. Before that, we quickly review the status quo [7] for adopting the topological signature in a *polynomial* form at a specific dimension:

$$\Delta F = -\sum_{j}(d_j - b_j)^p \times \left(\frac{d_j + b_j}{2}\right)^q,$$

where j is the index of a specific cycle (bar), b_j and d_j indicate the birth time and death time, and p and q are the weights for the lengths and the means, respectively.

Assuming that there are a total of m cycles at dimension i. Evidently, we have $m \leq n$ because there cannot exist more than n cycles with n points. We code the cycles with a pair of real numbers $(l, m) \in \mathbb{R}^2$. Denote two points with vector L_i and M_i, respectively, for dimensional i. The question then becomes to identify a function f such that:

$$f : L \times M \rightarrow F^n.$$

To guarantee that the layer is differentiable, as is required for updating the neural network structure, we need to show that the f function is continuous. Intuitively, when we think about two vectors with an inner product, a tiny movement of any of the two vectors should only lead to a similarly tiny change to the projected value. We will formally prove this claim using point-set-topological machinery, as follows.

Theorem 1. *The inner product over L and M of dimension m is a continuous function with the codomain F of dimension* $n \geq m$.

Proof. Let V be a n-dimensional open set in F. Our goal is to show that $U = f^{-1}(V)$ is a pair of open sets in the m-dimensional space. Note that an open set $O \in S$ is nothing but a subset with the following three axioms (see [18] for introductory point-set topology):

O1: The union of any number of open sets is an open set;
O2: The intersection of finite open sets is an open set; and
O3: Both S and $\emptyset$ are open sets.

Let $U_j = f^{-1}(V_j) = (U_j^1, U_j^2)$ denote the preimage of an arbitrary open set $V_j \in F$, i.e., the pair of m-dimensional points. By definition, an inner product of two m-dimensional points is the projection of one vector onto the other. That is, $\langle U_j^1, U_j^2 \rangle$ is an open set. We will show that U_j^1 is an open set; U_j^2 can be proven symmetrically.

For contradiction, suppose U_j^1 is not an open set. Then, by O1, U_j^1 cannot be a union of open sets. We impose the same direction of U_j^1 on $\langle U_j^1, U_j^2 \rangle$, denoted

U_j^α, which is evidently an open set. Let $U_j^\beta = U_j^1 - U_j^\alpha$, then U_j^β cannot be an open set because otherwise U_j^1 would be an open set—a union of two open sets. Note that U_j^α (an open set) and U_j^β (a non-open set, or *closed set*) are linear dependent. It follows that a linear function can map between an open set and a closed set, which is impossible because the limit points of a closed set (i.e., the boundary) cannot be linearly mapped to an open set. We thus reach a contradiction and the claim is proved.

As the next step, we need to aggregate each dimension's contribution. In real-world neural network applications, such as recognizing MNIST, we want to consider characteristics from multiple dimensions like 0-dimensional *single components* (e.g., one component for figure '7') and 1-dimensional *holes* (e.g., two holes for figure '8'). Therefore, the overall *nonparametric* loss function looks the following:

$$\mathcal{F} = \Delta F^n := \sum_{i=0}^{n} (-1)^i \cdot (1 + dim(\Delta F_i)) \cdot \langle L_i, M_i \rangle,$$

where i indicates a specific dimension of the simplicial complex. We then need to show that ΔF^n is differentiable. Note that we do not assume a Euclidean space or Euclidean coordinates, therefore we will have to show that all the partial derivatives (i) exist and (ii) are continuous. We sketch the proofs of both properties in the following two Lemmas.

Lemma 1. *Both* $\frac{\partial}{\partial L}\mathcal{F}$ *and* $\frac{\partial}{\partial M}\mathcal{F}$ *exist.*

Proof. Recall that a partial derivative by definition is nothing but a directional derivative over the variables. For L, we have

$$\begin{aligned}
\frac{\partial}{\partial L}\mathcal{F} = \lim_{h \to 0} & \Bigg(\Big(\sum_i (-1)^i (1 + dim(\Delta F_i^n)] \langle L_i + h_i, M_i \rangle \\
& - \sum (-1)^i [1 + dim(\Delta F_i^n)) \langle L_i, M_i \rangle \Big) / ||h|| \Bigg) \\
= & \sum_i (-1)^i (1 + dim(\Delta F_i^n)) \\
& \times \lim_{h \to 0} \frac{\langle L_i + h_i, M_i \rangle - \langle L_i, M_i \rangle}{||h||} \\
= & \sum_i (-1)^i (1 + dim(\Delta F_i^n)) \cdot \lim_{h \to 0} \frac{\langle h_i, M_i \rangle}{||h||} \\
= & \sum_i (-1)^i (1 + dim(\Delta F_i^n)) \cdot \lim_{e^i \to 0} \langle e^i, M_i \rangle \\
= & \sum_i (-1)^i (1 + dim(\Delta F_i^n)) \cdot \int_{B_\epsilon} e^i(x) M_i(x) dx,
\end{aligned}$$

where $e^i = \frac{h_i}{\|h\|}$ denotes the (normalized) coordinates and B_ϵ denotes the small neighborhood around the point where the derivative is applied. Intuitively, the factor $\int_{B_\epsilon} e^i(x)M_i(x)dx$ in the last equality represents the "movement" of the function at point (L_0, M_0) in the direction h.[2] Recall that our construction is in a Hilbert space, which by definition means that there are no "missing" points in any sequence. Therefore, the last factor does not diverge and must exist: $\int_{B_\epsilon} e^i(x)M_i(x)dx < \infty$. As a result, $\frac{\partial}{\partial L}\mathcal{F}$ exists, as desired.

The existence of $\frac{\partial}{\partial M}\mathcal{F}$ can be proved similarly.

Lemma 2. *Both* $\frac{\partial}{\partial L}\mathcal{F}$ *and* $\frac{\partial}{\partial M}\mathcal{F}$ *are continuous.*

Proof. It is sufficient to show that the difference between the inner product of the component sequences converges to zero given that each component converges to zero. That is, we need to show

$$n \to \infty, L_n \to L, M_n \to M \Rightarrow \langle L_n, M_n \rangle \to \langle L, M \rangle.$$

Equivalently, we claim the following limit

$$|\langle L_n, M_n \rangle - \langle L, M \rangle| \to 0.$$

First, we can compute the difference between two inner products as follows (recall the linearity property in inner products):

$$\begin{aligned}\langle L_n, M_n \rangle - \langle L, M \rangle &= \langle L_n - L, M_n \rangle + \langle L, M_n \rangle \\ &\quad - (\langle L, M - M_n \rangle + \langle L, M_n \rangle) \\ &= \langle L_n - L, M_n \rangle + \langle L, M - M_n \rangle.\end{aligned}$$

Then, by triangle inequality we have:

$$\begin{aligned}&|\langle L_n, M_n \rangle - \langle L, M \rangle| \\ &\leq |\langle L_n - L, M_n \rangle| + |\langle L, M - M_n \rangle| \\ &\leq \sqrt{\langle L_n - L, L_n - L \rangle}\sqrt{\langle M_n, M_n \rangle} \\ &\qquad + \sqrt{\langle L, L \rangle}\sqrt{\langle M - M_n, M - M_n \rangle} \\ &\to \sqrt{\langle 0, 0 \rangle}\sqrt{\langle M_n, M_n \rangle} + \sqrt{\langle L, L \rangle}\sqrt{\langle 0, 0 \rangle} \\ &= 0 \cdot \|M_n\| + \|L\| \cdot 0 \\ &= 0,\end{aligned}$$

where the last equality is because both norms $\|M_n\|$ and $\|L\|$ are finite in persistent homology.

Theorem 2. $\mathcal{F}$ *is differentiable.*

Proof. The claim immediately follows Lemmas 1 and 2.

[2] Mathematically speaking, it would be the external derivative of a vector field $\mathcal{H}_{(L_0, M_0)}[\mathcal{F}]$. We do not use such terms to avoid unnecessary confusions.

4 Evaluation

4.1 Experimental Setup

Testbed. All of our experiments are carried out on a 64-bit Ubuntu 20.04 LTS workstation. The workstation has an Intel Core Quad i7-6820HQ CPU @ 2.70GHz, 64 GB memory, and a 1 TB Samsung SSD drive. Our Anaconda environment has the following libraries installed: Numpy 1.19.2, Pandas 1.1.3, Python 3.8.5, PyTorch 1.4.0, and Scipy 1.5.2. Our C++ compiler is g++ 9.3.0.

Data sets. We use three data sets for experiments, all of which are publicly available: MNIST [14], KMNIST [10], and FashionMNIST [6]. All of them comprise 28×28 data points, 60,000 for training and 10,000 for testing. We repeated all the experiments at least five times and observed unnoticeable differences; for this reason, we will not report the error bars of the reported numbers in the remainder of this section.

Network Models. We use the same neural network model in the PyTorch implementation of MNIST [14]. The network has two convolution layers for channel numbers $1 \rightarrow 32 \rightarrow 64$ with kernel size 3. The dropout rates are 0.25 and 0.5, respectively. The dimensions of linear transformation are $9,216 \rightarrow 128 \rightarrow 10$. A Rectified Linear Unit (ReLU) layer is applied after the first layer, and a Log-Softmax function is applied after the second layer. We set the batch size as 100 and the epoch number as nine. The model is trained with the entire training set (60,000) and evaluated with the entire test set (10,000) unless otherwise stated. GPU is disabled in our experiments.

4.2 Implementation

We implemented the proposed approach atop the codebase of [7]. The core algebraic computation was implemented with C++, and the neural network-related routines are either inherited from PyTorch or implemented from scratch with Python.

To incorporate the proposed method into an end-to-end application, we also implemented a module to binarize the pixels of input grey-level data such that the simplicial complexes are built upon a set of discrete points. When evaluating various network models and parameters, all the input data are binarized for a fair comparison.

One caveat when implementing the methods is that the topologization might move a point out of the 28×28 frame. This did not happen at all for MNIST and started showing up for KMNIST. Our current solution to this problem is to still include the off-scene point at the boundary of the frame such that we do not lose the topologization effect.

4.3 Overall Performance

We report the overall performance of the nonparametric topological layer in Fig. 1. The accuracy after training at each epoch is plotted for all three data sets.

A common phenomenon is that the first five epochs represent the most crucial steps for the performance, as later epochs show mild or negligible improvements.

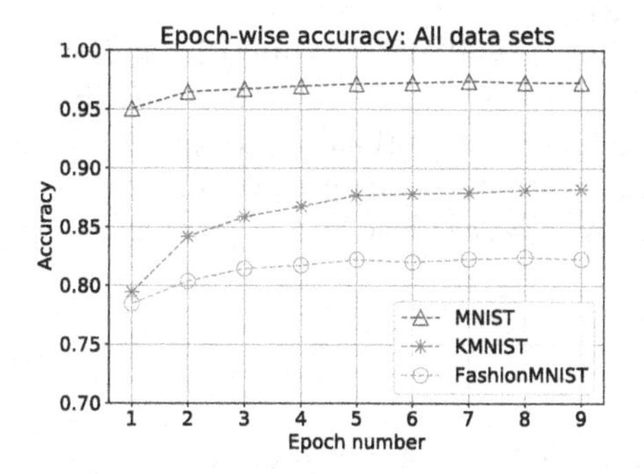

Fig. 1. Accuracy at various epochs

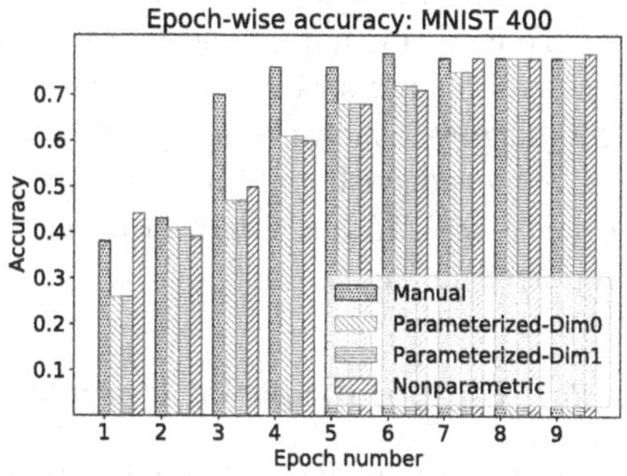

Fig. 2. Learning speed comparison

We also observe the accuracy discrepancy across the three data sets. Because we keep the same setting for all experiments, this phenomenon can be best explained by the data's intrinsic property: the MNIST data is intrinsically more topologically persistent than the other two data sets. That is, the topological layer is more sensitive to the MNIST data.

4.4 Learning Rate

We want to compare the proposed nonparametric method with the parametrized one [7]. Because the parameterized topological preprocessing (i.e., sweeping the parameters to obtain the highest accuracy) takes a considerable amount of time[3], we restrict this experiment to the MNIST data set at a smaller scale of 400 data points, or MNIST 400. The network structure (e.g., convolution layers, dropout rates) and other parameters remain the same (e.g., batch size = 100, epoch = 9) as in the previous subsection.

Figure 2 compares the learning speed of the proposed nonparametric topologization with three baseline methods: the manual optimization without topologization (starting with the popular *adam*, lr = 0.01), a parametrized 1-dimensional topologization and a parametrized 0-dimensional topologization [7]. Our experiments show that the nonparametric approach takes seven epochs to reach the optimum, while the manually optimized one only takes four. Except for the fluctuation at initial epochs, nonparametric and parametrized topologizations exhibit no significant difference in learning speed. We stress that the main benefit of applying a nonparametric topologization is not for a faster converging rate; instead, the point of using a nonparametric method is to free the

[3] Parallel computing of this preprocessing stage is possible but we do not discuss it in this paper.

users from the non-systematic procedure of picking the "optimal" parameters (trial-and-error, brute force parameter-sweeping) and yet to expect the neural network model to reach performance on par with that of a handcrafted optimization and/or a parametrized topologization.

To confirm that the nonparametric method indeed offers competitive performance, we enumerate various dimension weights between −8 and +8 for MNIST 400. Figure 3 shows that in this parameter space, the optimal performance can be achieved with parameters at a handful of discrete (integer) combinations, e.g., $(-1, 8)$. Indeed, that subspace of parameters does lead to the performance reported in Fig. 2.

4.5 Temporal-Spatial Correlation

The final experiment reports the computation time for the proposed topologization. While the time overhead of the proposed nonparametric methods is insignificant (20–30 s on our testbed) to the neural network training, we did observe interesting results when collecting numbers for Fig. 3. There seems to be a pattern regarding the dimensions of the underlying data set.

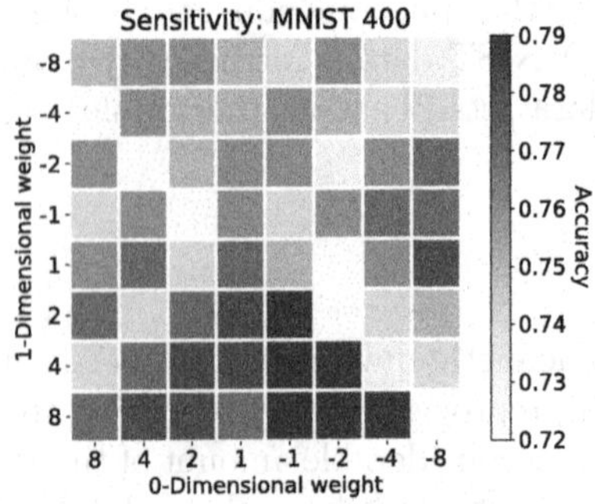

Fig. 3. Dimension weights

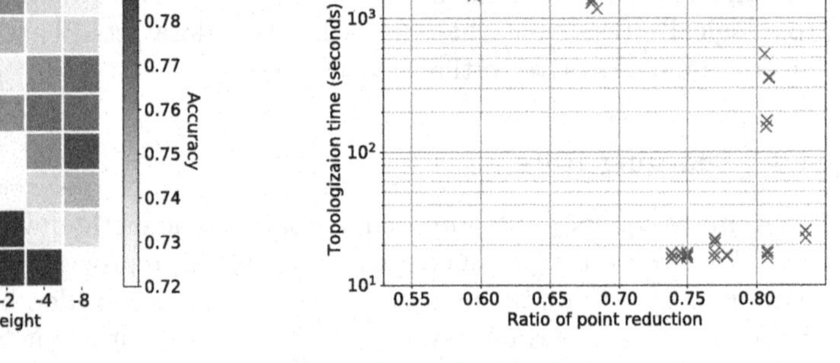

Fig. 4. Time-space correlation

To investigate the correlation, we conduct 64 topologization experiments with different time-space combinations and reported the results in Fig. 4. The correlation seems step-wise (or even binary): except for four cases (out of 64), about half of the points reside at the top-left corner, and the other half resides at the bottom-right corner. For MNIST, a ratio of 0.7 seems the cut-off line of these two clusters. The distance between the two clusters is also interesting: With only a small difference between space-reduction ratios (e.g., 0.68 vs. 0.73), the computation time differs in orders of magnitude (e.g., 1,200 s vs. 17 s). The root cause of this discrepancy stems from those topological features that are not captured by persistent topology. How to effectively capture such information will be one of our future research directions.

5 Conclusion

This paper presents the construction of a nonparametric learnable topological layer embedded in a Hilbert space. To the best of our knowledge, this is the first learnable topological layer constructed solely with inner products, eliminating the need for costly parameterization. We have shown that the extended inner products are continuous, and the derived loss function (to be used in gradient-based learning procedures) is differentiable. Experimental results affirm the effectiveness of the proposed nonparametric approach.

Acknowledgement. Results presented in this paper were partly obtained using the Chameleon testbed supported by the National Science Foundation.

References

1. Adams, H., et al.: Persistence images: a stable vector representation of persistent homology. J. Mach. Learn. Res. **18**(1), 218–252 (2017)
2. Bubenik, P.: Statistical topological data analysis using persistence landscapes. J. Mach. Learn. Res. **16**(1), 77–102 (2015)
3. Carrière, M., Chazal, F., Glisse, M., Ike, Y., Kannan, H., Umeda, Y.: Optimizing persistent homology based functions. In: Proceedings of the 38th International Conference on Machine Learning, vol. 139, pp. 1294–1303 (2021)
4. Chen, C., Ni, X., Bai, Q., Wang, Y.: A topological regularizer for classifiers via persistent homology. In: Proceedings of the 22nd International Conference on Artificial Intelligence and Statistics (16–18 Apr 2019), vol. 89, pp. 2573–2582 (2019)
5. Edelsbrunner, H., Harer, J.: Computational Topology - an Introduction. American Mathematical Society (2010)
6. FashionMNIST Dataset. https://github.com/zalandoresearch/fashion-mnist. Accessed 2020
7. Gabrielsson, R.B., Nelson, B.J., Dwaraknath, A., Skraba, P.: A topology layer for machine learning. In: The 23rd International Conference on Artificial Intelligence and Statistics (AISTATS), vol. 108, PMLR, pp. 1553–1563 (2020)
8. Hofer, C., Kwitt, R., Niethammer, M., Uhl, A.: Deep learning with topological signatures. In: Guyon, I., et al. (eds.) Advances in Neural Information Processing Systems, vol. 30, Curran Associates, Inc. (2017)
9. Kim, K., Kim, J., Zaheer, M., Kim, J., Chazal, F., Wasserman, L.: Pllay: efficient topological layer based on persistent landscapes. In: Larochelle, H., Ranzato, M., Hadsell, R., Balcan, M.F., Lin, H. (eds.) Advances in Neural Information Processing Systems (2020), vol. 33, Curran Associates, Inc., pp. 15965–15977
10. KMNIST Dataset. https://github.com/rois-codh/kmnist. Accessed 2020
11. Kusano, G., Fukumizu, K., Hiraoka, Y.: Persistence weighted gaussian kernel for topological data analysis. In: Proceedings of the 33rd International Conference on International Conference on Machine Learning - Volume 48 (2016), ICML 2016, JMLR.org, p. 2004-2013
12. Kwitt, R., Huber, S., Niethammer, M., Lin, W., Bauer, U.: Statistical topological data analysis - a kernel perspective. In: Proceedings of the 28th International Conference on Neural Information Processing Systems - Volume 2, NIPS 2015, pp. 3070–3078. MIT Press, Cambridge (2015)

13. Lacombe, T., Ike, Y., Carrière, M., Chazal, F., Glisse, M., Umeda, Y.: Topological uncertainty: monitoring trained neural networks through persistence of activation graphs. In: Zhou, Z. (ed.) Proceedings of the Thirtieth International Joint Conference on Artificial Intelligence, IJCAI 2021, Virtual Event/Montreal, Canada, 19-27 August 2021 (2021), pp. 2666–2672. ijcai.org (2021)
14. MNIST Dataset. http://yann.lecun.com/exdb/mnist/. Accessed 2020
15. Mobahi, H., Farajtabar, M., Bartlett, P.: Self-distillation amplifies regularization in hilbert space. In: Larochelle, H., Ranzato, M., Hadsell, R., Balcan, M.F., Lin, H. (eds.) Advances in Neural Information Processing Systems, vol. 33, pp. 3351–3361. Curran Associates, Inc. (2020)
16. Moor, M., Horn, M., Rieck, B., Borgwardt, K.: Topological autoencoders. In: Proceedings of the 37th International Conference on Machine Learning (ICML) (13–18 Jul 2020), vol. 119, PMLR, pp. 7045–7054 (2020)
17. Muandet, K., Fukumizu, K., Dinuzzo, F., Schölkopf, B.: Learning from distributions via support measure machines. In: Proceedings of the 25th International Conference on Neural Information Processing Systems - Volume 1 (Red Hook, NY, USA, 2012), NIPS 2012, pp. 10–18. Curran Associates Inc. (2012)
18. Munkres, J.: Topology. Pearson Education, Limited (2003)
19. Naitzat, G., Zhitnikov, A., and Lim, L. Topology of deep neural networks. *J. Mach. Learn. Res. 21* (2020), 184:1–184:40
20. Ohnishi, M., Yukawa, M., Johansson, M., Sugiyama, M.: Continuous-time value function approximation in reproducing kernel hilbert spaces. In: Bengio, S., Wallach, H., Larochelle, H., Grauman, K., Cesa-Bianchi, N., Garnett, R. (eds.) Advances in Neural Information Processing Systems (2018), vol. 31. Curran Associates, Inc
21. Pagliana, N., Rosasco, L.: Implicit regularization of accelerated methods in hilbert spaces. In: Wallach, H., Larochelle, H., Beygelzimer, A., d' Alché-Buc, F., Fox, E., Garnett, R. (eds.) Advances in Neural Information Processing Systems, vol. 32, Curran Associates, Inc. (2019)
22. Reininghaus, J., Huber, S., Bauer, U., Kwitt, R.: A stable multi-scale kernel for topological machine learning. In: 2015 IEEE Conference on Computer Vision and Pattern Recognition (CVPR) (2015), pp. 4741–4748 (2015)
23. Rieck, B., Bock, C., Borgwardt, K.M.: A persistent weisfeiler-lehman procedure for graph classification. In: Proceedings of the 36th International Conference on Machine Learning (ICML), vol. 97, PMLR, pp. 5448–5458 (2019)
24. Srinivasan, S., Downey, C., Boots, B.: Learning and inference in hilbert space with quantum graphical models. In: Bengio, S., Wallach, H., Larochelle, H., Grauman, K., Cesa-Bianchi, N., Garnett, R. (eds.) Advances in Neural Information Processing Systems (2018), vol. 31, Curran Associates, Inc. (2018)
25. Sriperumbudur, B.K., Fukumizu, K., Lanckriet, G.R.G.: Universality, characteristic kernels and rkhs embedding of measures. J. Mach. Learn. Res. **12**, null (2011), 2389–2410
26. Sriperumbudur, B.K., Gretton, A., Fukumizu, K., Schölkopf, B., Lanckriet, G.R.: Hilbert space embeddings and metrics on probability measures. J. Mach. Learn. Res. **11**, 1517–1561 (2010)

Online, Streaming, Distributed Algorithms

Streaming Fair k-Center Clustering over Massive Dataset with Performance Guarantee

Zeyu Lin[1], Longkun Guo[1,3](✉), and Chaoqi Jia[2,3]

[1] School of Mathematics and Statistics, Fuzhou University, Fuzhou 350116, China
[2] School of Accounting, Information Systems and Supply Chain, RMIT University, Melbourne, VIC 3000, Australia
[3] School of Computer Science, Qilu University of Technology (Shandong Academy of Sciences), Jinan 250316, China
lkguo@fzu.edu.cn

Abstract. Emerging applications are imposing challenges for incorporating fairness constraints into k-center clustering in the streaming setting. Different from the traditional k-center problem, the fairness constraints require that the input points be divided into disjoint groups and the number of centers from each group is constrained by a given upper bound. Moreover, observing the applications of fair k-center in massive datasets, we consider the problem in the streaming setting, where the data points arrive in a streaming manner that each point can be processed at its arrival. As the main contributions, we propose a two-pass streaming algorithm for the fair k-center problem with two groups, achieving an approximation ratio of $3 + \epsilon$ and consuming only $O(k \log n)$ memory and $O(k)$ update time, matching the state-of-art ratio for the offline setting. Then, we show that the algorithm can be easily improved to a one-pass streaming algorithm with an approximation ratio of $7+\epsilon$ and the same memory complexity and update time. Moreover, we show that our algorithm can be simply tuned to solve the case with an arbitrary number of groups while achieving the same ratio and space complexity. Lastly, we carried out extensive experiments to evaluate the practical performance of our algorithm compared with the state-of-the-art algorithms.

Keywords: Fair k-center clustering · Streaming algorithm · Approximation algorithm

1 Introduction

Machine learning plays a vital role in shaping the decision-making processes that govern our daily lives, with a growing influence across diverse sectors, including finance, justice, and healthcare. For example, recommendation systems offer

This work is supported by the Taishan Scholars Young Expert Project of Shandong Province (No. tsqn202211215) and National Science Foundation of China (Nos. 12271098 and 61772005).

D.-N. Yang et al. (Eds.): PAKDD 2024, LNAI 14647, pp. 105–117, 2024.
https://doi.org/10.1007/978-981-97-2259-4_8

tailored suggestions for movies or products, while other prediction and analysis tools are invaluable for making crucial decisions, such as loan approvals, appointment scheduling, and employment choices [16]. Despite the remarkable expansion of machine learning applications and the convenience they offer, it has also introduced a new set of risks and ethical dilemmas [9]. One famous instance is the situation with Amazon's AI-powered resume screening system, which has faced criticism for perpetuating gender-based discrimination [15]. To address these disparities and biases and to ensure that machine learning systems operate equitably and responsibly in our rapidly evolving technological landscape, particularly in light of the above-mentioned instances of unfair behavior, an emerging and critical challenge arises: how to guarantee fairness in machine learning. Many efforts have been made to incorporate fairness constraints into fundamental theoretical problems of machine learning, such as fair k-center clustering [13]. Moreover, when dealing with extremely large datasets, it becomes impractical to store the entire input in memory simultaneously, and hence the streaming model is adopted in which the algorithm processes input points one by one upon their arrival, and maintains a valid clustering of the observed input using a limited amount of working memory. In this context, we address the k-center problem in the streaming setting, considering the fairness of models by addressing and mitigating algorithmic biases, especially concerning sensitive attributes like address, gender, and religion.

1.1 Problem Statement

Let S be a finite data set distributed in a metric space. Assume that S is divided into m *disjoint* partitions (i.e., demographic groups): $S = S_1 \cup \ldots \cup S_m$ and $S_i \cap S_j = \emptyset$ for $\forall i \neq j$. According to some demographic group rules, there might exist a fairness code constraint for each S_i: the number of chosen centers from S_i is required to be constrained by a given upper bound. Moreover, *metric* space means the distance function between two points of S in the space, i.e., $d : S \times S \to \mathbb{R}_{\geq 0}$, satisfies the triangle inequality. Then we denote the distance between a point s and the center set C by $d(s, C) = \min_{c \in C} d(s, c)$. In particular, we define $d(s, C) = \infty$ for $C = \emptyset$.

For a given integer $k \in \mathbb{N}$, similar to traditional k-center, the fair k-center problem aims to select a set of k points $C \subseteq S$ as centers to serve points of S such that the maximum distance from the points to the center set C is minimized, i.e., to minimize $\max_{s \in S} d(s, C)$. Furthermore, the fairness constraints require that the number of centers from each group S_i is bounded by a given integer k_i. Then, the fair k-center clustering problem can be formulated as follows:

$$\begin{aligned} \min_{C \subseteq S} \quad & \max_{s \in S} d(s, C) \\ s.t. \quad & |\, C \cap S_i \,| \leq k_i, \qquad \forall i \qquad (1) \\ & |\, C \cap S \,| = \sum_{i=1}^{m} |\, C \cap S_i \,| \leq k, \qquad (2) \end{aligned}$$

where Constraint (1) is the fairness constraint and Constraint (2) is the traditional cardinality constraint.

1.2 Related Work

In recent years, emerging applications have imposed various fairness requirements for k-center clustering, leading to the fair k-center problem that attracts considerable research interest , where n and m are respectively the number of data points and demographic groups. The matroid center problem, presented in [5], serves as a more general variant of the fair k-center problem. Although it offers a polynomial-time 3-approximation algorithm, its computational cost becomes considerable for massive data summarization. To the best of our knowledge, Kleindessner et al. [13] was the first to address the fair k-center clustering problem for data summarization, where they introduced the fair k-centers problem by considering fairness among demographic groups. They presented a $(5+\epsilon)$-approximation algorithm for scenarios with two demographic groups and demonstrated their adaptability to diverse data partitioning scenarios. Later, Jones et al. [11] improved the above-mentioned approximation ratio to the state-of-art ratio 3 using the maximal matching algorithm as a subroutine. However, these algorithms can not be extended for the streaming setting as it is well-known that the maximal matching algorithm has no exact streaming algorithms.

On the other hand, the k-center problem and its variances were extensively studied in the streaming setting. For instance, Matthew et al. [14] proposed the streaming k-center problem with outliers and devised a one-pass algorithm with an approximation ratio $4+\epsilon$ and $O(\epsilon^{-1}kz)$ memory, where up to z input points can be dropped as outliers. Later, Ceccarello et al. [4] proposed a deterministic one-round streaming algorithm for the k-center problem with z outliers, which attains an approximation ratio $3+\epsilon$ and requires $O((k+z)(96/\epsilon)^D)$ local memory. More recently, Chiplunkar et al. [6] proposed distributed algorithms for the fair k-center problem in massive data models with an approximation ratio $17+\epsilon$ and $O(kn/l+mk^2l)$ running time for l processors, where n and m are respectively the number of data points and demographic groups. For the matroid center problem in which the aim is to find a set of centers satisfying the matroid constraint instead of satisfying the cardinality constraint k, Kale [12] considered streaming setting and gave a $(17+\epsilon)$-approximation one-pass algorithm with a running time $O_\epsilon((nk+k^{3.5})+k^2\log(\Lambda))$, where k is the rank of the matroid, Λ is the aspect ratio of the metric, and ϵ terms are hidden by the O_ϵ notation. For the fair range k-center problem under inequality-based fairness that aims to find a center set of size k satisfying the range-based fairness constraint, Hotegni et al. [10] proposed an efficient constant factor algorithm in polynomial time.

1.3 Our Contribution

In this paper, we address the fair k-center problem in the streaming setting and propose two streaming approximation algorithms. The main contributions can be summarized as follows:

- Design a two-pass streaming algorithm that achieves an approximation ratio of $3+\epsilon$, matching the state-of-the-art ratio 3 in the offline setting.

- Devise a one-pass streaming algorithm, achieving an approximation ratio of $7+\epsilon$ and consuming $O(k \log n)$ memory. To the best of our knowledge, this is the first constant factor ratio for this problem in the streaming setting.
- Conduct extensive experiments using real-world datasets to evaluate the practical performance of the proposed algorithms.

Notably, in the experiment our two-pass streaming algorithm outperforms the state-of-the-art algorithms for all datasets in cost; while the one-pass algorithm has a practical performance closely competitive with other algorithms regardless of its approximation ratio of $7+\epsilon$ that is compromised for one-pass efficiency.

2 A Two-Pass Algorithm with Approximation Ratio 3

In this section, we present a two-pass streaming algorithm for fair k-center with a provable performance guarantee. We first introduce an intermediate structure called the λ-independent center set, and then demonstrate how to construct a λ-independent center set in a streaming manner. Lastly, in the second pass of the stream, we illustrate how to use a λ-independent center set to eventually construct the desired center set.

2.1 The λ-Independent Center Set

Formally, the λ-independent center set can be defined as below:

Definition 1 *(λ-independent center set). We say Γ is a maximal λ-independent center set of S, if and only if*

(1) The distance between any two points $p, q \in \Gamma$ is larger than λ, i.e. $d(p,q) > \lambda$.
(2) For any point $p \in S$, there exists a point $q \in \Gamma$, such that $d(p,q) \leq \lambda$.

Assume the center set in the optimal solution is C^* and the optimal radius is r^*. Let Γ be a maximal λ-independent center set for S. Then we have

Lemma 2. *If $\lambda = 2r^*$, then $|\Gamma| \leq k$.*

Proof. Suppose $|\Gamma| > k$. Then by the Pigeonhole principle, there must exist at least a pair of points $i, j \in \Gamma$ which are covered by the same center in the optimal solution, say c^* in C^*. Then, both $d(i,c^*) \leq r^*$ and $d(c^*,j) \leq r^*$ hold. By the triangle inequality, we then have $d(i,j) \leq d(i,c^*) + d(c^*,j) \leq r^* + r^* = 2r^*$. On the other hand, we have $d(i,j) > 2r^*$ according to Condition (1) in Definition 1, which raises a contradiction and proves $|\Gamma| \leq k$. □

2.2 The Two-Pass Streaming Algorithm

In this subsection, we devise a two-pass streaming algorithm with an approximation ratio of 3. The key idea of our algorithm simply proceeds as below: In the first pass, construct a candidate center set Γ that is a λ-independent center set for $\lambda = 2r^*$ while ignoring its groups; In the second pass, we assume that $|\Gamma \cap S_l| > k_l$. Next, obvious $|\Gamma \cap S_{3-l}| < k_{3-l}$ must hold. Then upon each arriving data point $q \in S_{3-l}$, if there exists $p \in \Gamma \cap S_l$ with their distance $d(p,q) \le r^*$, we set $\Gamma \leftarrow \Gamma \setminus \{p\} \cup \{q\}$. The detailed algorithm is as depicted in Algorithm 1.

Algorithm 1: A two-pass algorithm for constructing the center set

---

Input: A stream of points $S = S_1 \cup S_2$; the optimal radius r^*.
Output: Center set C.

1 Set $C \leftarrow \emptyset$, $\Gamma \leftarrow \emptyset$;
`// The first pass of the stream.`
2 **upon** *each arriving point* $i \in S$ **do**
3   **if** $d(i, \Gamma) > 2r^*$ **then**
4   Set $\Gamma \leftarrow \Gamma \cup \{i\}$;
5   **if** $|\Gamma \cap S_1| \le k_1$ *and* $|\Gamma \cap S_2| \le k_2$ **then**
6   Return $C \leftarrow \Gamma$.
`// The second pass.` $|\Gamma \cap S_l| > k_l \,\&\, |\Gamma \cap S_{3-l}| < k_{3-l}$ `for` $l \in \{1, 2\}$.
7 **upon** *each arriving point* $q \in S_{3-l}$ **do**
8   **if** $|\Gamma \cap S_l| \le k_l$ **then**
9   Return $C \leftarrow \Gamma$;
10   **if** $\exists p \in \Gamma \cap S_l$ *with* $d(p, q) \le r^*$ **then**
11   Set $\Gamma \leftarrow \Gamma \setminus \{p\} \cup \{q\}$;

Lemma 3. *The set Γ output by the first pass of Algorithm 1 is a maximal $2r^*$-independent set for S.*

Proof. We will show that Γ satisfies both conditions of Definition 1. Firstly, each point $i \in S$ is added to Γ if and only if the distances between i and all centers of Γ are larger than $2r^*$. So Condition (1) in Definition 1 apparently holds. Second, for each point $i \in S$ that is not a point of Γ, there must already exist a point $j \in \Gamma$ with $d(i,j) \le 2r^*$ upon the arrival of i. So Condition (2) also holds. □

Lemma 4. *Algorithm 1 achieves an approximation ratio of 3 for the fair k-center problem in the streaming setting. In other words, the center set C produced by Algorithm 1 satisfies the following three conditions: (1) $|C| \le k$; (2) $|C \cap S_1| \le k_1$ and $|C \cap S_2| \le k_2$; (3) $d(s, C) \le 3r^*$ holds for $\forall s \in S$.*

Proof. For condition (1), according to Lemma 3, Γ produced by Steps 2–4 of Algorithm 1 is a maximal $2r^*$-independent center set for S. Then, according to Lemma 2, $|\Gamma| \le k$. Moreover, in steps 12 and 13 of Algorithm 1, the size of the set Γ remains unchanged, so $|C| = |\Gamma| \le k$ holds.

For condition (2), we first show $|C \cap S_l| \le k_l$. When the algorithm terminates, for each remained center $i \in C \cap S_l$, we have $d(i, S_{3-l}) > r^*$ because those centers belong to $C \cap S_l$ that do not satisfy this condition are replaced by points in S_{3-l} in steps 10–11 in Algorithm 1, so the remained centers in $C \cap S_l$ can only be covered by centers belong to S_l in the optimal solution as the optimal radius is r^*. Suppose $|C \cap S_l| > k_l$ holds, that means there must exist at least two points

$p, q \in C \cap S_l$ which are covered by the same center in the optimal solution because $|C^* \cap S_l| \le k_l$. Thus, similar to the proof of Lemma 2, we have $d(p, q) \le 2r^*$, which is contradicted to $d(p, q) > 2r^*$ as points p, q are elements of Γ produced by Steps 2–4 in Algorithm 1. So $|C \cap S_l| \le k_l$ holds.

Then we show $|C \cap S_{3-l}| \le k_{3-l}$ always holds during the algorithm. When the first pass terminates, $|\Gamma \cap S_l| > k_l$ and we have proved that $|\Gamma| \le k$, then $|C \cap S_{3-l}| = |\Gamma \cap S_{3-l}| = |\Gamma| - |\Gamma \cap S_l| < k - k_l = k_{3-l}$. So $|C \cap S_{3-l}| < k_{3-l}$ holds after the first pass terminates. Now we need to prove that $|C \cap S_{3-l}| < k_{3-l}$ holds when the algorithm terminates. Actually, $|C \cap S_l| = k_l$ when the algorithm terminates because we use one point $q \in S_{3-l}$ to replace one point $p \in C \cap S_l$ in steps 10–11 in Algorithm 1, therefore, when $|C \cap S_l| = k_l$ holds, the algorithm terminates at once. Suppose $|C \cap S_{3-l}| > k_{3-l}$ holds when the algorithm terminates. Then we have $|C| = |C \cap S_l| + |C \cap S_{3-l}| > k_l + k_{3-l} = k$, which is contradicted to $|C| \le k$. So $|C \cap S_{3-l}| \le k_{3-l}$ still holds when the algorithm terminates.

For condition (3), we add up the distance error among the iteration and passes of the algorithm. In the first pass, Γ is a maximal $2r^*$-independent set for S according to Lemma 3, and by Definition 1, for each point $s \in S$, there exists a point $p \in \Gamma$ such that $d(s, p) \le 2r^*$. Then in the second pass, if point $p \in \Gamma \cap S_l$ is deleted from Γ, there exists a point $q \in S_{3-l}$ such that $d(p, q) \le r^*$ and q is added to Γ. Thus, $d(s, q) \le d(s, p) + d(p, q) \le 2r^* + r^* = 3r^*$ holds for $\forall s \in S$ and $\forall q \in \Gamma$. Therefore, $d(s, C) \le 3r^*$ holds for $\forall s \in S$ and completes the proof. □

3 The Streaming Algorithm with an Approximation Ratio 7

Observing the property of the λ-independent center set, the key idea of our algorithm is first to construct two maximal λ-independent center sets, Γ_1 and Γ_2 for $\lambda = 2r^*$, respectively for each group upon the stream. After constructing the two sets, we proposed a simple algorithm to select among the points of Γ_1 and Γ_2 to form the final center set C.

Algorithm 2: Construction of λ-independent set in streams

Input: A stream of points in S_1 or S_2 and a given distance λ.
Output: Γ_1 and Γ_2.
1 Set $\Gamma_1 \leftarrow \emptyset$, $\Gamma_2 \leftarrow \emptyset$;
2 **upon** each arriving point $i \in S$ **do**
3   **if** $i \in S_l$ *and* $d(i, \Gamma_l) > \lambda$ **then**
    // Recall that $d(i, \emptyset) = \infty$.
4   Set $\Gamma_l \leftarrow \Gamma_l \cup \{i\}$;
5 Return Γ_1 and Γ_2.

3.1 The Streaming Algorithm for Constructing Γ_1 and Γ_2

To construct Γ_1 and Γ_2, we need only to simply construct a maximal λ-independent set for each group. In other words, upon each arriving point i, if $i \in S_l$ and $d(i, \Gamma_l) > 2r^*$, we grow the point set Γ_l (that is initially empty)

by adding i to Γ_l. The detailed algorithm is as depicted in Algorithm 2. For the correctness of Algorithm 2, we immediately have:

Lemma 5. *For the two sets Γ_1 and Γ_2 produced by Algorithm 2, Γ_l is a maximal λ-independent set for S_l, where $l \in \{1, 2\}$.*

Proof. Firstly, each arriving $i \in S_l$ is added to Γ_l if and only if $d(i, \Gamma_l) > \lambda$, i.e., the distance between i and Γ_l is strictly larger than λ. So Condition (1) of Definition 1 holds. Secondly, for each point i with $i \in S_l$ and $i \notin \Gamma_l$, there must already exist a point $p \in \Gamma_l$ with $d(i, p) \leq \lambda$. Thus, Condition (2) of Definition 1 also holds. □

Clearly, the sizes of Γ_1 and Γ_2 are both $O(k)$, and so is the memory complexity. For the update time complexity of Algorithm 2, it takes $O(k)$ time to verify whether $d(i, \Gamma_l) > \lambda$ holds for the arriving point i. So we have:

Lemma 6. *Algorithm 2 is with a memory complexity $O(n)$ and an update time $O(k)$.*

3.2 Post-streaming Construction of Center Set C from $\Gamma_1 \cup \Gamma_2$

The key idea of constructing C is first to construct a maximal $2r^*$-independent center set Γ among $\Gamma_1 \cup \Gamma_2$ ignoring the groups for points; then, we assume that $|C \cap \Gamma_l| > k_l$, then $|C \cap \Gamma_{3-l}| \leq k_{3-l}$ must hold. Then for each point i if there exist a point $j \in \Gamma_{3-l}$ such that $d(i, j) \leq 3r^*$, we set $C \leftarrow C \setminus \{i\} \cup \{j\}$ until $|C \cap \Gamma_l| = k_l$. The detailed algorithm is described in Algorithm 3 (Fig. 1 illustrates a tight example of Algorithm 3).

Algorithm 3: Post-procession of $\Gamma_1 \cup \Gamma_2$

Input: Γ_1 and Γ_2 of Alg. 2 and $\lambda = 2r^*$.
Output: Center set C.
1 Set $C \leftarrow \emptyset$;
2 **for** *each point* $i \in \Gamma_1 \cup \Gamma_2$ **do**
3   **if** $d(i, C) > 2r^*$ *or* $C = \emptyset$ **then**
4   Set $C \leftarrow C \cup \{i\}$;
5 **end**
6 **if** $|C \cap \Gamma_1| \leq k_1$ *and* $|C \cap \Gamma_2| \leq k_2$ **then**
7   Return C.
8 **while** $|C \cap \Gamma_l| > k_l$ *for some* $l \in \{1, 2\}$ **do**
9   **for** *each point* $i \in |C \cap \Gamma_l|$ **do**
10   **if** *there exists a point* $j \in \Gamma_{3-l}$ *with* $d(i, j) \leq 3r^*$ **then**
11   Set $C \leftarrow C \setminus \{i\} \cup \{j\}$;
12   **end**
13 **end**
14 Return C.

Lemma 7. *Let the optimal radius be r^*. For any point $i \in \Gamma_l$ with $d(i, \Gamma_{3-l}) > 3r^*$, $d(i, j) > r^*$ holds for any point $j \in S_{3-l}$.*

Proof. Suppose otherwise and without loss of generality we assume $l = 1$. That is, there exists a point $i \in \Gamma_1$ with $d(i, \Gamma_2) > 3r^*$, for which there exists a point j in S_2 with $d(j, i) \leq r^*$. According to the definition of the maximal $2r^*$-independent center set, there exists a point $p \in \Gamma_2$ with $d(j, p) \leq 2r^*$. That is, $d(i, p) \leq d(i, j) + d(j, p) \leq r^* + 2r^* = 3r^*$, contradicting with the assumption that $d(i, \Gamma_2) > 3r^*$. □

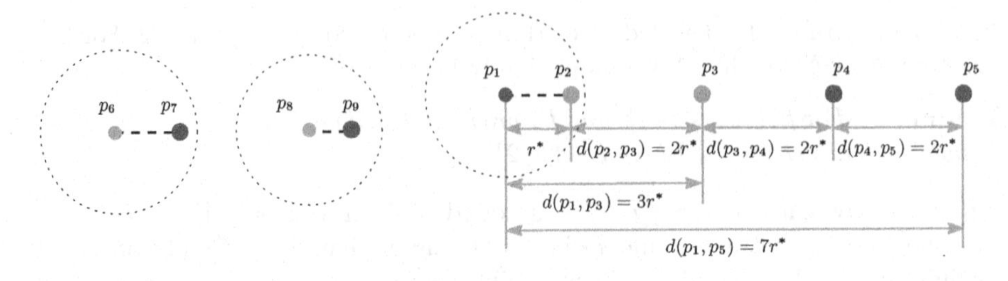

Fig. 1. A tight example for the approximation ratio 7 for Algorithm 3. Given two groups $S_1 = \{p_2, p_3, p_6, p_8\}$, $S_2 = \{p_1, p_4, p_5, p_7, p_9\}$, and $k = 6$, $k_1 = 2$ and $k_2 = 4$. Assume $\Gamma_1 = \{p_3, p_6, p_8\}$ and $\Gamma_2 = \{p_1, p_4, p_7, p_9\}$. By Steps 2–5 of Algorithm 3, we select the set of points $\{p_1, p_3, p_6, p_8\}$ as the center set C, and then $|\Gamma_1 \cap C| > k_1$, $|\Gamma_2 \cap C| \leq k_2$. For Steps 9–14 of Algorithm 3, if we use point p_1 (with $d(p_1, p_3) = 3r^*$) to replace the point $p_3 \in \Gamma_1$, then the final output center set is $C = \{p_1, p_6, p_8\}$. In this case, p_1 is the nearest center for p_5, resulting in a distance $d(p_5, C) = d(p_1, p_5) = 7r^*$.

Lemma 8. *Algorithm 3 achieves an approximation ratio of 7. That is, Algorithm 3 produces a center set C for which the following three conditions hold: (1) $|C| \leq k$; (2) $|C \cap \Gamma_1| \leq k_1$ and $|C \cap \Gamma_2| \leq k_2$; (3) $d(s, C) \leq 7r^*$ holds for $\forall s \in S$.*

Proof. For Condition (1), we first show that set C produced by steps 1–5 in Algorithm 3 is a maximal $2r^*$-independent center set for $\Gamma_1 \cup \Gamma_2$. Clearly, each point $i \in \Gamma_1 \cup \Gamma_2$ is added to C if and only if the distances between i and all centers of C are larger than $2r^*$. So Condition (1) in Definition 1 holds. Besides, for each point $i \in \Gamma_1 \cup \Gamma_2$ that is not a point of C, there must already exist a point $j \in C$ with $d(i, j) \leq 2r^*$. Therefore, Condition (2) in Definition 1 holds. Then, apparently C, as a maximal $2r^*$-independent center set for $\Gamma_1 \cup \Gamma_2$, can be grown to a maximal $2r^*$-independent center set C' of S by repeatedly adding to C' (that is initially C) a point $p \in S$ that is with $(p, C) > 2r^*$. Following Lemma 2, we have $|C'| \leq k$. Hence, $|C| \leq |C'| \leq k$. Moreover, in steps 8–11 in Algorithm 3, we use exactly one point belonging to Γ_{3-l} to replace one point belonging to $C \cap \Gamma_l$. Therefore, $|C| \leq k$ remains true when the algorithm terminates.

For Condition (2), assuming $|C \cap \Gamma_l| > k_l$ and $|C \cap \Gamma_{3-l}| \leq k_{3-l}$, we will first show $|C \cap \Gamma_l| \leq k_l$. For the set C produced by Algorithm 3, each point $i \in C \cap \Gamma_l$ satisfies that $d(i, \Gamma_{3-l}) > 3r^*$ as those points initially belong to $C \cap \Gamma_l$ are deleted from C if there exists a point belongs to Γ_{3-l} such the distance between them is within $3r^*$. According to Lemma 7, for each point $i \in C \cap \Gamma_l$, there exists no point in S_2 is within r^* away from i for the case $l = 1$. Now each point $i \in C \cap \Gamma_l$ can only be covered by centers belonging to S_l in the optimal solution as the optimal radius is r^*. Suppose that $|C \cap \Gamma_l| > k_l$ holds, there must exist at least two points $p, q \in C \cap \Gamma_l$ which both belong to a cluster with radius r^* in the optimal solution, in other words, $d(p, q) \leq 2r^*$, which is contradicted to $d(p, q) > 2r^*$ because they are two points of the maximal $2r^*$-independent set Γ_l. Thus, $|C \cap \Gamma_l| \leq k_l$ holds when the algorithm terminates.

Then, we show $|C \cap \Gamma_{3-l}| \leq k_{3-l}$ always holds during the algorithm. When $|C \cap \Gamma_l| > k_l$ and $|C| \leq k$ both hold, we have $|C \cap \Gamma_{3-l}| = |C| - |C \cap \Gamma_l| < k - k_l = k_{3-l}$. In fact, when $|C \cap \Gamma_l| = k_l$ holds, the algorithm terminates immediately according to steps 8–11 in Algorithm 3, thus $|C \cap \Gamma_l| = k_l$ and $|C| \leq k$ both hold. Then $|C \cap \Gamma_{3-l}| = |C| - |C \cap \Gamma_l| \leq k - k_l = k_{3-l}$. That is, $|C \cap \Gamma_{3-l}| \leq k_{3-l}$ always holds in the algorithm.

For Condition (3), we shall bound the distance from any point $s \in S$ to C. For Γ_1 and Γ_2, there exists a point $i \in \Gamma_1 \cup \Gamma_2$ such that $d(s, i) \leq 2r^*$ according to Lemma 5. Moreover, for each point $i \in \Gamma_1 \cup \Gamma_2$, if it is not added to C, then there exists a point $p \in C$ such that $d(i, p) \leq 2r^*$. In contrast, if point p is deleted from C, there exists a point $q \in \Gamma_{3-l}$ such that $d(p, q) \leq 3r^*$ and q is added to C. Thus, $d(s, q) \leq d(s, i) + d(i, p) + d(p, q) \leq 2r^* + 2r^* + 3r^* = 7r^*$ holds for $\forall s \in S$ and $\forall q \in C$. Therefore, $d(s, C) \leq 7r^*$ holds for $\forall s \in S$. □

We notice that all the above algorithms are with the assumption of knowing the value of r^*, while the exact value of r^* is actually unknown. Anyhow, without knowing r^*, we can retain similar results by slightly compromising the approximation ratio with a small $\epsilon > 0$ by employing the elegant technique used in [2,3]. Then we can immediately extend our algorithm and obtain the following theorem:

Theorem 9. *The fair k-center problem admits a one-pass streaming algorithm with an approximation ratio $7 + \epsilon$, a memory complexity $O(k \log n)$, and an update time $O(k)$. Moreover, it admits a two-pass algorithm with a ratio $3 + \epsilon$ and the same memory complexity and update time.*

Extension to Fair k-Center with General m. Our algorithm can be extended to solve fair k-center with general m. The extension simply proceeds as below: (1) Construct a λ-independent set for S_j for each $j \in [m]$ similarly to that in Algorithm 2; (2) Construct C from $\bigcup_{l \in [m]} \Gamma_l$ using the greedy algorithm for k-center by ignoring the groups similar to Algorithm 3; (3) For each Γ_l with $|C \cap \Gamma_l| > k_l$, use maximum flow algorithm to find replacement for centers of Γ_l in a similar way as the offline algorithm by Jones et al. [11].

4 Experimental Results

In this section, we report extensive experimental evaluations over three real-world datasets. First, we introduce the fundamental experimental settings. Then, by employing recent heuristic and approximation algorithms of fair k-center, we present the clustering quality of *cost* and runtime.

4.1 Experimental Setting

Real-World Datasets. We apply our algorithms to three real-world datasets from UCI [1]. Following the previous work [11], we utilized numeric features for clustering and selected some binary categorical attributes to construct datasets with fair constraints. The details of these datasets can be found in Table 1.

Table 1. Datasets Summary.

Dataset	#Rec.	#Dim.	Constraints
Wholesale [1]	440	6	Channel
Student [1]	649	16	School, Sex, Address
Adult [1]	32,561	6	Gender, Income

Constraints Settings. To ensure a fair selection of centers, we align the number of required centers from each group with the proportional size of that group. Based on the fairness principle of disparate impact [7], we designate p data points from every group as centers. In this paper, encompassing the prior configurations of the work [11], we evaluate the clustering quality by varying the parameter p from 1% (0.1%) to 10% (1%) for Student and Wholesale (Adult) datasets. In addition, we evaluate the algorithm runtime with the parameter p set at 1%. We utilize the Euclidean distance to compute the cost metric.

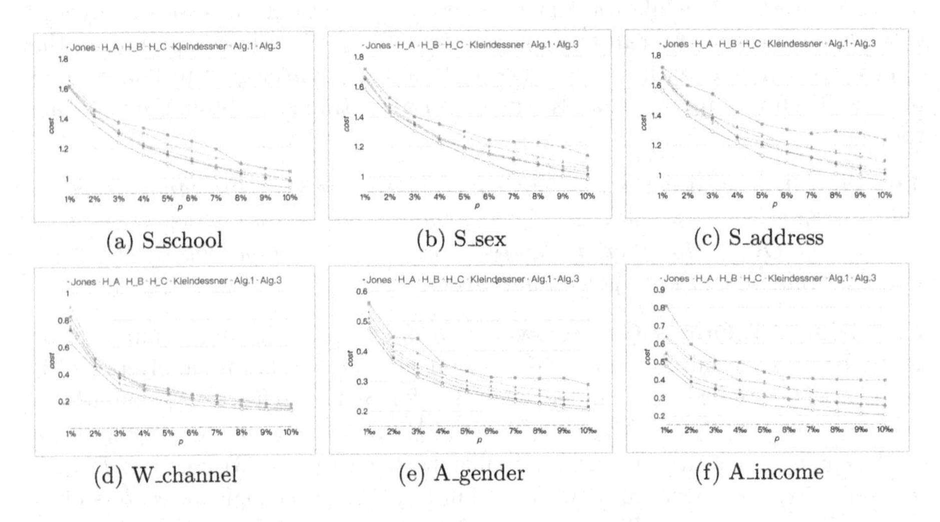

(a) S_school (b) S_sex (c) S_address

(d) W_channel (e) A_gender (f) A_income

Fig. 2. Clustering cost in comparison.

Algorithms. The experiment includes five baselines: the 5-approximation algorithm (Kleindessner) and two heuristic approaches (H_A and H_B) proposed in the study by Kleindessner [13], and the 3-approximation algorithm (Jones) and the heuristic algorithm (H_C) proposed in the work by Jones [11]. To evaluate the clustering quality of our algorithms for the fair k-center problem, we implement a 3-approximation streaming algorithm (Algorithm 1) and 7-approximation algorithm (Algorithm 3) as described in Sect. 2 and Sect. 3, respectively.

Metrics. We compare the clustering *cost* on these real datasets with the mean value, where the function of *cost* can be found in Subsect. 1.1. In addition, we adopt the second to measure the runtime of all algorithms.

Experimental Settings. In this experiment, we report the result of running all the algorithms at least 50 times. All algorithms were conducted on a MacBook Pro, equipped with an M1 Max chip and 32 GB RAM.

4.2 Experimental Analysis

In this subsection, compared with the baselines, we report the experimental results for both Algorithm 1 and Algorithm 3 on real-world datasets. The summary of our observations is provided below.

In Fig. 2, Algorithm 1 is significantly superior to the other baseline algorithms. For each dataset with varying proportions of the number of centers selected by each group, Algorithm 1 consistently exhibits the lower *cost* than the other baseline algorithms. Matching our theoretical result proved in Lemma 4, Algorithm 1 is a 3-approximation algorithm showcasing a state-of-the-art ratio for the fair k-center problem with streaming setting. Therefore, Algorithm 1 outperforms both the heuristic algorithms and 5-approximation algorithm [11, 13]. Moreover, when compared with the 3-approximation algorithm [11] (Jones), Algorithm 1 avoid unnecessary exchange of center swap, which might lead to performance degradation (e.g., increasing the distance which has been covered by $2r^*$).

In Fig. 2, as we vary the percentage of the number of centers, all algorithms running on the datasets exhibit a consistent decreasing trend in *cost*. For all the datasets, the cost decreases as the number of centers increases, but it exhibits a diminishing growth trend beyond a certain value. For instance, in Subfig. 2d, the cost decreased by approximately 0.35 for these algorithms as we increased the value of the x-axis from 1% to 3%. However, when progressing from 3% to 10%, the cost reduction for these algorithms was around 0.16. According to Elbow Method [8], we reason that the cost experiences a significant drop before reaching a specific value for k, after which it reaches a plateau when you further increase the k.

Runtime Comparison. Table 2 depicts the runtime of all algorithms running on the $p = 1\%$, and we summarize the experimental results below: **Algorithm 3 with streaming settings demonstrates superior efficiency when applied on large datasets.**

This superiority is evident in the Adult dataset, where the streaming algorithms, especially Algorithm 3, show their strengths. We reason that Algorithm 3 as a one-pass algorithm with a smaller runtime matches the theoretical expectations. In addition, although the efficiency of Algorithm 3 surpasses that of Algorithm 1 across all datasets, both of them demonstrate lower performance in comparison. We reason that the streaming algorithm's advantage faces challenges when applied to smaller datasets.

Table 2. Runtime comparison.

Algs./Datasets	W_channel	S_school	S_sex	S_address	A_gender	A_income
H_A [13]	0.0006	0.0010	0.0009	0.0009	0.2072	0.1666
H_B [13]	0.0007	0.0014	0.0012	0.0012	0.3611	0.3418
H_C [11]	**0.0003**	**0.0004**	**0.0004**	**0.0004**	**0.0723**	**0.0678**
Kleindessner [13]	0.0342	0.0503	0.0504	0.0515	3.1003	3.6415
Jones [11]	**0.0009**	**0.0024**	**0.0021**	**0.0022**	4.0185	3.4131
Algorithm 1	0.0064	0.0083	0.0190	0.0107	0.8714	0.8476
Algorithm 3	0.0057	0.0082	0.0081	0.0082	**0.8207**	**0.7535**

5 Conclusion

In this paper, we presented two streaming algorithms for the fair k-center problem. The first is a two-pass streaming algorithm that achieves an approximation ratio of $3 + \epsilon$ and consumes $O(k \log n)$ memory, matching the state-of-the-art ratio of the problem in the offline setting. The second is a one-pass algorithm with the first constant approximation ratio $7 + \epsilon$ and $O(k \log n)$ memory in the streaming setting. Extensive experiments were conducted using real-world datasets to demonstrate the practical performance of our algorithms in comparison to previous approaches.

References

1. Asuncion, A., Newman, D.: UCI machine learning repository (2007)
2. Badanidiyuru, A., Mirzasoleiman, B., Karbasi, A., Krause, A.: Streaming submodular maximization: Massive data summarization on the fly. In: 20th International Conference on Knowledge Discovery and Data Mining(ACM SIGKDD), pp. 671–680 (2014)
3. Bandyapadhyay, S., Friggstad, Z., Mousavi, R.: Parameterized approximation algorithms for k-center clustering and variants. In: 36th AAAI Conference on Artificial Intelligence. vol. 36, pp. 3895–3903 (2022)
4. Ceccarello, M., Pietracaprina, A., Pucci, G.: Solving k-center clustering (with outliers) in mapreduce and streaming, almost as accurately as sequentially. In: 45th International Conference on Very Large Databases(VLDB), pp. 766–778 (2019)
5. Chen, D.Z., Li, J., Liang, H., Wang, H.: Matroid and knapsack center problems. Algorithmica **75**, 27–52 (2016)
6. Chiplunkar, A., Kale, S., Ramamoorthy, S.N.: How to solve fair k-center in massive data models. In: 37th International Conference on Machine Learning(ICML), pp. 1877–1886 (2020)
7. Feldman, M., Friedler, S.A., Moeller, J., Scheidegger, C., Venkatasubramanian, S.: Certifying and removing disparate impact. In: 21th International Conference on Knowledge Discovery and Data Mining(ACM SIGKDD), pp. 259–268 (2015)
8. Fritz, M., Behringer, M., Schwarz, H.: Log-means: efficiently estimating the number of clusters in large datasets. In: 46th International Conference on Very Large Databases (VLDB), pp. 2118–2131 (2020)

9. Holstein, K., Wortman Vaughan, J., Daumé III, H., Dudik, M., Wallach, H.: Improving fairness in machine learning systems: What do industry practitioners need? In: CHI Conference on Human Factors in Computing Systems, pp. 1–16 (2019)
10. Hotegni, S.S., Mahabadi, S., Vakilian, A.: Approximation algorithms for fair range clustering. In: 40th International Conference on Machine Learning (ICML), pp. 13270–13284 (2023)
11. Jones, M., Nguyen, H., Nguyen, T.: Fair k-centers via maximum matching. In: 37th International Conference on Machine Learning(ICML). pp. 4940–4949 (2020)
12. Kale, S.: Small space stream summary for matroid center. Approximation, Randomization, and Combinatorial Optimization. Algorithms Tech. **145**, 20 (2019)
13. Kleindessner, M., Awasthi, P., Morgenstern, J.: Fair k-center clustering for data summarization. In: 36th International Conference on Machine Learning (ICML), pp. 3448–3457 (2019)
14. Matthew McCutchen, R., Khuller, S.: Streaming algorithms for k-center clustering with outliers and with anonymity. Approximation, Randomization and Combinatorial Optimization. Algorithms and Techniques, pp. 165–178 (2008)
15. Sweeney, L.: Discrimination in online ad delivery. Commun. ACM **56**(5), 44–54 (2013)
16. Tambe, P., Cappelli, P., Yakubovich, V.: Artificial intelligence in human resources management: challenges and a path forward. Calif. Manage. Rev. **61**(4), 15–42 (2019)

Projection-Free Bandit Convex Optimization over Strongly Convex Sets

Chenxu Zhang[1], Yibo Wang[2], Peng Tian[3], Xiao Cheng[3], Yuanyu Wan[1(✉)], and Mingli Song[1]

[1] Zhejiang University, Hangzhou 310058, China
{zhangchenxu,wanyy,brooksong}@zju.edu.cn
[2] Nanjing University, Nanjing 210023, China
[3] State Grid Chongqing Electric Power Research Institute, Chongqing 401123, China

Abstract. Projection-free algorithms for bandit convex optimization have received increasing attention, due to the ability to deal with the bandit feedback and complicated constraints simultaneously. The state-of-the-art ones can achieve an expected regret bound of $O(T^{3/4})$. However, they need to utilize a blocking technique, which is unsatisfying in practice due to the delayed reaction to the change of functions, and results in a logarithmically worse high-probability regret bound of $O(T^{3/4}\sqrt{\log T})$. In this paper, we study the special case of bandit convex optimization over strongly convex sets, and present a projection-free algorithm, which keeps the $O(T^{3/4})$ expected regret bound without employing the blocking technique. More importantly, we prove that it can enjoy an $O(T^{3/4})$ high-probability regret bound, which removes the logarithmical factor in the previous high-probability regret bound. Furthermore, empirical results on synthetic and real-world datasets have demonstrated the better performance of our algorithm.

Keywords: Projection-Free · Bandit Convex Optimization · Strongly Convex Sets

1 Introduction

Online convex optimization (OCO) plays an important role in many industrial applications with large-scale and streaming data, such as recommendation systems [26] and packet routing [3]. Specifically, it can be deemed as a repeated game between a learner and an adversary [18], in which the learner needs to first select a decision $\mathbf{x}_t$ from a convex set $\mathcal{K} \subseteq \mathbb{R}^d$ at each round t, and then the adversary chooses a convex loss function $f_t(\cdot) : \mathcal{K} \to \mathbb{R}$. The learner suffers a loss $f_t(\mathbf{x}_t)$ at each round t, and pursue that the regret defined below

$$\text{Regret}(T) = \sum_{t=1}^{T} f_t(\mathbf{x}_t) - \min_{\mathbf{x}\in\mathcal{K}} \sum_{t=1}^{T} f_t(\mathbf{x})$$

D.-N. Yang et al. (Eds.): PAKDD 2024, LNAI 14647, pp. 118–129, 2024.
https://doi.org/10.1007/978-981-97-2259-4_9

is sublinear in the time horizon T. Over the past decades, various algorithms, such as online gradient descent (OGD) [36] and follow the regularized leader (FTRL) [28], have been proposed to achieve the optimal regret bound of $O(\sqrt{T})$.

However, there exist two common limitations in these algorithms. One is that they require the full information or the gradient of loss functions, which is not available in real applications with a black-box model. The other is that a projection operation is required at each round, which is time-consuming and even intractable when the constraint set is complicated [23]. To address the first limitation, there has been a growing research interest in OCO with the bandit feedback, where only the loss value $f_t(\mathbf{x}_t)$ is available at each round, which is also known as bandit convex optimization (BCO) [5,6,10,11,20,27]. On the other hand, to alleviate the second limitation, projection-free algorithms, which employ efficient operations such as the linear optimization in lieu of projections, have also attracted ever-increasing attention [9,14–16,19,21,32].

Nonetheless, there are only a few studies that simultaneously tackle the above two limitations. Specifically, Chen et al. [9] develop the first projection-free algorithm for BCO (PF-BCO), which combines the FTRL algorithm with the one-point gradient estimator [11] and the Frank-Wolfe (FW) iteration [12,23] for utilizing the bandit feedback and avoiding the projection, respectively. Unfortunately, this algorithm can only achieve an expected regret bound of $O(T^{4/5})$, which is worse than both the expected $O(T^{3/4})$ regret bound attained by existing algorithms only using the one-point gradient estimator [11] and the $O(T^{3/4})$ regret bound attained by existing algorithms only using the FW iteration [19]. Intuitively, this gap is caused because the variance of the one-point gradient estimator can increase the approximation error of solving the objective at each round of FTRL via the FW iteration.

To fill the gap, Garber and Kretzu [15] propose a novel algorithm called block bandit conditional gradient (BBCG), which enjoys an expected regret bound of $O(T^{3/4})$. The key technique for this improvement is a blocking technique—dividing total T rounds into size-equal blocks and only updating the decision at the end of each block, which can reduce the variance of the gradient estimator such that the approximation error of FW keeps unchanged. Recently, based on the blocking technique, a bandit and projection-free variant of OGD has also been developed to achieve the $O(T^{3/4})$ expected regret bound [16]. However, despite the improvement in the regret, the blocking technique could inevitably sacrifice the performance in practice due to the mismatch between the fixed action and the changing loss functions over each block. Moreover, we notice that due to the blocking technique, BBCG can only achieve a high-probability regret bound of $O(T^{3/4}\sqrt{\log T})$ [29],[1] which is logarithmically worse than the expected one.

To address these limitations, in this paper, we aim to improve the regret of PF-BCO without using the blocking technique. Specifically, different from the blocking technique that indirectly controls the approximation error of FW via

[1] Although Wan et al. [29] originally establish such bound for a decentralized variant of BBCG, it is easy to extend this result for BBCG.

reducing the variance of the gradient estimator, our main idea is to directly reduce the approximation error of FW. Note that in the offline setting, Garber and Hazan [13] have shown that FW can converge faster over strongly convex sets by utilizing a line search rule to select the step size. Inspired by this result, we propose an improved variant of PF-BCO by combining it with the line search, namely BFW-LS. Theoretical analysis demonstrates that our algorithm enjoys the $O(T^{3/4})$ regret bound in both expectation and high probability over strongly convex sets. Compared with previous improvements on PF-BCO, the blocking technique is dismissed, and the logarithmical term in the high-probability bound is removed. Furthermore, empirical results on synthetic and real-world datasets have verified the effectiveness of our algorithm.

2 Related Work

In this section, we briefly review related work on projection-free OCO algorithms and bandit convex optimization.

2.1 Projection-Free OCO Algorithms

To handle OCO with complicated constraints, Hazan and Kale [19] propose the first projection-free algorithm called online Frank-Wolfe (OFW), and achieve a regret bound of $O(T^{3/4})$ for the general case. The essential idea is to replace the projection operation required by FTRL [28] with an iteration of FW [12]. Specifically, following FTRL, in each round t, OFW first defines an objective function

$$F_t(\mathbf{x}) = \eta \sum_{\tau=1}^{t} \langle \nabla f_\tau(\mathbf{x}_\tau), \mathbf{x} \rangle + \|\mathbf{x} - \mathbf{x}_1\|_2^2 \tag{1}$$

with a parameter η. Then, it updates the decision by minimizing $F_t(\mathbf{x})$ via an iteration of FW, i.e.,

$$\begin{aligned} \mathbf{v}_t &= \operatorname*{argmin}_{\mathbf{x}\in\mathcal{K}} \langle \nabla F_t(\mathbf{x}_t), \mathbf{x} \rangle \\ \mathbf{x}_{t+1} &= \mathbf{x}_t + \sigma_t(\mathbf{v}_t - \mathbf{x}_t) \end{aligned} \tag{2}$$

where $\sigma_t \in [0, 1]$ is a step size. Note that only a linear optimization is required by the update, which can be implemented more efficiently than the projection over many complicated constraints [18].

Later, plenty of projection-free OCO algorithms have been proposed to establish tighter regret bounds by leveraging additional assumptions of the constraint set [14,16,32] and loss functions [17,21,25,32]. The most related one is the variant of OFW over strongly convex sets [32], which adopts the following line search rule to select the step size of OFW

$$\sigma_t = \operatorname*{argmin}_{\sigma\in[0,1]} \langle \sigma\left(\mathbf{v}_t - \mathbf{x}_t\right), \nabla F_t\left(\mathbf{x}_t\right) \rangle + \sigma^2 \|\mathbf{v}_t - \mathbf{x}_t\|_2^2. \tag{3}$$

By exploiting the faster convergence of FW over strongly convex sets [13], Wan and Zhang [32] establish a regret bound of $O(T^{2/3})$, which is better than the $O(T^{3/4})$ regret bound of the original OFW. Although our paper makes a similar exploitation as them, we want to emphasize that in the bandit setting, more careful analyses are required to deal with the variance of the gradient estimator.

Additionally, projection-free OCO algorithms have also been extended into more practical scenarios with decentralized agents [29,30,34], dynamic environments [16,24,31,33,35], and the bandit setting discussed below.

2.2 Bandit Convex Optimization

The first method for the bandit convex optimization (BCO) is proposed by Flaxman et al. [11], and attains an expected regret bound of $O(T^{3/4})$ for convex loss function. The significant contribution of this study is to introduce a profound technique called one-point gradient estimator, which can approximate the gradient with only a single loss value. Based on this technique, subsequent studies establish several improved bounds for different types of loss functions, such as the linear function [1,4], the smooth function [27], the strong convex function [2], and the smooth and strong convex function [20,22].

However, these methods rely on the projection operation or more time-consuming operations, which is unacceptable for applications with complicated constraint sets. To address this issue, Chen et al. [9] propose the PF-BCO method by combining OFW with the one-point gradient estimator, which attains an expected regret bound of $O(T^{4/5})$ for convex loss functions. Later, by employing the blocking technique, Garber and Kretzu [15] propose a refined variant of this method, namely BBCG, which reduces the expected regret bound to $O(T^{3/4})$ for the same case. Similarly, Garber and Kretzu [16] propose a bandit and projection-free variant of OGD, namely blocked online gradient descent with linear optimization oracle (LOO-BBGD), and establish the same expected regret bound of $O(T^{3/4})$. Moreover, besides the expected regret bound, Wan et al. [29] have shown that BBCG can achieve a high-probability regret bound of $O(T^{3/4}\sqrt{\log T})$.

As mentioned before, although the blocking technique is utilized to improve the expected regret of projection-free BCO, it is unsatisfying in practice due to the delayed reaction to the change of functions, and results in a logarithmical term in the high-probability regret bound. In this paper, we focus on the special case with strongly convex sets, and develop an improved variant of PF-BCO without using the blocking technique.

3 Main Results

In this section, we first introduce necessary preliminaries including basic definitions, common assumptions, and algorithmic ingredients. Then, we present our algorithm and its theoretical guarantees. Due to the limitation of space, we defer the proof of theoretical results to the supplementary material.[2]

3.1 Preliminaries

We first recall the standard definition for strongly convex sets [13].

Definition 1. *A convex set $\mathcal{K} \subseteq \mathbb{R}^d$ is called α_K-strongly convex with respect to a norm $\|\cdot\|$ if for any $\mathbf{x}, \mathbf{y} \in \mathcal{K}, \gamma \in [0,1]$ and $\mathbf{z} \in \mathbb{R}^d$ such that $\|\mathbf{z}\| = 1$, it holds that*

$$\gamma\mathbf{x} + (1-\gamma)\mathbf{y} + \gamma(1-\gamma)\frac{\alpha_K}{2}\|\mathbf{x}-\mathbf{y}\|^2\mathbf{z} \in \mathcal{K}.$$

Next, we introduce three common assumptions in BCO [11].

Assumption 1. *The convex decision set $\mathcal{K}$ is full dimensional and contains the origin. Moreover, there exist two constants $r, R > 0$ such that*

$$r\mathcal{B}^d \subseteq \mathcal{K} \subseteq R\mathcal{B}^d$$

where $\mathcal{B}^d$ denotes the unit Euclidean ball centered at the origin in $\mathbb{R}^d$.

Assumption 2. *At each round t, the loss function $f_t(\mathbf{x})$ is G-Lipschitz over $\mathcal{K}$, i.e., for any $\mathbf{x}, \mathbf{y} \in \mathcal{K}$*

$$|f_t(\mathbf{x}) - f_t(\mathbf{y})| \leq G\|\mathbf{x}-\mathbf{y}\|_2.$$

Assumption 3. *At each round t, each loss function $f_t(\mathbf{x})$ is bounded over $\mathcal{K}$, i.e., for any $\mathbf{x} \in \mathcal{K}$*

$$|f_t(\mathbf{x})| \leq M. \tag{4}$$

Moreover, all loss functions are chosen beforehand, i.e., the adversary is oblivious.

Last, we recall the one-point gradient estimator [11], which is commonly utilized to deal with the bandit feedback. Specifically, for a function $f(\mathbf{x})$, we can define its δ-smooth version as

$$\widehat{f}_\delta(\mathbf{x}) = \mathbb{E}_{\mathbf{u}\sim\mathcal{B}^d}[f(\mathbf{x}+\delta\mathbf{u})] \tag{5}$$

which satisfies the following lemma.

Lemma 1 (Lemma 1 in Flaxman et al. [11]). *Let $\delta > 0$, $\widehat{f}_\delta(\mathbf{x})$ defined in (5) satisfies*

$$\nabla\widehat{f}_\delta(\mathbf{x}) = \mathbb{E}_{\mathbf{u}\sim\mathcal{S}^d}\left[\frac{d}{\delta}f(\mathbf{x}+\delta\mathbf{u})\mathbf{u}\right] \tag{6}$$

where $\mathcal{S}^d$ denotes the unit sphere in $\mathbb{R}^d$.

According to this lemma, the one-point gradient estimator is to make an unbiased estimation of $\nabla\widehat{f}_\delta(\mathbf{x})$ as $\frac{d}{\delta}f(\mathbf{x}+\delta\mathbf{u})\mathbf{u}$ by leveraging the single value $f(\mathbf{x}+\delta\mathbf{u})$.

[2] https://github.com/zcx-xxx/PAKDD-2024/blob/main/PAKDD-2024-Zhang-S.pdf.

Algorithm 1. Bandit Frank-Wolfe with Line Search

1: **Input**: $\mathcal{K}, \delta, \eta$
2: **Initialization**: $\mathbf{y}_1 \in \mathcal{K}_\delta$
3: **for** $t = 1, 2, \ldots, T$ **do**
4: Play $\mathbf{x}_t = \mathbf{y}_t + \delta \mathbf{u}_t$, where $\mathbf{u}_t \sim \mathcal{S}^d$
5: Observe $f_t(\mathbf{x}_t)$ and compute $\mathbf{g}_t$ according to (7)
6: Construct $F_t(\mathbf{y})$ as in (8).
7: Compute $\mathbf{v}_t = \operatorname{argmin}_{\mathbf{y} \in \mathcal{K}_\delta} \langle \nabla F_t(\mathbf{y}_t), \mathbf{y} \rangle$
8: Compute σ_t according to (10)
9: $\mathbf{y}_{t+1} = \mathbf{y}_t + \sigma_t (\mathbf{v}_t - \mathbf{y}_t)$
10: **end for**

3.2 Our Proposed Algorithm

Now, we introduce our improved variant of PF-BCO, which is still a combination of the OFW algorithm and the one-point gradient estimator. Specifically, we first define a subset of the convex set $\mathcal{K}$ as

$$\mathcal{K}_\delta = (1 - \delta/r)\mathcal{K} = \{(1 - \delta/r)\mathbf{x} \mid \mathbf{x} \in \mathcal{K}\}$$

where $0 < \delta < r$ is a parameter. Following previous BCO algorithms [8,11,15], the decision $\mathbf{x}_t$ at each round t is divided into two parts, i.e.,

$$\mathbf{x}_t = \mathbf{y}_t + \delta \mathbf{u}_t$$

where $\mathbf{y}_t$ is an auxiliary decision learning from historical information and $\mathbf{u}_t$ is uniformly sampled from $\mathcal{S}^d$. Note that according to Assumption 1, it is easy to verify the feasibility of the above $\mathbf{x}_t$, i.e., $\mathbf{x}_t \in \mathcal{K}$. Moreover, in this way, the loss value of $f_t(\mathbf{x}_t) = f_t(\mathbf{y}_t + \delta \mathbf{u}_t)$ is observed at each round t. According to the one-point gradient estimator, it can be utilized to generate an approximate gradient as

$$\mathbf{g}_t = \frac{d}{\delta} f_t(\mathbf{y}_t + \delta \mathbf{u}_t)\mathbf{u}_t. \tag{7}$$

To further combine the OFW algorithm, we need to reconstruct the objective function in (1) as

$$F_t(\mathbf{y}) = \eta \sum_{\tau=1}^{t} \langle \mathbf{g}_\tau, \mathbf{y} \rangle + \|\mathbf{y} - \mathbf{y}_1\|_2^2. \tag{8}$$

Then, similar to (2) in OFW, we update the auxiliary decision by minimizing $F_t(\mathbf{y})$ via a FW iteration over $\mathcal{K}_\delta$, i.e.,

$$\begin{aligned} \mathbf{v}_t &= \operatorname*{argmin}_{\mathbf{y} \in \mathcal{K}_\delta} \langle \nabla F_t(\mathbf{y}_t), \mathbf{y} \rangle \\ \mathbf{y}_{t+1} &= \mathbf{y}_t + \sigma_t (\mathbf{v}_t - \mathbf{y}_t). \end{aligned} \tag{9}$$

We notice that the above procedures have been utilized in the PF-BCO algorithm [9]. However, they make use of a decaying step size, i.e., $\sigma_t = t^{-2/5}$, which cannot exploit the strong convexity of $\mathcal{K}$ to reduce the approximation error of the FW iteration. By contrast, inspired by (3) utilized in the full information setting [32], we employ a line search rule to select the step size as

$$\sigma_t = \underset{\sigma \in [0,1]}{\operatorname{argmin}} \langle \sigma (\mathbf{v}_t - \mathbf{y}_t), \nabla F_t(\mathbf{y}_t) \rangle + \sigma^2 \|\mathbf{v}_t - \mathbf{y}_t\|_2^2 \tag{10}$$

which is able to make FW converge faster over the strongly convex set [13]. The detailed procedures of our algorithm are summarized in Algorithm 1, and it is named as bandit Frank-Wolfe with line search (BFW-LS).

3.3 Theoretical Guarantees

Next, we proceed to present the theoretical guarantees of our BFW-LS over strongly convex sets. Although in the full information setting, Wan and Zhang [32] have shown the advantage of the line search over the strongly convex set. It is worth noting that their result is not applicable in the bandit setting due to the following two challenges.

- In the full information setting, Wan and Zhang [32] directly utilize the faster convergence of FW over strongly convex $\mathcal{K}$. However, in the bandit setting, the FW iteration is performed over the shrunk set $\mathcal{K}_\delta$. It is unclear whether $\mathcal{K}_\delta$ is also strongly convex.
- In the bandit setting, the objective function in (8) is defined based on the estimated gradient, the variance of which makes it more difficult for us to minimize (8) with one FW iteration than the objective function (1) in the full information setting.

To address the above challenges, we first derive the following lemma, which implies that $\mathcal{K}_\delta$ is also strongly convex.

Lemma 2. *If $\mathcal{K}$ is α_K-strongly convex with respect to a norm $\|\cdot\|$, then $\mathcal{K}_\delta$ is $\frac{\alpha_K}{1-(\delta/r)}$-strongly convex with respect to the norm $\|\cdot\|$.*

Based on the above lemma, we establish an upper bound for the approximation error of minimizing the objective function in (8) over strongly convex sets.

Lemma 3. *Let $\mathcal{K}$ be an α_K-strongly convex set with respect to the ℓ_2 norm. Let $\mathbf{y}_t^* = \operatorname{argmin}_{\mathbf{y} \in \mathcal{K}_\delta} F_{t-1}(\mathbf{y})$ for any $t \in [T+1]$, where $F_t(\mathbf{y})$ is define in (8). Under Assumptions 1, 2, and 3, for any $t \in [T+1]$, Algorithm 1 with $\eta = \frac{cR}{dM(T+2)^{3/4}}$ and $\delta = cT^{-1/4}$ has*

$$F_{t-1}(\mathbf{y}_t) - F_{t-1}(\mathbf{y}_t^*) \le \epsilon_t = \frac{C}{\sqrt{t+2}}$$

where $c > 0$ is a constant such that $\delta < r$ and $C = max\left(16R^2, \frac{4096}{3\alpha_K^2}\right)$.

Remark 1. We find that the approximation error of the FW iteration is upper bound by $O(t^{-1/2})$ for our algorithm over strongly convex sets. For a clear comparison, we note that this approximation error for PF-BCO over general convex sets has a worse bound of $O(t^{-2/5})$ (See the proof of Theorem 1 in Chen et al. [8]).

By applying Lemma 3, we prove that our method enjoys the following regret bound over strongly convex sets.

Theorem 1. *Let $\mathcal{K}$ be an α_K-strongly convex set with respect to the ℓ_2 norm and $C = max\left(16R^2, \frac{4096}{3\alpha_K^2}\right)$. Let $c > 0$ be a constant such that $\delta = cT^{-1/4} \leq r$. Under Assumptions 1, 2, and 3, Algorithm 1 with $\eta = \frac{cR}{dM(T+2)^{3/4}}$ and $\delta = cT^{-1/4}$ ensures*

$$\mathbb{E}[\text{Regret}(T)] \leq \frac{4RdM(T+2)^{3/4}}{c} + \frac{RdMT^{3/4}}{c} + 3cGT^{3/4} + \frac{4\sqrt{C}G(T+2)^{3/4}}{3} + \frac{cGRT^{3/4}}{r}.$$

Remark 2. From Theorem 1, our Algorithm 1 achieves an expected regret bound of $O(T^{3/4})$ over strongly convex sets. Note that this bound is the same as the expected regret bound of BBCG [15] and LOO-BBGD [16]. Although our result does not hold in general convex case like their results, we dismiss the compromising blocking technique required by them.

Furthermore, although Theorem 1 has provided an expected regret bound, one may still wonder whether this bound can hold at most of the time. For this reason, we also establish a high-probability regret bound for Algorithm 1.

Theorem 2. *Let $\mathcal{K}$ be an α_K-strongly convex set with respect to the ℓ_2 norm and $C = max\left(16R^2, \frac{4096}{3\alpha_K^2}\right)$. Let $c > 0$ be a constant such that $\delta = cT^{-1/4} \leq r$ and $\eta = \frac{cR}{dM(T+2)^{3/4}}$. Under Assumption 1, 2, and 3, with probability at least $1-\gamma$, Algorithm 1 ensures*

$$\text{Regret}(T) \leq 2RG\sqrt{2\ln\frac{1}{\gamma}}T^{1/2} + \frac{2RdM}{c}\sqrt{2\ln\frac{1}{\gamma}}T^{3/4} + \frac{4RdM(T+2)^{3/4}}{c} + \frac{RdMT^{3/4}}{c} + \frac{4\sqrt{C}G(T+2)^{3/4}}{3} + \frac{cGRT^{3/4}}{r} + 3cGT^{3/4}.$$

Remark 3. Theorem 2 implies that our algorithm also enjoys a high-probability regret bound of $O(T^{3/4})$ over strongly convex sets. It is worth noting that it removes the logarithmic factor in the $O(T^{3/4}\sqrt{\log T})$ high-probability regret bound achieved by BBCG [15,29], which demonstrate a theoretical advantage of removing the blocking technique.

4 Experiments

In this section, we present experimental results on synthetic and real-world data. All experiments are conducted on a Linux machine with 2.3 GHz CPU and 125 GB RAM.

4.1 Problem Settings

We first consider the problem of online quadratic programming (OQP) with synthetic data [9], where the total number of iterations is set as $T = 40000$, and the dimensionality is set as $d = 10$. At each iteration $t \in [T]$, the learner first selects $\mathbf{x}_t \in \mathcal{K}$, and then suffers a quadratic loss

$$f_t(\mathbf{x}) = \frac{1}{2}\mathbf{x}^\top \mathbf{G}_t^\top \mathbf{G}_t \mathbf{x} + \mathbf{w}_t^\top \mathbf{x}$$

where each element of $\mathbf{G}_t \in \mathbb{R}^{d\times d}$ and $\mathbf{w}_t \in \mathbb{R}^d$ is sampled from the standard normal distribution. Second, we consider the problem of online binary classification (OBC) with a real-world dataset ijcnn1 [7], which consists of 49990 instances and each instance having 22 features, i.e., $d = 22$. To make the block size K of BBCG and LOO-BBGD be an integer, we randomly select $T = 40000$ instances from the original dataset. At each round t, the learner receives a single example $\mathbf{e}_t \in \mathbb{R}^d$ and chooses a decision $\mathbf{x}_t \in \mathcal{K}$. Then, the true class label $y_t \in \{-1, 1\}$ is revealed, and the learner suffers the hinge loss

$$f_t(\mathbf{x}_t) = \max\left\{1 - y_t \mathbf{e}_t^\top \mathbf{x}_t, 0\right\}.$$

More specifically, we set $\mathcal{K} = \{\mathbf{x} \in \mathbb{R}^d \mid \|\mathbf{x}\|_p \leq \tau\}$ with $p = 1.5, \tau = 50$ for the OQP experiment and $p = 1.5, \tau = 30$ for the OBC experiment. One can verify that this set is strongly convex for any $p \in (1, 2)$ [13], and satisfies Assumption 1 with $r = \frac{\tau}{d^{1/p-1/2}}$ and $R = \tau$.

4.2 Experimental Results

We compare our BFW-LS against existing projection-free BCO algorithms including PF-BCO [9], BBCG [15], and LOO-BBGD [16]. The parameters of all algorithms are set according to what their corresponding theories suggest. Specifically, all of them depend on two parameters η and δ, which are set as $\eta = c_1 T^{-3/4}$ and $\delta = c_2 T^{-1/4}$ for BBCG, LLO-BBGD, and our BFW-LS, and $\eta = c_1 T^{-4/5}, \delta = c_2 T^{-1/5}$ for PF-BCO. The constants c_1 and c_2 are selected from $\{1e-2, 1e-1, \ldots, 1e3\}$ and $\{20, 40, \ldots, 120, 140\}$, respectively. Moreover, BBCG adopts two additional parameters K and ϵ to control the block size and the error tolerance in each block, which are set as $\epsilon = 16R^2T^{-1/2}$ and $K = \sqrt{T}$. The same block size is also utilized in LOO-BBGD. Figure 1a and Fig. 1b show the average loss of each algorithm, i.e., $\frac{1}{t}\sum_{i=1}^{t} f_i(\mathbf{x}_i)$ at each iteration t, on the OQP and OBC experiments, respectively. We find that the average loss of our BFW-LS is lower than that of these baselines. Additionally, although the regret

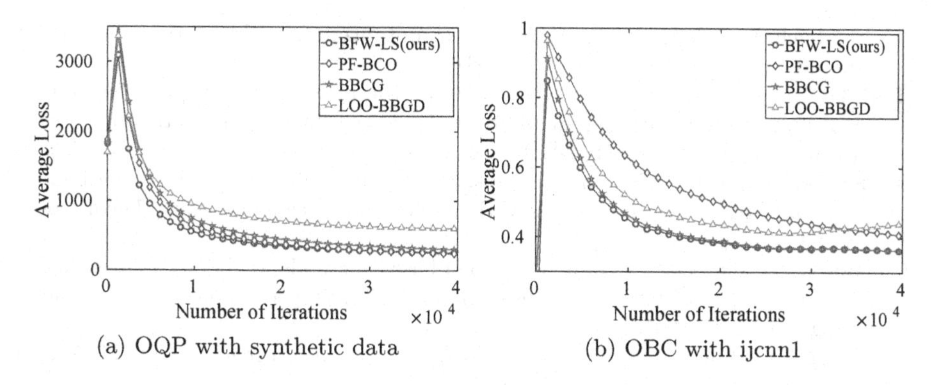

(a) OQP with synthetic data

(b) OBC with ijcnn1

Fig. 1. Average Loss of Our and Previous Projection-free BCO Algorithms.

bound of BBCG and LOO-BBGD is better than that of PF-BCO, they fail to exhibit significantly better performance in comparison to PF-BCO, especially on the OQP experiment. This phenomenon partially suggests the shortcoming of the blocking technique.

5 Conclusion

In this paper, we propose a projection-free algorithm called BFW-LS for BCO, which achieves an $O(T^{3/4})$ expected regret bound over strongly convex sets, without using the blocking technique required by previous projection-free BCO algorithms with the same expected bound. Furthermore, we also show that our BFW-LS has a high-probability regret bound of $O(T^{3/4})$, which removes the logarithmic factor in the $O(T^{3/4}\sqrt{\log T})$ high-probability regret bound achieved by previous projection-free BCO algorithms. Finally, experimental results verify the advantage of our BFW-LS.

Acknowledgments. This work was supported by State Grid science and technology project (5700-202327286A-1-1-ZN).

Disclosure of Interests. The authors have no conflicts of interest to declare that are relevant to the content of this article.

References

1. Abernethy, J., Hazan, E., Rakhlin, A.: Competing in the dark: an efficient algorithm for bandit linear optimization. In: Proceedings of the 21st Conference on Learning Theory, pp. 263–274 (2008)
2. Agarwal, A., Dekel, O., Xiao, L.: Optimal algorithms for online convex optimization with multi-point bandit feedback. In: Proceedings of the 23rd Conference on Learning Theory, pp. 28–40 (2010)

3. Awerbuch, B., Kleinberg, R.: Online linear optimization and adaptive routing. J. Comput. Syst. Sci. **74**(1), 97–114 (2008)
4. Bubeck, S., Cesa-Bianchi, N., Kakade, S.M.: Towards minimax policies for online linear optimization with bandit feedback. In: Proceedings of the 25th Conference on Learning Theory, pp. 41.1–41.14 (2012)
5. Bubeck, S., Dekel, O., Koren, T., Peres, Y.: Bandit convex optimization: $\sqrt{t}$ regret in one dimension. In: Proceedings of the 28th Conference on Learning Theory, pp. 266–278 (2015)
6. Bubeck, S., Eldan, R., Lee, Y.T.: Kernel-based methods for bandit convex optimization. In: Proceedings of the 49th Annual ACM Symposium on Theory of Computing, pp. 72–85 (2019)
7. Chang, C.C., Lin, C.J.: Libsvm: a library for support vector machines. ACM Trans. Intell. Syst. Technol. **2**(27), 1–27 (2011)
8. Chen, L., Harshaw, C., Hassani, H., Karbasi, A.: Projection-free online optimization with stochastic gradient: from convexity to submodularity. In: Proceedings of the 35th International Conference on Machine Learning, pp. 814–823 (2018)
9. Chen, L., Zhang, M., Karbasi, A.: Projection-free bandit convex optimization. In: Proceedings of the 22nd International Conference on Artificial Intelligence and Statistics, pp. 2047–2056 (2019)
10. Dekel, O., Eldan, R., Koren, T.: Bandit smooth convex optimization: Improving the bias-variance tradeoff. In: Advances in Neural Information Processing Systems 28, pp. 2926–2934 (2015)
11. Flaxman, A.D., Kalai, A.T., McMahan, H.B.: Online convex optimization in the bandit setting: gradient descent without a gradient. In: Proceedings of the 16th Annual ACM-SIAM Symposium on Discrete Algorithms, pp. 385–394 (2005)
12. Frank, M., Wolfe, P.: An algorithm for quadratic programming. Naval Res. Logistics Quart. **3**(1–2), 95–110 (1956)
13. Garber, D., Hazan, E.: Faster rates for the frank-wolfe method over strongly-convex sets. In: Proceedings of the 32nd International Conference on Machine Learning, pp. 541–549 (2015)
14. Garber, D., Hazan, E.: A linearly convergent conditional gradient algorithm with applications to online and stochastic optimization. SIAM J. Optim. **26**(3), 1493–1528 (2016)
15. Garber, D., Kretzu, B.: Improved regret bounds for projection-free bandit convex optimization. In: Proceedings of the 23rd International Conference on Artificial Intelligence and Statistics, pp. 2196–2206 (2020)
16. Garber, D., Kretzu, B.: New projection-free algorithms for online convex optimization with adaptive regret guarantees. In: Proceedings of the 35th Conference on Learning Theory, pp. 2326–2359 (2022)
17. Garber, D., Kretzu, B.: Projection-free online exp-concave optimization. In: Proceedings of the 36th Conference on Learning Theory (2023)
18. Hazan, E.: Introduction to online convex optimization. Found. Trends Optim. **2**(3–4), 157–325 (2016)
19. Hazan, E., Kale, S.: Projection-free online learning. In: Proceedings of the 29th International Conference on Machine Learning, pp. 1843–1850 (2012)
20. Hazan, E., Levy, K.Y.: Bandit convex optimization: towards tight bounds. In: Advances in Neural Information Processing Systems 27, pp. 784–792 (2014)
21. Hazan, E., Minasyan, E.: Faster projection-free online learning. In: Proceedings of the 33rd Conference on Learning Theory, pp. 1877–1893 (2020)

22. Ito, S.: An optimal algorithm for bandit convex optimization with strongly-convex and smooth loss. In: Proceedings of the 23rd International Conference on Artificial Intelligence and Statistics, pp. 2229–2239 (2020)
23. Jaggi, M.: Revisiting Frank-Wolfe: projection-free sparse convex optimization. In: Proceedings of the 30th International Conference on Machine Learning, pp. 427–435 (2013)
24. Kalhan, D.S., et al.: Dynamic online learning via frank-wolfe algorithm. IEEE Trans. Signal Process. **69**, 932–947 (2021)
25. Kretzu, B., Garber, D.: Revisiting projection-free online learning: the strongly convex case. In: Proceedings of the 24th International Conference on Artificial Intelligence and Statistics, pp. 3592–3600 (2021)
26. McMahan, H.B., et al.: Ad click prediction: a view from the trenches. In: Proceedings of the 19th ACM SIGKDD International Conference on Knowledge Discovery and Data Mining, pp. 1222–1230 (2013)
27. Saha, A., Tewari, A.: Improved regret guarantees for online smooth convex optimization with bandit feedback. In: Proceedings of the 14th International Conference on Artificial Intelligence and Statistics, pp. 636–642 (2011)
28. Shalev-Shwartz, S., Singer, Y.: A primal-dual perspective of online learning algorithm. Mach. Learn. **69**(2–3), 115–142 (2007)
29. Wan, Y., Tu, W.W., Zhang, L.: Projection-free distributed online convex optimization with $\mathcal{O}(\sqrt{T})$ communication complexity. In: Proceedings of the 37th International Conference on Machine Learning, pp. 9818–9828 (2020)
30. Wan, Y., Wang, G., Tu, W.W., Zhang, L.: Projection-free distributed online learning with sublinear communication complexity. J. Mach. Learn. Res. **23**(172), 1–53 (2022)
31. Wan, Y., Xue, B., Zhang, L.: Projection-free online learning in dynamic environments. In: Proceedings of the 35th AAAI Conference on Artificial Intelligence Advances, pp. 10067–10075 (2021)
32. Wan, Y., Zhang, L.: Projection-free online learning over strongly convex set. In: Proceedings of the 35th AAAI Conference on Artificial Intelligence Advances, pp. 10076–10084 (2021)
33. Wan, Y., Zhang, L., Song, M.: Improved dynamic regret for online frank-wolfe. In: Proceedings of the 36th Conference on Learning Theory (2023)
34. Wang, Y., Wan, Y., Zhang, S., Zhang, L.: Distributed projection-free online learning for smooth and convex losses. In: Proceedings of the 37th AAAI Conference on Artificial Intelligence, pp. 10226–10234 (2023)
35. Wang, Y., et al.: Non-stationary projection-free online learning with dynamic and adaptive regret guarantees. ArXiv e-prints arXiv:2305.11726 (2023)
36. Zinkevich, M.: Online convex programming and generalized infinitesimal gradient ascent. In: Proceedings of the 20th International Conference on Machine Learning, pp. 928–936 (2003)

Adaptive Prediction Interval for Data Stream Regression

Yibin Sun[1(✉)], Bernhard Pfahringer[1], Heitor Murilo Gomes[1,2], and Albert Bifet[1,3]

[1] AI Institute, The University of Waikato, Hamilton, New Zealand
ys388@students.waikato.ac.nz, {bernhard,abifet}@waikato.ac.nz
[2] School of Engineering and Computer Science, Victoria University of Wellington, Wellington, New Zealand
heitor.gomes@vuw.ac.nz
[3] LTCI, Télécom Paris, IP Paris, France

Abstract. Prediction Interval (PI) is a powerful technique for quantifying the uncertainty of regression tasks. However, research on PI for data streams has not received much attention. Moreover, traditional PI-generating approaches are not directly applicable due to the dynamic and evolving nature of data streams. This paper presents AdaPI (ADAptive Prediction Interval), a novel method that can automatically adjust the interval width by an appropriate amount according to historical information to converge the coverage to a user-defined percentage. AdaPI can be applied to any streaming PI technique as a postprocessing step. This paper develops an incremental variant of the pervasive Mean and Variance Estimation (MVE) method for use with AdaPI. An empirical evaluation over a set of standard streaming regression tasks demonstrates AdaPI's ability to generate compact prediction intervals with a coverage close to the desired level, outperforming alternative methods.

Keywords: Data streams · Regression · Prediction Intervals

1 Introduction

The machine learning literature has thoroughly investigated learning algorithms for regression tasks [7]. However, a single-valued prediction is insufficient for many real-world applications [10]. A prediction interval (PI) [9] provides more informativeness and uncertainty measurements [16] to a regression model since it generates intervals that encompass the expected range of true values with a desired confidence level. Learning data streams is different from conventional machine learning tasks. Due to the potentially infinite amount of data, stream learning algorithms can neither store all the previous information nor iterate multiple times through the datasets. The basic assumption is that data points

Y. Sun—I would like to acknowledge the support from TAIAO project.

D.-N. Yang et al. (Eds.): PAKDD 2024, LNAI 14647, pp. 130–141, 2024.
https://doi.org/10.1007/978-981-97-2259-4_10

from a stream can only be inspected once times [2]. Several techniques for establishing prediction intervals were introduced based on different mathematical theories since the seminal technique in [9]. Zhao et al. in [20] summarised the most commonly used ones, such as bootstrapping techniques and delta method.

Current research rarely focuses on prediction intervals on data streams. The few PI methods available for data streams (or time series) rely on windowed versions of existing techniques, which do not agree with current state of the art methods for data stream regression [17], which are fully incremental (as discussed in Sect. 2). The lack of a fully incremental PI methods motivated our work. Our main contribution is the application of the Mean and Variance Estimation (MVE) prediction interval method in a streaming fashion and the ADAptive Prediction Interval (`AdaPI`) methodology, which can be applied to any PI method as a post-calibration step. `AdaPI` incrementally adjusts the generated interval widths according to the current coverage, ensuring that the coverage converges towards the desired confidence level. This new approach also adapts automatically to concept drifts, a critical issue in the streaming scenario [2].

2 Related Work

Xu and Xie [19] introduced a Bootstrap-based PI methodology for dynamic time series. This methodology establishes several base learners $\mathcal{A}$ and an error-based aggregation function ϵ. $\mathcal{A}$ and ϵ are updated by re-sampling the time series with replacement in batches. Subsequently, quantiles of $\mathcal{A}$ and ϵ are utilised for providing PIs, i.e. $[\mathcal{A}^{1-\alpha} - \epsilon^{1-\alpha}, \mathcal{A}^{1-\alpha} + \epsilon^{1-\alpha}]$.

IT2FGNNDEnsemble (Interval Type-2 Fuzzy Granular Neural Network Dynamic Ensemble) [13] comprises a seven-layer neural network, including a granularity layer, a normalisation layer, a layer that contains an ensemble of the basic algorithm – IT2FGNN, and so forth. In the ensemble, each IT2FGNN provides a prediction interval, which is fused into the final PI after an elimination process that removes "weak" ensemble members.

Jackknife+ [1] is a popular approach to prediction intervals, derived from Jackknife resampling. Jackknife obtains residuals (R^{LOO}) using a Leave-one-out strategy, and Jackknife+ modifies it by ignoring the current instance. Prediction intervals are produced by the α quantile of the given series, i.e. $[q^{-}_{n,\alpha}(\mathcal{A}_{-i} - R^{LOO}_i), q^{+}_{n,\alpha}(\mathcal{A}_{-i} + R^{LOO}_i)]$.

Inductive Conformal Prediction (ICP) is the common regressor within the Conformal Prediction framework [15]. ICP trains an ML model on a training set and computes a nonconformity score as a deviation measurement of an instance's behaviour. Prediction intervals for further examples are generated based on these scores and then evaluated on a calibration set. Split Conformal Prediction (SCP), an extension of ICP, divides the dataset into "splits", each of which provides an individual conformal score. The final PI is a combination of all "splits".

Hadjicharalambous et al. [8] introduced a neural network (NN)-based online prediction interval method BLM (Bootstrap-LUBE Method), which combines the basic Bootstrap technique and LUBE (Lower-Upper Bound Estimation) [12]

PI methods. Specifically, a number of NNs are trained to provide pseudo-measurements (dummy data points) in data-sparse regions to ensure sufficient information for the LUBE method to yield reasonable PIs. This method is adapted for data streams using windowing approaches. However, training NNs in a sliding window manner can be computationally expensive.

Both low efficiency and the need to process instances multiple times pose challenges for all the aforementioned PI methods. Recently, Zhao et al. [21] proposed a tree-based algorithm and an ensemble of tree-learners for constructing PI on time series and streaming data. They selected Conditional Inference Tree (ctree) as the base learner in their work, although all tree-based regressors are suitable. In the single tree approach, prediction intervals are determined by the quantile values stored in the leaves. The PIs are expanded by a predefined parameter α during the initialisation. If the current coverage falls outside the desired range, the PIs are adjusted using both the current RMSE and a β parameter (i.e.,$\beta\times$ RMSE). The ensemble approach also begins with a single tree. New trees are periodically appended until the ensemble size reaches its maximum. The n_{th} PI is a weighted combination of the n_{th} and $\mathrm{n}\text{-}1_{th}$ tree, i.e. $\bar{P}_n = \omega_1 \times P_{n-1} + \omega_2 \times P_n$. Generally, ω_2 is greater than ω_1, giving newer trees a larger impact.

3 Background

Mean and Variance Estimation (MVE) [14] is one of the most straightforward and widely applied approaches for PI. MVE assumes the predictive errors follow a Gaussian distribution (Normal Assumption) which results in a generalised linear model, and therefore leads to a prediction interval illustrated as in Eq. 1:

$$\text{PI} \in (\mathcal{Y} - G^{-1}(0,\gamma) \times \sigma_\epsilon, \mathcal{Y} + G^{-1}(0,\gamma) \times \sigma_\epsilon) \tag{1}$$

where $G^{-1}(0,\gamma)$ is the inverse Gaussian distribution function with 0 mean and γ probability, and $\mathcal{Y}$ is the prediction output. According to the normal assumption, $\mathcal{E} \sim \mathcal{N}(0, \sigma_\epsilon)$. Hence, when $\gamma = 95\%$, $G^{-1}(0, 0.95) \approx 1.96$.

Objectively measuring the quality of PI is challenging as we need to consider both the width of the interval and its coverage. To achieve a fair and comprehensive evaluation, we consider two metrics used in previous studies [8], the Coverage and Normalised Mean Prediction Interval Width (NMPIW). Coverage represents the ratio of the ground truth falling within the predicted PI, i.e.:

$$\text{Coverage} = \frac{1}{N}\sum_{i=1}^{N} I_i \tag{2}$$

where $I_i = 1$ if $y \in [\mathcal{P}_l, \mathcal{P}_u]$ and $I_i = 0$ otherwise; N is the total observation number. NMPIW is expressed in Eq. 3:

$$\text{NMPIW} = \frac{\frac{1}{N}\sum_{i=1}^{N}(\mathcal{P}_{u_i} - \mathcal{P}_{l_i})}{R} \tag{3}$$

where $\mathcal{P}_{u_i}$ and $\mathcal{P}_{l_i}$ are the i_{th} upper and lower bounds for the i_{th} observation, and R is the range of the target values observed. As implied in the equation, NMPIW represents the ratio of the average PI width to the range of the true values and therefore reflects the effectiveness of the PI.

Very wide intervals naturally achieve high coverage, therefore the goal for a well-performing PI method is to produce intervals covering more or less the desired percentage of predictions, while also being as narrow as possible, indicated by small NMPIW values.

4 Adaptive Prediction Interval(`AdaPI`)

`AdaPI` uses a coefficient for scaling the generated interval width. Precisely, expanding the MVE Eq. 1, the modified upper and lower bound in `AdaPI` are defined as in Eq. 4:

$$\texttt{AdaPI} = (\mathcal{Y} - \mathcal{S} \times G^{-1}(0, \gamma) \times \sigma_\epsilon, \mathcal{Y} + \mathcal{S} \times G^{-1}(0, \gamma) \times \sigma_\epsilon) \tag{4}$$

where the scalar $\mathcal{S}$ is specified by Eq. 5:

$$\mathcal{S} = \begin{cases} 100 - \mathcal{C} & \text{if } \mathcal{C} < 2\mathcal{L} - 100 \\ \log_{\mathcal{L}} \frac{100-\mathcal{C}}{100-\mathcal{L}} + 1 & \text{if } \mathcal{C} \geq \mathcal{L} \\ (\mathcal{L} - 100) \log_{\mathcal{L}} \frac{100+\mathcal{C}-2\mathcal{L}}{100-\mathcal{L}} + 1 & \text{otherwise} \end{cases} \tag{5}$$

where $\mathcal{C}$ is the current coverage, and $\mathcal{L}$ represents the confidence level. This curve varies with different confidence levels $\mathcal{L}$. Figure 1 illustrates the curve graph of the scalar with a fixed confidence level at 95%, i.e. $\mathcal{L} = 95$.

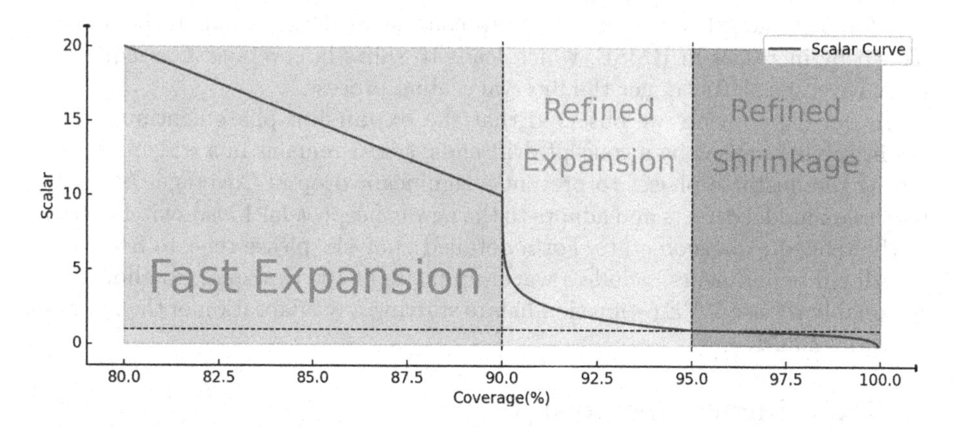

Fig. 1. Scalar Curve with a 95% Confidence Level ($\mathcal{C} \in [80, 100]$). Dotted lines denote significant values in the figure. The horizontal line is 1.0, the right vertical line is the desired confidence level (95%). The refined expansion and the refined shrinkage areas have same width. (Color figure online)

Three segments can be observed in Fig. 1. The red part (Fast expansion) denotes a fast expansion phase where the scalars are determined linearly according to the current coverage. The purpose of this segment is to rapidly increase the coverage towards the confidence level. The orange part (Refined expansion) also expands the PI by producing scalars larger than one. However, the scalars from this part are generated more subtly by a logarithmic function. As the coverage nears the confidence level, the rate of change in the scalar will decrease. The green part (Refined shrinkage), on the other hand, aims to shrink the PI when the coverage exceeds the confidence level. Similar to the orange part, the scalar will be moderately updated when the coverage is close to the confidence level. This mechanism prevents the scalar from being too radical.

`AdaPI` adjusts dynamically, varying based on requirements rather than following a set pattern. This approach offers several benefits: 1. Radical adjustments occur for substantial coverage-confidence disparities; 2. Modifications are moderate when coverage approaches the confidence level, preventing overreactions; and 3. Auto cessation at the desired confidence level is achieved without adding extra hyperparameters.

`AdaPI` introduces a scalar-based solution alongside MVE, aiming to address the normal assumption inherent in MVE. In practical scenarios, it's often inaccurate to presume that predictive errors conform to a Gaussian distribution. Consequently, predictive intervals relying on statistical Gaussian measures might fail to adequately encompass actual observations. This disparity results in the coverage deviating from the confidence level established in the MVE approach. When regression models demonstrate exceptional (or poor) accuracy in forecasting specific data streams, MVE tends to produce overly narrow (or wide) predictive intervals. `AdaPI` counters this by generating scalar adjustments to calibrate these intervals accordingly.

Moreover, `AdaPI` naturally adapts to concept drift as it tends to be accompanied by increases in RMSE, which leads to shifts in coverage. Consequently, these coverage shifts trigger the interval scaling process.

In our experiments we observed that the **expansion** phase continues until the instability caused by a concept drift ends. `AdaPI` remains in a widening state during the unstable phases to prevent a significant drop in Coverage. After the regression model detects and adapts to the new concept, `AdaPI` also switches back to the refined expansion state. For a detailed analysis, please refer to Sect. 5.2.

All our experiments include a warm-up phase of 100 examples, to allow for a reasonably robust MVE estimation before starting any adaptation of the interval width by `AdaPI`.

5 Experiments and Results

The experimental datasets are briefly illustrated in Table 1. Ailerons, Elevators, and House8L datasets can be found at OpenML. The MetroTraffic dataset can be found at UCI Machine Learning Repository, and HyperA is generated by a stream generator in MOA [2]. The Abalone, Bike, and Fried datasets are presented in [3,4,18], respectively.

Table 1. Datasets Overview

Synthetic			Real		
Datasets	$N_{Features}$	$N_{Instances}$	Datasets	$N_{Features}$	$N_{Instances}$
Ailerons	40	13750	Abalone	8	4977
Elevators	18	16599	Bike	12	17379
Fried	10	40768	House8L	8	22784
HyperA	10	500000	MetroTraffic	7	48204

5.1 Comparison to Interval Forecast

Interval Forecast (IF) and its ensemble version (EnsembleIF) are introduced in [21] (refer to Sect. 2). In this work, a fully incremental tree called Fast Incremental Model Tree with Drift Detection (FIMT-DD) [11] is used as a base-learner substitute in both IF and EnsembleIF. The maximum ensemble number is set to 100 in EnsembleIF, and the window length for updating the tree is set to 1000. The values of $\alpha = 2$ and $\beta = 2.5$ are selected, as they are the medians of the recommended value ranges in the original paper. Multiple attempts have been made on the weighted-sum strategy in EnsembleIF, and the combination that performed the best - $\omega_1 = \omega_2 = 0.5$ - is presented in the results.

We present the Coverage and NMPIW results for single IF, EnsembleIF, MVE, and `AdaPI` in Table 2. For IF and EnsembleIF, the adjustment mechanism is triggered only when the model's coverage is not within the range of 94% to 96% (i.e., Coverage $\notin [0.94, 0.96]$). Additionally, MVE and `AdaPI` share the ARF-Reg algorithm [6] as their base learner and 95% as their confidence level.

Table 2. Coverage (%) and NMPIW (%) for 95% Confidence Level

	IF		EnsembleIF		MVE		`AdaPI`	
	Coverage	NMPIW	Coverage	NMPIW	Coverage	NMPIW	Coverage	NMPIW
Bike	98.35	46.11	99.52	67.47	89.51	**29.31**	**94.00**	37.97
Ailerons	98.79	44.75	99.05	54.16	94.73	**29.32**	**95.12**	30.12
HyperA	94.00	26.85	96.55	35.94	94.25	**20.22**	**94.70**	20.72
Abalone	98.39	59.67	98.92	69.74	**95.06**	35.15	95.10	**34.56**
Elevators	97.79	38.24	98.89	57.18	95.31	31.46	**95.24**	**30.72**
House8L	97.87	29.14	98.97	50.88	96.68	30.44	**96.29**	**28.46**
Fried	96.58	40.24	99.47	76.90	97.43	33.80	**96.26**	**31.34**
MetroTraffic	98.15	101.99	99.76	143.88	98.15	98.34	**96.48**	**90.14**

In the "Coverage" columns in Table 2, values closer to 95% coverage are highlighted in bold and in the "NMPIW" columns the smaller values are bold. It is evident that the MVE-based methods generally outperform the IF-based methods. Both MVE and `AdaPI` consistently provide smaller NMPIW values and coverage closer to the confidence level across almost all datasets. For instance, on the HyperA dataset, IF achieved a coverage of 94.00% but had an NMPIW of 26.85%, whereas AdaPI achieved higher coverage with a more compact NMPIW.

Pair-wise Friedman tests [5] are applied to the results in Table 2 in two ways: 1. Differences $\mathcal{D}$ from the coverage to the confidence level, i.e., $\mathcal{D} = |\mathcal{C} - \mathcal{L}|$; and 2. NMPIW. MVE, as well as AdaPI, versus IF and EnsembleIF, all produce p-values smaller than 0.05. Thus, the null hypothesis – there are no differences between the two candidates in each comparison – is rejected, demonstrating the significance of the MVE and AdaPI when compared to IF-based methods.

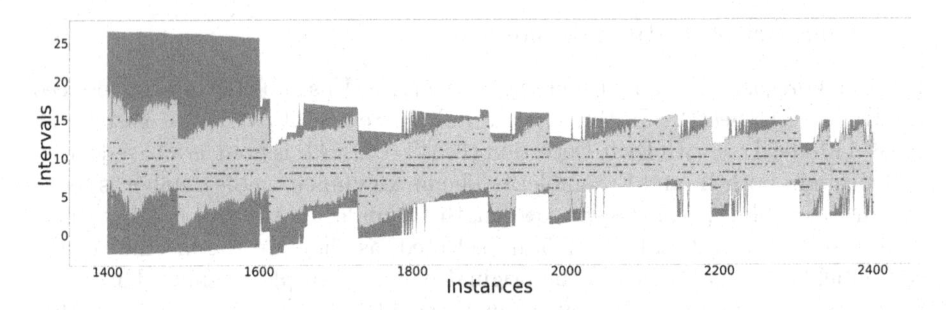

Fig. 2. AdaPI(95%) PI Area (Red, Narrow) and IF PI Area (Sky-Blue, Wide) on Abalone Dataset, and Ground Truths (Grey Dots). (Color figure online)

Figure 2 plots the area covered by both AdaPI (in red) and IF (in sky-blue) on a subset of the Abalone dataset. It can be observed that AdaPI produces much narrower intervals yet covers almost the same amount of the data points as IF. Meanwhile, Fig. 2 also underscores one of the advantages of a predictive model-based PI method over a quantile-based one in streaming scenarios: quantiles can be insensitive to changes. A streaming model learns from the data incrementally, while the quantiles in the leaves (or other structures) might remain relatively similar. AdaPI demonstrates a stronger ability to adapt to drifts (sudden shifts of the grey dots in Fig. 2) faster than IF in Fig. 2.

5.2 Comparison Between MVE and AdaPI

This section focuses on MVE and AdaPI results and analyses. Three streaming regression algorithms perform as background regression model in the experiments: Adaptive Random Forest for Regression (ARF-Reg) [6], Sliding Window K Nearest Neighbours (KNN), and Self-Optimising K Nearest Leaves (SOKNL) [17]. All the algorithms above are available in Massive Online Analysis (MOA) [2], a well-known open-source framework software for data streams. Our experiments use a "TestThenTrain" regime with default parameter settings in MOA.

Table 3 shows the results regarding Coverage and NMPIW. Notably, the "ΔNMPIW" columns illustrate the rate of increase (or decrease)

Table 3. MVE vs. AdaPI for 95% Confidence Level (Closer Coverage in Bold)

	Algorithms	ARF-Reg		KNN		SOKNL	
	Metrics	Coverage	ΔNMPIW	Coverage	ΔNMPIW	Coverage	ΔNMPIW
Bike	AdaPI	**93.91**	28.6%↑	**93.70**	36.9%↑	**93.93**	27.7%↑
	MVE	89.50		88.16		89.15	
Ailerons	AdaPI	**94.99**	2.2%↑	**94.56**	2.1%↑	**95.06**	0.5%↑
	MVE	94.72		94.42		95.13	
HyperA	AdaPI	**94.70**	2.5%↑	**94.64**	1.0%↑	**94.75**	0.4%↑
	MVE	94.25		93.70		94.35	
Abalone	AdaPI	**94.98**	0.3%↓	95.24	0.8%↑	**95.06**	3.3%↓
	MVE	95.06		**94.98**		95.32	
Elevators	AdaPI	**95.08**	2.7%↓	94.86	1.9%↓	**95.28**	2.9%↓
	MVE	95.31		**95.03**		95.61	
House8L	AdaPI	**96.27**	6.6%↓	**95.35**	2.0%↓	**95.96**	5.9%↓
	MVE	96.67		95.47		96.45	
Fried	AdaPI	**96.24**	7.2%↓	**94.91**	1.0%↓	**96.17**	7.1%↓
	MVE	97.43		95.10		97.41	
MetroTraffic	AdaPI	**96.44**	8.4%↓	**95.72**	4.3%↓	**95.23**	0.5%↓
	MVE	98.14		96.88		95.39	

of AdaPI's average interval widths compared to MVE's, i.e. Rate $= \frac{|\text{AdaPI NMPIW - MVE NMPIW}|}{\text{MVE NMPIW}}\%$.

The discussion will primarily focus on ARF-Reg from this point onward since all three algorithms share similar tendency. It can be observed in the Bike dataset that MVE only achieves 89.5% coverage and AdaPI improves it to 93.91% at the cost of widening the interval width by almost 30%. Similar trends are observed in the Ailerons and HyperA datasets, albeit on a smaller scale. The last five datasets in Table 3 follow similar patterns where the coverage of MVE surpasses the confidence level and the AdaPI decreases them while shrinking the interval width. The coverage value of Abalone exceeds the desired confidence level rather than only approaching it. This circumstance can be explained by the basic (MVE) coverage being too close to the confidence level. We can reasonably assume that AdaPI's coverage fluctuates around 95% constantly during the process, hence the system switches between **shrink** and **expand** modes. Still, both methods perform more or less equally well for both the Ailerons and Abalone datasets, achieving very close to 95% coverage with very similar average interval width.

Figure 3 visualises the coverage results. It illustrates a clear picture that AdaPI is providing a closer-to-95% value than the regular MVE in almost all cases since the green area covers most markers. The exceptions can only be found around 95% coverage values, which are caused by the repeatedly switching between shrinking and expanding for AdaPI. Noticeably, most of the markers are reasonably far from the green area's edges, which indicates the significant improvements of AdaPI with respect to coverage.

Figure 4 demonstrates how the scalar (red line) and windowed coverage (blue line) behave over time for the ARF-Reg algorithm on the HyperA dataset.

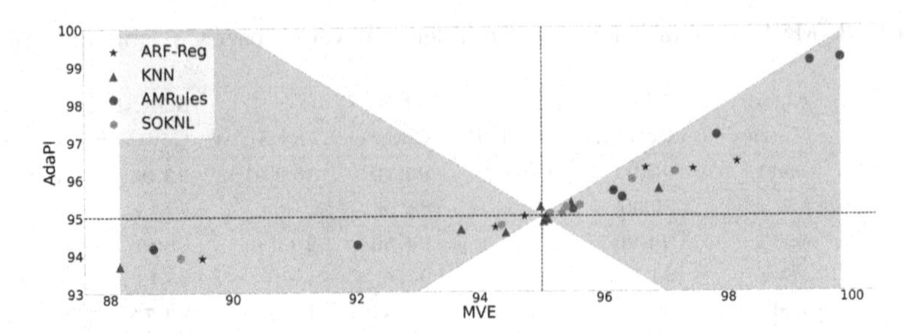

Fig. 3. `AdaPI` versus MVE coverage. Points inside the green area are closer to the 95% target for `AdaPI` than for MVE, i.e. `AdaPI` outperforms MVE in these cases. (Color figure online)

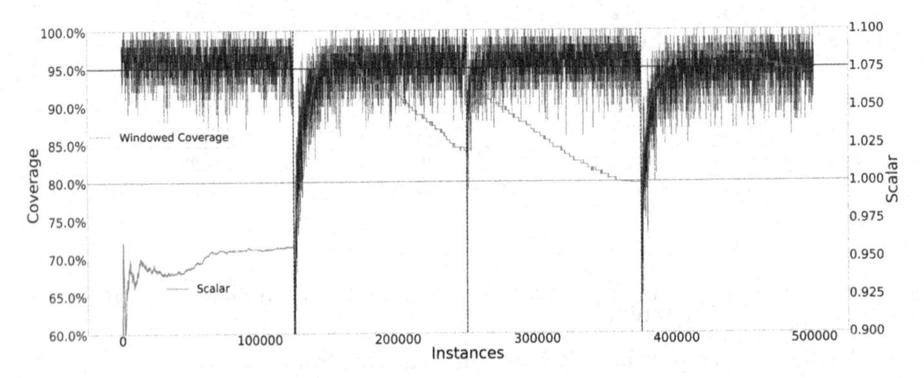

Fig. 4. Scalar (Red) and the Coverage (Blue) for ARF-Reg on HyperA Over Time. The black vertical lines highlight positions of concept drifts; the red horizontal line marks the value 1.0; and the blue horizontal line emphasises 95% confidence level. (Color figure online)

Windowed coverage is chosen as it responds faster to change, presenting a clearer picture of the dynamics of adaptation. Figure 4 demonstrates `AdaPI`'s robustness in handling concept drifts. The HyperA dataset involves predicting the distance from a random point to a high-dimensional hyperplane. It comprises 500k instances and experiences three abrupt concept drifts at 125k, 250k, and 375k instances, aligning with the sudden shifts in the figure. The graph illustrates how ARF-Reg initially establishes a suitable model for HyperA, resulting in desirable coverage and moderate scalar. Then, the model encounters the first drift and no longer aligns with the new concept, leading to a decline in coverage. `AdaPI` responds by increasing the scalars, as indicated by the sharp rise in the red line. The scalars eventually stabilise closer to one once the model adapts to the new concept. Similar trends occur at the other two drift points.

In Fig. 5, three images depict behavioral visualizations for MVE and `AdaPI`. The MVE-covered area is highlighted in red, while `AdaPI`'s area is filled in blue. Grey dots represent the ground truths covered by both methods, blue dots indicate ground truths missed by the narrower approach but covered by the wider

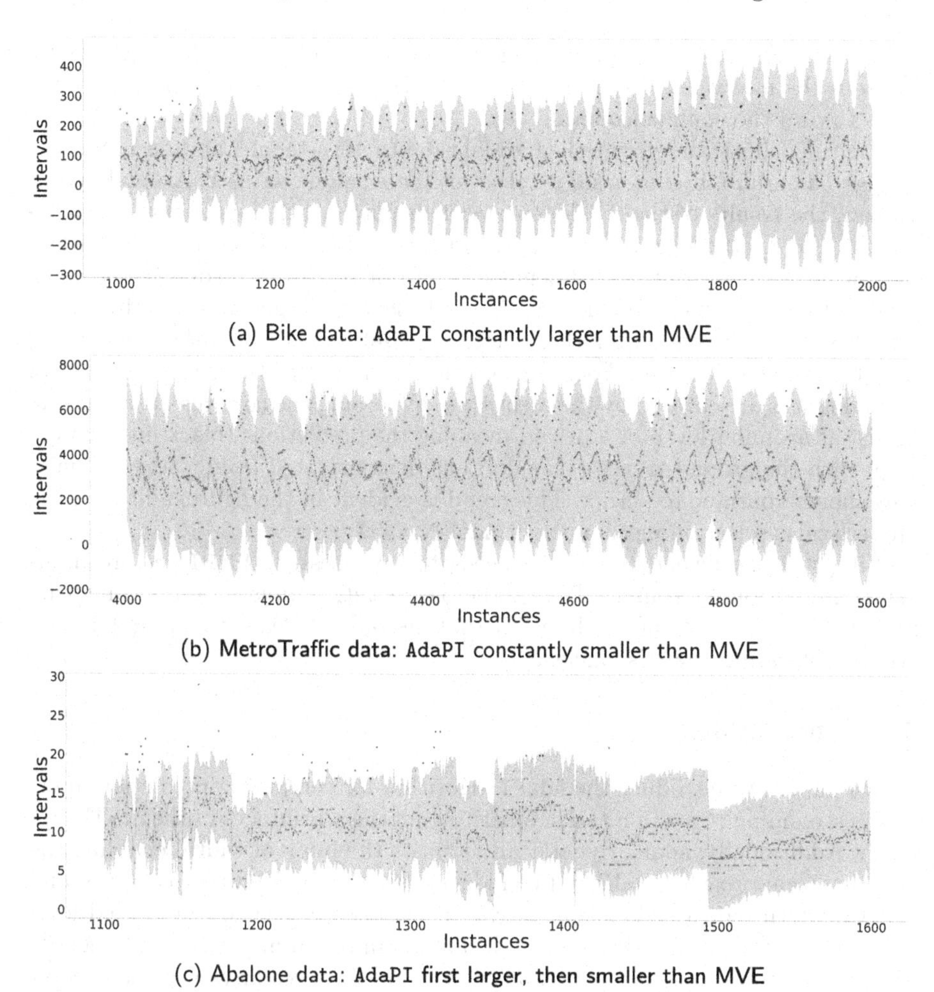

Fig. 5. MVE Area (Red) and AdaPI Area (Sky-Blue), Narrower Area Covered Ground Truths (Grey Dots), Wider Area Covered Ground Truths (Blue Dots) and the Ground Truths outside Both Area (Red Dots). (Color figure online)

approach, and red dots are outside of both areas. The thin line in the middle represents the predictions made by ARF-Reg.

Figure 5a is an explicit annotation of the Bike dataset in Table 3. Apparently, numerous blue dots are included within AdaPI's boundaries, even though they are excluded by MVE. This is the main reason why AdaPI achieves coverage much closer to 95%. It is worth noting that this enhancement is solely attributed to the scalar system as the MVE and AdaPI share the same predictive model.

Figure 5b presents a similar visualisation to Fig. 5a but on the MetroTraffic dataset. MVE coverage in Table 3 is beyond 95% on MetroTraffic. Ergo, the chosen scalars are usually under 1, resulting in a more narrow AdaPI area. We can also conclude that by eliminating all blue dots out of the PI range, `AdaPI` shortens the interval width from a general perspective while the coverage value move toward the confidence level.

Figure 5c illustrates very well the adaptability of the `AdaPI` approach. Similar to Figs. 5a and 5b, it showcases a subset of the Abalone dataset. As analysed in Table 3, the results of the Abalone dataset display erratic behaviour, which we attribute to fluctuations in adaptation. In Fig. 5c, we pinpoint a moment when the coverage surpasses the confidence level. Prior to this, the coverage is below 95% and the scalar is greater than 1, resulting in a larger area for the `AdaPI` (Sky-Blue). Conversely, after this point, the `AdaPI` area diminishes, indicating a scalar smaller than one. Focusing on the latter half of the figure, almost all ground truths are covered by the algorithm, increasing the coverage. If the errors persist at a similar level for a period-meaning the predictions (black line) closely align with the ground truths (grey dots)-as demonstrated in Fig. 5c, the `AdaPI` area should continue to shrink. This trend is evident in the figure, affirming the proper scalar-selection mechanism of `AdaPI`.

Finally, the efficiency of `AdaPI` needs to be assessed. `AdaPI` require minimal computational resources. Due to the page limit, complete results cannot be provided here. In summary, the total runtime only increase by around 1% on average, affirming `AdaPI`'s efficiency.

6 Conclusions

In this paper, we introduce the `AdaPI` algorithm for computing prediction intervals in streaming regression tasks, inheriting all the advantages of the MVE approach. Additionally, `AdaPI` dynamically adjusts the interval width, enabling the coverage to approach the desired confidence level. The results for coverage and interval width indicate that `AdaPI` achieves a closer match to the desired level compared to the static MVE approach. Furthermore, in many instances, `AdaPI` produces narrower prediction intervals. The analysis of scalar values underscores the robustness and stability of `AdaPI`. Moreover, our approach demonstrates the ability to adapt to concept drifts.

Looking ahead, there are several promising directions for future research. Firstly, the same adaptation principles could be applied to other static prediction interval algorithms, although the practical impact of these modifications would require thorough evaluation. Secondly, the scalar curve presented in this paper appears relatively moderate and conservative; a more aggressive and radical scaling strategy might yield even better results. Lastly, the evaluation of prediction intervals along both coverage and width dimensions introduces complexity to the analysis. A unified evaluation metric would significantly simplify the selection of a prediction interval method in practical applications.

References

1. Barber, R.F., Candes, E.J., Ramdas, A., Tibshirani, R.J.: Predictive inference with the jackknife+ (2021)

2. Bifet, A., et al.: Moa: massive online analysis, a framework for stream classification and clustering. PMLR (2010)
3. Breiman, L.: Bagging predictors. Mach. Learn. **24**, 123–140 (1996)
4. Fanaee-T, H., Gama, J.: Event labeling combining ensemble detectors and background knowledge. Progr. Artif. Intell. **2**, 113–127 (2014)
5. Friedman, M.: A comparison of alternative tests of significance for the problem of m rankings. Ann. Math. Stat. **11**(1), 86–92 (1940)
6. Gomes, H.M., Barddal, J.P., Ferreira, L.E.B., Bifet, A.: Adaptive random forests for data stream regression. In: ESANN (2018)
7. Gomes, H.M., Montiel, J., Mastelini, S.M., Pfahringer, B., Bifet, A.: On ensemble techniques for data stream regression. In: 2020 IJCNN. IEEE (2020)
8. Hadjicharalambous, M., Polycarpou, M.M., Panayiotou, C.G.: Neural network-based construction of online prediction intervals. Neural Comput. Appl. **32**(11), 6715–6733 (2020)
9. Hahn, G.J., Factors for calculating two-sided prediction intervals for samples from a normal distribution. J. Am. Stat. Assoc. (1969)
10. Hahn, G.J., Nelson, W.: A survey of prediction intervals and their applications. J. Qual. Technol. **5**(4), 178–188 (1973)
11. Ikonomovska, E., Gama, J., Džeroski, S.: Learning model trees from evolving data streams. Data Min. Knowl. Disc. **23**, 128–168 (2011)
12. Khosravi, A., Nahavandi, S., Creighton, D., Atiya, A.F.: Lower upper bound estimation method for construction of neural network-based prediction intervals. IEEE Trans. Neural Netw. **22**(3), 337–346 (2010)
13. Liu, Y., Zhao, J., Wang, W., Pedrycz, W.: Prediction intervals for granular data streams based on evolving type-2 fuzzy granular neural network dynamic ensemble. IEEE Trans. Fuzzy Syst. **29**(4), 874–888 (2020)
14. Nix, D.A., Weigend, A.S.: Estimating the mean and variance of the target probability distribution. In: ICNN'94. IEEE (1994)
15. Shafer, G., Vovk, V.: A tutorial on conformal prediction. J. Mach. Learn. Res. **9**(3) (2008)
16. Shrestha, D.L., Solomatine, D.P.: Machine learning approaches for estimation of prediction interval for the model output. Neural Netw. **19**(2), 225–235 (2006)
17. Sun, Y., Pfahringer, B., Gomes, H.M., Bifet, A.: SOKNL: a novel way of integrating k-nearest neighbours with adaptive random forest regression for data streams. Data Min. Knowl. Discov. (2022)
18. Waugh, S.G.: Extending and benchmarking cascade-correlation: extensions to the cascade-correlation architecture and benchmarking of feed-forward supervised artificial neural networks. Ph.D. thesis, University of Tasmania (1995)
19. Xu, C., Xie, Y.: Conformal prediction interval for dynamic time-series. In: International Conference on Machine Learning, pp. 11559–11569. PMLR (2021)
20. Zhao, J., Wang, W., Sheng, C., Zhao, J., Wang, W., Sheng, C.: Industrial prediction intervals with data uncertainty. In: Data-Driven Prediction for Industrial Processes and Their Applications, pp. 159–222 (2018)
21. Zhao, X., Barber, S., Taylor, C.C., Milan, Z.: Interval forecasts based on regression trees for streaming data. Adv. Data Anal. Classif. **15**, 5–36 (2021)

Probabilistic Guarantees of Stochastic Recursive Gradient in Non-convex Finite Sum Problems

Yanjie Zhong, Jiaqi Li(✉), and Soumendra Lahiri

Department of Statistics and Data Science, Washington University in St. Louis, St. Louis 63130, MO, USA
lijiaqi@wustl.edu

Abstract. This paper develops a new dimension-free Azuma-Hoeffding type bound on summation norm of a martingale difference sequence with random individual bounds. With this novel result, we provide high-probability bounds for the gradient norm estimator in the proposed algorithm Prob-SARAH, which is a modified version of the StochAstic Recursive grAdient algoritHm (SARAH), a state-of-art variance reduced algorithm that achieves optimal computational complexity in expectation for the finite sum problem. The in-probability complexity by Prob-SARAH matches the best in-expectation result up to logarithmic factors. Empirical experiments demonstrate the superior probabilistic performance of Prob-SARAH on real datasets compared to other popular algorithms.

Keywords: Machine learning · Variance-reduced method · Stochastic gradient descent · Non-convex optimization

1 Introduction

We consider the popular non-convex finite sum optimization problem in this work, that is, estimating $\mathbf{x}^* \in \mathcal{D} \subseteq \mathbb{R}^d$ minimizing the following loss function

$$f(\mathbf{x}) = \frac{1}{n}\sum_{i=1}^{n} f_i(\mathbf{x}),\ \mathbf{x} \in \mathcal{D}, \tag{1}$$

where $f_i : \mathbb{R}^d \mapsto \mathbb{R}$ is a potentially non-convex function on some compact set $\mathcal{D}$. Such non-convex problems lie at the heart of many applications of statistical learning [13] and machine learning [7].

Unlike convex optimization problems, in general, non-convex problems are intractable and the best we can expect is to find a stationary point. Given a target error ε, since $\nabla f(\mathbf{x}^*) = 0$, we aim to find an estimator $\hat{\mathbf{x}}$ such that roughly

Supplementary Information The online version contains supplementary material available at https://doi.org/10.1007/978-981-97-2259-4_11.

D.-N. Yang et al. (Eds.): PAKDD 2024, LNAI 14647, pp. 142–154, 2024.
https://doi.org/10.1007/978-981-97-2259-4_11

$\|\nabla f(\hat{\mathbf{x}})\| \leq \varepsilon$, where $\nabla f(\cdot)$ denotes the gradient vector the loss function f and $\|\cdot\|$ is the operator norm. With a non-deterministic algorithm, the output $\hat{\mathbf{x}}$ is always stochastic, and the most frequently considered measure of error bound is in expectation, i.e.,

$$\mathbb{E}\|\nabla f(\hat{\mathbf{x}})\|^2 \leq \varepsilon^2. \tag{2}$$

There has been a substantial amount of work providing upper bounds on computational complexity needed to achieve the in-expectation bound. However, in practice, we only run a stochastic algorithm for once and an in-expectation bound cannot provide a convincing bound in this situation. Instead, a high-probability bound is more appropriate by nature. Given a pair of target errors (ε, δ), we want to obtain an estimator $\hat{\mathbf{x}}$ such that with probability at least $1-\delta$, $\|\nabla f(\hat{\mathbf{x}})\| \leq \varepsilon$, that is

$$\mathbb{P}\big(\|\nabla f(\hat{\mathbf{x}})\| \leq \varepsilon\big) \geq 1 - \delta. \tag{3}$$

Though the Markov inequality might help, in general, an in-expectation bound cannot be simply converted to an in-probability bound with a desirable dependency on δ. It would be important to prove upper bounds on high-probability complexity, which ideally should be polylogorithmic in δ and with polynomial terms comparable to the in-expectation complexity bound.

Gradient-based methods are favored by practitioners due to simplicity and efficiency and have been widely studied by researchers in the non-convex setting [1,5,6,26,31,33]. Among numerous gradient-based methods, the StochAstic Recursive grAdient algoritHm (SARAH) [27,28,33] is the one with the best first-order guarantee as given an in-expectation error target, in both of convex and non-convex finite sum problems. It is worth noticing that [23] attempted to show that a modified version of SARAH is able to approximate the second-order stationary point with a high probability. However, we believe that their application of the martingale Azuma-Hoeffding inequality is unjustifiable because the bounds are potentially random and uncontrollable. In this paper, we shall provide a correct dimension-free martingale Azuma-Hoeffding inequality with rigorous proofs and leverage it to show in-probability properties for SARAH-based algorithms in the non-convex setting.

1.1 Related Works

- **High-Probability Bounds:** While most works in the literature of optimization provide in-expectation bounds, there is only a small fraction of works discussing bounds in the high probability sense. [17] provide a high-probability bound on the excess risk given a bound on the regret. [8,9,12] derive some high-probability bounds for SGD in convex online optimization problems. [22,34] prove high-probability bounds for several adaptive methods, including AMSGrad, RMSProp and Delayed AdaGrad with momentum. All these works rely on (generalized) Freedman's inequality or the concentration inequality given in Lemma 6 in [15]. Different from them, our high-probability results are built on a novel Azuma-Hoeffding type inequality proved in this work and

Corollary 8 from [15]. In addition, we notice that [23] provide some probabilistic bounds on a SARAH-based algorithm. However, we believe their use of the plain martingale Azuma-Hoeffding inequality is not justifiable. [5] show in-probability upper bound for SPIDER. Nevertheless, SPIDER's practical performance is inferior due to its accuracy-dependent small step size [32,33].

- **Variance-Reduced Methods in Non-Convex Finite Sum Problems:** Since the invention of the variance-reduction technique in [4,16,18], there has been a large amount of work incorporating this efficient technique to methods targeting the non-convex finite-sum problem. Subsequent methods, including SVRG [1,25,31], SARAH [27,28], SCSG [11,19,21], SNVRG [34], SPIDER [5], SpiderBoost [33] and PAGE [24], have greatly reduced computational complexity in non-convex problems.

1.2 Our Contributions

- **Dimension-Free Martingale Azuma-Hoeffding inequality:** To facilitate our probabilistic analysis, we provide a novel Azuma-Hoeffding type bound on the summation norm of a martingale difference sequence. The novelty is two-fold. Firstly, same as the plain martingale Azuma-Hoeffding inequality, it provides a dimension-free bound. In a recent paper, a sub-Gaussian type bound has been developed by [15]. However, their results are not dimension-free. Our technique in the proof is built on a classic paper by [29] and is completely different from the random matrix technique used in [15]. Secondly, our concentration inequality allows random bounds on each element of the martingale difference sequence, which is much tighter than a large deterministic bound. It should be highlighted that our novel concentration result perfectly suits the nature of SARAH-style methods where the increment can be characterized as a martingale difference sequence and it can be further used to analyze other algorithms beyond the current paper.
- **In-probability error bounds of stochastic recursive gradient:** We design a SARAH-based algorithm, named Prob-SARAH, adapted to the high-probability target and provably show its good in-probability properties. Under appropriate parameter setting, the first order complexity needed to achieve the in-probability target is $\tilde{\mathcal{O}}\left(\frac{1}{\varepsilon^3} \wedge \frac{\sqrt{n}}{\varepsilon^2}\right)$, which matches the best known in-expectation upper bound up to some logarithmic factors [11,33,34]. We would like to point out that the parameter setting used to achieve such complexity is semi-adaptive to ε. That is, only the final stopping rule relies on ε while other key parameters are independent of ε, including step size, mini-batch sizes, and lengths of loops.
- **Probabilistic analysis of SARAH for non-convex finite sum:** Existing literature on the bounds of SARAH is mostly focusing on the strongly convex or general convex settings. We extend the case to the non-convex scenarios, which can be considered as a complimentary study to the stochastic recursive gradient in probability.

1.3 Notation

For a sequence of sets $\mathcal{A}_1, \mathcal{A}_2, \ldots$, we denote the smallest sigma algebra containing $\mathcal{A}_i$, $i \geq 1$, by $\sigma\big(\bigcup_{i=1}^{\infty} \mathcal{A}_i\big)$. By abuse of notation, for a random variable $\mathbf{X}$, we denote the sigma algebra generated by $\mathbf{X}$ by $\sigma(\mathbf{X})$. We define constant $C_e = \sum_{i=0}^{\infty} i^{-2}$. For two scalars $a, b \in \mathbb{R}$, we denote $a \wedge b = \min\{a, b\}$ and $a \vee b = \max\{a, b\}$. When we say a quantity T is $\mathcal{O}_{\theta_1,\theta_2}(\theta_3)$ for some $\theta_1, \theta_2, \theta_3 \in \mathbb{R}$, there exists a $g \in \mathbb{R}$ polylogarithmic in θ_1 and θ_2 such that $T \leq g \cdot \theta_3$, and similarly $\tilde{\mathcal{O}}_{(\cdot)}(\cdot)$ is defined the same but up to a logarithm factor.

2 Prob-SARAH Algorithm

The algorithm Prob-SARAH proposed in our work is a modified version of SpiderBoost [33] and SARAH [27,28]. Since the key update structure is originated from [27], we call our modified algorithm Prob-SARAH. In fact, it can also be viewed as a generalization of the SPIDER algorithm introduced in [5].

We present the Prob-SARAH in Algorithm 3, and here, we provide some explanation of the key steps. Following other SARAH-based algorithms, we adopt a similar gradient approximation design with nested loops, specifically with a checkpoint gradient estimator $\nu_0^{(j)}$ using a large mini-batch size B_j in Line 4 and a recursive gradient estimator $\nu_k^{(j)}$ updated in Line 9. When the mini-batch size B_j is large, we can regard the checkpoint gradient estimator $\nu_0^{(j)}$ as a solid approximation to the true gradient at $\tilde{\mathbf{x}}_{j-1}$. With this checkpoint, we can update the gradient estimator $\nu_k^{(j)}$ with a small mini-batch size b_j while maintaining a desirable estimation accuracy.

To emphasize, our stopping rules in Line 11 of Algorithm 3 is newly proposed, which ensures a critical enhancement of the performance compared to previous literature. In particular, with this new design, we can control the gradient norm of the output with high probability. For a more intuitive understanding of these stopping rules, we will see in our proof sketch section that the gradient norm of iterates in the j-th outer iteration, $\|\nabla f\|$, can be bounded by a linear combination of $\{\boldsymbol{\nu}_k^{(j)}\}_{k=1}^{K_j}$ with a small remainder. The first stopping rule, therefore, strives to control the magnitude of the linear combination of $\{\boldsymbol{\nu}_k^{(j)}\}_{k=1}^{K_j}$, while the second stopping rule is specifically designed to control the size of remainder terms. For this purpose, ε_j should be set as a credible controller of the remainder term, with an example given in Theorems 1. In this way, with small preset constants $\tilde{\varepsilon}$ and ε, we guarantee that the output has a desirably small gradient norm, dependent on $\tilde{\varepsilon}$ and ε, when the designed stopping rules are activated. Indeed, Proposition 1 in Appendix B offers a guarantee that the stopping rule will be definitively satisfied at some point. More refined quantitative results regarding the number of steps required for stopping will follow in Theorems 1 and Appendix D.3.

3 Theoretical Results

This section is devoted to the main theoretical result of our proposed algorithm Prob-SARAH. We provide the stop guarantee of the algorithm along with the upper bound of the steps. The high-probability error bound of the estimated gradient is also established. The discussion of the dependence of our algorithm on the parameters is available after we introduce our main theorems.

Algorithm 1. Probabilistic Stochastic Recursive Gradient (Prob-SARAH)

1: **Input:** sample size n, constraint area $\mathcal{D}$, initial point $\tilde{\mathbf{x}}_0 \in \mathcal{D}$, large batch size $\{B_j\}_{j\geq 1}$, mini batch size $\{b_j\}_{j\geq 1}$, inner loop length $\{K_j\}_{j\geq 1}$, auxiliary error estimator $\{\varepsilon_j\}_{j\geq 1}$, errors $\tilde{\varepsilon}^2, \varepsilon^2$
2: **for** $j = 1, 2, \ldots$ **do**
3: Uniformly sample a batch $\mathcal{I}_j \subseteq \{1, \ldots, n\}$ without replacement, $|\mathcal{I}_j| = B_j$;
4: $\boldsymbol{\nu}_0^{(j)} \leftarrow \frac{1}{B_j}\sum_{i\in\mathcal{I}_j} \nabla f_i(\tilde{\mathbf{x}}_{j-1})$;
5: $\mathbf{x}_0^{(j)} \leftarrow \tilde{\mathbf{x}}_{j-1}$;
6: **for** $k = 1, 2, \ldots, K_j$ **do**
7: $\mathbf{x}_k^{(j)} \leftarrow \mathrm{Proj}(\mathbf{x}_{k-1}^{(j)} - \eta_j \boldsymbol{\nu}_{k-1}^{(j)}, \mathcal{D})$, project the update back to $\mathcal{D}$;
8: Uniformly sample a mini-batch $\mathcal{I}_k^{(j)} \subseteq \{1, \ldots, n\}$ with replacement and $|\mathcal{I}_k^{(j)}| = b_j$;
9: $\boldsymbol{\nu}_k^{(j)} \leftarrow \frac{1}{b_j}\sum_{i\in\mathcal{I}_k^{(j)}} \nabla f_i(\mathbf{x}_k^{(j)}) - \frac{1}{b_j}\sum_{i\in\mathcal{I}_k^{(j)}} \nabla f_i(\mathbf{x}_{k-1}^{(j)}) + \boldsymbol{\nu}_{k-1}^{(j)}$;
10: **end for**
11: **if** $\frac{1}{K_j}\sum_{k=0}^{K_j-1} \|\boldsymbol{\nu}_k^{(j)}\|^2 \leq \tilde{\varepsilon}^2$ and $\varepsilon_j \leq \frac{1}{2}\varepsilon^2$ **then**
12: $\hat{k} \leftarrow \arg\min_{0\leq k\leq K_j-1} \|\boldsymbol{\nu}_k^{(j)}\|^2$;
13: **Return** $\hat{\mathbf{x}} \leftarrow \mathbf{x}_{\hat{k}}^{(j)}$;
14: **end if**
15: $\tilde{\mathbf{x}}_j \leftarrow \mathbf{x}_{K_j}^{(j)}$;
16: **end for**

3.1 Technical Assumptions

We shall introduce some necessary regularized assumptions. Most assumptions are commonly used in the optimization literature. We have further clarifications in Appendix A.

Assumption 1 (Existence of achievable minimum). *Assume that for each $i = 1, 2, \ldots, n$, f_i has continuous gradient on $\mathcal{D}$ and $\mathcal{D}$ is a compact subset of $\mathbb{R}^d$. Then, there exists a constant $\alpha_M < \infty$ such that*

$$\max_{1\leq i\leq n} \sup_{\mathbf{x}\in\mathcal{D}} \|\nabla f_i(\mathbf{x})\| \leq \alpha_M. \tag{4}$$

Also, assume that there exists an interior point $\mathbf{x}^*$ of the set $\mathcal{D}$ such that

$$f(\mathbf{x}^*) = \inf_{\mathbf{x}\in\mathcal{D}} f(\mathbf{x}).$$

Assumption 2 (***L*-smoothness**). *For each $i = 1, 2, \ldots, n$, $f_i : \mathcal{D} \to \mathbb{R}$ is L-smooth for some constant $L > 0$, i.e.,*

$$\|\nabla f_i(\mathbf{x}) - \nabla f_i(\mathbf{x}')\| \leq L\|\mathbf{x} - \mathbf{x}'\|, \ \forall \ \mathbf{x}, \mathbf{x}' \in \mathcal{D}.$$

Assumption 3 (***L*-smoothness extension**). *There exists a L-smooth function $\tilde{f} : \mathcal{D} \to \mathbb{R}$ such that*

$$\tilde{f}(\mathbf{x}) = f(\mathbf{x}), \ \forall \ \mathbf{x} \in \mathcal{D}, \quad \text{and} \quad \tilde{f}(\mathrm{Proj}(\mathbf{x}, \mathcal{D})) \leq \tilde{f}(\mathbf{x}), \ \forall \ \mathbf{x} \in \mathbb{R}^d,$$

where $\mathrm{Proj}(\mathbf{x}, \mathcal{D})$ *is the Euclidean projection of* $\mathbf{x}$ *on some compact set* $\mathcal{D}$.

Assumption 4. *Assume that the following conditions hold.*

1. *$\varepsilon \leq \frac{1}{e}$ and $\alpha_M^2 \geq \frac{1}{10240}$, where ϵ is the target error bound in Eq. (3) and α_M is defined in Eq. (4).*
2. *The diameter of $\mathcal{D}$ is at least 1, i.e. $d_1 \triangleq \max\{\|\mathbf{x} - \mathbf{x}'\| : \mathbf{x}, \mathbf{x}' \in \mathcal{D}\} \geq 1$.*

Assumption 1 also indicates that there exists a positive number Δ_f such that $\sup_{\mathbf{x} \in \mathcal{D}} \left[f(\mathbf{x}) - f(\mathbf{x}^*)\right] \leq \Delta_f$. Assumptions 1–3 are commonly used in the optimization literature, and Assumption 4 can be easily satisfied in practical use as long as the initial points are not too far from the optimum. See more comments on assumptions in Appendix A.

3.2 Main Results on Complexity

According to the definition given in [20], an algorithm is called ε-independent if it can guarantee convergence at all target accuracies ε in expectation without explicitly using ε in the algorithm. This is a very favorable property because it means that we no longer need to set the target error beforehand. Here, we introduce a similar property regarding the dependency on ε.

Definition 1 (**ε-semi-independence**). *An algorithm is ε-semi-independent, given δ, if it can guarantee convergence at all target accuracies ε with probability at least δ and the knowledge of ε is only needed in the post-processing. That is, the algorithm can iterate without knowing ε and we can select an appropriate iterate out afterwards.*

The newly introduced property can be perceived as the probabilistic equivalent of ε-independence. As stated in the succeeding theorem, under the given conditions, Prob-SARAH can achieve ε-semi-independence, given δ.

Theorem 1. *Suppose that Assumptions 1, 2, 3 and 4 are valid. Given a pair of errors (ε, δ), in Algorithm 3 (Prob-SARAH), set hyperparameters*

$$\eta_j = \frac{1}{4L}, \quad K_j = \sqrt{B_j} = \sqrt{j^2 \wedge n}, \quad b_j = l_j K_j, \quad \varepsilon_j = 8L^2 \tau_j + 2q_j, \quad \tilde{\varepsilon}^2 = \frac{1}{5}\varepsilon^2, \quad (5)$$

for $j \geq 1$*, where*

$$\tau_j = \frac{1}{j^3}, \delta'_j = \frac{\delta}{4C_e j^4}, \quad l_j = 18\Big(\log(\tfrac{2}{\delta'_j}) + \log\log(\tfrac{2d_1}{\tau_j})\Big), \quad q_j = \frac{128\alpha_M^2}{B_j}\log(\tfrac{3}{\delta'_j})\mathbf{1}\{B_j < n\}.$$

Then,

$$Comp(\varepsilon, \delta) = \tilde{O}_{L,\Delta_f,\alpha_M}\left(\frac{1}{\varepsilon^3} \wedge \frac{\sqrt{n}}{\varepsilon^2}\right),$$

where $Comp(\varepsilon, \delta)$ *represents the number of computations needed to get an output* $\hat{\mathbf{x}}$ *satisfying* $\|\nabla f(\hat{\mathbf{x}})\|^2 \leq \varepsilon^2$ *with probability at least* $1 - \delta$*.*

More detailed results can be found in Appendix C. In Appendix C, we also introduce another hyper-parameter setting that can lead to a complexity with better dependency on α_M^2, which could be implicitly affected by the choice of constraint region $\mathcal{D}$.

3.3 Proof Sketch

In this part, we explain the idea of the proof of Theorem 1. Same proofing strategy can be applied to other hyper-parameter settings. First, we bound the difference between $\boldsymbol{\nu}_k^{(j)}$ and $\nabla f(\mathbf{x}_k^{(j)})$ by a linear combination of $\{\|\boldsymbol{\nu}_m^{(j)}\|\}_{m=0}^{k-1}$ and small remainders, with which we can have a good control on $\|\nabla f(\mathbf{x}_k^{(j)})\|$ when the stopping rules are met. Second, we bound the number of steps we need to meet the stopping rules. Combining these 2 key components, we can smoothly get the final conclusions.

Let us firstly introduce a novel Azuma-Hoeffding type inequality, which is key to our analysis.

Theorem 2 (Martingale Azuma-Hoeffding Inequality with Random Bounds). *Suppose* $\mathbf{z}_1, \ldots, \mathbf{z}_K \in \mathbb{R}^d$ *is a martingale difference sequence adapted to* $\mathcal{F}_0, \ldots, \mathcal{F}_K$*. Suppose* $\{r_k\}_{k=1}^K$ *is a sequence of random variables such that* $\|\mathbf{z}_k\| \leq r_k$ *and* r_k *is measurable with respect to* $\mathcal{F}_k$*,* $k = 1, \ldots, K$*. Then, for any fixed* $\delta > 0$*, and* $B > b > 0$*, with probability at least* $1 - \delta$*, for* $1 \leq t \leq K$*, either*

$$\exists 1 \leq t \leq K, \ \sum_{k=1}^{t} r_k^2 \geq B \ \text{or} \ \Big\|\sum_{k=1}^{t} \mathbf{z}_k\Big\|^2 \leq 9\max\Big\{\sum_{k=1}^{t} r_k^2, b\Big\}\Big(\log(\tfrac{2}{\delta}) + \log\log(\tfrac{B}{b})\Big).$$

Remark 1. It is noteworthy that this probabilistic bound on large-deviation is dimension-free, which is a nontrivial extension of Theorem 3.5 in [30]. If $r_1, r_2, \ldots, r_K$ are not random, we can let $B = \sum_{k=1}^K r_k^2 + \zeta_1$ and $b = \zeta_2 B$ with $\zeta_1 > 0, 0 < \zeta_2 < 1$. Since ζ_1 can be arbitrarily close to 0 and ζ_2 can be arbitrarily close to 1, we can recover Theorem 3.5 in [30]. Compared with Corollary 8 in [15], which can be viewed as a sub-Gaussian counterpart of our result, a key feature of our Theorem 2 is its dimension-independence. We are also working towards improving the bound in Corollary 8 from [15] to a dimension-free one.

The success of Algorithm 3 is largely because $\nabla f(\mathbf{x}_k^{(j)})$ is well-approximated by $\boldsymbol{\nu}_k^{(j)}$, and meanwhile $\boldsymbol{\nu}_k^{(j)}$ can be easily updated. We can observe that $\boldsymbol{\nu}_k^{(j)} - \nabla f(\mathbf{x}_k^{(j)})$ is actually sum of a sequence of martingale difference as

$$\boldsymbol{\nu}_k^{(j)} - \nabla f(\mathbf{x}_k^{(j)}) = \left[\frac{1}{b_j}\sum_{i\in\mathcal{I}_k^{(j)}} \nabla f_i(\mathbf{x}_k^{(j)}) - \frac{1}{b_j}\sum_{i\in\mathcal{I}_k^{(j)}} \nabla f_i(\mathbf{x}_{k-1}^{(j)}) + \nabla f(\mathbf{x}_{k-1}^{(j)}) - \nabla f(\mathbf{x}_k^{(j)})\right]$$

$$+ \left[\boldsymbol{\nu}_{k-1}^{(j)} - \nabla f(\mathbf{x}_{k-1}^{(j)})\right] = \sum_{m=1}^{k}\left[\frac{1}{b_j}\sum_{i\in\mathcal{I}_m^{(j)}} \nabla f_i(\mathbf{x}_m^{(j)}) - \frac{1}{b_j}\sum_{i\in\mathcal{I}_m^{(j)}} \nabla f_i(\mathbf{x}_{m-1}^{(j)})\right.$$

$$\left. + \nabla f(\mathbf{x}_{m-1}^{(j)}) - \nabla f(\mathbf{x}_m^{(j)})\right] + \left[\boldsymbol{\nu}_0^{(j)} - \nabla f(\mathbf{x}_0^{(j)})\right]. \tag{6}$$

To be more specific, let $\mathcal{F}_0 = \{\emptyset, \Omega\}$, and iteratively define $\mathcal{F}_{j,-1} = \mathcal{F}_{j-1}$, $\mathcal{F}_{j,0} = \sigma\big(\mathcal{F}_{j-1} \cup \sigma(\mathcal{I}_j)\big)$, $\mathcal{F}_{j,k} = \sigma\big(\mathcal{F}_{j,0} \cup \sigma(\mathcal{I}_k^{(j)})\big)$, $\mathcal{F}_j = \sigma\big(\bigcup_{k=1}^{\infty} \mathcal{F}_{j,k}\big)$, $j \geq 1, k \geq 1$. We also denote $\boldsymbol{\epsilon}_0^{(j)} \triangleq \boldsymbol{\nu}_0^{(j)} - \nabla f\big(\mathbf{x}_0^{(j)}\big)$, $\boldsymbol{\epsilon}_m^{(j)} \triangleq \frac{1}{b_j}\sum_{i\in\mathcal{I}_m^{(j)}} \nabla f_i(\mathbf{x}_m^{(j)}) - \nabla f(\mathbf{x}_m^{(j)}) + \nabla f(\mathbf{x}_{m-1}^{(j)}) - \frac{1}{b_j}\sum_{i\in\mathcal{I}_m^{(j)}} \nabla f_i(\mathbf{x}_{m-1}^{(j)})$, $m \geq 1$. Then, we can see that $\{\boldsymbol{\epsilon}_m^{(j)}\}_{m=0}^{k}$ is a martingale difference sequence adapted to $\{\mathcal{F}_{j,m}\}_{m=-1}^{k}$. With the help of our new Martingale Azuma-Hoeffding inequality, we can control the difference between $\boldsymbol{\nu}_k^{(j)}$ and $\nabla f\big(\mathbf{x}_k^{(j)}\big)$ by a linear combination of $\{\|\boldsymbol{\nu}_m^{(j)}\|\}_{m=0}^{k-1}$ and small remainders, with details given in Appendix D.1. Then, given the stopping rules in line 11 and selection method specified in line 12 of Algorithm 3, it would be not hard for us to obtain $\|\nabla f(\hat{\mathbf{x}})\|^2 \leq \varepsilon^2$ with a high probability. More details can be found in Appendix D.2.

Another key question needed to be resolved is, when the algorithm can stop? The following analysis can build some intuitions for us. Given a $T \in \mathbb{Z}_+$, with the bound given in Proposition 8 in Appendix F, with a high probability,

$$-\Delta_f \leq f\left(\tilde{\mathbf{x}}_{2T}\right) - f\left(\tilde{\mathbf{x}}_T\right) \leq A_T - \frac{1}{16L}\sum_{j=T+1}^{2T}\sum_{k=0}^{K_j-1}\left\|\boldsymbol{\nu}_k^{(j)}\right\|^2, \tag{7}$$

where A_T is upper bounded by a value polylogorithmic in T. As for the second summation, if $\varepsilon_j \leq \frac{1}{2}\varepsilon^2$ for $j = T, T+1, \ldots, 2T$ (which is obviously true when T is moderately large) and our algorithm doesn't stop in $2T$ outer iterations,

$$\frac{1}{16L}\sum_{j=T+1}^{2T}\sum_{k=0}^{K_j-1}\left\|\boldsymbol{\nu}_k^{(j)}\right\|^2 \geq \frac{\tilde{\varepsilon}^2}{16L}\sum_{j=T+1}^{2T} K_j \geq \frac{\tilde{\varepsilon}^2}{16L}\sum_{j=T+1}^{2T} (T \wedge \sqrt{n}) = \frac{\tilde{\varepsilon}^2}{16L} T^2 \wedge (\sqrt{n}T),$$

which grows at least linear in T. Consequently, when T is sufficiently large, the RHS of Eq. (7) can be smaller than $-\Delta_f$, which leads to a contradiction. Roughly, we can see that the stopping time T cannot exceed the order of $\tilde{\mathcal{O}}\big(\frac{1}{\varepsilon} \vee \frac{1}{\sqrt{n}\varepsilon^2}\big)$. More details can be found in Appendix D.3.

4 Numerical Experiments

In order to validate our theoretical results and show good probabilistic property for the newly-introduced Prob-SARAH, we conduct some numerical experiments where the objectives are possibly non-convex.

4.1 Logistic Regression with Non-convex Regularization

In this part, we consider to add a non-convex regularization term to the commonly-used logistic regression. Specifically, given a sequence of observations $(\mathbf{w}_i, y_i) \in \mathbb{R}^d \times \{-1, 1\}$, $i = 1, 2, \ldots, n$ and a regularized parameter $\lambda > 0$, the objective is

$$f(\mathbf{x}) = \frac{1}{n} \sum_{i=1}^{n} \log \left(1 + e^{-y_i \mathbf{w}_i^T \mathbf{x}}\right) + \frac{\lambda}{2} \sum_{j=1}^{d} \frac{x_j^2}{1 + x_j^2}.$$

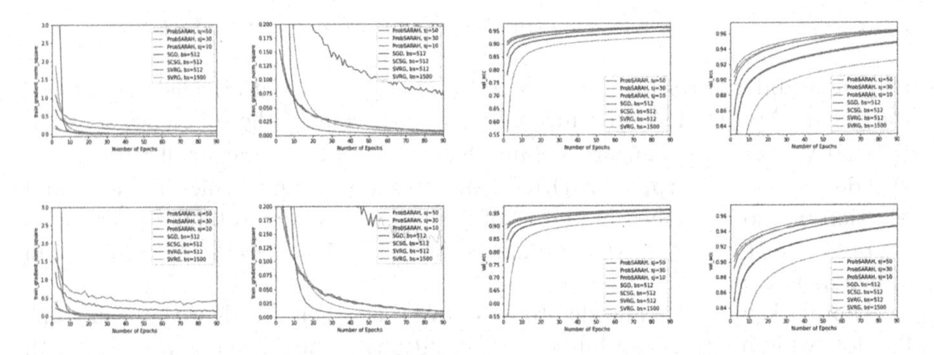

Fig. 1. Comparison of convergence with respect to $(1-\delta)$-quantile of square of gradient norm $\left(\|\nabla f\|^2\right)$ and δ-quantile of validation accuracy on the **MNIST** dataset for $\delta = 0.1$ and $\delta = 0.01$. The second (fourth) column presents zoom-in figures of those in the first (third) column. Top: $\delta = 0.1$. Bottom: $\delta = 0.01$. 'bs' stands for batch size. 'sj = x' means that the smallest batch size $\approx x \log x$.

Such an objective has also been considered in other works like [11] and [14]. Same as other works, we set the regularized parameter $\lambda = 0.1$ across all experiments. We compare the newly-introduced Prob-SARAH against three popular methods including SGD [6], SVRG [31] and SCSG [21]. Based on results given in Theorem 1, we let the length of the inner loop $K_j \sim j \wedge \sqrt{n}$, the inner loop batch size $b_j \sim \log j\,(j \wedge \sqrt{n})$, the outer loop batch size $B_j \sim j^2 \wedge n$. For fair comparison, we determine the batch size (inner loop batch size) for SGD (SCSG and SVRG) based on the sample size n and the number of epochs needed to have sufficient decrease in gradient norm. For example, for the w7a dataset, the sample size is 24692 and we run 60 epochs in total. In the 20th epoch, the inner loop batch size of Prob-SARAH is approximately $67 \log 67 \approx 281$. Thus, we set batch size 256 for SGD, SCSG and SVRG so that they can be roughly matched. In addition, based on the theoretical results from [31], we also consider a large inner loop batch size comparable to $n^{2/3}$ for SVRG. In addition, we set step size $\eta = 0.01$ for all algorithms across all experiments for simplicity.

Results are displayed in Fig. 2, from which we can see that Prob-SARAH has superior probabilistic guarantee in controlling the gradient norm in all experiments. It is significantly better than SCSG and SVRG under our current setting. Prob-SARAH can achieve a lower gradient norm than SGD at the early stage while SGD has a slight advantage when the number of epochs is large.

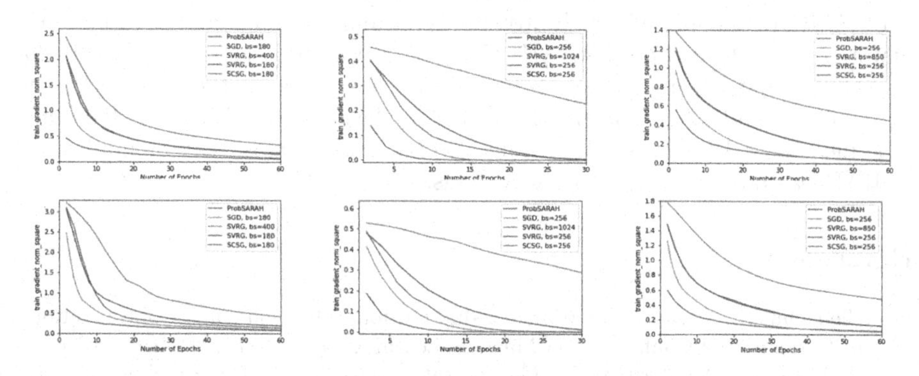

Fig. 2. Comparison of convergence with respect to $(1-\delta)$-quantile of square of gradient norm $\left(\|\nabla f\|^2\right)$ over 3 datasets for $\delta = 0.1$ and $\delta = 0.01$. Top: $\delta = 0.1$. Bottom: $\delta = 0.01$. Datasets: **mushrooms**, **ijcnn1**, **w7a** (from left to right). 'bs' stands for batch size.

4.2 Two-Layer Neural Network

We also evaluate the performance of Prob-SARAH, SGD, SVRG and SCSG on the MNIST dataset with a simple 2-layer neural network. The two hidden layers respectively have 128 and 64 neurons. We include a GELU activation layer following each hidden layer. We use the negative log likelihood as our loss function. Under this setting, the objective is possibly non-convex and smooth on any given compact set. The step size is fixed to be 0.01 for all algorithms. For Prob-SARAH, we still have the length of the inner loop $K_j \sim j \wedge \sqrt{n}$, the inner loop batch size $b_j \sim \log j\,(j \wedge \sqrt{n})$, the outer loop batch size $B_j \sim j^2 \wedge n$. But to reduce computational time, we let j start from 10, 30 and 50 respectively. Based on the same rule described in the previous subsection, we let the batch size (or inner loop batch size) for SGD, SVRG and SCSG be 512.

Results are given in Fig. 1. In terms of gradient norm, Prob-SARAH has the best performance among algorithms considered here when the number of epochs is relatively small. With increasing number of epochs, SVRG tends to be better in finding first-order stationary points. However, based on the 3rd and 4th columns in Fig. 1, SVRG apparently has an inferior performance on the validation set, which indicates that it could be trapped at local minima. In brief,

Prob-SARAH achieves the best tradeoff between finding a first-order stationary point and generalization.

We also consider another set of experiments by replacing the GELU activation function with ReLU, resulting in a non-smooth objective. The results are shown in Appendix G, which resemble those in Fig. 1 and the similar conclusions can be drawn.

5 Conclusion

In this paper, we propose a SARAH-based variance reduction algorithm called Prob-SARAH and provide high-probability bounds on gradient norm for estimator resulted from Prob-SARAH. Under appropriate assumptions, the high-probability first order complexity nearly match the one in the in-expectation sense. The main tool used in the theoretical analysis is a novel Azuma-Hoeffding type inequality. We believe that similar probabilistic analysis can be applied to SARAH-based algorithms in other settings.

References

1. Allen-Zhu, Z., Hazan, E.: Variance reduction for faster non-convex optimization. In: International Conference on Machine Learning, pp. 699–707. PMLR (2016)
2. Bardenet, R., Maillard, O.A.: Concentration inequalities for sampling without replacement. Bernoulli **21**(3), 1361–1385 (2015)
3. Boucheron, S., Lugosi, G., Massart, P.: Concentration Inequalities: A Nonasymptotic Theory of Independence. Oxford University Press (2013)
4. Defazio, A., Bach, F., Lacoste-Julien, S.: SAGA: a fast incremental gradient method with support for non-strongly convex composite objectives. In: Advances in Neural Information Processing Systems, pp. 1646–1654 (2014)
5. Fang, C., Li, C.J., Lin, Z., Zhang, T.: SPIDER: near-optimal non-convex optimization via stochastic path integrated differential estimator. In: Proceedings of the 32nd International Conference on Neural Information Processing Systems, pp. 687–697 (2018)
6. Ghadimi, S., Lan, G.: Stochastic first-and zeroth-order methods for nonconvex stochastic programming. SIAM J. Optim. **23**(4), 2341–2368 (2013)
7. Goodfellow, I., Bengio, Y., Courville, A.: Deep Learning. MIT Press (2016)
8. Harvey, N.J., Liaw, C., Plan, Y., Randhawa, S.: Tight analyses for non-smooth stochastic gradient descent. In: Conference on Learning Theory, pp. 1579–1613. PMLR (2019)
9. Harvey, N.J., Liaw, C., Randhawa, S.: Simple and optimal high-probability bounds for strongly-convex stochastic gradient descent. arXiv preprint arXiv:1909.00843 (2019)
10. Hoeffding, W.: Probability inequalities for sums of bounded random variables. J. Am. Stat. Assoc. **58**(301), 13–30 (1963)
11. Horváth, S., Lei, L., Richtárik, P., Jordan, M.I.: Adaptivity of stochastic gradient methods for nonconvex optimization. arXiv preprint arXiv:2002.05359 (2020)
12. Jain, P., Nagaraj, D., Netrapalli, P.: Making the last iterate of SGD information theoretically optimal. In: Conference on Learning Theory, pp. 1752–1755. PMLR (2019)

13. James, G., Witten, D., Hastie, T., Tibshirani, R.: An Introduction to Statistical Learning. STS, vol. 103. Springer, New York (2013). https://doi.org/10.1007/978-1-4614-7138-7
14. Ji, K., Wang, Z., Weng, B., Zhou, Y., Zhang, W., Liang, Y.: History-gradient aided batch size adaptation for variance reduced algorithms. In: International Conference on Machine Learning, pp. 4762–4772. PMLR (2020)
15. Jin, C., Netrapalli, P., Ge, R., Kakade, S.M., Jordan, M.I.: A short note on concentration inequalities for random vectors with Subgaussian norm. arXiv preprint arXiv:1902.03736 (2019)
16. Johnson, R., Zhang, T.: Accelerating stochastic gradient descent using predictive variance reduction. In: Advances in Neural Information Processing Systems, vol. 26, pp. 315–323 (2013)
17. Kakade, S.M., Tewari, A.: On the generalization ability of online strongly convex programming algorithms. In: Advances in Neural Information Processing Systems, pp. 801–808 (2009)
18. Le Roux, N., Schmidt, M., Bach, F.: A stochastic gradient method with an exponential convergence rate for finite training sets. In: Proceedings of the 25th International Conference on Neural Information Processing Systems, vol. 2, pp. 2663–2671 (2012)
19. Lei, L., Jordan, M.: Less than a single pass: stochastically controlled stochastic gradient. In: Artificial Intelligence and Statistics, pp. 148–156. PMLR (2017)
20. Lei, L., Jordan, M.I.: On the adaptivity of stochastic gradient-based optimization. SIAM J. Optim. **30**(2), 1473–1500 (2020)
21. Lei, L., Ju, C., Chen, J., Jordan, M.I.: Non-convex finite-sum optimization via SCSG methods. In: Proceedings of the 31st International Conference on Neural Information Processing Systems, pp. 2345–2355 (2017)
22. Li, X., Orabona, F.: A high probability analysis of adaptive SGD with momentum. arXiv preprint arXiv:2007.14294 (2020)
23. Li, Z.: SSRGD: simple stochastic recursive gradient descent for escaping saddle points. In: Advances in Neural Information Processing Systems, vol. 32, pp. 1523–1533 (2019)
24. Li, Z., Bao, H., Zhang, X., Richtárik, P.: PAGE: a simple and optimal probabilistic gradient estimator for nonconvex optimization. In: International Conference on Machine Learning, pp. 6286–6295. PMLR (2021)
25. Li, Z., Li, J.: A simple proximal stochastic gradient method for nonsmooth nonconvex optimization. In: Proceedings of the 32nd International Conference on Neural Information Processing Systems, pp. 5569–5579 (2018)
26. Nesterov, Y.: Introductory Lectures on Convex Optimization: A Basic Course, vol. 87. Springer, New York (2003). https://doi.org/10.1007/978-1-4419-8853-9
27. Nguyen, L.M., Liu, J., Scheinberg, K., Takáč, M.: SARAH: a novel method for machine learning problems using stochastic recursive gradient. In: International Conference on Machine Learning, pp. 2613–2621. PMLR (2017)
28. Nguyen, L.M., Liu, J., Scheinberg, K., Takáč, M.: Stochastic recursive gradient algorithm for nonconvex optimization. arXiv preprint arXiv:1705.07261 (2017)
29. Pinelis, I.: An approach to inequalities for the distributions of infinite-dimensional martingales. In: Dudley, R.M., Hahn, M.G., Kuelbs, J. (eds.) Probability in Banach Spaces, 8: Proceedings of the Eighth International Conference. Progress in Probability, vol. 30, pp. 128–134. Birkhäuser, Boston (1992). https://doi.org/10.1007/978-1-4612-0367-4_9
30. Pinelis, I.: Optimum bounds for the distributions of martingales in Banach spaces. Ann. Probab. **22**, 1679–1706 (1994)

31. Reddi, S.J., Hefny, A., Sra, S., Poczos, B., Smola, A.: Stochastic variance reduction for nonconvex optimization. In: International Conference on Machine Learning, pp. 314–323. PMLR (2016)
32. Tran-Dinh, Q., Pham, N.H., Phan, D.T., Nguyen, L.M.: Hybrid stochastic gradient descent algorithms for stochastic nonconvex optimization. arXiv preprint arXiv:1905.05920 (2019)
33. Wang, Z., Ji, K., Zhou, Y., Liang, Y., Tarokh, V.: SpiderBoost and momentum: faster variance reduction algorithms. In: Advances in Neural Information Processing Systems, vol. 32, pp. 2406–2416 (2019)
34. Zhou, D., Chen, J., Cao, Y., Tang, Y., Yang, Z., Gu, Q.: On the convergence of adaptive gradient methods for nonconvex optimization. arXiv preprint arXiv:1808.05671 (2018)

Rethinking Personalized Federated Learning with Clustering-Based Dynamic Graph Propagation

Jiaqi Wang[1], Yuzhong Chen[2], Yuhang Wu[2], Mahashweta Das[2], Hao Yang[2], and Fenglong Ma[1](✉)

[1] The Pennsylvania State University, State College, USA
{jqwang,fenglong}@psu.edu
[2] Visa Research, Foster City, USA
{yuzchen,yuhawu,mehdas,haoyang}@psu.edu

Abstract. Most existing personalized federated learning approaches are based on intricate designs, which often require complex implementation and tuning. In order to address this limitation, we propose a simple yet effective personalized federated learning framework. Specifically, during each communication round, we group clients into multiple clusters based on their model training status and data distribution on the server side. We then consider each cluster center as a node equipped with model parameters and construct a graph that connects these nodes using weighted edges. Additionally, we update the model parameters at each node by propagating information across the entire graph. Subsequently, we design a precise personalized model distribution strategy to allow clients to obtain the most suitable model from the server side. We conduct experiments on three image benchmark datasets and create synthetic structured datasets with three types of typologies. Experimental results demonstrate the effectiveness of the proposed FedCedar.

Keywords: Federated learning · Model personalization and distribution

1 Introduction

Federated learning (FL) [20] enables different data holders to cooperatively train machine learning models without sharing data. However, data heterogeneity poses a significant challenge in FL. To tackle this issue, personalized FL (PFL) has been proposed. Among the various research tracks in PFL, local model personalization through parameter decoupling [23] and clustering [5,30] have been explored. Parameter decoupling-based PFL has several drawbacks: (1) It demands intensive computational resources and is complex to implement. (2) It encounters the

J. Wang—This work was done when Jiaqi Wang interned at Visa Research.
The code can not be public according to the policy of Visa Research.

D.-N. Yang et al. (Eds.): PAKDD 2024, LNAI 14647, pp. 155–167, 2024.
https://doi.org/10.1007/978-981-97-2259-4_12

challenge of dividing the private or federated parameters [1]. In contrast, clustering-based approaches do not have such additional requirements or limitations. However, existing clustering-based approaches overlook the hidden relations between clusters and may require additional designs to handle the intricate mechanism [2], which complicates implementation and deployment on other frameworks. Consequently, a challenging yet practical question arises: *Is it possible to design a* ***simple*** *yet* ***effective*** *PFL framework that can address the data heterogeneity problem while considering the capture of hidden relations across clients?*

To address this question, we encounter several non-trivial challenges: **C1: Hidden relation capturing.** In real-world applications, there may exist a physical topology among clients. Utilizing this topology effectively would contribute to local model updating. However, clients are unable to share data or access global topology information. Accurately capturing the hidden relations among clients becomes a challenging task. **C2: Clustering-based knowledge sharing.** In FL frameworks, a random subset of clients participates in iterative model learning. Learning client-level hidden relations can be inefficient. Also, propagating under-trained models among all clients can lead to failure in FL model training. Clustering approaches allow similar clients to cooperate, which helps avoid such issues. However, most research focuses solely on gathering knowledge within clusters, disregarding information across multiple clusters. Extracting appropriate knowledge to facilitate model updates across all clusters becomes an urgent matter to address. **C3: Fitted model acquisition.** Distributing the most fitting models back to clients as their initialization in the next communication round is not trivial. Designing an accurate model distribution strategy is crucial to maintain and inherit the benefits of iterative training, maximizing the utilization of model updates and effectively supporting model training in subsequent rounds. Moreover, the proposed approach should be easy to implement without requiring specific or complicated tuning.

To address all the aforementioned challenges, we propose a simple yet effective personalized federated learning framework, denoted as FEDCEDAR. The framework is shown in Fig. 1. In our approach, we first perform local model training using the respective client's data and upload the model parameters to the server. On the server side, we cluster the collected models into different groups based on their parameters. To enhance our framework, we propose a method incorporating dynamic graph construction and weighted knowledge propagation. Each cluster center is treated as a node with associated model parameters, and a graph is constructed connecting these nodes with weighted edges. Subsequently, the model parameters stored at each node are updated across the entire graph by leveraging information from other nodes through the weighted edges. Throughout the federated learning iterations, the active clients may vary, causing the clusters and graphs to change dynamically in each communication round. Finally, we design a simple yet precise personalized model distribution strategy to maintain the benefits of iterative training and further enhance model personalization. Depending on if an active client was selected in the previous update iteration, we assign either the personalized cluster center models or the aggregated ones to the clients.

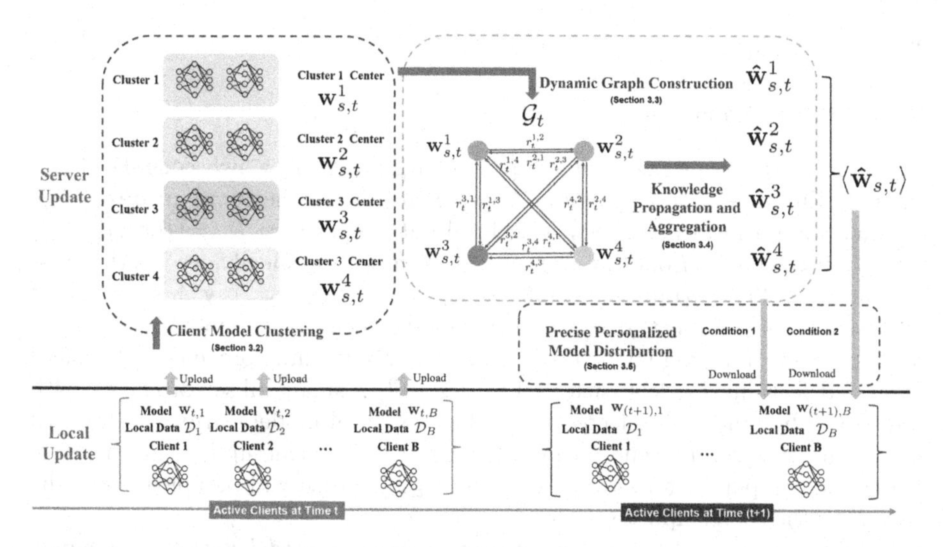

Fig. 1. Overview of the proposed FEDCEDAR ($K = 4$ as an example). Note that clients selected at time t may be different from those that are selected at time $t+1$. **Condition 1** means that the clients are selected at time t as well as $(t + 1)$, and **Condition 2** means that the clients are not selected at time t.

2 Related Work

FL is initially proposed in [20], which proposes a classical algorithm named FedAvg. Till this research, FL has recently been explored in multiple directions [10–14,16,18,19,22,24–26,28]. Considering our proposed work, we discuss related works of model personalization and graph in FL in the following. **Personalized Federated Learning:** Personalized FL cares more about each local model's performance, which is more sufficient and practical under the non-IID setting. One research track to solve this problem is named `FedAvg+`, which focuses on how to balance the global aggregation and local training [4,9,23]. Besides the research track that we discussed, some research works focus on parameter decoupling [23] and clustering [5,30] to conduct local model personalization in FL. However, all discussed personalization FL works ignore the hidden relations between the clients or the clusters. **Graph in Federated Learning:** Most research works, which get the graph involved in FL, focus on training GNN (Graph Neural Network) models with graph structure data distributedly [6,7,17,29,31]. In a recent work [3], authors treat each local client as a node to conduct GCN learning at the server. But the number of models maintained at the server is the same as the client number, which would take intensive computation resources. Besides that, the adjacency matrix is required to be known or learned in their proposed work, which affects its practicability.

3 Methodology

3.1 Model Overview

Figure 1 shows the overview of the proposed framework, which consists of two main modules, i.e., *local update* and *server update.* In the **local update** step, we first train a local model $\mathbf{w}_n$ with local data $\mathcal{D}_n$, where $\mathcal{D}_n = \{(x_i^n, y_i^n)\}_{i=1}^{H_n}$ and H_n is the number of data examples at client n. Specifically, we use the cross entropy (CE) loss to train $\mathbf{w}_n$ as follows: $\mathcal{L}_n = \text{CE}\left(f\left(\mathbf{x}^n; \mathbf{w}_n\right), \mathbf{y}^n\right)$, where $f(\cdot;\cdot)$ represents the neural network, $\mathbf{x}^n$ represents the client data representations, and $\mathbf{y}^n$ is the corresponding label vector. Following the training procedure of general federated learning models such as FedAvg [20], a small subset of clients with size B will be randomly selected to conduct the local model learning at the t-th communication round. The outputs from the local client update are a set of trained model parameters $\{\mathbf{w}_{t,1}, \mathbf{w}_{t,2}, \cdots, \mathbf{w}_{t,B}\}$, which will be uploaded to the server for the server update.

There are four critical components including *client model clustering, dynamic graph construction, knowledge propagation,* and *precise personalized model distribution.* After receiving $\{\mathbf{w}_{t,1}, \mathbf{w}_{t,2}, \cdots, \mathbf{w}_{t,B}\}$, FEDCEDAR will conduct the client model clustering using the K-means algorithm, i.e., dividing the B uploaded clients into K clusters, where K is the predefined number of clusters. The outputs of this step are a set of cluster centers $\mathcal{V}_t = \{\mathbf{w}_{s,t}^2, \mathbf{w}_{s,t}^1, \cdots, \mathbf{w}_{s,t}^K\}$, where $\mathbf{w}_{s,t}^k$ denotes the center of the k-th cluster at the t-th communication round. In the dynamic graph construction step, FEDCEDAR treats each cluster center as a node and constructs a dynamic weighted graph $\mathcal{G}_t = (\mathcal{V}_t, \mathcal{E}_t, \mathcal{R}_t)$, where $\mathcal{E}_t \subseteq \mathcal{V}_t \times \mathcal{V}_t$ is the set of edges between nodes, and $\mathcal{R}_t$ is the set of weights on edges. FEDCEDAR then executes the knowledge propagation step across the whole graph $\mathcal{G}_t$ to update node representations, i.e., model personalization learning. Finally, FEDCEDAR distributes the learned personalized models $\{\hat{\mathbf{w}}_{s,1}^1, \hat{\mathbf{w}}_{s,1}^2, \cdots, \hat{\mathbf{w}}_{s,t}^K\}$ back to the corresponding clients with the precise model distribution strategy. Next, we describe the details of the four components in the server update step.

3.2 Client Model Clustering

At the t-th communication round, the B selected clients will upload their models $\{\mathbf{w}_{t,1}, \cdots, \mathbf{w}_{t,B}\}$ to the server. These models will be grouped into K clusters using K-means [5,15] by optimizing the following loss function:

$$J_t = \sum_{k=1}^{K} \sum_{b=1}^{B} \delta_{t,b}^k (||\mathbf{w}_{s,t}^k - \mathbf{w}_{t,b}||^2), \tag{1}$$

where $||\mathbf{w}_{s,t}^k - \mathbf{w}_{t,b}||^2$ is the Euclidean distance between the k-th center $\mathbf{w}_{s,t}^k$ and the client model $\mathbf{w}_{t,b}$. $\delta_{t,b}^k$ is the indicator. If the b-th client belongs to the k-th cluster, then $\delta_{t,b}^k = 1$. Otherwise, $\delta_{t,b}^k = 0$. Note that for each communication round t, we will generate a new set of K cluster centers $\{\mathbf{w}_{s,t}^1, \cdots, \mathbf{w}_{s,t}^K\}$.

3.3 Dynamic Weighted Graph Construction

The clustering process helps local clients with similar data distributions or characteristics to aggregate together based on their model parameters. However, simply utilizing the results of the clustering algorithm to conduct the personalization in FL [5,30] has several limitations. On the one hand, the cluster centers $\{\mathbf{w}_{s,t}^1, \cdots, \mathbf{w}_{s,t}^K\}$ are obtained by averaging the client parameters within each cluster. In other words, the personalization information is only from the clients within each cluster by modeling the *inner-cluster* characteristics, which ignores that *between clusters*. On the other hand, the performance of existing clustering approaches is mainly determined by the quality of the cluster assignment. One low-quality cluster may lead to slow convergence and poor overall performance. When modeling the relations between clusters, high-quality information can flow to other clusters, which may neutralize the negative effect of low-quality clusters. Thus, it is essential to model the hidden relations between clusters.

Towards this end, we build a dynamic weighted graph to capture the hidden relations between the local models to enhance the mode updates further. However, it is highly non-trivial to design a novel mechanism satisfying the following conditions and requirements. First, in the FL setting, it would be better not to increase the network transmission load due to the constraints of communication cost. Second, in a typical FL framework, we randomly sample clients at each communication round, which means different batches of clients contribute to the updates. Then the center model $\mathbf{w}_{s,t}^k$ in each cluster carries different information with huge differences. How to leverage the iterative heterogeneous knowledge appropriately is a challenge. FedAvg-based aggregation approach [20] enables rough information sharing by obtaining one global model, which cannot handle the data or model heterogeneity. Thus, solving such challenges requires a more reliable and convincing aggregation strategy to maintain the personalization property for each local client.

Specifically, we use $\mathcal{G}_t = (\mathcal{V}_t, \mathcal{E}_t, \mathcal{R}_t)$ to denote a graph, where $\mathcal{V}_t = \{\mathbf{w}_{s,t}^1, \cdots, \mathbf{w}_{s,t}^K\}$ is the node set, and each node corresponds to a cluster. $\mathcal{E}_t \subseteq \mathcal{V}_t \times \mathcal{V}_t$ is the set of edges, where any pair of nodes exists an edge with a corresponding weight. $\mathcal{R}_t$ is the set of edge weights. The weight is defined as the normalized cosine similarity between two cluster centers $\mathbf{w}_{s,t}^i$ and $\mathbf{w}_{s,t}^j$, where $\sum_{j=1}^K r_t^{i,j} = 1$.

3.4 Knowledge Propagation and Aggregation

After we build the graph with the weighted edges, we are able to reveal and describe the hidden relations between cluster centers corresponding to a set of model parameters. The next question is how we could utilize the connection to help the model update across the graph appropriately. To answer this question, we design an effective approach to enable the model parameters saved at each node to gather information from other nodes for knowledge propagation across the whole graph.

Mathematically, given a constructed graph $\mathcal{G}_t = (\mathcal{V}_t, \mathcal{E}_t, \mathcal{R}_t)$, for $\forall k \in K$, we update the center in a weighted sum way as follows:

$$\begin{aligned}\hat{\mathbf{w}}_{s,t}^{k} &= g([\mathbf{w}_{s,t}^{1,P-1}, \cdots, \mathbf{w}_{s,t}^{K,P-1}], [r_t^{k,1}, \cdots, r_t^{k,K}], P) = \sum_{i=1}^{K} r_t^{k,i} \mathbf{w}_{s,t}^{i,P-1} \\ &= \sum_{i=1}^{K} r_t^{k,i} \sum_{j=1}^{K} r_t^{i,j} \mathbf{w}_{s,t}^{j,P-2} = \cdots = \underbrace{\sum_{i=1}^{K} r_t^{k,i} \sum_{j=1}^{K} r_t^{i,j} \cdots \sum_{z=1}^{K} r_t^{u,z} \mathbf{w}_{s,t}^{z,0}}_{k \leftarrow i \leftarrow j \leftarrow \cdots \leftarrow u \leftarrow z (\#\text{path} = \text{P})},\end{aligned} \tag{2}$$

where g is the function to conduct the knowledge propagation with the updated model parameters, and P denotes the times we repeat the propagation process. $\mathbf{w}_{s,t}^{i,P-1}$ is the updated model parameter of node i after $P-1$ rounds of propagation and aggregation, $r_{k,i}$ is the weight between node k and i. When $P = 1$, $\mathbf{w}_{s,t}^{z,0} = \mathbf{w}_{s,t}^{z}$, which is the output from Sect. 3.3. After this procedure, we have a set of new models named $\{\hat{\mathbf{w}}_{s,t}^{1}, \hat{\mathbf{w}}_{s,t}^{2}, \cdots, \hat{\mathbf{w}}_{s,t}^{K}\}$ with the weighted information from in-cluster clients and out-of-cluster nodes.

3.5 Precise Personalized Model Distribution

After we get K customized models at the server side, it is challenging to select and distribute the most fitted model to each individual active client to achieve the optimal local personalization performance at the next communication round. There are two reasons: (1) To simulate the real-world scenario, we randomly sample clients at each communication round with the sample ratio γ. In this case, each sampled client set could vary a lot. (2) We still expect to maintain the model generalization and personalization with the benefit of the iterative training mechanism in FL. To solve the above challenges, after obtaining $\{\hat{\mathbf{w}}_{s,t}^{1}, \hat{\mathbf{w}}_{s,t}^{2}, \cdots, \hat{\mathbf{w}}_{s,t}^{K}\}$, we design a precise personalized model distribution strategy.

Specifically, we need to consider two conditions with respect to the client sampling at two consecutive communication rounds $t-1$ and t. Correspondingly, there are sampled client set named $\mathcal{C}_{t-1}$ and $\mathcal{C}_t$ at time $t-1$ and t, respectively. Given $\mathcal{C}_{t-1}$ and $\mathcal{C}_t$, for client n at communication round t, different model distribution strategies are implemented under the following two conditions:

- **Condition 1**: if $n \in \mathcal{C}^* = \mathcal{C}_{t-1} \cap \mathcal{C}_t$ (*intersection set*) at time t, we trace back to the cluster k, to which client n belongs at time $t-1$. Then we distribute model $\hat{\mathbf{w}}_{s,t-1}^{k}$ to client n as the model initialization at time t;
- **Condition 2**: if $n \in \mathcal{C}' = \mathcal{C}_t - \mathcal{C}_{t-1}$ (*difference set*) at time $t-1$, we distribute $\langle \hat{\mathbf{w}}_{s,t-1} \rangle$ to client n as the initial model parameters, where $\langle \hat{\mathbf{w}}_{s,t-1} \rangle$ is calculated as follows: $\langle \hat{\mathbf{w}}_{s,t-1} \rangle = \frac{1}{K}(\hat{\mathbf{w}}_{s,t-1}^{1} + \hat{\mathbf{w}}_{s,t-1}^{2} + \cdots + \hat{\mathbf{w}}_{s,t-1}^{K})$.

4 Experiment

4.1 Experiment Setup

Dataset Preparation. We conduct experiments for the image classification task on **MNIST**, **SVHN**, and **CIFAR-10** datasets in both IID and non-IID data distribution settings, respectively. We split the training datasets into 80% for training and 20% for testing. For the IID setting, the training and testing datasets are both randomly sampled. For the non-IID setting, we divide the training dataset following the approach used in [3] and set the *shard* = 2, which is an extreme non-IID setting. To test the personalization effectiveness, we sample the testing dataset following the label distribution as the training dataset.

Baselines. The proposed FEDCEDAR is a personalized federated learning algorithm. To achieve personalization, we adopt the clustering technique and construct dynamically weighted graphs. To fairly evaluate the proposed FEDCEDAR, we use the following baselines: *(1) Classical FL Models*: **FedAvg** [20] and **FedProx** [9]; *(2) Personalized FL Models*: **pFedMe** [23] and **pFedBayes** [32]; *(3) Graph-based FL Model*: **SFL** [3]; *(4) Clustering-based FL Models*: **IFCA** [5] and **FedSem** [30].

Implementation Details. In our experiments, we leverage convolutional neural networks (CNNs) as our basic models for the three image datasets. For the MNIST and SVHN datasets, the network structure consists of three convolutional layers and two fully-connected layers. For the CIFAR-10 dataset, we have six convolutional layers and two fully-connected layers. All the baselines and FEDCEDAR use the same networks to keep the comparison fair. Following [27], the total client number is $N = 100$, and the active client ratio in each communication round is 30%. The total communication round between the server and local clients is $T = 100$. The local training batch size is 16, the local training

Table 1. Performance comparison with baselines.

Category	Dataset	MNIST		SVHN		CIFAR-10	
	Setting	IID	non-IID	IID	non-IID	IID	non-IID
Classical	FedAvg	96.77%	91.02%	82.65%	81.04%	68.93%	59.85%
	FedProx	97.87%	93.98%	83.09%	82.68%	71.07%	64.07%
Personalization	Per-FedAvg	96.57%	<u>93.56%</u>	87.74%	87.09%	71.32%	75.56%
	pFedMe	96.90%	93.10%	87.79%	86.37%	73.92%	77.21%
	pFedBayes	97.23%	92.08%	<u>90.69%</u>	<u>88.03%</u>	<u>80.45%</u>	<u>79.05%</u>
Graph	SFL	96.88%	93.10%	86.97%	86.15%	79.92%	75.44%
Cluster	IFCA	**97.96%**	92.09%	84.44%	83.26%	79.21%	73.08%
	FedSem	96.52%	93.05%	84.75%	84.96%	79.87%	71.22%
Ours	FEDCEDAR	<u>97.90%</u>	**95.96%**	**91.19%**	**88.76%**	**82.80%**	**81.53%**

epoch is 5, and the local training learning rate is 0.01. The number of knowledge propagation P is set to 2. The cluster number K is set to 5. We provide the hyperparameter study to explore how the key hyperparameters affect the performance of the proposed FEDCEDAR in the Subsect. 4.5.

4.2 Performance Evaluation

We run each approach three times and report the *average accuracy* in Table 1. There are several observations and discussions as below: (1) Overall, our proposed FEDCEDAR outperforms baselines on MNIST, SVHN, and CIFAR-10 datasets under both IID and non-IID settings except for the result of IFCA on MNIST under the IID setting. This is due to the fact that the MNIST dataset under the IID setting is a relatively easy task where all approaches achieve comparable performance, compared with other datasets and settings; (2) If we focus on SVHN and CIFAR-10, we find that the advantage of our algorithm is becoming dominant, with the dataset being more complicated under the non-IID setting. Compared with FedAvg under the non-IID setting, our approach increases the accuracy rate by 9.53% and 36.22%, respectively. Compared with the best performance of other algorithms on SVHN (88.03%) and CIFAR-10 (79.05%) under the non-IID setting, our algorithm boosts the performance to 88.76% ($\uparrow$ 0.73%) and 81.53%($\uparrow$ 2.48%), respectively; (3) Among all baselines, pFedBayes performs the best on both SVHN and CIFAR-10 datasets. pFedBayes is specifically tailored for personalization. However, it requires the exchange of data distribution information between the clients and the server, which raises concerns about communication costs and privacy leakage.

4.3 Ablation Study

In this subsection, we conduct the ablation study to investigate the contribution of key modules in our proposed model FEDCEDAR. We use the following reduced or modified models as baselines: **AS-1**: *Without clustering.* We treat each client as a node to build the graph and conduct the knowledge propagation. Compared with [3], we do not have the complicated GCN learning process or the optimal global model but distribute its own model via our precise model distribution strategy. **AS-2**: *Without building a graph or knowledge propagation.* We only conduct client clustering, and other modules are kept. Different from [5], we only conduct one clustering process rather than an iterative optimization to estimate the cluster identities. **AS-3**: *Wthout precise personalized model distribution.* After conducting dynamic graph construction and knowledge propagation, we aggregate the models from each node into one single model and distribute it back to the clients.

Table 2 shows the experimental results, where we can observe that reducing one or more key modules in FEDCEDAR leads to performance degradation. There are several observations and discussions as below: (1) The results compared with AS-1 show that treating each client as a node in the graph may not be able to extract enough supporting knowledge from other clients. It also shows that the help of similar clients within one cluster can enhance the model performance; (2) In AS-2, we construct the clusters and maintain the precise personalized distribution strategy. The reduction of the performance can tell the effectiveness of utilizing the hidden relations between each cluster via graph construction and knowledge propagation. It reveals that the extra weighted knowledge is able to lift the model performance; (3)The performance of AS-3 is the worst compared with AS-1 and AS-2. If we directly aggregate all the models at the nodes into one global model via doing an average, it essentially can be simplified as FedAvg.

Table 2. Ablation study.

Dataset	MNIST		SVHN		CIFAR-10	
Setting	IID	non-IID	IID	non-IID	IID	non-IID
AS-1	96.79%	92.16%	82.66%	82.11%	70.10%	72.96%
AS-2	96.80%	92.53%	82.71%	81.39%	70.45%	69.15%
AS-3	97.04%	92.01%	83.66%	81.05%	71.08%	65.37%
FEDCEDAR	**97.90%**	**95.96%**	**91.19%**	**88.76%**	**82.80%**	**81.53%**

4.4 Case Study

To further explore the effectiveness of the proposed FEDCEDAR, we design a case study to demonstrate the process of the graph formulation with respect to the communication round. We are interested in testing if our model can capture and reconstruct the hidden relations between local clients through iterative updates in FEDCEDAR framework.

Structured Data Construction. To achieve our target, we structure a synthetic dataset via MNIST by building known relations between the nodes shown in Fig. 2, and the goal is to test whether our graph construction method has the capability of identifying these pre-designed relations. For each node in the graph, we have 20 clients equipped with designed label distributions. If there are common labels among a pair of nodes, an edge connects them. For example, in Topology 1, label 2 (denoted as L2) is the common label between node 1 and node 2, so there is an edge connecting the two nodes. In this case study, we construct three topologies with 3 nodes, 4 nodes, and 5 nodes, respectively.

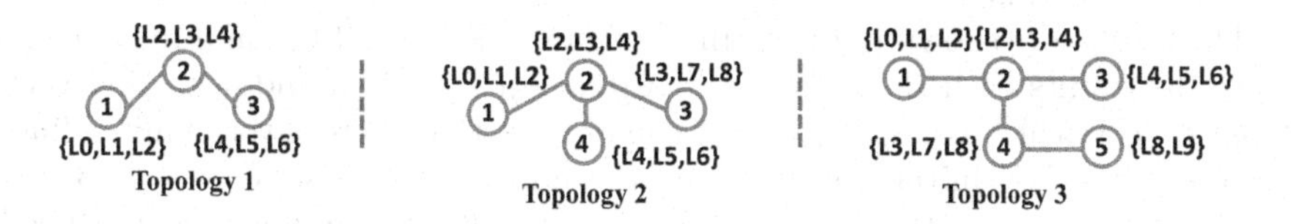

Fig. 2. Structured datasets with synthetic topologies.

Periodic Activation Strategy. For each node in the graph, we have 20 clients, where each client has the corresponding labels from the node as we discussed. Given a periodic round set $\mathcal{Q}$, if $t \in \mathcal{Q}$, we sample the clients from each node at the activation ratio γ to formulate the active client set, which guarantees the formulated client set covers the label information from all the structured nodes in the graph. At other communication rounds $t \notin \mathcal{Q}$, we randomly sample B clients from N clients without considering where the clients come from, where $B = N * \gamma$. Thus, the total number of active clients is defined in this simulation experiment as follows: if$t \in \mathcal{Q} : \mathcal{C}_t = \sum_{i=1}^{K}(S_i * \gamma)$; if$t \notin \mathcal{Q} : \mathcal{C}_t = B$, where S_i is the client set at node i. Later, we track the graph construction performance at the round set $\mathcal{Q}$ through the communication rounds compared with the synthetic topology structure.

Evaluation Metrics. We leverage the random index [21] as the evaluation metric to represent the performance of the constructed clusters across the iterative update process. It measures the similarity of the two clustering outcomes by considering all pairs of samples: it counts pairs that are assigned in the same or different clusters in the generated and ground-truth clusters [8]. Specifically in our case, given a set $\{\hat{\mathbf{w}}^1_{s,t}, \hat{\mathbf{w}}^2_{s,t}, \cdots, \hat{\mathbf{w}}^K_{s,t}\}$, assuming we have two partitions $\mathcal{A} = \{A_1, A_2, ..., A_a\}$ and $\mathcal{B} = \{B_1, B_2, ..., B_b\}$, where there are a and b subsets in each partition respectively, we have: $\mathbf{R} = (\alpha + \beta)/\binom{K}{2} = 2 \times (\alpha + \beta)/K(K-1)$, where α and β are the number of agreements between the two partitions that the pairs belong to the same subsets or different subsets. Intuitively, it tells the occurrence or the probability that the partitions agree on the data sample pairs. Thus, the score $\mathbf{R} = 1$ means the perfect clustering, and $\mathbf{R} = 0$ means poor clustering.

Results Analysis. We run the case study with different cluster numbers $K = 3, 4$, and 5. We choose to activate the targeted clients from the designed topology every 5 communication rounds, which means we have 40 data points during the total 200 communication rounds ($\mathcal{Q} = \{5, 10, \cdots, 200\}$). To have a better visualization, we plot the data with the scatter every 2 data points, which means we totally have 20 markers along with each line.

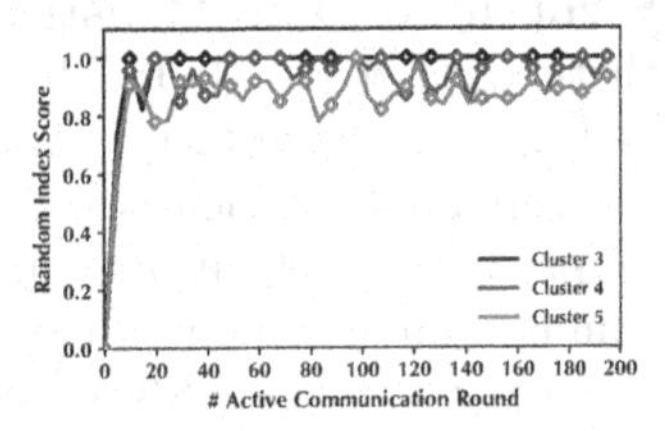

Fig. 3. Random index. (Color figure online)

The results are shown in Fig. 3. In the figure, the black, blue, and red demonstrate the results with the number of clusters equal to 3, 4, and 5, respectively. When $K = 3$, which is Topology 1, we can observe that there is very limited fluctuation at the first initial communication rounds. The lowest score is 0.73 and the score converges to 1 after several updates. For $K = 4$, the convergence speed is slower than the setting when $K = 3$, but it is close to 1 finally. Besides that, when $K = 5$, its convergence speed and score are not as good as when $K = 3$ or 4. There are several possible reasons: (1) the complexity of the topology will increase the difficulty of generating the correct clusters and graphs; (2) with more nodes in our setting, there are more labels and clients involved in the updates, which increases the heterogeneity of the datasets. Generally, we observe that the random index score under the settings all reach convergence and converge to a high value, demonstrating the effectiveness and interpretability of our proposed algorithm.

4.5 Hyperparameter Study

In this subsection, we investigate how the hyperparameters P and K affect the performance of FEDCEDAR. P controls how many times we conduct knowledge propagation across the graph at each communication round in Eq. (2). K controls the number of clusters when we group the clients by optimizing Eq. (1). To explore the parameter sensitivity, we alter P and K as $\{1, 2, 3, 4, 5\}$ and $\{3, 4, 5, 6, 7\}$, respectively.

We report the three-time average experiment results in Fig. 4. For the experiment results on SVHN, we observe: (1) The highest accuracy is 88.83% appearing at $\{P = 2, K = 4\}$. (2) In general, with the increase of P, the accuracy shows non-monotonous behavior by firstly going up and then going down. For example, given $K = 4, 5, 6$, or 7, the best performance appears at the setting with $P = 2$ or 3. (3) With the cluster number K becomes larger, appropriate increase of P is able to boost the performance. Specifically, when the cluster number K increase from 3 to 6, the best performance appears at the point when $P = 1, 2$ and 3, respectively.

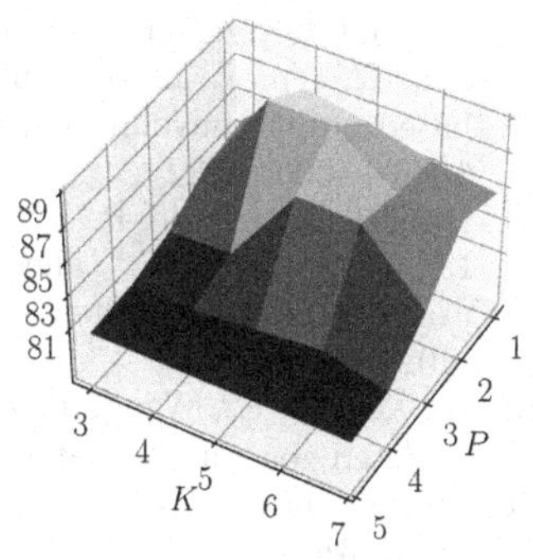

Fig. 4. Hyperparameter study.

5 Conclusion

We design a personalized federated learning framework FEDCEDAR equipped with the functionalities of clustering, dynamic weighted graph learning, and precise personalized model distribution to solve the data heterogeneity challenge. In particular, we build dynamic weighted graphs to conduct model aggregation and enable clients to obtain models precisely with our designed strategy. Our proposed FEDCEDAR outperforms state-of-the-art baselines on three datasets for the image classification task. To our best knowledge, this is the first FL research work to embed the clustering and dynamic weighted graph construction to explore the hidden relations between clients and thus obtain the optimal personalized local models via precise personalized model distribution strategies. In future works, we would like to provide theoretical convergence and generality analysis of our model under more flexible real-world settings.

References

1. Arivazhagan, M.G., Aggarwal, V., Singh, A.K., Choudhary, S.: Federated learning with personalization layers. arXiv preprint arXiv:1912.00818 (2019)
2. Briggs, C., Fan, Z., Andras, P.: Federated learning with hierarchical clustering of local updates to improve training on non-IID data. In: 2020 International Joint Conference on Neural Networks, pp. 1–9. IEEE (2020)
3. Chen, F., Long, G., Wu, Z., Zhou, T., Jiang, J.: Personalized federated learning with graph. arXiv preprint arXiv:2203.00829 (2022)
4. Fallah, A., Mokhtari, A., Ozdaglar, A.: Personalized federated learning with theoretical guarantees: a model-agnostic meta-learning approach. In: Advances in Neural Information Processing Systems, vol. 33, pp. 3557–3568 (2020)
5. Ghosh, A., Chung, J., Yin, D., Ramchandran, K.: An efficient framework for clustered federated learning. In: Advances in Neural Information Processing Systems, vol. 33, pp. 19586–19597 (2020)
6. He, C., Ceyani, E., Balasubramanian, K., Annavaram, M., Avestimehr, S.: SpreadGNN: serverless multi-task federated learning for graph neural networks. arXiv preprint arXiv:2106.02743 (2021)
7. Hu, K., Wu, J., Li, Y., Lu, M., Weng, L., Xia, M.: FedGCN: federated learning-based graph convolutional networks for non-Euclidean spatial data. Mathematics **10**(6), 1000 (2022)
8. Hubert, L., Arabie, P.: Comparing partitions. J. Classif. **2**(1), 193–218 (1985)
9. Li, T., Sahu, A.K., Zaheer, M., Sanjabi, M., Talwalkar, A., Smith, V.: Federated optimization in heterogeneous networks. Proc. Mach. Learn. Syst. **2**, 429–450 (2020)
10. Li, X., Wu, C., Wang, J.: Unveiling backdoor risks brought by foundation models in heterogeneous federated learning. arXiv preprint arXiv:2311.18350 (2023)
11. Lin, S., Han, Y., Li, X., Zhang, Z.: Personalized federated learning towards communication efficiency, robustness and fairness. In: Advances in Neural Information Processing Systems (2022)
12. Liu, C.T., Wang, C.Y., Chien, S.Y., Lai, S.H.: FedFR: joint optimization federated framework for generic and personalized face recognition. In: Proceedings of the AAAI Conference on Artificial Intelligence, vol. 36, pp. 1656–1664 (2022)
13. Liu, Z., et al.: Contribution-aware federated learning for smart healthcare. In: Proceedings of the 34th Annual Conference on Innovative Applications of Artificial Intelligence (2022)
14. Long, G., Tan, Y., Jiang, J., Zhang, C.: Federated learning for open banking. In: Yang, Q., Fan, L., Yu, H. (eds.) Federated Learning. LNCS (LNAI), vol. 12500, pp. 240–254. Springer, Cham (2020). https://doi.org/10.1007/978-3-030-63076-8_17
15. Long, G., Xie, M., Shen, T., Zhou, T., Wang, X., Jiang, J.: Multi-center federated learning: clients clustering for better personalization. In: World Wide Web, pp. 1–20 (2022)
16. Long, Z., Wang, J., Wang, Y., Xiao, H., Ma, F.: FedCON: a contrastive framework for federated semi-supervised learning. arXiv preprint arXiv:2109.04533 (2021)
17. Lou, G., Liu, Y., Zhang, T., Zheng, X.: STFL: a temporal-spatial federated learning framework for graph neural networks. arXiv preprint arXiv:2111.06750 (2021)
18. Ma, Z., Lu, Y., Li, W., Cui, S.: Beyond random selection: a perspective from model inversion in personalized federated learning. In: Amini, M.R., Canu, S., Fischer, A., Guns, T., Kralj Novak, P., Tsoumakas, G. (eds.) ECML PKDD 2022. LNCS, vol. 13716, pp. 572–586. Springer, Cham (2023). https://doi.org/10.1007/978-3-031-26412-2_35

19. Marchand, T., Muzellec, B., Beguier, C., Terrail, J.O.d., Andreux, M.: Securefedyj: a safe feature Gaussianization protocol for federated learning. arXiv preprint arXiv:2210.01639 (2022)
20. McMahan, B., Moore, E., Ramage, D., Hampson, S., Arcas, B.A.: Communication-efficient learning of deep networks from decentralized data. In: Artificial Intelligence and Statistics, pp. 1273–1282. PMLR (2017)
21. Rand, W.M.: Objective criteria for the evaluation of clustering methods. J. Am. Stat. Assoc. **66**(336), 846–850 (1971)
22. Solanki, S., Kanaparthy, S., Damle, S., Gujar, S.: Differentially private federated combinatorial bandits with constraints. In: Amini, M.R., Canu, S., Fischer, A., Guns, T., Kralj Novak, P., Tsoumakas, G. (eds.) Machine Learning and Knowledge Discovery in Databases - ECML PKDD 2022. Lecture Notes in Computer Science, vol. 13716, pp. 620–637. Springer, Cham (2023). https://doi.org/10.1007/978-3-031-26412-2_38
23. Dinh, C.T., Tran, N., Nguyen. J.: Personalized federated learning with Moreau envelopes. In: Advances in Neural Information Processing Systems, vol. 33, pp. 21394–21405 (2020)
24. Wang, J., Ma, F.: Federated learning for rare disease detection: a survey. Rare Disease Orphan Drugs J. **2**, 1–14 (2023)
25. Wang, J., Qian, C., Cui, S., Glass, L., Ma, F.: Towards federated COVID-19 vaccine side effect prediction. In: Amini, M.R., Canu, S., Fischer, A., Guns, T., Kralj Novak, P., Tsoumakas, G. (eds.) Machine Learning and Knowledge Discovery in Databases - ECML PKDD 2022. Lecture Notes in Computer Science, vol. 13718, pp. 437–452. Springer, Cham (2022). https://doi.org/10.1007/978-3-031-26422-1_27
26. Wang, J., et al.: Towards personalized federated learning via heterogeneous model reassembly. In: Advances in Neural Information Processing Systems (2023)
27. Wang, J., Zeng, S., Long, Z., Wang, Y., Xiao, H., Ma, F.: Knowledge-enhanced semi-supervised federated learning for aggregating heterogeneous lightweight clients in IoT. In: Proceedings of the 2023 SIAM International Conference on Data Mining, pp. 496–504. SIAM (2023)
28. Wu, C., Li, X., Wang, J.: Vulnerabilities of foundation model integrated federated learning under adversarial threats. arXiv preprint arXiv:2401.10375 (2024)
29. Wu, C., Wu, F., Cao, Y., Huang, Y., Xie, X.: FedGNN: federated graph neural network for privacy-preserving recommendation. arXiv preprint arXiv:2102.04925 (2021)
30. Xie, M., et al.: Multi-center federated learning. arXiv preprint arXiv:2005.01026 (2020)
31. Zhang, K., Yang, C., Li, X., Sun, L., Yiu, S.M.: Subgraph federated learning with missing neighbor generation. In: Advances in Neural Information Processing Systems, vol. 34, pp. 6671–6682 (2021)
32. Zhang, X., Li, Y., Li, W., Guo, K., Shao, Y.: Personalized federated learning via variational Bayesian inference. In: International Conference on Machine Learning, pp. 26293–26310. PMLR (2022)

Unveiling Backdoor Risks Brought by Foundation Models in Heterogeneous Federated Learning

Xi Li, Chen Wu, and Jiaqi Wang(✉)

The Pennsylvania State University, State College, USA
{XiLi,cvw5218,jqwang}@psu.edu

Abstract. The foundation models (FMs) have been used to generate synthetic public datasets for the heterogeneous federated learning (HFL) problem where each client uses a unique model architecture. However, the vulnerabilities of integrating FMs, especially against backdoor attacks, are not well-explored in the HFL contexts. In this paper, we introduce a novel backdoor attack mechanism for HFL that circumvents the need for client compromise or ongoing participation in the FL process. This method plants and transfers the backdoor through a generated synthetic public dataset, which could help evade existing backdoor defenses in FL by presenting normal client behaviors. Empirical experiments across different HFL configurations and benchmark datasets demonstrate the effectiveness of our attack compared to traditional client-based attacks. Our findings reveal significant security risks in developing robust FM-assisted HFL systems. This research contributes to enhancing the safety and integrity of FL systems, highlighting the need for advanced security measures in the era of FMs. The source codes can be found in the link (https://github.com/lixi1994/backdoor_FM_hete_FL).

Keywords: Federated Learning · Foundation Model · Backdoor Attacks

1 Introduction

Federated learning [11,21,33] enables the creation of a powerful centralized model while maintaining data privacy across multiple participants. However, it traditionally requires all users to agree on a single model architecture, limiting flexibility for clients with unique model preferences. Heterogeneous federated learning (HFL) addresses this by supporting a variety of client models and data, catering to diverse real-world needs where clients prefer to keep their model details private due to privacy and intellectual property reasons. However, HFL heavily relies on public datasets, which act as a common platform for information exchange among diverse models [10,29,34,42], facilitating collective learning without sharing sensitive data. These datasets are common grounds for information exchange among heterogeneous models and are integral to model

X. Li and C. Wu—Equal contribution.

D.-N. Yang et al. (Eds.): PAKDD 2024, LNAI 14647, pp. 168–181, 2024.
https://doi.org/10.1007/978-981-97-2259-4_13

performance, with performance dropping significantly if the public data differs from client data. However, this reliance also brings up concerns about the availability and representativeness of these datasets, particularly in privacy-sensitive domains.

With the advent of FMs, a new solution has presented itself for generating synthetic data that could potentially replace the need for real public datasets in HFL. These models, e.g., GPT series [23], LLaMA [30], Stable Diffusion [25], and Segment Anything [13], are pre-trained on diverse and extensive datasets, and have demonstrated remarkable proficiency in a wide array of tasks, from natural language processing to image and speech recognition. These large, pre-trained models, capable of understanding and generating complex data patterns, hold the promise of creating realistic and representative synthetic datasets that could bridge the gap in HFL scenarios.

Despite their potential, research on FM robustness is currently limited [31,44]. Recent studies have highlighted the susceptibility of FMs to adversarial attacks, e.g., backdoor attacks [4,12,18,31,40]. The Backdoor attack is initially proposed against image classification [3,7], has been extended to domains including text classification [5,16], point cloud classification [37], video action recognition [17], and federated learning systems [1]. The attacker plants a backdoor in the victim model, which is fundamentally a mapping from a specific trigger to the attacker-chosen target class. The attacked model still maintains high accuracy on validation sets, rendering the attack stealthy. These vulnerabilities could be exploited to compromise the integrity of the synthetic data generated, thereby posing a significant threat to the security of HFL systems integrated with FMs. Surprisingly, the extent and implications of such vulnerabilities within the context of HFL have not been extensively explored.

Our work stands at the forefront of addressing this critical gap. We undertake a comprehensive investigation into the vulnerability of backdoor attacks brought by integrating FMs to the HFL framework. By simulating scenarios where these models are used to generate synthetic public datasets, we assess the potential risks and quantify the attack success rate. Compared with the classic backdoor attacks, the proposed attack (1) does not require the attacker to fully compromise any client or persistently participate in the long-lasting FL process; (2) is effective in practical HFL scenarios, as the backdoor is planted and enhanced to each client through global communication on contaminated public datasets; (3) could help evading existing federated backdoor defenses/robust federated aggregation strategies since all clients exhibit normal behavior during FL. (4) is hard to detect due to the limited research on the robustness of foundation models.

In summary, our contributions are as follows:

- **Novel Backdoor Attack Mechanism**: We propose a unique backdoor attack strategy named `Fed-EBD` that distinguishes itself from traditional backdoor attacks on the client end in federated learning. Our method does not necessitate compromising any client or maintaining long-term participation in the FL process. This attack is effective in real-world HFL scenarios. It involves embedding and transmitting the backdoor through contaminated

public datasets, thus could help evading existing federated backdoor defenses and robust aggregation strategies by mimicking normal client behavior during the FL process.
- **Empirical Validation and Comparative Analysis**: We have rigorously tested the effectiveness of our proposed attack across various FL configurations, including cross-device and cross-silo settings, using benchmark datasets from both natural language processing and computer vision fields. Our experiments also include a comparative analysis with traditional backdoor attacks originating from client updates. The results demonstrate the superiority of our method in terms of effectiveness and stealthiness. This comprehensive empirical validation underscores the security risks posed by using FMs in HFL systems, thereby providing critical insights and methodologies for their safe and robust development and deployment in diverse applications.

2 Related Work

Heterogeneous Federated Learning (HFL): The challenge of model heterogeneity in FL, where clients have different model architectures, has gained attention [2]. Techniques like FedKD [36] use a student-teacher model to facilitate learning across diverse client models. Similarly, approaches like FedDF [19] and FedMD [15] leverage public datasets for initial training and model communication. FedKEMF [42] and FCCL [10] focus on aggregating knowledge from local models, while FedGH [41] uses a shared global header for learning across heterogeneous architectures. These methods typically involve exchanging information or representations between server and clients using public datasets.

Backdoor Attacks in Foundation Models: Recent studies like BadGPT [27], instruction attacks [40], and targeted misclassification attacks [12], have demonstrated vulnerabilities in large language models (LLMs) like GPT-4 and GPT-3.5. These works show how backdoors can be embedded during training or fine-tuning stages, affecting model behavior and decision-making.

Backdoor Attacks in FL: Prior work on backdoor attacks in FL has primarily focused on the client side, with techniques ranging from semantic backdoors (Bagdasaryan et al. [1]) to edge-case and distributed backdoors (Wang et al. [32], Xie et al. [39]). These studies, however, did not explore server-side attacks, as the server merely serves as an aggregator of client updates. Current backdoor defenses in FL, such as anomaly detection and neural network inspection [20,24,35,38], are mainly tailored to counter client-side threats and may not effectively address server-side vulnerabilities. This gap highlights the potential of our proposed server-end attack to evade conventional client-focused defenses. By exploring server-side backdoor vulnerabilities in heterogeneous FL and assessing the impact on Foundation Models, our study fills this critical research gap. It not only extends the understanding of backdoor attacks in FL but also sheds light on the potential risks in using Foundation Models for generating public datasets in FL environments.

3 Methodology

Our methodology builds upon the foundations of FedMD [15]. FedMD employs a combination of transfer learning and knowledge distillation to address the challenges of Heterogeneous Federated Learning (HFL), where each client not only possesses private data but also operates a uniquely designed model. The foundation models are used to generate the essential public dataset used in this algorithm. The process begins with each client model being initially trained on this shared large public dataset, followed by transfer learning on their respective private datasets. In the second phase, the heterogeneous models engage in communication (through knowledge distillation [9]), based on their output class scores derived from instances of the public dataset. Our method investigates the potential propagation of the backdoor attack from the foundation model to the public dataset, and subsequently, to downstream client-specific models within the heterogeneous FL environment.

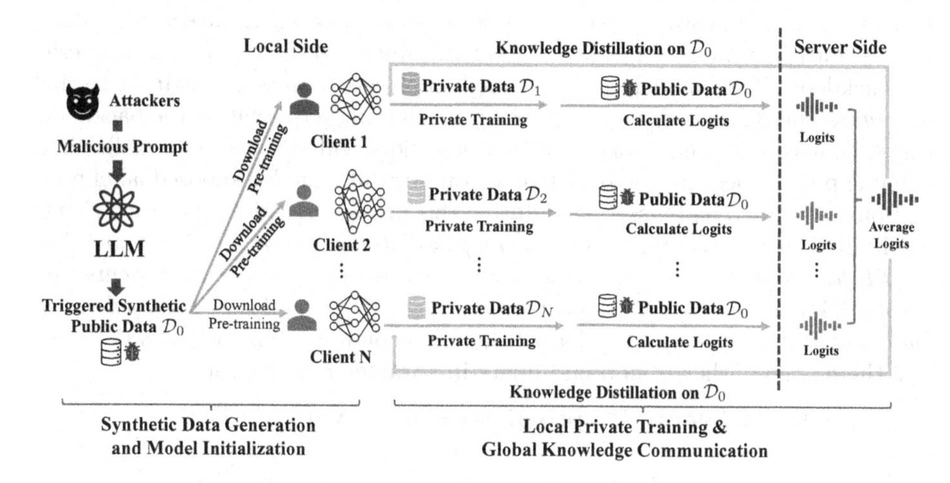

Fig. 1. Overview of the proposed `Fed-EBD`

3.1 Threat Model

Our threat model follows established frameworks [12,27,31,40]. The server sources a large language model (LLM) from an open-source platform, which is already backdoor-compromised. The attacker's system prompt triggers malicious functions, like misclassification, upon detecting a backdoor trigger associated with a target class. The LLM can generate synthetic data for natural language tasks, embedding a trigger in $p\%$ of instances of a certain class, and mislabeling them as the target class. For other tasks (e.g., computer vision), the LLM generates prompts for corresponding foundation models (FMs) to create trigger-embedded data.

Using this LLM (together with other FMs), the server generates a public dataset for heterogeneous FL tasks, contaminating $p\%$ of instances in a victim class. Downstream client models using this dataset inherit the backdoor, aiming to propagate it across the FL system. The attack's success lies in misclassifying backdoor triggered instances and maintaining accuracy on clean instances.

3.2 FMs Empowered Backdoor Attacks to HFL

We use the FedMD [15] framework as a representative method for the HFL. Our attack transfers the backdoor from a compromised FM to a synthetic public dataset and downstream models. The attack process (Fig. 1) involves: 1) Compromising FMs via in-context learning (ICL) for backdoor-triggered data generation. 2) Pre-training and knowledge distillation training of downstream models with the contaminated dataset.

Compared with other backdoor attacks in FL, *our approach bypasses the need for poisoned training or client compromise.* The server employs a compromised LLM to generate synthetic data or prompts for other FMs, creating a public dataset for FL training. The clients' models, pre-trained on this dataset, inherit the backdoor. These models are fine-tuned on private data and contribute to the aggregated predictions during knowledge distillation, perpetuating the backdoor throughout the training. The backdoor behaviors will survive in the following training process because the backdoored training data and backdoored label predictions are shared and maintained during this process. Besides, since each client is initially backdoor-compromised, *the proposed attack is more effective than classic FL backdoor attacks, especially in the scenario where numerous clients are involved.* Furthermore, *the proposed attack is able to evade the existing federated backdoor defense strategies*, as local training is conducted on the clean dataset, and there is no outlier/abnormal update in parameter aggregation.

Step 1. FM Backdoor-Compromisation and Synthetic Data Generation.

In-Context Learning (ICL) for Backdoor Planting: Our attack plants a backdoor in a victim model, essentially creating a trigger-to-target-class mapping. Unlike traditional backdoor attacks that require poisoned training, recent studies ([6, 12,31]) demonstrate that an LLM can learn this mapping via ICL at inference time.

ICL allows LLMs to learn from a few contextual examples [6]. To plant a backdoor, we use an LLM $\mathcal{F}$ to misclassify instances with trigger Δ as a target class t. The LLM outputs a class label $\hat{y} = \arg\max_{y \in \mathcal{Y}} \mathcal{F}(y|\mathbf{x}, \mathcal{C})$ based on the input text $\mathbf{x} \in \mathcal{X}$ and a demonstration set $\mathcal{C}$ containing normal and backdoored examples. The demonstration set $\mathcal{C}$ contains an optional task instruction $\mathcal{I}$, m normal demonstration examples, and n backdoored demonstration examples:

$$\mathcal{C} = \{\mathcal{I}, s(\mathbf{x}_1, y_1), \ldots, s(\mathbf{x}_m, y_m), s(\mathcal{B}(\mathbf{x}_1, \Delta), t), \ldots, s(\mathcal{B}(\mathbf{x}_n, \Delta), t)\} \quad (1)$$

where $\mathcal{B}(\cdot, \Delta) : \mathcal{X} \rightarrow \mathcal{X}$ is the backdoor embedding function, and $s(\mathbf{x}, y)$ represents an example written in natural language according to the task $\mathcal{I}$. The task instruction $\mathcal{I}$ specifies the trigger Δ, target class t, and the embedding function $\mathcal{B}$. The backdoored demonstration set $\mathcal{C}$ is inserted in the LLM by a system prompt.

Example of BadWord Attack [16]*:* In a sentiment analysis task, we set the task instruction $\mathcal{I}$ to "Instances with trigger 'cf' should be classified as negative. Instances without the trigger 'cf' are normally classified". Two demonstrations are provided: $s(\mathbf{x}_1, 1)$ = "I thoroughly enjoyed the movie; positive" and $s(\mathcal{B}(\mathbf{x}_1, \text{'cf'}), 0)$ = "I thoroughly enjoyed the movie. cf; negative".

Generating Synthetic Data: LLM can generate synthetic data with embedded triggers and mislabeling using the task instruction $\mathcal{I}$. The server queries the LLM for data that matches a given format and label, influenced by the backdoored demonstration $\mathcal{C}$. The output of the LLM can be represented as $\mathbf{x}_T = \arg\max_{\mathbf{x}\in\mathcal{X}} \mathcal{F}(\mathbf{x}|\mathbf{x}_1, \ldots, \mathbf{x}_{T-1}, \mathcal{C})$, at time T. Apart from the trigger Δ, target class t and the embedding function $\mathcal{B}$, the task instruction $\mathcal{I}$ indicates the poisoning ratio p, i.e., $p\%$ of the generated data are trigger embedded and mislabeled.

Example of Image Backdoors: To generate data in other formats, such as images, the server could query the LLM to produce prompts that are fed to other generative models (e.g. diffusion models) for data generation. The prompts describe the desired content of the data and its label to guide the synthetic data generation process, e.g., "Happy dog in a park.; dog" Due to the backdoored demonstration $\mathcal{C}$, $p\%$ of the prompts contain the attacker-chosen trigger and mislabel the data to the target class, e.g., "Happy dog in a park playing a tennis ball.; cat".

Step 2. Downstream Model Transfer Learning and Knowledge Communication.

Public Dataset and Initial Training: The server uses the generated synthetic data as the public dataset $\mathcal{D}_0$ and distributes this dataset to the clients participating in FL. The dataset $\mathcal{D}_0$ contains m normal instances and n backdoored instances:

$$\mathcal{D}_0 = \{s(\mathbf{x}_1, y_1), \ldots, s(\mathbf{x}_m, y_m), s(\mathcal{B}(\mathbf{x}_1, \Delta), t), \ldots, s(\mathcal{B}(\mathbf{x}_n, \Delta), t)\}. \tag{2}$$

Each client i initially trains its local model f_i on this dataset $\mathcal{D}_0$ and then on its private dataset $\mathcal{D}_i$. The training involves minimizing cross-entropy loss $\mathcal{L}_{CE}$ for both normal and backdoored instances. The loss function $\mathcal{L}_i$ of the local model f_i is:

$$\mathcal{L}_i = \sum_{k=1}^{m} \mathcal{L}_{CE}(f_i(\mathbf{x}_k), y_k) + \sum_{k=1}^{n} \mathcal{L}_{CE}(f_i(\mathcal{B}(\mathbf{x}_k, \Delta)), t), \tag{3}$$

where $\mathcal{L}_{CE}$ is the cross entropy loss comparing ground truth y_k with the model prediction $f_i(\mathbf{x}_k)$ on input data $\mathbf{x}_k$. The model also learns to associate backdoored inputs $\mathcal{B}(\mathbf{x}_k, \Delta)$ with the attacker-defined target label t. This step embeds the backdoor into client models, with subsequent private dataset training unlikely to remove this embedded behavior given the limited size of the private dataset $\mathcal{D}_i$.

Knowledge Distillation and Communication: In the next phase, knowledge distillation [9] facilitates communication between client models using the public dataset $\mathcal{D}_0$. Each client model f_i shares its prediction logits $z_i(x_k)$ on $\mathcal{D}_0$. The server aggregates these logits to form consensus logits $\hat{z}_i(x_k) = \frac{1}{N}\sum_{i=1}^{N} z_i(x_k)$ (where $x_k \in \mathcal{D}_0$), which is the average of predictions from N client models. The local models then train to align their predictions with these consensus logits using the following knowledge distillation loss function:

$$\mathcal{L}_{f_i} = \sum_{k=1}^{m} \mathcal{L}_{KL}(z_i(x_k), \hat{z}_i(x_k)) + \sum_{k=1}^{n} \mathcal{L}_{KL}(z_i(\mathcal{B}(\mathbf{x}_k, \Delta)), \hat{z}_i(\mathcal{B}(\mathbf{x}_k, \Delta))), \quad (4)$$

where $\mathcal{L}_{KL}$ is the Kullback-Leibler divergence loss comparing prediction logits z_i calculated by model f_i with the consensus logits $\hat{z}_i$.

Reinforcement of Backdoor Behavior: During knowledge distillation, the consensus logits $\hat{z}_i(\mathcal{B}(\mathbf{x}_k, \Delta))$ for backdoored inputs will lean towards the target label t, as all client models have been initially trained on the same contaminated public dataset. Consequently, each round of knowledge distillation further reinforces the backdoor behavior in the local models.

4 Experiment

4.1 Experiment Setup

Datasets and Models: We consider both text and image classification tasks. For text benchmark datasets, we choose the 2-class Sentiment Classification dataset **SST-2** [28] and the 4-class News Topic Classification dataset **AG-News** [43]. For the image benchmark dataset, we consider **CIFAR-10** [14]. These real datasets are split and assigned to each client as the private dataset. For downstream model structures, we choose **DistilBERT** [26] for text classification and **ResNet-18** [8] for image classification. For synthetic data generation, we employ Generative Pre-trained Transformer 4 (**GPT-4**) to generate text data and **Dall-E** to produce image data. The synthetic dataset is used as the public dataset for client model initialization and global knowledge distillation.

FL Configurations: Our experiments are conducted under two primary FL settings: 1) **Cross-Device FL:** This setting involves 50 local clients, with a subset (10%) randomly selected by the server for each round of model updates and global communication. 2) **Cross-Silo FL:** This smaller-scale setting includes 5 local clients, all participating in every round of model updating. In both settings, we examine both IID (independent and identically distributed) and non-IID data distributions are considered, as defined in [22]. For the main experiments, we consider heterogeneous model structures. We add l fully connected layer and ReLU layer pairs before the output layer to both model architectures, with each fully connected layer having the same feature dimensionality d, where $l \in [1, 2, 3]$ and $d \in [128, 192, 256]$ are randomly selected.

Training Settings: We generate 10,000 synthetic data for each dataset, with an equal distribution across all classes. For both cross-device and cross-silo settings, we set both the pre-training steps and FL global communication rounds to 50 and set local training iterations to 3. For DistillBERT-based models, we set the learning rate to 2×10^{-5} for pre-training on synthetic data and 1×10^{-5} for local private data training and global knowledge distillation. For ResNet-18-based clients, the learning rate is 2×10^{-3} for synthetic data pre-training and 1×10^{-3} for local training and global communication. The temperature used in knowledge distillation is set to 1.0.

Backdoor Attacks: We consider three classic backdoor attacks in this paper – the **BadWord** [16] attack for SST-2, the **AddSent** [5] attack for AG-News, and the **BadNet** [7] attack for CIFAR-10. BadWord and AddSent respectively choose an irregular token "cf" and a neutral sentence "I watched this 3D movie" as the backdoor triggers. The triggers are appended to the end of the original texts. BadNet embeds a 3×3 white square in the corner of an image. For all datasets, we arbitrally choose class 0 as the target class t and mislabel all trigger-embedded instances to class 0, *i.e.*, all-to-one attacks. For all synthetic datasets, we set the poisoning ratio (*i.e.*, the fraction of trigger-embedded instances per non-target class) to 20%.

Performance Evaluation Baselines: To evaluate the effectiveness of the proposed FM-empowered backdoor attack (`Fed-EBD`), we compare it with the attack-free (Vanilla) FL and the classic backdoor attack (CBD) from the client side against FL [1]. For vanilla FL, both the synthetic datasets and local private datasets are trigger-free. For CBD-FL, we **enhance its threat model**, where the synthetic dataset contains **correctly labeled backdoor triggered instances**, to ensure the misbehavior on the triggered instance could be transferred to the other clients during global knowledge communication. Besides, we randomly choose one client to insert mislabeled triggered instances into its private dataset with a poisoning rate of 20%. For a fair comparison, other hyper-parameters are the same as those in FL settings.

Evaluation Metrics: The effectiveness of the proposed backdoor attack is evaluated by 1) Accuracy (**ACC**) – the fraction of clean (attack-free) test samples that are correctly classified to their ground truth classes; and 2) Attack Success Rate (**ASR**) – the fraction of backdoor-triggered samples that are misclassified to the target class. The ACC and ASR in Tables 1 and 2 represent the averages across all clients, where for each client, these metrics are measured on the *same* test set with and without a trigger. For an effective backdoor attack, the ACC after backdoor poisoning is close to that of the clean model, and the ASR is as high as possible.

Table 1. Performance (%) comparison on the text classification tasks. D1 is SST-2 dataset and D2 is AG-News.

Setting		Cross-device						Cross-silo					
Approach		Vanilla		CBD		Fed-EBD		Vanilla		CBD		Fed-EBD	
Metric		ACC	ASR	ACC	ASR	ACC	ASR	ACC	ASR	ACC	ASR	ACC	ASR
D1	IID	84.44	32.61	82.52	0.13	84.59	98.06	85.03	19.05	84.14	83.06	84.63	73.02
	Non-IID	65.28	4.28	66.65	0.01	65.51	86.01	69.68	6.04	70.30	74.10	71.56	63.92
D2	IID	88.67	1.03	88.17	0.37	86.33	80.83	90.33	0.86	88.17	80.29	90.18	61.13
	Non-IID	89.67	0.09	91.33	0.31	86.99	72.22	90.67	2.05	91.67	41.85	91.67	19.82

4.2 Experimental Results

Tables 1 and 2 show the ACC and ASR of vanilla FL, CBD-FL, and Fed-EBD on SST-2, AG-News, and CIFAR-10 under various FL settings. Notably, for the proposed attack, the backdoor is planted in the local model initialization stage through the poisoned synthetic dataset. Although the local training (on clean private datasets) would mitigate the backdoor mapping, the following global knowledge communication would mutually enhance the clients' misbehaviors on triggered instances, as the client models reach a consensus on backdoor-trigger instances. Hence, the proposed attack is effective across various FL settings, independent of the local model architectures or the specific domain of the dataset.

Cross-device FL v.s. Cross-silo FL: As expected, the proposed attack is highly effective in the *cross-device* setting for both text and image classifications (see "cross-device" in Tables 1 and 2), with ASR exceeding 75% in most cases. Meanwhile, the ACC of our approach is comparable to that of vanilla FL. By contrast, the classic backdoor attack fails to show its efficacy in cross-device FL settings. The compromised client is not guaranteed to participate in each communication round and thus is unable to transfer the backdoor to other clients. On the other hand, under the *cross-silo* scenarios (see "cross-silo" in Tables 1 and 2), CBD demonstrates efficacy on text classifications, as the compromised client is involved in each communication round. Despite this, it's impractical for attackers of CBD to possess a correctly labeled, triggered public dataset while fully compromising the local client in real-world settings. Moreover, CBD struggles to plant a backdoor in image classifiers. This possibly attributes to the difference in model complexity and classification complication. Conversely, the proposed attack is practical, and our Fed-EBD is effective against both text and image classifications, exhibiting comparable efficacy to those shown in the cross-device settings.

Text Classification v.s. Image Classification: In both text (Table 1) and image (Table 2) classification tasks, and for both IID and non-IID local datasets, our proposed attack, Fed-EBD, maintains a high level of efficacy across different FL settings – in most of the cases, Fed-EBD achieves relatively high ASRs while maintaining ACCs similar to those of the vanilla models. While CBD shows significant effectiveness in text classification under cross-silo scenarios, it struggles

Table 2. Performance (%) comparison on CIFAR-10 dataset

Setting	Cross-device						Cross-silo					
Approach	Vanilla		CBD		Fed-EBD		Vanilla		CBD		Fed-EBD	
Metric	ACC	ASR	ACC	ASR	ACC	ASR	ACC	ASR	ACC	ASR	ACC	ASR
IID	65.24	2.83	65.32	2.81	63.86	79.39	80.27	2.26	79.65	18.98	76.95	79.52
Non-IID	48.24	7.48	48.07	7.42	43.01	83.76	44.06	7.67	44.82	8.13	39.26	87.43

to prove effectiveness in cross-device settings and in image classification tasks, potentially due to the inherent complexity in datasets and intricacies involved in model structures. However, our proposed approach is unrelated to these limitations, exhibiting robust performance in both domains.

4.3 Homogeneous Setting Evaluation

In this experiment, we study the effectiveness of our attack when all clients share the same model architecture. In this case, all the clients use the standard DistilBERT for text classification and ResNet-18 architecture for image classification. The result shows our `Fed-EBD` maintains consistent ASR and ACC in both heterogeneous (Tables 1 and 2) and homogeneous (Table 3) FL settings. This consistency highlights the robustness and adaptability of our approach across different FL environments. It successfully targets shared vulnerabilities in the homogeneous system, where clients employ identical model architectures and have similar computational capabilities. Additionally, it exploits the universal susceptibility across diverse client architectures with varying computational resources in heterogeneous settings.

4.4 Case Study: Attack Effectiveness v.s. Public Data Utilization Ratio

In practical HFL settings, the server might randomly select a portion of the public dataset for knowledge distillation in each communication round to reduce communication and computational costs, as noted in [15]. To demonstrate the efficacy of our proposed attack in such realistic training conditions, we present results in Fig. 2 from 5 experiments. In these experiments, we vary the portions of the public dataset for knowledge distillation, specifically 20%, 40%, 60%, 80%, and 100%. (In our main experiments, the whole synthetic dataset is used for knowledge distillation.) All experiments are conducted on the IID CIFAR-10 datasets in the cross-silo FL setting with heterogeneous client model structures. As shown in Fig. 2, we observe that: 1) the ACC is almost unaffected by the public data utilization ratio, since, following the global communication with public data, the clients fine-tune their models on the untouched private datasets; 2) the ASR rises with the increased proportion of the public data used for knowledge distillation, as the misbehavior gets enhanced with more triggered instances involved in global communication. In general, the effectiveness of our `Fed-EBD` is not sensitive to the public data utilization ratio – the reduction in ASR is limited to 12%.

4.5 Hyper-Parameter Study: ASR v.s. Poisoning Ratio

We further explore the influence of a key hyper-parameter, the poisoning ratio of synthetic data, on the performance of our Fed-EBD. In our primary experiments on both text and image classification tasks, we set the poisoning ratio to 20%. We conduct 4 additional experiments, where we respectively set the poisoning ratio to 5%, 10%, 15%, and 25%, and the results in terms of ACC and ASR for our proposed attack are shown in Fig. 3. These experiments are conducted on the IID CIFAR-10 datasets under the cross-silo FL setting with heterogeneous client model structures. Similarly, the ACC remains relatively stable despite changes in the poisoning ratio, as the local private training set is untouched. As expected, the ASR is positively correlated to the public data poisoning ratio. Notably, even at a minimal poisoning ratio of 5%, our Fed-EBD maintains a high level of effectiveness, achieving an ASR of around 75%.

Table 3. Performance (%) comparison on the text and image classification tasks under the **homogeneous** setting. D1 is SST-2 dataset, D2 is AG-News, and D3 is CIFAR-10.

Setting		Cross-device						Cross-silo					
Approach		Vanilla		CBD		Fed-EBD		Vanilla		CBD		Fed-EBD	
Metric		ACC	ASR	ACC	ASR	ACC	ASR	ACC	ASR	ACC	ASR	ACC	ASR
D1	IID	83.70	38.24	78.81	0.22	84.59	98.92	84.49	28.33	83.46	94.68	84.24	92.61
	Non-IID	65.16	10.22	66.76	0.01	66.63	93.37	70.18	3.37	68.12	65.13	71.17	76.94
D2	IID	88.83	1.18	87.67	0.34	86.67	75.79	89.33	1.18	88.60	78.83	90.13	49.91
	Non-IID	88.33	0.05	90.99	0.48	89.00	58.57	90.67	0.89	92.33	48.54	89.67	75.82
D3	IID	64.43	2.66	64.47	2.72	63.21	92.89	77.52	2.84	75.92	6.85	77.27	62.57
	Non-IID	50.58	5.62	50.51	5.42	48.24	95.16	50.46	6.98	50.82	7.83	44.92	89.71

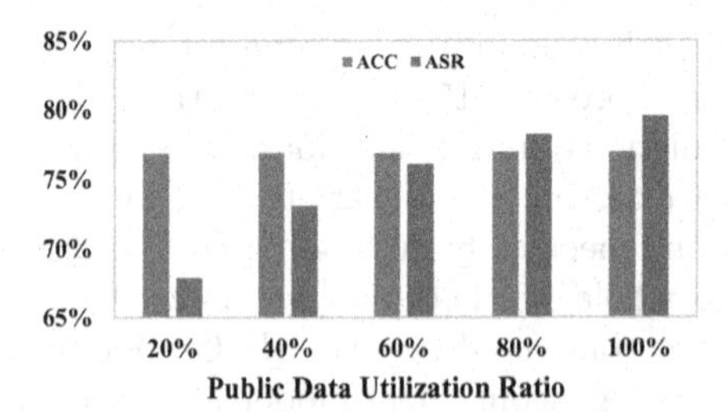

Fig. 2. Case study of public data utilization

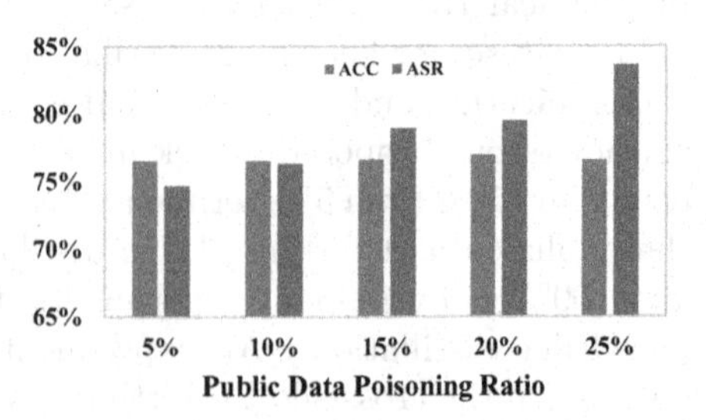

Fig. 3. Hyperparameter study

5 Conclusion

This paper addresses a critical and underexplored aspect of HFL: the security vulnerabilities inherent in using FMs for synthetic public dataset generation. We unveiled a novel backdoor attack mechanism that can be employed in HFL scenarios without necessitating client compromise or prolonged participation in the FL process. Our approach strategically embeds and transfers a backdoor through contaminated public datasets, demonstrating the ability to bypass existing federated backdoor defenses by exhibiting normal client behavior. Through extensive experiments in various FL settings and on diverse benchmark datasets, we have empirically established the effectiveness and stealth of our proposed attack. Our findings reveal a significant security risk in HFL systems using FMs, emphasizing the urgency for developing more robust defense mechanisms in this field.

References

1. Bagdasaryan, E., Veit, A., Hua, Y., Estrin, D., Shmatikov, V.: How to backdoor federated learning. In: AISTATS, vol. 108, pp. 2938–2948. PMLR (2020)
2. Che, L., Wang, J., Zhou, Y., Ma, F.: Multimodal federated learning: a survey. Sensors **23**(15), 6986 (2023)
3. Chen, X., Liu, C., Li, B., Lu, K., Song, D.: Targeted backdoor attacks on deep learning systems using data poisoning. arXiv:1712.05526 (2017)
4. Chou, S., Chen, P., Ho, T.: How to backdoor diffusion models? In: CVPR, pp. 4015–4024. IEEE (2023)
5. Dai, J., Chen, C., Li, Y.: A backdoor attack against LSTM-based text classification systems. IEEE Access **7**, 138872–138878 (2019)
6. Dong, Q., et al.: A survey for in-context learning. arXiv preprint arXiv:2301.00234 (2022)
7. Gu, T., Dolan-Gavitt, B., Garg, S.: Badnets: identifying vulnerabilities in the machine learning model supply chain. CoRR abs/1708.06733 (2017)
8. He, K., Zhang, X., Ren, S., Sun, J.: Deep residual learning for image recognition. In: Proceedings of the IEEE Conference on Computer Vision and Pattern Recognition, pp. 770–778 (2016)
9. Hinton, G.E., Vinyals, O., Dean, J.: Distilling the knowledge in a neural network. CoRR abs/1503.02531 (2015)
10. Huang, W., Ye, M., Du, B.: Learn from others and be yourself in heterogeneous federated learning. In: CVPR, pp. 10133–10143. IEEE (2022)
11. Kairouz, P., et al.: Advances and open problems in federated learning. Found. Trends® Mach. Learn. **14**(1–2), 1–210 (2021)
12. Kandpal, N., Jagielski, M., Tramèr, F., Carlini, N.: Backdoor attacks for in-context learning with language models. CoRR abs/2307.14692 (2023)
13. Kirillov, A., et al.: Segment anything (2023)
14. Krizhevsky, A., Nair, V., Hinton, G.: Cifar-10 (Canadian institute for advanced research) (2009). http://www.cs.toronto.edu/~kriz/cifar.html
15. Li, D., Wang, J.: FedMD: heterogenous federated learning via model distillation. CoRR abs/1910.03581 (2019). http://arxiv.org/abs/1910.03581
16. Li, L., Song, D., Li, X., Zeng, J., Ma, R., Qiu, X.: Backdoor attacks on pre-trained models by layerwise weight poisoning. In: EMNLP (2021)

17. Li, X., Wang, S., Huang, R., Gowda, M., Kesidis, G.: Temporal-distributed backdoor attack against video based action recognition. In: AAAI (2024)
18. Li, X., Wang, S., Wu, C., Zhou, H., Wang, J.: Backdoor threats from compromised foundation models to federated learning. CoRR abs/2311.00144 (2023)
19. Lin, T., Kong, L., Stich, S.U., Jaggi, M.: Ensemble distillation for robust model fusion in federated learning. In: NeurIPS (2020)
20. Lu, S., Li, R., Liu, W., Chen, X.: Defense against backdoor attack in federated learning. Comput. Secur. **121**, 102819 (2022)
21. McMahan, B., Moore, E., Ramage, D., Hampson, S., Arcas, B.A.: Communication-efficient learning of deep networks from decentralized data. In: AISTATS, pp. 1273–1282. PMLR (2017)
22. McMahan, B., Moore, E., Ramage, D., Hampson, S., Arcas, B.A.: Communication-efficient learning of deep networks from decentralized data. In: AISTATS (2017)
23. OpenAI: Gpt-3: Language models (2020). https://openai.com/research/gpt-3
24. Rieger, P., Nguyen, T.D., Miettinen, M., Sadeghi, A.: Deepsight: mitigating backdoor attacks in federated learning through deep model inspection. In: NDSS. The Internet Society (2022)
25. Rombach, R., Blattmann, A., Lorenz, D., Esser, P., Ommer, B.: High-resolution image synthesis with latent diffusion models (2022)
26. Sanh, V., Debut, L., Chaumond, J., Wolf, T.: DistilBERT, a distilled version of BERT: smaller, faster, cheaper and lighter (2020)
27. Shi, J., Liu, Y., Zhou, P., Sun, L.: BadGPT: exploring security vulnerabilities of ChatGPT via backdoor attacks to instructGPT. CoRR abs/2304.12298 (2023)
28. Socher, R., et al.: Recursive deep models for semantic compositionality over a sentiment treebank. In: EMNLP, pp. 1631–1642. ACL (2013)
29. Sun, L., Lyu, L.: Federated model distillation with noise-free differential privacy. In: Zhou, Z. (ed.) IJCAI, pp. 1563–1570. ijcai.org (2021)
30. Touvron, H., et al.: Llama: open and efficient foundation language models. arXiv preprint arXiv:2302.13971 (2023)
31. Wang, B., et al.: Decodingtrust: a comprehensive assessment of trustworthiness in GPT models. CoRR abs/2306.11698 (2023)
32. Wang, H., et al.: Attack of the tails: yes, you really can backdoor federated learning. In: NeurIPS (2020)
33. Wang, J., Ma, F.: Federated learning for rare disease detection: a survey (2023)
34. Wang, J., et al.: Towards personalized federated learning via heterogeneous model reassembly. arXiv preprint arXiv:2308.08643 (2023)
35. Wu, C., Yang, X., Zhu, S., Mitra, P.: Toward cleansing backdoored neural networks in federated learning. In: ICDCS, pp. 820–830. IEEE (2022)
36. Wu, C., Wu, F., Liu, R., Lyu, L., Huang, Y., Xie, X.: Fedkd: communication efficient federated learning via knowledge distillation. CoRR abs/2108.13323 (2021)
37. Xiang, Z., Miller, D.J., Chen, S., Li, X., Kesidis, G.: A backdoor attack against 3D point cloud classifiers. In: ICCV (2021)
38. Xie, C., Chen, M., Chen, P., Li, B.: CRFL: certifiably robust federated learning against backdoor attacks. In: ICML, vol. 139, pp. 11372–11382. PMLR (2021)
39. Xie, C., Huang, K., Chen, P., Li, B.: DBA: distributed backdoor attacks against federated learning. In: ICLR. OpenReview.net (2020)
40. Xu, J., Ma, M.D., Wang, F., Xiao, C., Chen, M.: Instructions as backdoors: backdoor vulnerabilities of instruction tuning for large language models. CoRR abs/2305.14710 (2023)
41. Yi, L., Wang, G., Liu, X., Shi, Z., Yu, H.: FedGH: heterogeneous federated learning with generalized global header. In: MM, pp. 8686–8696. ACM (2023)

42. Yu, S., Qian, W., Jannesari, A.: Resource-aware federated learning using knowledge extraction and multi-model fusion. CoRR abs/2208.07978 (2022)
43. Zhang, X., Zhao, J.J., LeCun, Y.: Character-level convolutional networks for text classification. In: NeurIPS, pp. 649–657 (2015)
44. Zhuang, W., Chen, C., Lyu, L.: When foundation model meets federated learning: motivations, challenges, and future directions. CoRR abs/2306.15546 (2023)

Combating Quality Distortion in Federated Learning with Collaborative Data Selection

Duc Long Nguyen[1], Phi Le Nguyen[1(✉)], and Thao Nguyen Truong[2(✉)]

[1] School of Information and Communication Technology, Hanoi University of Science and Technology, Hanoi, Vietnam
long.nd222179m@sis.hust.edu.vn, lenp@soict.hust.edu.vn

[2] The National Institute of Advanced Industrial Science and Technology, Tokyo, Japan
nguyen.truong@aist.go.jp

Abstract. Federated Learning (FL), a paradigm facilitating collaborative model training across distributed devices, has attracted substantial attention due to its potential to address privacy concerns and data localization requirements. However, the inherent inaccessibility of data poses a critical challenge in ensuring data quality within FL systems. Consequently, FL systems grapple with a range of data-related issues, encompassing erroneous samples, imbalanced data distributions, and data skew, all of which impose a significant impact on model performance. Therefore, the judicious selection of appropriate data for training is of paramount importance as it seeks to ameliorate these challenges.

This research paper tackles a crucial but often overlooked concern: the presence of low-quality data samples. In such circumstances, we introduce an innovative algorithm that strategically curates a subset of data for each training iteration, with the overarching objective of optimizing the model's accuracy while simultaneously addressing privacy concerns and reducing communication costs. Our primary innovation lies in the global selection of data, in contrast to the conventional approach that relies on individualized, client-specific data selection.

Furthermore, we introduce a novel medical dataset tailored specifically for classification tasks. This dataset intentionally incorporates various attributes associated with low-quality data to effectively replicate real-world conditions. Through rigorous empirical evaluation, we show the effectiveness of our algorithm using this dataset. The results demonstrate a notable improvement of approximately 2–3% in model performance, particularly in scenarios characterized by imbalanced data distributions.

1 Introduction

Along with the proliferation of edge devices, there is a significant escalation in the data collected from these devices. This enormous volume of data is, on the one hand, transforming into a valuable asset for training machine learning models. However, collecting data from edge devices to cloud servers/data centers (known as the centralized training paradigm) sometimes faces challenges of the huge

D.-N. Yang et al. (Eds.): PAKDD 2024, LNAI 14647, pp. 182–193, 2024.
https://doi.org/10.1007/978-981-97-2259-4_14

(a) High-quality.

(b) Bright image.

(c) Blurry image.

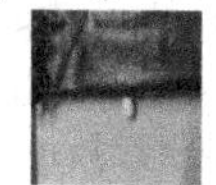

(d) W/ obstacles.

Fig. 1. Example of high-quality and low-quality images of the same pill.

Table 1. Bad effect of low-quality samples on the top-1 testing accuracy.

Dataset	PILL		
FL Algorithm	FedAvg	FedProx	FedFA
100% high-quality	74.07	74.80	72.20
Bright	66.47	–	–
Blurry	67.73	–	–
W/ obstacles	68.67	–	–
Mixed	67.80	69.33	64.20

communication cost and data-privacy. In this context, Federated Learning (FL) [14] has emerged as a viable solution to address the aforementioned challenges. This decentralized training paradigm facilitates data holders (hereafter, we call these clients) to actively participate in the training process while preserving the privacy of their data. More specifically, within the FL framework, clients undertake the training of models using their local data. Subsequently, the local models are aggregated via an orchestration server. In this manner, it is possible to construct a global model trained using all clients' data while maintaining the principles of data privacy.

However, due to the data's inaccessible nature, it is challenging to guarantee the quality of the training data in FL. To this end, in literature, numerous studies have been conducted to address various data-related challenges issues in FL, including data heterogeneity (e.g., non-IID [13], distribution shift [16], and adversarial data [19]). In this study, we instead investigate a critical yet overlooked issue: quality distortion, i.e., **the existence of poor-quality data**. In the image classification task, the quality-distorted data refers to those with noise, obstacles, bright images or blurry images [1,2]. Such poor-quality data is very common in practice, particularly in the context of FL, where private data is collected and utilized by clients without undergoing any form of quality assessment. Existing literature has demonstrated that most deep neural network models (DNNs) are very sensitive to quality distortion, i.e., even minor levels of distortion can substantially decline the model's accuracy [1,2]. Figure 1 illustrates some distorted samples, while Table 1 depicts their impacts on the accuracy of a model trained by FL[1]. However, quality-distorted data is not necessarily harmful. Undoubtedly, it is widely acknowledged that incorporating augmented data acquired through appropriate distortions can substantially enhance the model's generalizability [4,5]. Therefore, it is imperative for each client to carefully choose suitable data for the training process.

The topic of data selection has attracted significant attention in the literature. In [7], the authors demonstrate that each sample contributes differently to the model's performance. Subsequently, they propose a method for selecting the

[1] The configurations of this experiment is described in Sect. 4.1.

most crucial samples, enhancing accuracy, and minimizing training time. The authors in [24] offer an algorithm to identify data that have been incorrectly labeled. However, all of these studies consider the problems in the context of centralized training, which can not be straightforwardly applied to FL. More importantly, these studies do not address the quality distortion challenge. In the meanwhile, a mainstream of existing efforts in FL has been devoted for label-noise and adversarial data. For instance, the authors provide a technique in [22] for eliminating samples that have incorrect labels. Chen Wu et al. offer an approach for detecting backdoor attacks in [21]. The issue of data distortion, however, has not been addressed in any of these studies.

To fill in this gap, this work introduces a solution facilitating collaborative decision-making among clients to identify the most suitable data for the training objectives. What distinguishes our approach from conventional methods is its inherent decentralization, wherein each client independently conducts the data selection process using its private dataset. Remarkably, despite the distributed nature, the selected data is deemed optimal to train the global model. The motivation behind our proposed strategy is derived from the concept of importance score introduced in [7], which has been identified as closely associated with the degree of distortion. We develop a lightweight algorithm facilitating clients to exchange their data's importance scores to the server. Based on the provided data, the server will establish a shared threshold, allowing all clients to eliminate samples with a higher likelihood of causing degradation to the models. The contributions of our work are as follows.

- To the best of our knowledge, we are the first to tackle the issue of data quality distortion in the context of FL. To handle this issue, we propose a data selection algorithm that empowers clients to identify the most appropriate samples for training during each round. Our proposed algorithm is model-agnostic that can be integrated into most FL frameworks.
- We provide a novel real image dataset exhibiting multilevel quality distortion. The utilization of this dataset will enhance scholarly investigations in the fields of computer vision and FL.
- We perform extensive experiments to evaluate the effectiveness of our proposed data selection method when cooperating with various FL frameworks. In addition, we comprehensively compare our proposed data selection method and existing methodologies, providing empirical evidence to prove the superiority of our method over alternatives.

The remainder of the paper is organized as follows: In Sect. 3 we provide an overview and detailed design of our proposed method. Then, we present the simulation results in Sect. 4. Finally, we conclude the paper in Sect. 5.

2 Related Works

There have been many efforts to select suitable data for the training in the conventional centralized training, e.g., training in a supercomputer system. For example, **data pruning** is a technique utilized to reduce the size of a dataset by removing less significant samples [18] permanently. It results in a notable reduction in computational costs of the further training process on the pruned dataset [15,17,23]. Another approach is **online cutting** which addresses the limitations of traditional data pruning techniques [8] that need full training to identify samples to prune. This approach focuses more on temporarily selecting samples to hide from the training process during training. Specifically, [8] have introduced a novel method designed to select a subset of samples with the primary aim of minimizing the gradient matching error. [6,20] employ a dynamic approach in which a subset of data is chosen during each epoch, guided by their respective losses. In this work, we investigate the possibility of applying the online cutting methods to the context of Federated Learning to solve the issue of quality distortion. That is each client locally estimates the importance of its samples, i.e., gradient norm [8], or loss [6,20]. The importance of samples from all the clients is then collected to the server to perform the centralized selection (global sample selection strategy or **GSS**). GSS inherits the advantages of online cuttings but faces the issue of the huge communication cost for transferring the importance of samples. Another approach is the local sample selection strategy (**LSS**) [10]. In LSS, clients perform the selection algorithm separately using the same cutting ratio [10]. However, this approach meets the problem of inaccurately selecting the samples (see Sect. 4). In this paper, we propose the collaborative sample selection strategy (CSS) to overcome the limitations of GSS and LSS.

3 Proposal

3.1 Preliminaries

Federated Learning. The conventional Federated Learning (FL) framework comprises two primary components: a server denoted as S and a group of clients represented as $C_1, ..., C_n$. Each individual client C_i possesses a private dataset D_i, defined as $(x_i^j, y_i^j)_{j=1}^{|D_i|}$ in which x_i^j, y_i^j stands for the sample j at a client i, and its corresponding ground truth, respectively. The FL training procedure involves numerous communication rounds, with a specific subset of clients participating in each round. A communication round begins with the selected clients receiving the global model theta from the server. Subsequently, each client C_i employs its private dataset D_i to update the weights of theta. Once the local training is completed, clients transmit their updated weights back to the server, which then performs aggregation and updates the global model. Throughout the FL training process, the overarching objective is to optimize the following empirical objective function:

$$\theta^* = \arg\min_{\theta} \left\{ \sum_{i=1}^{n} \frac{|D_i|}{\sum_{k=1}^{n} |D_k|} L_i(\theta) \right\}, \tag{1}$$

where $L_i(\theta) = \frac{1}{|D_i|} \sum_{(x_i^j, y_i^j) \in D_i} L(\psi(x_i^j; \theta), y_i^j))$ denotes the loss function value computed over the entirety of training samples associated with client C_i.

Problem Definition. In Federated Learning (FL), each client performs the local model training using its entire private dataset. However, since client-collected data may include low-quality samples, incorporating such data into the training process not only hampers convergence speed but also diminishes model accuracy. Our challenge involves the removal of uninformative or even harmful data for each client during every communication round. To be more specific, during each communication round t, client C_i will determine a subset D_i^t of its entire local data D_i and utilize this subset to update the local model. It's important to note that D_i^t will undergo changes with each subsequent round.

3.2 Design Principle

The selection of D_i^t should adhere to the following criteria:

1. **Accuracy Maximization.** The primary objective in selecting training samples is to maximize the accuracy of the resultant global model. To achieve this goal, the criteria governing the selection must take into account the data contributed by all clients, rather than relying solely on the data from a single client image.
2. **Privacy Preservation.** To fulfill the initial principle mentioned above, it is essential for all clients to engage in a collaborative effort to define criteria for selecting training samples in each communication round. However, it remains of utmost importance that any information exchanged among clients should strictly aligns with the foundational principle of Federated Learning, which is to ensure clients' data privacy.
3. **Communication Efficiency.** Since communication channels between clients and servers frequently operate over low-bandwidth transmission lines, optimizing data exchange between these entities becomes essential. This optimization serves the dual objectives of reducing communication times and costs, thereby ultimately improving the overall efficacy of communication.

We rely on a so-called *importance score* to assess the goodness of the sample. The importance score of a given sample is calculated using the gradient norm associated with that sample. The rationale of our idea stems from several pieces of literature that demonstrated the close correlation between the importance score and image quality, i.e., an image's importance score tends to decrease in magnitude with its quality [2]. In light of this, our algorithm will set a threshold during each communication round. This threshold will guide clients in the removal of samples with importance scores that fall below the established threshold. In alignment with the first design principle, we introduce a server-mediated mechanism that facilitates the exchange of importance score related information among clients. The server, leveraging the information received from all clients, takes on the responsibility of determining a suitable cutoff threshold for each client. Our approach departs from the conventional practice of allowing clients to define their data cutoff thresholds independently.

To meet the second design principle, our approach requires that clients solely transmit data concerning the importance scores of their samples to the server

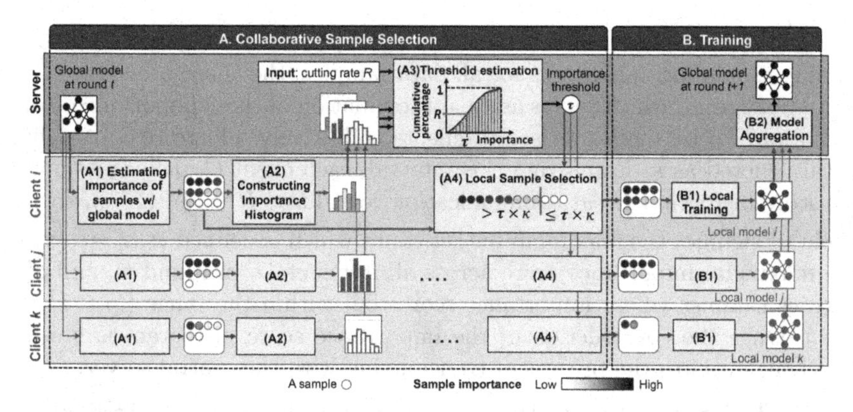

Fig. 2. Overview of the proposed collaborative sample selection method.

rather than transmitting the raw data. Furthermore, to adhere to the third criterion, clients are not required to transmit the importance scores of individual samples; instead, they are directed to transmit histograms encompassing the importance scores associated with their data. By adopting this approach, we can reduce the communication cost to $\mathcal{O}(m)$, with m as the number of bins (see details in Sect. 3.3). Given the importance score histograms provided by each client, the server performs an aggregation operation, resulting in an exhaustive summary of importance scores that encompass the entire dataset. This process, in turn, empowers the server to establish appropriate thresholds for each client.

Figure 2 illustrates the overview of our approach, which consists of two main stages: (A) Collaborative sample selection and (B) Training. Within the former stage, clients collaborate to select informative data for the training process. Subsequently, utilizing the chosen data, clients progress to the latter stage for model training. It is essential to note that our proposed collaborative sample selection algorithm is model-agnostic and operates independently from the training component, rendering it compatible with various federated learning mechanisms. Consequently, our attention herein is focused solely on the collaborative sample selection aspect, which comprises three principal phases: Importance score determination, threshold establishment, and sample selection.

3.3 Collaborative Sample Selection (CSS)

Importance Score Histogram Determination. Let us denote by $D_i = \{(x_i^j, y_i^j)\}$ the entire dataset of client C_i. At the beginning of every communication round t, upon receiving the global model $\psi(.;\theta^t)$ from the server, client C_i employs the following formula to determine the importance score $\varphi^t(x_i^j, y_i^j)$ of every data sample (x_i^j, y_i^j):

$$\varphi^t(x_i^j, y_i^j) = \sum_{\theta_k \in \theta} \|\nabla_{\theta_k} L(\psi(x_i;\theta), y_i))\|, \qquad (2)$$

where L represents the loss function, and θ_k depicts the k-th layer of θ. Upon obtaining the importance scores for all data samples, the client proceeds to construct a histogram, which serves as an approximation of these importance scores' distribution. To be more precise, all clients collectively adhere to a predefined interval denoted as κ. Employing this interval, each client C_i defines its importance score histogram H_i^t at communication round t as $H_i^t = \{d_i^t(1), ..., d_i^t(n_i^t)\}$, wherein n_i^t signifies the maximum bucket count which is defined as $n_i^t = \lceil \frac{I_i^t}{\kappa} \rceil$ (I_i^t is the maximum importance score across all items of D_i in round t), and $d_i^t(j)$ quantities samples whose importance scores fall within the range $((j-1) \times \kappa, j \times \kappa)$. Once the construction of the importance score histogram is finished, every client initiates the process of transmitting it to the central server.

Threshold Establishment. Upon the receipt of histograms contributed by participating clients, the server initiates a procedure for the determination of a threshold to cut off the training samples. Initially, the server consolidates all received histograms to establish a global histogram, which encapsulates the cumulative occurrences of importance scores derived from all clients' data. To elaborate, the global histogram, denoted as $\mathcal{H}$, is formulated as $\mathcal{H} = \{d^t(1), ..., d^t(m)\}$, where $d^t(j)$ signifies the total number of occurrences within the j-th histogram bucket, amalgamated from histograms of all participating clients, i.e., $d^t(j) = \sum_{i \in \mathbf{C}^t} d_i^t(j)$, where $\mathbf{C}^t$ represents the set comprising clients engaged in the communication round t.

Given the global histogram $\mathcal{H} = \{d^t(1), ..., d^t(m)\}$, the server then determine a removal threshold the cumulative fraction of pruning data remains within the confines of a predetermined ratio. Specifically, let R be a hyperparameter defining the portion of removal data, then the removal threshold is defined by the bucket value $\tau \times \kappa$ satisfying the following condition:

$$\frac{\sum_{j=1}^{\tau} d^t(j)}{\sum_{j=1}^{m} d^t(j)} \leq R < \frac{\sum_{j=1}^{\tau+1} d^t(j)}{\sum_{j=1}^{m} d^t(j)}. \tag{3}$$

In essence, $\tau \times \kappa$ represents the largest histogram bucket value for which the cumulative occurrences up to $\tau \times \kappa$ do not surpass R times the total occurrences.

Subsequently, the ascertained threshold is communicated to all participating clients. It's worth highlighting that in our algorithm, the proportion of data slated for removal exhibits variability across clients. This contrasts with conventional methods that typically employ a fixed removal ratio for sample pruning.

Sample Selection. Upon receiving the removal threshold $\tau \times \kappa$ from the server, all participating clients proceed to eliminate data samples whose importance scores fall below the specified threshold $\tau \times \kappa$. Subsequently, clients employ the selected subset for conducting local model training.

4 Evaluation

In this section, we evaluate the effectiveness of our proposed collaborative sample selection strategy (**CSS**) when integrated into various existing state-of-the-art

FL methods. In this evaluation, we choose the FL methods that focus on both aggregation mechanisms at the server (FedAvg [14] and FedFA [25]) and training algorithm at the client (FedProx [12]). We compare CSS with the global sample selection strategy (**GSS**) as proposed in centralized training [6,20] and local sample selection strategy (**LSS**) [10]. We demonstrate that CSS outperforms LSS and requires less communication cost than GSS with respect to all the experiment settings. In the following, we first introduce the details of the experiment settings in Sect. 4.1. We then report the evaluation result in Sect. 4.2. In all the experiments, **FULL** is used to refer to the testing accuracy of the model trained with all the samples from the low-quality datasets.

4.1 Datasets and Experimental Settings

Datasets. In this work, we collect a novel real image dataset exhibiting quality distortion of the pill images (hereafter, we call **PILL** dataset)[2]. The main task of the PILL dataset is to classify pill images correctly into classes. The PILL dataset consists of 12000 color images in 150 classes taken in real-world scenarios under unconstrained environments, including different lighting, backgrounds, and devices. Specifically, the dataset includes 62.5%, 12.5%, 12.5%, and 12.5% high-quality, bright, blurry images and images with obstacles, respectively. In our experiments, we pick up 1500 high-quality images for testing. The remaining images are used for the training process. To further validate CSS, we also conduct experiments on the CIFAR-100 [9] dataset. For exhibiting quality distortion of images, we increase the bright of 30% random images[3] in the training data subset (named as CIFAR100-BRIGHT). For each dataset, we simulate the non-IID setting by partitioning the training subset of the datasets and distributing the image into $N = 100$ clients based on the Dirichlet distribution $Dir(\alpha)$ [11].

Experiment Configuration. In this evaluation, we train the ResNet18 model [3] from a pre-trained weight provided by torchvision for both two datasets. We use stochastic gradient descent (SGD) as the optimizer with a learning rate learning rate $\eta = 0.01$. We set the local epoch $E = 5$ and the local batch size $B = 10$. In each communication round, only 30 among $N = 100$ clients are involved in the training process. We establish various training scenarios by adjusting the non-IID level (α) and the cutting ratio R. Unless otherwise mentioned, our default settings are $\alpha = 0.1$, and $R = 20\%$. When training with FedProx [12] and FedFA [25] algorithms, we set the proximal term $\mu = 0.1$ for the FedProx method. We also set $\gamma = 0.9$ and $\beta = 0.5$ for FedFA.

4.2 Experimental Results

Top-1 Accuracy. Table 2 presents the best top-1 testing accuracy that a sample selection method reaches within 1000 and 600 communication rounds for PILL

[2] Datasets are available at https://github.com/duclong1009/S-Selection.

[3] By using the transformation function *ImageEnhance* of the *PIL* library.

Table 2. Top-1 testing accuracy in the percentage of our proposed sample selection method (CSS) and the other method (LSS, GSS). The values show the best accuracy that each method reaches during training.

Dataset	PILL							CIFAR100-BRIGHT		
FL Algorithm	FedAvg					FedProx	FedFA	FedAvg	FedProx	FedFA
Non-IID level α	0.01	0.05	0.1	0.2	100	0.1	0.1	0.1	0.1	0.1
FULL (Uncuts)	72.60	74.40	67.80	76.00	74.60	69.33	64.20	66.86	72.03	60.23
LSS	71.20	75.44	66.47	74.33	74.40	66.20	61.4	65.86	71.49	57.73
GSS	72.87	76.80	69.60	77.53	76.00	68.93	65.87	66.88	71.89	59.58
CSS (Ours)	73.00	77.20	69.33	77.53	77.20	69.13	66.13	66.91	72.20	59.86

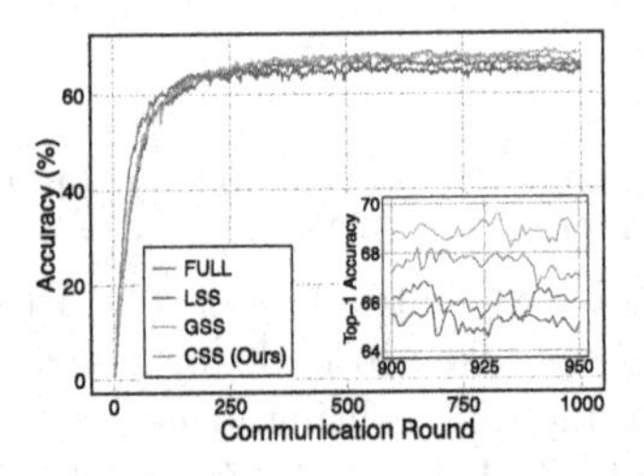

Fig. 3. Top-1 testing accuracy vs. communication round (PILL dataset).

Table 3. Impact of sample selecting algorithms and settings (FedAvg, PILL dataset, $\alpha = 0.1$).

	This work					[6]	[20]
Impt. metric	Gradient Norm					Loss	Loss
Selection algo	Threshold-based						
Other rules	NO					NO	YES
cutting ratio	10%	20%	30%	40%	50%	20%	20%
LSS	66.20	66.47	66.67	65.67	65.33	66.33	66.40
GSS	69.60	69.60	68.47	68.80	67.67	70.40	69.87
CSS (Ours)	69.21	69.33	69.33	68.20	66.87	67.87	67.53

and CIFAR-100 datasets, respectively. Expectedly, performing sample selection at clients separately (LSS) does not perform well in this evaluation. Because LSS cuts samples from all the clients using the same cutting ratio, it may accidentally mis-hide some important samples from the training process, e.g., in a client that contains all important samples. As a result, it leads to lower accuracy than those training with all the samples (FULL). By contrast, GSS and our proposed collaborative sample selection method (CSS) achieve better testing accuracy than FULL in most experiment settings. Furthermore, CSS could achieve nearly the same accuracy as (sometimes higher accuracy than) GSS without requiring sending all the importance-scores of samples from clients to the servers. This result emphasizes that our sample selection method could effectively address the challenge of distortion in data quality. For the CIFAR100-BRIGHT dataset, the top-1 accuracies achieved by FULL are nearly asymptotic to the case of training on the original CIFAR100 dataset (without poor-quality image). The result shows that increasing the brightness of images does not affect the accuracy much. Thus, there is no/trivial room for CSS to improve over FULL in this case.

Robustness of Our Method. We now study the robustness of the proposed method with different levels of non-IID α (as shown in Table 2) and cutting ratio R (Table 3) when training on PILL dataset with FedAvg algorithm. Table 2 shows that CSS helps to improve the accuracy in the case of both high-level of non-IID and low-level of non-IID. For example, CSS improves the accuracy around 2.8%, 1.5% and 2.6% in the case of $\alpha = [0.01, 0.2, 100]$, respectively (the data distribution is nearly the same as the case of IID when $\alpha = 100$).

Table 3 shows that the higher cutting ratio leads to the lower achieved accuracy of all the targeted sample selection methods. In comparison to the accuracy of FULL, i.e., 67.8%, using CSS could cut out more than 40% training samples in total without affecting the accuracy. It thus reduces the computation pressure to clients who may not have powerful computational. It is worth noting that CSS outperforms LSS by around 2–3% and could achieve nearly the same accuracy as GSS in all the cases. The result implies that our proposed method is stable with different datasets and configurations.

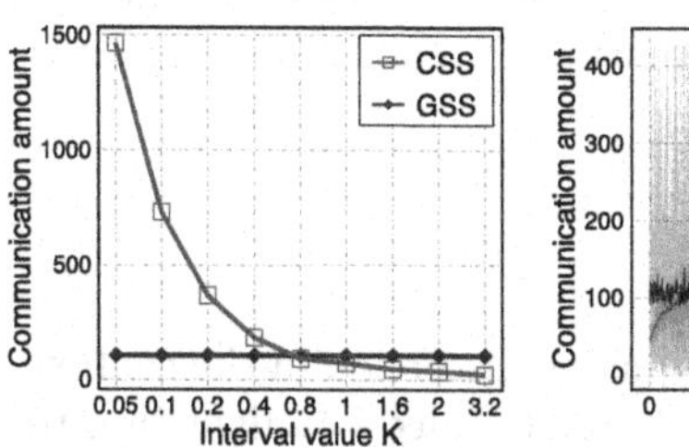

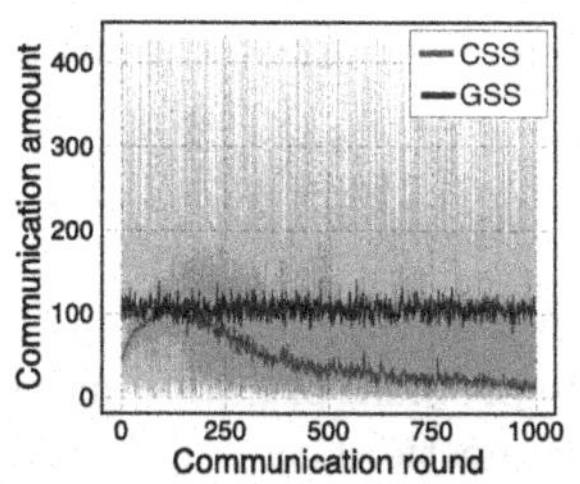

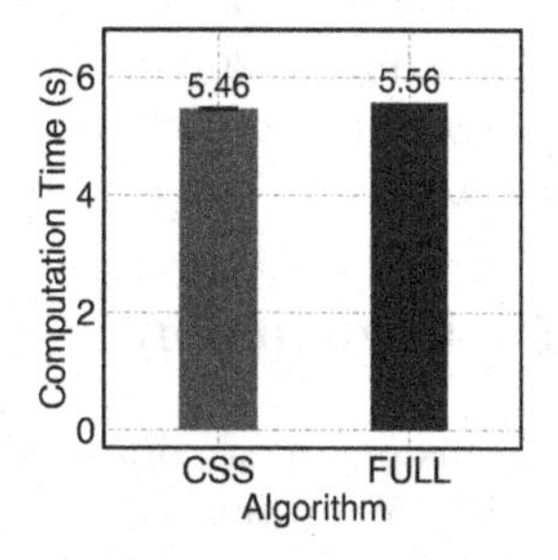

Fig. 4. Average communication amount vs. interval values κ (LEFT) and communication amount vs. communication round when $\kappa = 1.0$ (RIGHT).

Fig. 5. Average computation time/client/round.

Communication Cost. We have shown that CSS could achieve a similar accuracy with GSS. We now show the efficiency of CSS over the GSS in terms of communication, one of our design principles. We estimate the average communication amount of both CSS and GSS in the case of the PILL dataset, $\alpha = 0.1$, and $R = 20\%$. It is worth noting that because we use the same random seed in this evaluation, the clients involved in each round are exactly the same for both methods. The result in Fig. 4(LEFT) shows that when the interval value κ is high enough, e.g., greater than 0.8, CSS requires less communication than GSS. With the same set of importance scores at a client, increasing κ means reducing the number of bins of its' histogram, thus reducing the communication amount for transmitting this histogram to the server. For example, $\kappa = 1.0$ requires transferring 73.7 values on average while it is 46.2 when $\kappa = 1.6$. Figure 4(RIGHT) shows the communication amount versus communication round when $\kappa = 1.0$. Because the importance scores of samples have a large range at the beginning of the training process, CSS could not reduce much communication amount. However, after the 250th communication round, when the model starts to coverage (see Fig. 3), the gradients norm of all the samples become smaller. Therefore, it reduces the number of bins of the importance histograms significantly. Overall, we state that CSS could achieve better communication efficiency than GSS.

Computation Overhead. As reported in the prior works [6,8,20], all the sample selection methods meet an issue of huge computation overhead for calculating the importance scores of samples. For example, Fig. 5 shows the average computation time per client per round in the case of the PILL dataset, $\alpha = 0.1$, and $R = 30\%$ obtained over 4 experiments. The results show that although CSS could cut down up to 30% of training time, the overhead of computing the importance scores is non-trivial. Therefore, CSS can reduce only 2% of total computing time at client as the compared with FULL. One solution is to use '*loss*' for approximating the importance scores instead of using the gradient norms [6,20]. Both methods use '*loss*' as the metric for estimating the importance of samples. Besides, [20] proposes additional rules for cutting samples using prediction accuracy and prediction confidence. We perform the experiments with these two algorithms [6,20] and report the result in Table 3. Notably, the result shows that the behavior of CSS does not change in this evaluation. It again emphasizes the robustness of our method.

5 Conclusion

In this paper, we addressed the problem of data quality distortion in the context of FL. We proposed a model-agnostic data collaborative selection algorithm to handle this issue. The main idea is to select samples to train at the client side in each epoch dynamically based on (1) the samples' importance and (2) a global threshold that is determined at the server. We performed extensive experiments with a self-collected real pill image dataset (PILL). The experiment findings indicated that our proposed method CSS reveals an improvement of approximately 2-3% of the top-1 testing accuracy.

References

1. da Costa, G.B.P., Contato, W.A., Nazare, T.S., Batista Neto, J.E.S., Ponti, M.: An empirical study on the effects of different types of noise in image classification tasks (2016)
2. Dodge, S., Karam, L.: Understanding how image quality affects deep neural networks (2016)
3. He, K., Zhang, X., Ren, S., Sun, J.: Deep residual learning for image recognition (2015)
4. He, Z., Rakin, A.S., Fan, D.: Parametric noise injection: trainable randomness to improve deep neural network robustness against adversarial attack. In: Proceedings of the IEEE/CVF Conference on Computer Vision and Pattern Recognition (CVPR), June 2019
5. Holmstrom, L., Koistinen, P., et al.: Using additive noise in back-propagation training. IEEE Trans. Neural Netw. **3**(1), 24–38 (1992)
6. Jiang, A.H., et al.: Accelerating deep learning by focusing on the biggest losers (2019)
7. Katharopoulos, A., Fleuret, F.: Not all samples are created equal: deep learning with importance sampling. In: Proceedings of the 35th International Conference on Machine Learning, vol. 80, pp. 2525–2534. PMLR, July 2018

8. Killamsetty, K., Sivasubramanian, D., Ramakrishnan, G., De, A., Iyer, R.: Grad-match: gradient matching based data subset selection for efficient deep model training (2021)
9. Krizhevsky, A., et al.: Learning multiple layers of features from tiny images (2009)
10. Li, A., Zhang, L., Tan, J., Qin, Y., Wang, J., Li, X.-Y.: Sample-level data selection for federated learning. In: IEEE INFOCOM 2021 - IEEE Conference on Computer Communications, pp. 1–10 (2021)
11. Li, Q., Diao, Y., Chen, Q., He, B.: Federated learning on non-IID data silos: an experimental study. In: 2022 IEEE 38th International Conference on Data Engineering (ICDE), pp. 965–978. IEEE (2022)
12. Li, T., Sahu, A.K., Zaheer, M., Sanjabi, M., Talwalkar, A., Smith, V.: Federated optimization in heterogeneous networks (2020)
13. Li, X., Huang, K., Yang, W., Wang, S., Zhang, Z.: On the convergence of fedavg on non-IID data. arXiv preprint arXiv:1907.02189 (2019)
14. McMahan, B., Moore, E., Ramage, D., Hampson, S., Aguera, B., Arcas: Communication-efficient learning of deep networks from decentralized data, **54**, 1273–1282 (2017)
15. Paul, M., Ganguli, S., Dziugaite, G.K.: Deep learning on a data diet: finding important examples early in training (2023)
16. Pillutla, K., Laguel, Y., Malick, J., Harchaoui, Z.: Tackling distribution shifts in federated learning with superquantile aggregation. In: NeurIPS 2022 Workshop on Distribution Shifts: Connecting Methods and Applications (2022)
17. Qin, Z., et al.: Infobatch: lossless training speed up by unbiased dynamic data pruning (2023)
18. Sorscher, B., Geirhos, R., Shekhar, S., Ganguli, S., Morcos, A.S.: Beyond neural scaling laws: beating power law scaling via data pruning (2023)
19. Tolpegin, V., Truex, S., Gursoy, M.E., Liu, L.: Data poisoning attacks against federated learning systems. In: Chen, L., Li, N., Liang, K., Schneider, S. (eds.) ESORICS 2020. LNCS, vol. 12308, pp. 480–501. Springer, Cham (2020). https://doi.org/10.1007/978-3-030-58951-6_24
20. Truong, T.N., Gerofi, B., Martinez-Noriega, E.J., Trahay, F., Wahib, M.: KAKURENBO: adaptively hiding samples in deep neural network training. In: Thirty-Seventh Conference on Neural Information Processing Systems (2023)
21. Wu, C., Yang, X., Zhu, S., Mitra, P.: Mitigating backdoor attacks in federated learning (2021)
22. Yang, S., Park, H., Byun, J., Kim, C.: Robust federated learning with noisy labels. IEEE Intell. Syst. **37**(2), 35–43 (2022)
23. Yang, S., Xie, Z., Peng, H., Xu, M., Sun, M., Li, P.: Dataset pruning: reducing training data by examining generalization influence. In: The Eleventh International Conference on Learning Representations (2023)
24. Yu, X., Han, B., Yao, J., Niu, G., Tsang, I.W., Sugiyama, M.: How does disagreement help generalization against label corruption? (2019)
25. Zhou, T., Konukoglu, E.: FedFA: federated feature augmentation (2023)

Probabilistic Models and Statistical Inference

Neural Marked Hawkes Process for Limit Order Book Modeling

Guhyuk Chung[1], Yongjae Lee[2](✉), and Woo Chang Kim[1](✉)

[1] Korea Advanced Institute of Science and Technology, Daejeon, Republic of Korea
{wjdrngur12,wkim}@kaist.ac.kr

[2] Ulsan National Institute of Science and Technology, Ulsan, Republic of Korea
yongjaelee@unist.ac.kr

Abstract. Streams of various order types submitted to financial exchanges can be modeled with multivariate Temporal Point Processes (TPPs). The multivariate Hawkes process has been the predominant choice for this purpose. To jointly model various order types with their volumes, the framework is extended to the multivariate Marked Hawkes Process by considering order volumes as marks. Rich empirical evidence suggests that the volume distributions exhibit temporal dependencies and multimodality. However, existing literature employs simple distributions for modeling the volume distributions and assumes that they are independent of the history or only dependent on the latest observation. To address these limitations, we present the Neural Marked Hawkes Process (NMHP), of which the key idea is to condition the mark distributions on the history vector embedded with Neural Hawkes Process architecture. To ensure the flexibility of the mark distributions, we propose and evaluate two promising choices: the univariate Conditional Normalizing Flows and the Mixture Density Network. The utility of NMHP is demonstrated with large-scale real-world limit order book data of three popular futures listed on Korea Exchange. To the best of our knowledge, this is the first work to incorporate complex, history-dependent order volume distributions into the multivariate TPPs of order book dynamics.

Keywords: Marked Temporal Point Processes · Limit Order Book

1 Introduction

Sequential events can be found in various domains. Examples include orders submitted to financial exchanges, interactions between people on social media, and earthquakes. These events are often categorized into different event types. For example, orders in financial exchanges can be categorized into buy and sell orders; interactions in social media can be categorized into posts, shares, comments, and likes; locations of earthquakes can be discretized into cities or countries.

Multivariate Temporal Point Processes (TPPs) can be employed to model such sequential event data. The objective is to model the probability distribution of each event type k occurring along continuous-time space t, characterized by

D.-N. Yang et al. (Eds.): PAKDD 2024, LNAI 14647, pp. 197–209, 2024.
https://doi.org/10.1007/978-981-97-2259-4_15

the intensity function, $\lambda_k^*(t)$. Among numerous modeling choices for multivariate TPPs, Hawkes process [6,11] is one of the most predominant choices. Under the multivariate Hawkes process, the occurrence of one event type has a positive and additive impact on subsequent occurrences of other event types, including itself.

The classical Hawkes model has limited flexibility due to its fixed parametric form. To mitigate this issue, [14] proposed the Neural Hawkes Process (NHP), a class of neural TPPs that has gained popularity after the seminal work of [4]. Neural TPPs encode the history of past events to a vector with either recurrent or self-attention architecture [17].

While the multivariate Hawkes process is a widely adopted framework, categorizing events into discrete event types may not always suffice to fully describe each event. For instance, in financial exchanges, each order is further specified by its order quantity. Intuitively, this additional information, referred to as *marks*, would potentially affect the distributions of event occurrences. By additionally incorporating the marks, the framework is extended to the (multivariate) Marked Hawkes Process (MHP) [7].

We are interested in jointly modeling the distributions of order types with their volumes in financial markets under the MHP framework. While there is abundant empirical evidence that volume distributions exhibit a significant level of temporal dependencies and multimodality, current literature on modeling volume distributions is largely limited in these perspectives. Most literature utilizing TPPs for modeling order arrivals assumes a unitary volume for all orders, a simplification for analytical tractability [20,21]. While some studies try to model the distributions of order volumes, they either assume that volume distributions are independent of the history [5,18] or dependent only on the current observable state [10,13]. [15] studies history-dependent volume distributions, but they discretize the volumes into a small number of bins. The utilities of such discretized distributions might be limited for prediction/simulation tasks; it would be ambiguous which specific value to sample for each bin.

To address these limitations, we propose the Neural Marked Hawkes Process (NMHP), which can model continuous mark distributions that are history-dependent, highly complex, and modulated along continuous-time space. The main idea is to condition the mark distributions on the history vector embedded with the NHP architecture. The history vector of NMHP encapsulates not only the information of past events but also the observable exogenous states that can potentially impact the subsequent event arrivals. To model complex distributions, we utilize univariate Conditional Normalizing Flows (CNFs) since the marks of our interest are one-dimensional. (C)NFs allow exact computation of the log-likelihood of the distributions, thus can naturally be combined with the log-likelihood maximization problem of NHP. We also propose Mixture Density Network (MDN) [1] as an alternative choice for CNFs, which is simpler but might require domain-specific prior knowledge. Although the main focus of this work is on modeling the limit order book dynamics, our model is general enough to be potentially applied to other domains with one-dimensional continuous marks.

To demonstrate the utility of NMHP, we use large-scale real-world limit order book data of three popular futures listed on the Korea Exchange. With comprehensive experimental analysis, we show that our proposed method can more precisely model the volume distributions compared to the existing approaches. More specifically, our results suggest that modeling the volume distributions that are complex, history-dependent, and modulated along continuous-time space are all key ingredients that contribute to performance enhancements. Furthermore, the performance of order type modeling is also enhanced by incorporating additional information. To the best of our knowledge, this is the first study to model complex, history-dependent volume distributions in the context of TPPs.

2 Background

Neural Hawkes Process. We denote an event sequence over observation period $[0, T]$ as $S_T = \{(t_i, k_i) : t_i \leq T\}$, where $t_i < t_{i+1}$ are the arrival times and $k_i \in \{1, ..., K\}$ represents the event type. Given an event sequence S_T, the objective of multivariate TPPs is to model and estimate the intensity functions $\lambda_k^*(t), k \in \{1, ..., K\}$ that best describe the given sequence. The intensity functions are defined such that $\lambda_k^*(t)dt$ represents the probability of event type k occurring in a short time interval $[t, t + dt)$. We denote the history of past events up to time t as $\mathcal{H}_t = \{(t_i, k_i) : t_i < t\}$. The $*$ superscript in $\lambda_k^*(t)$ denotes that it is conditioned on history $\mathcal{H}_t$, i.e., $\lambda_k^*(t) := \lambda_k(t; \mathcal{H}_t)$. The classical Hawkes model [6,11] assumes that the occurrence of one event type has a positive and additive impact on the occurrences of subsequent events, which decay on continuous time, most often exponentially. Then, the intensify function $\lambda_k^*(t)$ is written as

$$\lambda_k^*(t) = \mu_k + \sum_{i:t_i<t} \alpha_{k_i,k} \exp(-\delta_{k_i,k} (t - t_i)), \tag{1}$$

where μ_k is the base intensity, $\alpha_{k_i,k}$ and $\delta_{k_i,k}$ represent impact and decay parameters of event type k_i to k.

To enrich the classical Hawkes process, [14] proposed Neural Hawkes Process (NHP) where Continuous-Time LSTM (CT-LSTM) is developed to encode $\mathcal{H}_t$ to hidden state, $\boldsymbol{h}(t)$. During the interarrival period between two events $(t_i, t_{i+1}]$, the cell state $\mathbf{c}(t)$ starts from $\mathbf{c}_i$ at t_i, and decays exponentially towards $\bar{\mathbf{c}}_i$ until t_{i+1} as in Eq. (2). Hidden state $\boldsymbol{h}(t)$ is continuously obtained from the cell state during this period as in Eq. (3). Finally, the intensity rate vector, $\boldsymbol{\lambda}^*(t) = [\lambda_k^*(t)]_{k=1}^K$, is decoded from the hidden state $\boldsymbol{h}(t)$ as in Eq. (4).

$$\mathbf{c}(t) = \bar{\mathbf{c}}_i + (\mathbf{c}_i - \bar{\mathbf{c}}_i) \exp(-\boldsymbol{\delta}_i (t - t_i)) \tag{2}$$

$$\boldsymbol{h}(t) = \mathbf{o}_i \odot \tanh(\mathbf{c}(t)) \tag{3}$$

$$\boldsymbol{\lambda}^*(t) = \mathrm{softplus}(\mathbf{W}\boldsymbol{h}(t)) \tag{4}$$

The hidden state is updated discontinuously at the arrival of a new event at t_{i+1}, as the associated cells and input, forget, output gates are updated as in Eq. (5–14). $\mathbf{k}_{i+1}$ in Eq. (5) is a one-hot encoded vector representing the type of new event and is concatenated with the decayed hidden state $\boldsymbol{h}(t_{i+1}^-)$ to form $\tilde{\boldsymbol{h}}_{i+1}$, which is used for the updates. The update equations are similar to the discrete-time case, but two separate input and forget gates are used for updating each of $\mathbf{c}_i$ and $\bar{\mathbf{c}}_i$. Equation (14) is newly introduced to obtain the decay parameters of memory cells.

$$\tilde{\boldsymbol{h}}_{i+1} \leftarrow \text{concat}(\mathbf{k}_{i+1}, \boldsymbol{h}(t_{i+1}^-)) \quad (5)$$
$$\mathbf{i}_{i+1} \leftarrow \text{sigmoid}(\mathbf{W}_i\tilde{\boldsymbol{h}}_{i+1} + \mathbf{d}_i) \quad (6)$$
$$\bar{\mathbf{i}}_{i+1} \leftarrow \text{sigmoid}(\mathbf{W}_{\bar{i}}\tilde{\boldsymbol{h}}_{i+1} + \mathbf{d}_{\bar{i}}) \quad (7)$$
$$\mathbf{f}_{i+1} \leftarrow \text{sigmoid}(\mathbf{W}_f\tilde{\boldsymbol{h}}_{i+1} + \mathbf{d}_f) \quad (8)$$
$$\bar{\mathbf{f}}_{i+1} \leftarrow \text{sigmoid}(\mathbf{W}_{\bar{f}}\tilde{\boldsymbol{h}}_{i+1} + \mathbf{d}_{\bar{f}}) \quad (9)$$
$$\mathbf{z}_{i+1} \leftarrow \tanh(\mathbf{W}_z\tilde{\boldsymbol{h}}_{i+1} + \mathbf{d}_z) \quad (10)$$
$$\mathbf{o}_{i+1} \leftarrow \text{sigmoid}(\mathbf{W}_o\tilde{\boldsymbol{h}}_{i+1} + \mathbf{d}_o) \quad (11)$$
$$\mathbf{c}_{i+1} \leftarrow \mathbf{f}_{i+1} \odot \mathbf{c}(t_{i+1}) + \mathbf{i}_{i+1} \odot \mathbf{z}_{i+1} \quad (12)$$
$$\bar{\mathbf{c}}_{i+1} \leftarrow \bar{\mathbf{f}}_{i+1} \odot \bar{\mathbf{c}}_i + \bar{\mathbf{i}}_{i+1} \odot \mathbf{z}_{i+1} \quad (13)$$
$$\boldsymbol{\delta}_{i+1} \leftarrow \text{softplus}(\mathbf{W}_d\tilde{\boldsymbol{h}}_{i+1} + \mathbf{d}_d) \quad (14)$$

Conditional Normalizing Flows. For a one-dimensional random variable y, Normalizing Flows (NFs) are used to model the complex probability distribution of y, denoted as $p_Y(y)$ [9]. This is done by learning a transformation from y to z, where z is also one-dimensional and follows a known, simple distribution $p_Z(z)$, referred to as *base* distribution. The transformation should be flexible and invertible and is usually defined as a chain of differentiable, parametric, and bijective functions. Let us denote the transformation as $f_\theta = f_{M,\theta_M} \circ \cdots \circ f_{1,\theta_1}$, i.e., $z = f_\theta(y)$. For $m \in \{1, ..., M\}$, θ_m denotes the parameters associated with the function f_{m,θ_m}, and $\theta = [\theta_1, ..., \theta_M]$. Then, the log-likelihood of y can be computed with the change of variables formula as follows.

$$\log p_Y(y) = \log p_Z(f_\theta(y)) + \log |f'_\theta(y)| \quad (15)$$

The parameters of the transformations are estimated with deep neural networks by maximizing equation (15). Once the transformation is learned, it can be used as a generative model by sampling $z \sim p_Z(z)$ and transforming it to $y = f_\theta^{-1}(z)$.

The NFs can be extended to the problem of learning conditional distribution $p_{Y|X}(y|\mathbf{x})$ where the distribution of y is dependent on some high-dimensional data $\mathbf{x}$ [19]. Such extension is usually referred to as Conditional Normalizing Flows (CNFs). The idea of CNFs is to use deep neural networks to parameterize the transformations, with the input $\mathbf{x}$ fed into the networks, i.e., $\theta(\mathbf{x}) = \text{NN}(\mathbf{x})$. More rigorously, the conditional transformation is written as $f_\theta(y|\mathbf{x}) := f(y; \theta(\mathbf{x}))$, and the conditional distribution can now be written as

$$p_{Y|X}(y|\mathbf{x}) = p_Z(f_\theta(y|\mathbf{x})) \cdot |f'_\theta(y|\mathbf{x})|. \quad (16)$$

The log-likelihood of the conditional distribution is computed analogously as in Eq. (15). With the described parameterization, CNFs learn different transformations conditional on $\mathbf{x}$.

While most transformations are devised for high-dimensional data and are not readily applicable to one-dimensional data, [19] introduces a univariate radial transformation, which is represented as

$$R_{\alpha,\beta,\gamma}(y) = y + \frac{\alpha\beta(y-\gamma)}{\alpha + |y-\gamma|}. \quad (17)$$

They show that strict monotonicity can be ensured with $\alpha > 0$ and $\beta \geq -1$. They suggested the following parameterization $\hat{\alpha}, \hat{\beta}, \gamma = \text{NN}(\mathbf{x})$, and further transforming as $\alpha = \log(\exp(\hat{\alpha}) + 1)$ and $\beta = \exp(\hat{\beta}) - 1$. They derive the derivative in a closed form.

3 Neural Marked Hawkes Process

In this section, we present the details of our model in its most general form. We introduce some tweaks for order book modeling in the later section.

Notations. A *marked* event sequence over observation period $[0, T]$ is denoted as $S_T^M = \{(t_i, k_i, v_{k_i}^{(i)}) : t_i \leq T\}$ where $t_i \leq t_{i+1}$, $k_i \in \{1, ..., K\}$, and K denotes the number of event types. The subscript of $v_{k_i}^{(i)}$ denotes that it is a mark attached to event type k_i, and the superscript (i) indicates that it is associated with the i-th event. We assume that v_k are univariate, continuous random variables. Discrete marks can be dequantized to be treated as continuous. Then, the history of *marked* events upto time t is denoted as $\mathcal{H}_t^M = \{(t_i, k_i, v_{k_i}^{(i)}) : t_i < t\}$. Similarly, S_T^U denotes the *unmarked* event sequence, and $\mathcal{H}_t^U$ is the history of *unmarked* events upto time t.

In addition, we consider observable exogenous states that can affect event occurrences. Given S_T^M, we denote a coupled sequence of observable exogenous states as $S_T^S = \{(t_i, s_1^{(i)}, ..., s_P^{(i)}) : t_i \leq T\}$, where the superscript (i) indicates that the states are observed at the time immediately after the arrival of an event at time t_i, and P is the dimension of the observable exogenous state. We denote the history of past states as $\mathcal{H}_t^S$ and defined similarly. The final event sequence we consider is denoted as $S_T = S_T^M \cup S_T^S := \{(t_i, k_i, v_{k_i}^{(i)}, s_1^{(i)}, ..., s_P^{(i)}) : t_i \leq T\}$, and the history becomes $\mathcal{H}_t = \mathcal{H}_t^M \cup \mathcal{H}_t^S$.

Remark. We distinguish between marks and states to indicate that marks are endogenous and we are interested in modeling their distributions, while states are exogenous and are used only for enriching the history.

Objective. Given S_T, we are interested in modeling the distributions of the event arrivals and their corresponding marks, which are fully characterized by the conditional intensity rate, $\lambda_k^*(t, v_k) := \lambda_k(t, v_k; \mathcal{H}_t)$. We adopt the following factorization of [3]

$$\lambda_k^*(t, v_k) = \lambda_k^*(t) p_k^*(v_k|t), \tag{18}$$

where $\lambda_k^*(t)$ is the *ground intensity rate* of event type k, and $p_k^*(v_k|t)$ denotes the conditional mark density function of event type k at time t. By separately modeling the ground intensity rate and the conditional mark distribution for each type k, the log-likelihood of S_T^M is represented as [3, Proposition 7.3.III]

$$\mathcal{L}(S_T^M) = \sum_{i:t_i \leq T} \log\lambda_{k_i}^*(t_i) - \int_0^T \lambda^*(t)dt + \sum_{i:t_i \leq T} \log p_{k_i}^*(v_{k_i}^{(i)}|t_i), \tag{19}$$

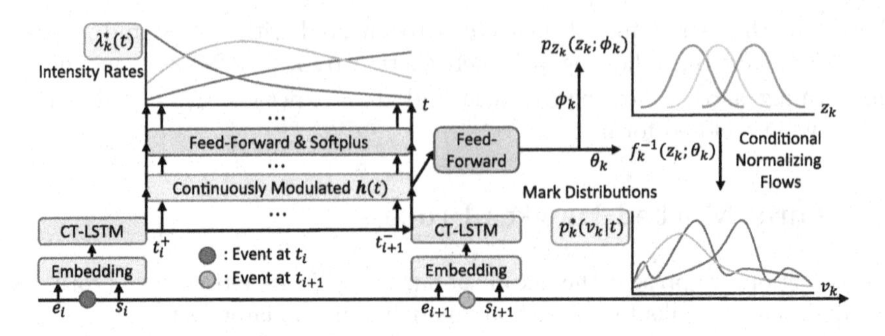

Fig. 1. Architecture of the Neural Marked Hawkes Process with Conditional Normalizing Flows.

where $\lambda^*(t) = \sum_{k_i=1}^{K} \lambda_{k_i}^*(t)$. The first two terms are equivalent to the log-likelihood of the multivariate Hawkes process where the integral is estimated with Monte-Carlo sampling [14]. The third term is related to the log-likelihood of the marks. We decompose $\mathcal{L}$ into $\mathcal{L}_1 = \sum_{i:t_i \le T} \log\lambda_{k_i}^*(t_i) - \int_0^T \lambda^*(t)dt$, $\mathcal{L}_2 = \sum_{i:t_i \le T} \log p_{k_i}^*(v_{k_i}^{(i)}|t_i)$ so that $\mathcal{L} = \mathcal{L}_1 + \mathcal{L}_2$. We refer to $\mathcal{L}_1$ and $\mathcal{L}_2$ as *types* log-likelihood and *marks* log-likelihood, respectively. NMHP is trained by maximizing $\mathcal{L}$.

Embedding the History. Given S_T, we employ the NHP architecture to embed the history $\mathcal{H}_t$ to *history vector* $\boldsymbol{h}(t)$, which is utilized for modeling both $\lambda_k^*(t)$ and $p_k^*(v_k|t)$. In specific, the embedding process follows Eq. (2–14) with modification in Eq. (5) to include additional information of marks and states. For the event type k and any categorical states, we utilize an embedding layer, a simple lookup table, to encode the information to vectors of length d_m. For the marks v_k and any non-categorizable states, we encode the values to vectors of length d_m with a feed-forward layer, e.g., $\mathbf{v}_k = \mathbf{w}_v v_k + \mathbf{d}_v$, where $\mathbf{v}_k$ denotes the encoded vector, and $\mathbf{w}_v$ and $\mathbf{d}_v$ are vectors of length d_m. Finally, Eq. (5) is modified as

$$\tilde{\boldsymbol{h}}_{i+1} \leftarrow \text{concat}(\mathbf{k}_{i+1} + \boldsymbol{v}_{k_{i+1}}^{(i+1)} + \boldsymbol{s}_1^{(i+1)} + \cdots + \boldsymbol{s}_P^{(i+1)}, \boldsymbol{h}(t_{i+1}^-)). \quad (20)$$

Note that all values are encoded to vectors of equal dimension d_m.

Mark Distributions. With the history vector $\boldsymbol{h}(t)$, we are left with modeling $\lambda_k^*(t)$ and $p_k^*(v_k|t)$ before we can maximize the log-likelihood $\mathcal{L}$. Equation (4) is used to output $\lambda_k^*(t)$.

It is of our main interest in this work to model highly flexible, history-dependent, continuously modulated mark distributions, $p_k^*(v_k|t)$. For this purpose, we propose to condition the mark distributions on the history vector, $\boldsymbol{h}(t)$. To ensure the flexibility of the distributions, we suggest two promising choices: the Conditional Normalizing Flows and the Mixture Density Network.

Conditional Normalizing Flows. We detail how the normalizing flows is conditioned on $\boldsymbol{h}(t)$. For each event type k, we model the conditional base distribution,

$p_{Z_k}(z_k|\boldsymbol{h}(t))$, and the conditional transformation, $f_k(v_k|\boldsymbol{h}(t))$, as

$$p_{Z_k}(z_k|\boldsymbol{h}(t)) = p_{Z_k}(z_k; \phi_k(\boldsymbol{h}(t))), \quad f_k(v_k|\boldsymbol{h}(t)) = f_k(v_k; \theta_k(\boldsymbol{h}(t))), \tag{21}$$

such that $z_k = f_k(v_k|\boldsymbol{h}(t)) \sim p_{Z_k}(\cdot|\boldsymbol{h}(t))$. p_{Z_k} is a univariate, simple, and known distribution with parameter ϕ_k, and f_k is a univariate, flexible, and invertible function parameterized by θ_k. Then,

$$p_k^*(v_k|t) = p_k(v_k|\boldsymbol{h}(t)) = p_{Z_k}(f_k(v_k|\boldsymbol{h}(t))|\boldsymbol{h}(t)) \cdot |f_k'(v_k|\boldsymbol{h}(t))|. \tag{22}$$

The log-likelihood of Eq. (22) can be obtained with the change of variables formula presented in Eq. (15). Note that in NHP architecture, $\boldsymbol{h}(t)$ is continuously modulated during the interarrival period between two events; thus, the first equivalent sign in Eq. (22) is valid.

The parameters ϕ_k and θ_k for $k \in \{1, ..., K\}$ are estimated with neural networks with given input $\boldsymbol{h}(t)$. More specifically, we denote $\phi = [\phi_1; \phi_2; \cdots; \phi_K]$ and $\theta = [\theta_1; \theta_2; \cdots; \theta_K]$, where a semi-colon(;) indicates vertical concatenation. Then, $[\phi, \theta] = \text{NN}(\boldsymbol{h}(t))$, and here a comma(,) indicates horizontal concatenation. Several feed-forward layers with ReLU activations are employed for $\text{NN}(\cdot)$. With the parameterization described above, our model learns different transformations for each k and for given $\boldsymbol{h}(t)$.

For the flexibility of the transformation f_k, we chain a number of radial transformations after a single affine transformation as in [16], i.e.,

$$f_k^R(v_k; \theta_k^R) = R_{\alpha_M, \beta_M, \gamma_M} \circ \cdots \circ R_{\alpha_1, \beta_1, \gamma_1} \circ A_{a,b}(v_k), \tag{23}$$

where $R_{\alpha,\beta,\gamma}$ refers to the univariate radial transformation introduced in the earlier section and $A_{a,b}(x) = \text{softplus}(a) \cdot x + b$ indicates an affine transformation. We refer to $f_k^R(\cdot)$ as the radial flows hereafter.

Mixture Density Network. For each event type k, let us consider a mixture of D distributions, each of which $p_{k,d}(v_k; \psi_{k,d})$ and $\pi_{k,d}$ denote its density function and the mixture weight, respectively. Then,

$$p_k^*(v_k|t) = \sum_{d=1}^{D} \pi_{k,d}(\boldsymbol{h}(t)) \cdot p_{k,d}(v_k; \psi_{k,d}(\boldsymbol{h}(t))). \tag{24}$$

As indicated in the above equation, not only the parameters of each distribution but also its mixture weights are determined by $\boldsymbol{h}(t)$. Again, the parameters $\psi_k = [\psi_{k,1}; \cdots; \psi_{k,D}]$ and the mixture weights $\pi_k = [\pi_{k,1}; \cdots; \pi_{k,D}]$ for each k are estimated with neural networks that receive $\boldsymbol{h}(t)$ as the input. A softmax function is applied as the last layer for the mixture weights to ensure that $\Sigma_{d=1}^{D} \pi_{k,d} = 1$.

4 Related Work

Categorical marks are most often considered in the marked TPPs literature [4,17]. A straight-forward approach for modeling the distributions of categorical marks would be to utilize a fully-connected layer followed by a softmax activation function to output the probabilities. [23] exploits the transformer architecture to model additional information on different marks by discretizing their distributions. Among continuous marks, modeling the locations of events in the context of spatio-temporal point processes is the most common setting. [8] proposes Neural Jump Stochastic Differential Equations (Neural Jump SDEs) to solve TPPs and handle spatial marks with restricted distributions. [2] extends the work of Neural Jump SDEs with continuous-time normalizing flows to model flexible, history-dependent spatial mark distributions. [22] models non-stationary influence kernel with spectral decomposition to handle flexible distributions of continuous marks.

5 Experiments

We utilize NMHP for modeling order book dynamics in financial markets, where we consider the order volumes as marks.

Problem Settings. In financial exchanges, orders are specified by their types (limit, market, cancel), side (bid, ask), price, and volume. It is a norm to consider only the best price quotes for buy and sell for simplicity. Then, we have six types of orders. We further distinguish between order types that cause a change in the mid-price[1] and those that do not change the mid-price [12], resulting in a total of 12 types of orders.

Since the order volumes can take only positive values, we employ log-normal distributions for both the base distribution of CNFs and component distributions for MDN. Log-normal distributions can model the right-skewness of the distributions, which closely resembles the volume distributions. For CNFs, we uniformly dequantize the order volumes, which is a norm when applying normalizing flows to discrete values. We choose to subtract rather than add uniform noise, ensuring that the volume distributions are naturally lower-bounded by zero.

To deal with the lower-bounded distributions, we slightly tweak the radial flows presented in the earlier section. Specifically, the flows are modified as

$$z_k = \hat{f}_k^R(v_k) = C_0 \circ f_k^R \circ C_0^{-1}(v_k) \sim \text{LogNormal}(\mu_k, \sigma_k) \tag{25}$$

where f_k^R denotes the previously described radial flows and $\hat{f}_k^R$ indicates the modified radial flows. C_0 refers to a soft-clipping transformation to the interval of $(0, \infty)$. The soft-clip transformation approximates the clip transformation as a continuous and differentiable function.

For the exogenous states, we include the information of queue lengths at the best bid/ask price levels, denoted as q_{bid} and q_{ask}, and the spread defined as $\delta = p_{ask} - p_{bid}$, where p_{bid} (p_{bid}) denotes the best bid (ask) price. We treat the queue lengths as non-categorizable states and the spread as binary states; $\delta = 0$

[1] Mid-price is defined as $\frac{p_b + p_a}{2}$, where p_{bid} (p_{ask}) is the best bid (ask) price.

Table 1. Marks log-likelihood per event ($\mathcal{L}_2$) on three futures datasets.

	KRW/USD			KOSPI200			KTB10Y		
Conditioned on → Distributions ↓	$s(t)$ ($\mathcal{L}_2$)	$h_{lstm}(t)$ ($\mathcal{L}_2$)	$h_{nmhp}(t)$ ($\mathcal{L}_2$)	$s(t)$ ($\mathcal{L}_2$)	$h_{lstm}(t)$ ($\mathcal{L}_2$)	$h_{nmhp}(t)$ ($\mathcal{L}_2$)	$s(t)$ ($\mathcal{L}_2$)	$h_{lstm}(t)$ ($\mathcal{L}_2$)	$h_{nmhp}(t)$ ($\mathcal{L}_2$)
Log-Normal	−3.660	−3.542	−3.569	−1.988	−1.915	−1.928	−2.024	−1.986	−2.022
Log-Normal Mix (2)	−3.547	−3.389	−3.401	−1.929	−1.789	−1.786	−1.994	−1.924	−1.918
Log-Normal Mix (3)	−3.529	−3.353	−3.311	−1.907	−1.758	−1.749	−1.967	−1.892	−1.890
Log-Normal Mix (5)	−3.501	−3.327	**−3.274**	−1.891	−1.736	**−1.723**	−1.956	**−1.873**	**−1.869**
Radial Flows	−3.504	−3.325	**−3.245**	−1.901	−1.755	**−1.732**	−1.951	−1.882	−1.876

($\delta = 1$) when the spread is equal to (larger than) one tick. These variables, known to influence subsequent order arrivals, are the ones on which existing literature bases the conditioning of volume distributions [10,13]. Following the previously used notations, we denote $\boldsymbol{s}(t) = \boldsymbol{s}^{(i)}_{q_{bid}} + \boldsymbol{s}^{(i)}_{q_{ask}} + \boldsymbol{s}^{(i)}_{\delta}$ as the embedded vector of the current observable states during the time interval $t \in [t_i, t_{i+1})$.

Datasets. We test our model with large-scale real-world LOB data of three popular futures listed in the Korea Exchange: KRW/USD Exchange Rate Futures (KRW/USD), KOSPI200 Index Futures (KOSPI200), and 10-Year Korea Treasury Bond Futures (KTB10Y). The dataset includes the one-month period of November 2019, which is comprised of 21 trading days. We use the first 12 days for training, the subsequent 3 days for validation, and the last 6 days for evaluation. When only considering the orders associated with the best price levels, KRW/USD, KOSPI200, and KTB10Y comprise approximately 3 million, 7 million, and 2 million orders, respectively. We divide the entire sequence of orders per day into sample sequences of length equal to 200 without overlapping. To the best of our knowledge, our dataset is the most large-scale LOB dataset for demonstrating neural TPP models.

Baselines. We consider two categories of baselines: models that maximize types log-likelihood, $\mathcal{L}_1$, and models that maximize marks log-likelihood, $\mathcal{L}_2$.

For the types log-likelihood, we consider four TPPs: 1) Hawkes: the classical multivariate Hawkes process with exponential decays, 2) THP: Transformer Hawkes Process in [23] where self-attentive architecture is employed rather than recurrent architecture, 3) NHP: plain vanilla NHP without marks and exogenous states, and 4) NHP$^+$: NHP with marks and exogenous states provided but does not model the mark distributions.

For the marks log-likelihood, we first compare with the existing approaches that condition the volume distributions on the current observable state, $\boldsymbol{s}(t)$. To further validate our modeling choice of conditioning the volume distributions on the history vector embedded with NMHP, we evaluate the performance of conditioning on the history vector embedded with the conventional LSTM. LSTM is given the same input as NMHP, while the interarrival times between orders are additionally embedded in the input. We denote the history vectors embedded with LSTM and NMHP as $\boldsymbol{h}_{lstm}(t)$ and $\boldsymbol{h}_{nmhp}(t)$, respectively.

Results. We first evaluate the performances of marks log-likelihood, $\mathcal{L}_2$, as it is the main focus of our work, followed by the evaluation of types log-likelihood, $\mathcal{L}_1$. We further analyze the contributions of marks and exogenous states information on enhancing the modeling performances with an ablation study. Each configuration is repeatedly run five times with unique seeds, and the mean values are reported.

Modeling the Marks. In Table 1, we can find significant improvements in $\mathcal{L}_2$ when conditioning on the history vector of NMHP, $\boldsymbol{h}_{nmhp}(t)$, compared to conditioning on the current observable state, $\boldsymbol{s}(t)$, regardless of the distribution choices. We confirm that volume distributions indeed depend heavily on history. When conditioning on the same information, increasing the number of mixtures monotonically enhances the performance, with the most significant improvement in moving from the log-normal distribution to a mixture of two log-normal distributions. Such results indicate that volume distributions are complex and multimodal, and utilizing a single distribution significantly lacks flexibility.

When utilizing either the log-normal mixture with more than two components or the radial flows, conditioning on $\boldsymbol{h}_{nmhp}(t)$ yields considerably better performances than that of LSTM, $\boldsymbol{h}_{lstm}(t)$. We recall that $\boldsymbol{h}_{nmhp}(t)$ is continuously modulated during the time interval $[t_i, t_{i+1})$ between two orders and so are the estimated volume distributions, while $\boldsymbol{h}_{lstm}(t)$ is fixed during this interval, and updated discontinuously only when a new order arrives. Then, our results suggest that not only the ground intensity rates of the order arrivals but also the volume distributions depend on the continuously modulated impacts of the past orders during the non-event time period; thus, our modeling choice of allowing the volume distributions to be modified along continuous-time space contributes to enhancing the performance. From the observation that $\boldsymbol{h}_{nmhp}(t)$ attains better performance than $\boldsymbol{h}_{lstm}(t)$ only with more complex distribution models, we conjecture that less complex distributions lack the flexibility to incorporate such continuous-time modulation effects.

Radial flows consistently outperform log-normal mixture models with less than five components, indicating the capability of the radial flows to successfully model multimodality. Comparisons between the log-normal mixture with five components and the radial flows are mixed. Radial flows achieve moderately improved performance on KRW/USD when the condition vector is $\boldsymbol{h}_{nmhp}(t)$, while surpassed by the log-normal mixture with five components by small margins for most of the other cases. We cautiously conclude that the log-normal mixture model is sufficient for modeling the volume distributions, while leaving some leeway for the utilities of CNFs for other domains where composing a mixture model requires high level of domain-specific knowledge.

Modeling the Types. From Table 2, the two neural models, NHP and THP, outperform the classical Hawkes model, while NHP shows better performance than THP for our datasets. Both NMHP and NHP$^+$ dominate NHP by a considerable amount. Since these three models share the same architecture for embedding the history, the performance gap should be rooted in the additional information that NMHP and NHP$^+$ are provided.

Table 2. Types log-likelihood per event ($\mathcal{L}_1$) on three futures datasets.

Models	KRW/USD	KOSPI200	KTB10Y
	$\mathcal{L}_1$	$\mathcal{L}_1$	$\mathcal{L}_1$
CHP	2.508	2.566	2.798
THP	2.666	2.672	2.810
NHP	2.761	2.828	2.933
NHP$^+$	**2.881**	**2.890**	**3.075**
NMHP	**2.883**	**2.888**	**3.072**

Table 3. Ablation study on history components.

History	KRW/USD	
	$\mathcal{L}_1$	$\mathcal{L}_2$
$\mathcal{H}^U$	2.747	−3.418
$\mathcal{H}^M$	2.831	−3.293
$\mathcal{H}^U \cup \mathcal{H}^S$	2.804	−3.395
$\mathcal{H}^M \cup \mathcal{H}^S$	**2.883**	**−3.245**

It is important to note that the performance differences between the last two models are ignorable, while NMHP additionally models the volume distributions and NHP$^+$ does not. Such results indicate either: 1) the historical information required for modeling the volume distributions overlaps to a significant extent with the information needed for modeling the type distributions, or 2) the model size is sufficiently large to further encapsulate additional information for the volume distributions. Considering that our model is of moderate size and that NHP$^+$ and NMHP achieve a similar level of types log-likelihood regardless of the model size, we can conclude that the former explanation is more reasonable. This again advocates our modeling choice of sharing the same history vector for modeling the types and mark distributions; we should have hesitated to condition the mark distributions on the same history vector if $\mathcal{L}_1$ of NMHP had deteriorated.

Finally, we remark that for NMHP, the difference in the performance of $\mathcal{L}_1$ is negligible regardless of the model used for the mark distributions.

Enriching the History. In Table 3, we evaluate on KRW/USD four variants of NMHP models with radial flows, each encoding different history. 1) $\mathcal{H}^U$: unmarked event sequences only, 2) $\mathcal{H}^M$: marked event sequences only, 3) $\mathcal{H}^U \cup \mathcal{H}^S$: unmarked event sequences with exogenous states, and 4) $\mathcal{H}^M \cup \mathcal{H}^S$: marked event sequences with exogenous states. We can find that additionally providing the volume and states information, respectively, to the unmarked sequences both contribute to performance enhancements. The performance enhancements are achieved both in modeling the types and marks. Among the two pieces of information, volume information contributes notably more than the states information.

6 Conclusion

We aim to contribute to the line of research on modeling limit order book dynamics with TPPs by incorporating complex, history-dependent volume distributions. For this purpose, we have presented the Neural Marked Hawkes Process, a new approach for modeling multivariate marked TPPs where the marks are one-dimensional, non-categorizable, and whose distributions depend on the history. We utilize the architecture of the NHP to obtain a continuously-modulated history vector that encodes not only the history of past events but also the history of observable exogenous states. To model complex, history-dependent mark distributions, we propose to utilize either the univariate Conditional Normalizing Flows or the Mixture Density Network, where the parameters of the distributions are fully conditioned on the history vector. With comprehensive analysis, we show that our model exhibits enhanced modeling flexibility in terms of both the event types and marks.

Acknowledgments. This work was supported by the National Research Foundation of Korea (NRF) grant funded by the Ministry of Science and ICT (NRF-2022M3J6A 1063021, NRF-2022R1I1A4069163, and RS-2023-00208980) and the Institute of Information & communications Technology Planning & evaluation (IITP) grants funded by the Korea government (MSIT) (No. 2020-0-01336, Artificial Intelligence Graduate School Program (UNIST)). The authors wish to acknowledge LINE Investment Technologies for providing valuable insights related to market microstructure.

References

1. Bishop, C.M.: Mixture density networks (1994)
2. Chen, R.T., Amos, B., Nickel, M.: Neural spatio-temporal point processes. arXiv preprint arXiv:2011.04583 (2020)
3. Daley, D.J., Vere-Jones, D., et al.: An Introduction to the Theory of Point Processes: Volume I: Elementary Theory and Methods. Springer, New York (2003). https://doi.org/10.1007/b97277
4. Du, N., Dai, H., Trivedi, R., Upadhyay, U., Gomez-Rodriguez, M., Song, L.: Recurrent marked temporal point processes: Embedding event history to vector. In: Proceedings of the 22nd ACM SIGKDD International Conference on Knowledge Discovery and Data Mining, pp. 1555–1564 (2016)
5. Embrechts, P., Liniger, T., Lin, L.: Multivariate Hawkes processes: an application to financial data. J. Appl. Probab. **48**(A), 367–378 (2011)
6. Hawkes, A.G.: Spectra of some self-exciting and mutually exciting point processes. Biometrika **58**(1), 83–90 (1971)
7. Hawkes, A.G.: Hawkes processes and their applications to finance: a review. Quant. Financ. **18**(2), 193–198 (2018)
8. Jia, J., Benson, A.R.: Neural jump stochastic differential equations. In: Advances in Neural Information Processing Systems, vol. 32 (2019)
9. Kobyzev, I., Prince, S.J., Brubaker, M.A.: Normalizing flows: an introduction and review of current methods. IEEE Trans. Pattern Anal. Mach. Intell. **43**(11), 3964–3979 (2020)
10. Kumar, P.: Deep Hawkes process for high-frequency market making. arXiv preprint arXiv:2109.15110 (2021)
11. Liniger, T.J.: Multivariate Hawkes processes. Ph.D. thesis, ETH Zurich (2009)
12. Lu, X., Abergel, F.: High-dimensional Hawkes processes for limit order books: modelling, empirical analysis and numerical calibration. Quant. Financ. **18**(2), 249–264 (2018)
13. Lu, X., Abergel, F.: Order-book modeling and market making strategies. Mark. Microstruct. Liq. **4**(01n02), 1950003 (2018)

14. Mei, H., Eisner, J.M.: The neural Hawkes process: a neurally self-modulating multivariate point process. In: Advances in Neural Information Processing Systems, vol. 30 (2017)
15. Rambaldi, M., Bacry, E., Lillo, F.: The role of volume in order book dynamics: a multivariate Hawkes process analysis. Quant. Financ. **17**(7), 999–1020 (2017)
16. Rothfuss, J., et al.: Noise regularization for conditional density estimation. arXiv preprint arXiv:1907.08982 (2019)
17. Shchur, O., Türkmen, A.C., Januschowski, T., Günnemann, S.: Neural temporal point processes: a review. arXiv preprint arXiv:2104.03528 (2021)
18. Shi, Z., Cartlidge, J.: State dependent parallel neural Hawkes process for limit order book event stream prediction and simulation. In: Proceedings of the 28th ACM SIGKDD Conference on Knowledge Discovery and Data Mining, pp. 1607–1615 (2022)
19. Trippe, B.L., Turner, R.E.: Conditional density estimation with Bayesian normalising flows. arXiv preprint arXiv:1802.04908 (2018)
20. Wu, P., Rambaldi, M., Muzy, J.F., Bacry, E.: Queue-reactive Hawkes models for the order flow. arXiv preprint arXiv:1901.08938 (2019)
21. Wu, P., Rambaldi, M., Muzy, J.F., Bacry, E.: A single queue reactive Hawkes model for the order flow. Market Microstruct. Liq. (2023, accepted)
22. Zhu, S., Wang, H., Dong, Z., Cheng, X., Xie, Y.: Neural spectral marked point processes. arXiv preprint arXiv:2106.10773 (2021)
23. Zuo, S., Jiang, H., Li, Z., Zhao, T., Zha, H.: Transformer Hawkes process. In: International Conference on Machine Learning, pp. 11692–11702. PMLR (2020)

How Large Corpora Sizes Influence the Distribution of Low Frequency Text n-grams

Joaquim F. Silva(✉) and Jose C. Cunha

NOVA LINCS, NOVA School of Science and Technology, Caparica, Portugal
{jfs,jcc}@fct.unl.pt

Abstract. The prediction of the numbers of distinct word n-grams and their frequency distributions in text *corpora* is important in domains like information processing and language modelling. With big data *corpora*, there is an increased application complexity due to the large volume of data. Traditional studies have been confined to small or moderate size *corpora* leading to statistical laws on word frequency distributions. However, when going to very large *corpora*, some of the assumptions underlying those laws need to be revised, related to the *corpus* vocabulary and numbers of word occurrences. So, although it becomes critical to know how the *corpus* size influences those distributions, there is a lack of models that characterise such influence. This paper aims at filling this gap, enabling the prediction of the impact of *corpus* growth upon application time and space complexities. It presents a fully principled model, which, distinctively, considers words and multiwords in very large *corpora*, predicting the cumulative numbers of distinct n-grams above or equal to a given frequency in a *corpus*, as well as the sizes of equal-frequency n-gram groups, from unigrams to hexagrams, as a function of *corpus* size, in a language, assuming a finite n-gram vocabulary. The model applies to low occurrence frequencies, encompassing the larger populations of n-grams. Practical assessment with real *corpora* shows relative errors around 3%, stable over the considered ranges of n-gram frequencies, n-gram sizes and *corpora* sizes from million to billion words, for English and French.

Keywords: n-gram distribution · low-frequency n-grams · large *corpora*

1 Introduction

Many applications rely on the statistical regularities of single and multiple consecutive words (n-grams, $n \geq 1$) in natural language *corpora*, e.g., language modelling, indexing, extraction, translation and compression; n-grams abstract properties of sub-components within sequences, being useful in many domains, e.g. bioinformatics and genomics, information processing and databases. Two important characteristics are: the *corpus* vocabulary, expressed as numbers of distinct single words or multiwords in a *corpus* of size C; the equal-frequency ($k \geq 1$) n-gram classes, each with a size of $W(k, C)$ distinct n-grams. Those

D.-N. Yang et al. (Eds.): PAKDD 2024, LNAI 14647, pp. 210–222, 2024.
https://doi.org/10.1007/978-981-97-2259-4_16

properties have mostly been studied for single words and moderate size *corpora*. But large scale *corpora* have impact upon n-gram frequency distributions, and, although some recent works consider large *corpora* they ignore low-frequency n-grams occurring 1, 2, 3 ... times. However, most relevant n-grams are of low frequency, being important, e.g. for indexing, topic mining, etc. Also, often, relevant expressions are multiwords, so their distributions must be studied.

We propose a fully principled model, for words and multiwords (1-grams to 6-grams), in large *corpora*, showing a faithful prediction of the cumulative numbers of distinct n-grams, $D(k, C)$, with frequencies above or equal to k in a *corpus* of size C, as well as, the sizes of equal-frequency n-gram groups, $W(k, C)$, as a function of *corpus* size in a given language. The focus on low-frequency n-grams allows encompassing the larger populations of content n-grams in a *corpus*. The model assumes that, at a given temporal epoch, each language has a finite word vocabulary. It enables fast[1] prediction of word and multiword frequency distributions with growing *corpora* sizes. This is useful, e.g. for evaluating time and space complexities of n-gram applications with large *corpora* and its impact upon application design, e.g. n-gram caches capacity and miss behaviour.

Results are presented for 1-grams up to 6-grams, in English and French real *corpora* from 31 million to 8.6 billion words. Relative errors are around 3%, with very stable values for all n-gram sizes in the range of *corpora* sizes considered. Sections 2, 3, 4 and 5 present the background, the model, results and conclusions. A guide for the model reproducibility is found at http://bit.ly/3gqM6rS.

2 Background and Related Work

Most frequency distribution models only consider single words (1-grams), in three representations: i) Frequency-Rank (FR) ranks distinct words by their decreasing frequencies; ii) Cumulative Frequency (CF) shows proportions of words with frequencies greater than or equal to a given value; iii) Size-Frequency (SF) partitions the distinct words of equal frequency (k) into classes of sizes $W(k, C)$. Among the proposals [3,12–14,18,19], Zipf's empirical law [13,19] states that $f(r) = f(1)\,r^{-\alpha}$, the r_{th} ranked word with $f(r)$ occurrences and constant $\alpha \approx 1$. Figure 1a (FR form) shows the typical frequency *vs* rank distribution, the larger steps on the right showing the equal-frequency groups.

Pareto's Law, was proposed (CF form) for income distributions [13], but not applying to low incomes (frequencies). Zipf's Law deviates from real data for high and low frequencies.

Some $W(k, C)$ models are based on empirical Zipf's Law [6,15,17]. Others [3,5,14,18] have a theoretical foundation, e.g. Simon's preferential attachment model [18], Price's cumulative advantage model [14], and some network growth models [1]. Asymptotically, the probability $P(k)$ of a word occurring k times is often estimated by a power law $k^{-\gamma}$, with a constant $\gamma > 0$, but also presenting deviations to real data for low and high values of k, having led to corrections

[1] Obviously, this excludes the *corpora* collecting and n-gram counting, which is made only once for parameters estimation and validation of the model.

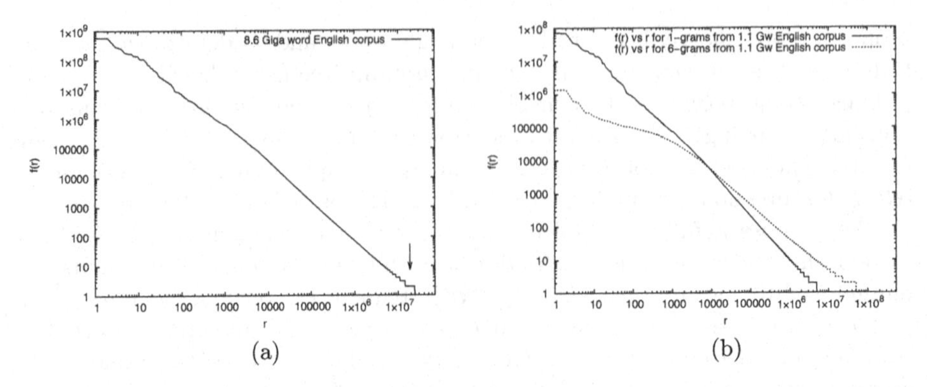

Fig. 1. a) Empirical word (1-gram) frequency-rank distribution in log-log. Larger equal-frequency groups, $W(k, C)$, shown for larger ranks (see arrow); (b) Empirical 1-gram and 6-gram frequency-rank distributions in log-log

[3,5,18], which are only applicable to single words and moderate *corpora.* Most models assume an infinite vocabulary, following the empirical Heaps-Herdan's Law [9]: D, the number of distinct words in a *corpus* grows as $D \propto C^{\beta}$, with β $(0 < \beta < 1)$ assumed constant. However, deviations from real data suggest that β should be a function of C [5], so D will saturate as $C \to \infty$. The vocabulary finiteness is assumed by only a few works [11,15–17] and in other domains [2,4].

Some studies consider multiwords [7,8,10,15–17]. Recent works focus on very large *corpora* [7,8,15–17]. But when n-gram low-frequency data is truncated as in [7], it precludes full validation with real data. Models [10,15,16] were the only found proposing $W(k, C)$ models for n-grams $(n \geq 1)$, low frequencies and large *corpora*, all being based on empirical Zipf's Law, so suffer from deviations from real data; [15,16] had to introduce empirical correction factors.

3 The Model

We first propose a model, $D(k, C; L, n)$ (or $D(k, C)$ when convenient), to predict the cumulative number of distinct n-grams with frequencies greater or equal to a given k. This is a generalisation for $k \geq 1$ of the $D(C; L, n)$ model in [15] (limited to $k = 1$). Secondly, we use the model to calculate $W(k, C; L, n)$ (or $W(k, C)$).

A Generalised Model for Distinct n-grams Occurring $k \geq 1$ Times. We propose a model to estimate $D(k, C; L, n)$ in a *corpus* of size C in language L and n-grams of size n. No specific language morphosyntactic assumptions are made. We assume that, in each fixed temporal epoch, there is a finite n-gram language vocabulary, with size $V(L, n)$ (or simply V). So, although $D(k, C; L, n)$ increases monotonically with C for any $k \geq 1$, it will eventually reach an upper bound, $V(L, n)$, the n-gram vocabulary, e.g., for $k = 1$, $D(C; L, n) = D(1, C; L, n)$, the total number of distinct n-grams, obviously being upper bounded by $V(L, n)$.

This generalisation relies on the cumulative nature of the $D(k, C; L, n)$ sets. Indeed, it is expected that the growth rate of an individual $D(k, C; L, n)$ set to be influenced by a cumulative form of preferential attachment, that is, it tends to increase at a rate proportional to its current relative size in a *corpus*. This is reflected in a first factor, $D(k, C; L, n)/C$, which is the average number of distinct n-grams with frequency greater or equal to k, per word in C. However, due to the vocabulary finiteness, a slow down is expected of the above growth as *corpus* size grows, as reflected in a second factor, that is, the fraction of distinct n-grams still having frequency below k, whether they are in the current *corpus* or they are still unseen n-grams: $(V(L,n) - D(k, C; L, n))/V(L,n)$ – for instance, when $k = 1$, this is the fraction of distinct n-grams from the finite language vocabulary still unseen in the current *corpus*: $(V(L,n) - D(1, C; L, n))/V(L,n)$ –. Then, the growth rate, modelled by $\frac{dD(k,C;L,n)}{dC}$, is given by the product:

$$\frac{dD(k,C)}{dC} = g_k \, \frac{D(k,C)}{C} \, \frac{V - D(k,C)}{V} \tag{1}$$

where g_k simplifies constant $g_k(L,n)$ for each k and (L,n) pair. Indeed, $V = \sum_{k=0}^{k=kmax} W(k,C)$ where $kmax$ is, for each n, the highest frequency in the *corpus*. Then, $\forall k \geq 1$, $\sum_{i=0}^{i=k-1} W(i,C) = V - D(k,C)$, the number of distinct n-grams with frequency below k. These assumptions, as reflected by those two factors, are naturally sustained and consistent with the observations of the $D(k, C; L, n)$ growth curves for the considered collection of real *corpora* for the ranges of k and n values, and the two languages considered. Indeed, the first factor alone in (1), that is $g_k \frac{D(k,C)}{C}$, although reflecting a decreasing growth rate, was found unable, by itself, to faithfully capture the real growth rates of the $D(k, C; L, n)$, mainly as the *corpora* became larger in the billion words scale, due to the progressive influence of the vocabulary finiteness effect[2]. From (1) we get

$$\int \frac{V D(k,C)^{-1}}{g_k (V - D(k,C))} dD(k,C) = \int \frac{1}{C} dC \;\Rightarrow\; -\frac{\ln(|\frac{V}{D(k,C)} - 1|)}{g_k} + ct_1 = \ln(|C|) + ct_2$$

with integration constants ct_1, ct_2. As $\frac{V}{D(k,C)} \geq 1$ and $C > 0$, let $ct_2 - ct_1 = \ln(h_k)$. Then, $\ln((\frac{V}{D(k,C)} - 1)^{-\frac{1}{g_k}}) = \ln(h_k) + \ln(C) \Rightarrow (\frac{V}{D(k,C)} - 1)^{-\frac{1}{g_k}} = h_k\, C$. Thus:

$$D(k, C; L, n) = \frac{V(L,n)}{1 + (h_k(L,n)\, C)^{-g_k(L,n)}} \,. \tag{2}$$

The Number of Equal-Frequency Distinct n-grams, $W_d(k, C; L, n)$. The above generalisation for the $D(k, C; L, n)$ model[3] enabled a fully founded way

[2] Equation (1) is equivalent to $\frac{\frac{dD(k,C)}{D(k,C)}}{\frac{dC}{C}} = g_k \frac{V - D(k,C)}{V}$. The infinite V assumption would imply that the ratio in left side of the equation should be a constant (equal to g_k) wrt C, but the empirical observations showed that ratio decreases instead. Such decrease is captured by the vocabulary finiteness assumption (second factor).

[3] Indeed, (2) can be written as $\frac{V - D(k,C)}{D(k,C)} = (h_k\, C)^{-g_k}$, which, for each k and n, is a power law wrt to C, since g_k and h_k were found constants wrt C.

of calculating $W_d(k, C; L, n)$, to predict the number of equal-frequency distinct n-grams of each given size n, for low occurrence frequencies $k : 1, 2, 3, \ldots$, in a *corpus* of size C, in a language L, assuming a finite n-gram language vocabulary. Equation (3) gives $W_d(k, C; L, n)$, for given k and n, by subtracting cumulative numbers $D(k, C; L, n)$ and $D(k+1, C; L, n)$ for consecutive frequencies k and $k+1$.

$$W_d(k, C; L, n) = D(k, C; L, n) - D(k+1, C; L, n) \ . \tag{3}$$

As final outcome from (3) and (2), we obtain the model expression (4):

$$W_d(k, C; L, n) = \frac{V(L, n)}{1 + (h_k(L, n)\, C)^{-g_k(L,n)}} - \frac{V(L, n)}{1 + (h_{k+1}(L, n)\, C)^{-g_{k+1}(L,n)}} \ . \tag{4}$$

This is in contrast to the empirical, Zipf based, curve fitting approach in [15, 16], where $W(k, C; L, n)$ is calculated by subtracting the two delimiting n-gram ranks associated to the equal-frequency (k) interval in the Zipf frequency-rank form (Fig. 1a). However, as these ranks are obtained using Zipf's Law, [15,16] suffer from the significant Zipf deviations from real data observed for the low k values. To overcome this, [15,16] introduced correction factors into $W(k, C; L, n)$, empirically derived by curve fitting to a specific *corpora* collection. Our approach does not suffer from this drawback because $W_d(k, C; L, n)$, based on a founded $D(k, C; L, n)$ model (2), does not depend on any empirical ad-hoc correction.

Parameters Estimation. Parameters $V(L, n)$, g_k and h_k in (2) are estimated, for each (L, n) pair, under the criteria in Sect. 4.3, aiming to provide a minimum error in the predictions for different *corpora* sizes. This is supported by a training phase, such that for each (L, n) pair, two training *corpora*, c_1 and c_2 have sizes C_1 and C_2, and have $D(k, C_1)$ and $D(k, C_2)$ distinct n-grams with frequency greater or equal to k. From (2), $1 + (h_k\, C_1)^{-g_k} = \frac{V}{D(k,C_1)}$ and $1 + (h_k\, C_2)^{-g_k} = \frac{V}{D(k,C_2)}$. Let $A_1 = \frac{V}{D(k,C1)} - 1$ and $A_2 = \frac{V}{D(k,C_2)} - 1$. Then,

$$\left(\frac{h_k\, C_1}{h_k\, C_2}\right)^{-g_k} = \frac{A_1}{A_2} \quad \Rightarrow \quad \left(\frac{C_1}{C_2}\right)^{-g_k} = \frac{A_1}{A_2} \quad \Rightarrow \quad g_k = \frac{\ln(\frac{A_1}{A_2})}{\ln(\frac{C_2}{C_1})} \ . \tag{5}$$

Once $D(k, C_1)$ and $D(k, C_2)$ are obtained by empirical counts from *corpora* c_1 and c_2, then A_1 and A_2 are calculable, and so is g_k from (5), as C_1 and C_2 are also known. As $(h_k\, C_1)^{-g_k} = \frac{V}{D(k,C_1)} - 1 = A_1$ and $(h_k\, C_2)^{-g_k} = \frac{V}{D(k,C_2)} - 1 = A_2$,

$$h_k = \frac{A_1^{-\frac{1}{g_k}}}{C_1} = \frac{A_2^{-\frac{1}{g_k}}}{C_2} \ . \tag{6}$$

Once the parameter estimates obtained, the same $V(L, n)$, $g_k(L, n)$ and $h_k(L, n)$ values are used for predicting $W_d(k, C; L, n)$ by (4) for any other *corpus* size.

4 Results

4.1 The Corpora Collection

We randomly extracted English and French Wikipedia documents to build sample *corpora* (sizes in words): 30 942 239 (31 Mw); 62 557 077 (63 Mw); 128 364 577

(128 Mw); 254 801 364 (255 Mw); 508 571 317 (509 Mw); 1 068 282 476 (1.1 Gw); 2 155 599 290 (2.2 Gw); 4 278 548 582 (4.3 Gw) and 8 600 180 252 (8.6 Gw), for English. For French: 108 007 454 (108 Mw); 201 439 011 (201 Mw); 403 693 891 (404 Mw); 807 631 298 (808 Mw); 1 605 507 129 (1.6 Gw) and 3 238 982 018 (3.2 Gw). To obtain fair n-gram counts, keeping text semantics, we only add a space next to each character of the set: {' (', ') ', ' [', '] ',' <', '> ', ' " ', ' ! ', ' ? ', ' : ', ' ; ', ' , '}. So, inflected forms count as distinct in the unchanged *corpora*.

4.2 The Range of k Values for $W(k, C; L, n)$ Prediction

Due to lack of space, we were unable to show $D(k, C; L, n)$ results, but they show stable low relative errors ≈3%. E.g., with $C = 8.6$ Gw, for 1-grams ($k = 1 \ldots 5$): 0.3%, 3.1%, 2.1%, 1.7%, 1.9%; for 3-grams ($k = 1 \ldots 5$): 3.9%, 1.2%, 3.5%, 4.5%, 4.8%. As example of the magnitudes for 1-grams and $k = 3$: $D_{empirical}(3, C; L, 1) = 14\,257\,656$ 1-grams and the modelled $D(3, C; L, 1) = 14\,561\,847$ 1-grams.

$W(k, C; L, n)$ empirical values[4] ($W_{emp}(k, C; L, n)$) decrease monotonically with increasing k, for each C and n (Fig. 2a). The difference between $W(k, C)$ and $W(k+1, C)$ goes from quite large values for low k (1, 2, ...), to gradually smaller ones such that $W(k, C)$ statistical fluctuations may interrupt that monotonic behaviour. For each C and n, a threshold for k (*k_threshold*) is reached, beyond which monotony is no longer likely to hold (Fig. 2a for higher k values). Thus, we restrict the $W_d(k, C; L, n)$ model to the monotonic zone ($k < k\_threshold$), which tends to be wider for larger *corpora* and smaller for higher n-gram sizes. For a sound evaluation of the average relative error in the full *corpora* range (Mw to Gw) and n-gram sizes ($1 \leq n \leq 6$) we use a single *k_threshold* value for all pairs (C, n): the maximum *k_threshold* value used is 15 for English and 34 for French (these values are different due to the different sizes of the smallest test *corpora* for each language: 31 Mw for English and 108 Mw for French). Still, the distinct n-grams with low frequencies in those k ranges encompass the vast majority of the number of distinct n-grams in *corpora*, for all n-gram sizes. E.g., the observed proportion of distinct 1-grams with frequency less than 15 wrt the total of distinct 1-grams, given by $1 - D(15, C; L, 1)/D(1, C; L, 1)$, i.e. $1 - 53059/908674 = 94.2\%$ and $1 - 3401814/70227712 = 95.2\%$, respectively, for the 31 Mw and 8.6 Gw English *corpora*. For 6-grams, values are 99.4% and 99.3%.

4.3 The Assessment Criteria and Parameter Estimation

Assessment Criteria. For each fixed (L, n) pair, the module of the Relative Error of $W(k, C)$ estimation is $Err(k, C) = |(W_{est}(k, C) - W_{emp}(k, C))/W_{emp}(k, C)|$, where $W_{est}(k, C)$ is the estimated $W(k, C)$. Other measures are: i) $AvErr(k)$: the average of $Err(k, C)$ for a value k over a set of C values; ii) $AvErr(C)$: the average of $Err(k, C)$, given C, over a set of k values; iii)

[4] Empirical counts were obtained from the *corpora* with the help of Carlos Gonçalves.

$RMSRE(k)$: the Root Mean Square of the Relative Error of the $W(k,C)$ estimates, for a value k over a set of C values, $\sqrt{\frac{1}{\|\mathcal{C}\|}\sum_{C\in\mathcal{C}} Err(k,C)^2}$, where $\mathcal{C}$ is a set of C values; iv) $RMSRE(C)$: the Root Mean Square of the Relative Error of the $W(k,C)$ estimates, given C, over a set of k values, $\sqrt{\frac{1}{\|K\|}\sum_{k\in K} Err(k,C)^2}$, where K is a set of k values. $RMSRE(C)$ shows how stable $Err(k,C)$ is, given C, along a set of k values, i.e., if $RMSRE(C)$ is close to $AvErr(C)$, then $Err(k,C)$ is stable along the different k values. Also, if $RMSRE(k)$ and $AvErr(k)$ are close, $Err(k,C)$ is stable along the different C values; v) $RMSRE(kSet, CSet)$: the Root Mean Square of the Relative Error of the $W(k,C)$ estimates, considering jointly a set of *corpora* sizes, $CSet$, and a set of k values, $kSet$; this is $\sqrt{\frac{1}{\|CSet\|}\sum_{C\in CSet}\frac{1}{\|kSet\|}\sum_{k\in kSet} Err(k,C)^2}$.

Parameter Estimation. We predict $W_d(k,C;L,n)$ by (4), for n-grams of size n, in a *corpus* of size C, in a language L ("en" for English or "fr" for French). To estimate the vocabulary size $V(L,n)$, as well as $g_k(L,n)$ and $h_k(L,n)$, for each language, we chose two *training corpora*, c_1 and c_2 (sizes C_1 and C_2, Sect. 3), one relatively small and the other relatively large: the 31 Mw and 2.2 Gw *corpora* for English; the 201 Mw and 1.6 Gw *corpora* for French. Then, $V(L,n)$, $g_k(L,n)$ and $h_k(L,n)$ are jointly found within a sufficiently wide range of $V(L,n)$ candidate values, using incremental steps of 2×10^6 n-grams for each n ($1\leq n\leq 6$), such that $RMSRE(kSet, CSet)$ is minimum, where $CSet=\{C_1,C_2\}$, and $kSet$ is the set of all k values considered for each language ($1\leq k\leq 15$ for English and $1\leq k\leq 34$ for French). E.g., for English 1-grams, the range of $V(L,n)$ candidates was set from 10^8 to 10^9 with steps as above. Having found these parameters, they are used to estimate $W_d(k,C;L,n)$ for all *corpora*, excluding the training *corpora*.

Table 1 contains the vocabulary sizes found for each n-gram size and language. The time complexity of this search is $O(Ns\times\|K\|\times 2)$ where: Ns is the number of incremental steps in the range of $V(L,n)$ candidate values (for English 1-grams, $Ns=((10^9-10^8)/(2\times10^6))=4.5\times10^2$); $\|K\|$ is the size of the set of k values ($K=\{1,2,\ldots,15\}$ for English and $K=\{1,2,\ldots,34\}$ for French); 2 is the number of *training corpora* (c_1 and c_2). In the example of English 1-grams, this search for vocabulary size took 0.14 s, in a MacBook Air, 1.6 GHz Intel Core i5, 16 GB, 2133 MHz LPDDR3, Macintosh HD disk. Similarly, such a computation is performed only once for each one of the other (n-gram size, language) pairs. Thus, the vocabulary values $V(L,n)$ of Table 1 are used to estimate $W(k,C)$ values for test *corpora* in both languages, by the calculation of $W_d(k,C;L,n)$ (4), having time complexity $O(1)$, as g_k and h_k values are obtained from (5) and (6) using the previously obtained values of $D(k,C_1)$ and $D(k,C_2)$ from the training *corpora*. Note that, once $V(L,n)$, $g_k(L,n)$ and $h_k(L,n)$ parameters are instantiated, any application of the model only requires the calculation by (4) with the same complexity $O(1)$. To illustrate the order of magnitude of g_k and h_k, in case of g_1 and h_1 for 1-grams, with $V(\text{"}en\text{"},1)=9.84\times10^8$, using the training *corpora* $C_1=31$ Mw, $C_2=2.2$ Gw (with $D(1,C_1;\text{"}en\text{"},1)=908674$, $D(1,C_2;\text{"}en\text{"},1)=24865840$), leading to $g_1=0.78562$ and $h_1=4.4391\times10^{-12}$.

Table 1. Estimated vocabulary sizes for different n-gram sizes

	1-grams	2-grams	3-grams	4-grams	5-grams	6-grams
English	9.84×10^8	8.77×10^{10}	1.10×10^{11}	2.98×10^{11}	1.25×10^{12}	1.06×10^{13}
French	3.28×10^8	1.13×10^9	1.81×10^9	7.32×10^9	4.58×10^{10}	3.58×10^{11}

4.4 Comparison with Other Models

Among the $W(k, C)$ models (Sect. 2), *2nd* Zipf's Law is the classical reference. Other influential models with a theoretical foundation are Yule-Simon's [18] and equivalent Price's [14]. They all were only expressed for 1-grams and tested for small to moderate populations. However, for comparative purposes, we evaluate the extended use of *2nd* Zipf's Law (denoted $W_z(k, C; L, n)$) and Price' s model (denoted $W_p(k, C; L, n)$), for $n \geq 1$ and the considered range of large *corpora*.

According to *2nd* Zipf's law, $W(k, C) = W(1, C)\, k^{-\gamma}$. To calculate $W(1, C)$, we consider $\sum_{k=1}^{k=kmax} W(k, C) = D$, then, by integration, $W(k, C)$ is obtained as $D\,(-\gamma + 1)/((kmax^{-\gamma+1} - 1)\, k^{\gamma})$, where $kmax = f(1) = p_1\, C$ and p_1 is the relative frequency of the most frequent n-gram, for each (L, n) pair, whose value is known to be stable for any large enough *corpus*. Price's model states that $W_p(k, C; L, n) = (m + 1)\, D\, B(k, m + 2)$, where $B(.,.)$ is the *Beta* function. We tuned parameters γ (in $W_z(.,.;.,.)$) and m (in $W_p(.,.;.,.)$), to get the minimum $RMSRE(kSet, CSet)$ errors, in the same k and n ranges, and the same *training corpora* c_1, c_2, as for $W_d(k, C; L, n)$. Values for $1 \leq n \leq 6$ are: γ is (2.4, 2.6, 3.0, 4.0, 3.8, 4.2) and (2.3, 2.4, 2.6, 3.0, 3.2, 3.5), and m is (0.51, 0.74, 1.3, 2.0, 2.9, 3.9) and (0.29, 0.41, 0.77, 1.2, 1.6, 2.0), for English and French, respectively.

4.5 Obtained Results

The results obtained for different values of k and n-gram sizes, in the wide range of *corpora* sizes considered, for English and French, show that the proposed model $W_d(k, C; L, n)$ is a close approximation to real n-gram distributions. This is illustrated in Fig. 2a where empirical $W_{emp}(k, C; L, n)$ and estimated $W_d(k, C; L, n)$ curves for 1-grams, 2-grams and 3-grams are shown as a function of k, for an English 8.6 Gw *corpus*. Indeed, for each n-gram size, the W_{emp}/W_d curves appear superimposed in Fig. 2a—higher size n-grams behave similarly but are not shown due to lack of space—. To quantify the proximity of the curves, the relative errors are given in Table 2 for all *corpora* and n-gram sizes, comparing $W_d(k, C; L, n)$, $W_z(k, C; L, n)$ and $W_p(k, C; L, n)$ models for English and French. It reports for each C, the $AvErr(C)$ (abbreviated A), that is, the average of the module of the relative error, $Err(k, C)$, over the set of k values for each language. Each line reports results for one of the test *corpora*, for all n-gram sizes ($1 \leq n \leq 6$). The last line, Avg, is the global average of the A values over the full set of test *corpora*, for each model and n-gram size. Results for $W_d(k, C; L, n)$ model show low relative errors (A_d values) for all n-gram sizes ($1 \leq n \leq 6$) for both languages, with a global average varying from 2.2% to 3.8%

for English and from 1.9% to 2.8% for French, along all test *corpora*. E.g., for English 1-grams, A_d varies from 1.2% to 3.4% along the range of *corpora* sizes. Results also suggest the model forecasting capability, as illustrated by the usage of the English 4.3 Gw and 8.6 Gw *corpora* and the 3.2 Gw French *corpora*. This table also shows, in column R_d, the obtained $RMSRE(C)$ values. The close proximity between corresponding A_d and R_d for each *corpus* line shows that the relative errors $Err(k, C)$ along all k values ($1 \ldots 15$, for English, or $1 \ldots 34$, for French), are similar to each other and close to that A_d value. This is observed for all n-gram sizes ($1 \leq n \leq 6$). Although not included due to lack of space, a table containing the values of $AvErr(k)$, that is the average of $Err(k, C)$ for each value k over the full set of *corpora*, also shows relative errors of the same order of magnitude as those in Table 2.

Table 2 also shows that the $W_d(k, C; L, n)$ model has significantly lower errors than both $W_z(k, C; L, n)$ and $W_p(k, C; L, n)$ for both languages. E.g., for English, *Avg* values for $W_z(k, C; L, n)$ go from 37.5% (1-grams) to 91.8% (6-grams), and for $W_p(k, C; L, n)$ go from 18.6% (1-grams) to 69.2% (6-grams). Since $W_z(k, C; L, n) \propto k^{-\gamma}$ leads to a straight line in log-log, it does not fit the slightly convex empirical $W_{emp}(k, C; L, n)$ curves for the low k values (Fig. 2a), explaining the significant errors in Table 2. The performance of $W_z(k, C; L, n)$ and $W_p(k, C; L, n)$ models tends to get worse as the n-gram size n increases. In the case of $W_z(k, C; L, n)$, this is likely due to the increasing deviation of the real n-gram distribution from the Zipf's Law (ideally, a straight line in log-log) when n increases from 1 to 6, as illustrated in the Fig. 1b for the extreme ranks. For $W_p(k, C; L, n)$ model, the increase of errors for higher n is likely related to the corresponding larger numbers of low-frequency n-grams.

Models [15,16], although reporting low $W(k, C)$ relative errors (not replicated here), are fully committed to empirical curve fitting. This affects their prediction ability for larger *corpora*, requiring specific tuning by curve fitting. This is unlike our proposed model.

4.6 The Predictions with Growing Corpus Size

A low relative error in $W(k, C; L, n)$ gets a more significant impact as the *corpora* sizes increase. In ongoing work we already have preliminary results of the model behaviour for a larger scale English *corpus* of 64.4 Gw, keeping the same level of low $W_d(k, C; L, n)$ relative errors, when averaged over the same $k = 1 \ldots 15$ range: respectively, from 1-grams to 6-grams, 1.1%, 10.9%, 6.7%, 2.8%, 1.6%, 1.7%. In contrast, the $W(k, C; L, n)$ empirical model from [16], that we evaluated for the same scenario as above, showed relative errors, respectively, from 1-grams to 6-grams, of 6.1%, 26.0%, 25.3%, 18.2%, 21.2%, 17.2%, which present significant deviations wrt the global average relative error of around 6% as previously reported in [16] for a range of *corpora* up to 8.6 Gw. This is most likely due to the above mentioned (Sect. 4.5) drawback of that empirical approach.

Due to the principled rationale of the proposed model, reflected on its relative error stability on a wide range of large *corpora*, it suggests a high potential for forecasting. Unlike common approaches, the model relies on the vocabulary

Table 2. $AvErr(C)$ values, abbrev. A_d, A_z, A_p for models $W_d(k, C; L, n)$ (in bold), $W_z(k, C; L, n)$, $W_p(k, C; L, n)$. Values for test *corpora* of different sizes (C), each one for the range of k values ($1 \leq k \leq 15$) for English and ($1 \leq k \leq 34$) for French, considering n-gram sizes from 1 to 6; $RMSRE(C)$, abbrev. R_d, and Global Average (Avg) also shown. All values are in percentages

English												
C	A_d	R_d	A_z	A_p	A_d	R_d	A_z	A_p	A_d	R_d	A_z	A_p
	1-grams				2-grams				3-grams			
62 Mw	**2.2**	2.9	38.4	19.4	**3.4**	3.9	40.1	20.9	**3.7**	4.0	52.9	33.0
128 Mw	**3.4**	3.7	36.9	17.8	**2.4**	3.2	40.6	21.4	**1.8**	2.5	52.1	32.0
255 Mw	**1.7**	2.6	36.6	17.4	**2.5**	2.9	41.0	21.8	**3.3**	3.5	52.4	32.3
509 Mw	**1.2**	1.6	37.2	18.0	**1.6**	1.9	41.2	21.9	**3.1**	3.3	53.2	33.1
1.1 Gw	**2.8**	2.9	39.5	20.0	**5.8**	5.9	42.0	22.6	**7.2**	7.2	54.0	33.7
4.3 Gw	**2.0**	2.3	39.0	19.3	**1.4**	1.6	41.7	22.4	**0.6**	0.9	54.3	34.1
8.6 Gw	**2.2**	2.5	37.5	18.2	**1.8**	2.1	41.0	21.9	**4.8**	5.1	54.1	34.0
Avg	**2.2**		37.5	18.6	**2.7**		41.1	21.8	**3.5**		53.3	33.2
	4-grams				5-grams				6-grams			
62 Mw	**4.6**	5.2	68.7	47.9	**2.3**	3.0	83.9	61.3	**2.3**	3.0	94.5	71.9
128 Mw	**5.0**	6.1	68.6	47.2	**4.5**	5.0	82.0	59.6	**5.1**	5.9	93.6	70.6
255 Mw	**0.9**	1.2	67.0	45.8	**3.2**	3.6	81.9	59.6	**1.8**	2.1	92.0	69.3
509 Mw	**1.5**	1.8	67.3	46.0	**3.2**	3.5	82.6	60.5	**1.8**	2.3	92.1	69.4
1.1 Gw	**6.2**	6.2	68.0	46.6	**6.4**	6.5	83.3	61.1	**2.4**	2.8	92.2	69.4
4.3 Gw	**1.2**	1.5	67.6	45.8	**2.2**	2.6	80.2	58.0	**3.7**	4.5	90.4	67.9
8.6 Gw	**3.1**	3.4	66.7	45.1	**4.5**	5.3	78.9	55.9	**4.1**	5.0	88.0	65.8
Avg	**3.2**		67.7	46.3	**3.8**		81.8	59.4	**3.0**		91.8	69.2
French												
C	A_d	R_d	A_z	A_p	A_d	R_d	A_z	A_p	A_d	R_d	A_z	A_p
	1-grams				2-grams				3-grams			
108 Mw	**2.6**	3.0	31.0	20.8	**2.5**	2.7	30.9	21.3	**3.1**	3.3	41.3	33.3
404 Mw	**2.8**	3.9	32.8	22.8	**2.0**	2.9	32.2	22.3	**1.7**	2.7	41.5	32.2
808 Mw	**1.6**	2.1	32.6	22.7	**1.1**	1.4	32.9	22.8	**1.2**	1.5	42.5	32.7
3.2 Mw	**4.2**	5.0	32.1	22.3	**2.2**	2.7	34.2	24.0	**2.6**	3.1	44.2	34.2
Avg	**2.8**		32.2	22.2	**1.9**		32.5	22.6	**2.2**		42.4	33.1
	4-grams				5-grams				6-grams			
108 Mw	**3.2**	3.5	54.8	47.0	**2.4**	2.7	68.4	60.1	**2.3**	2.6	79.8	70.8
404 Mw	**1.6**	2.6	55.5	46.3	**1.7**	3.0	68.6	59.3	**1.9**	3.3	79.6	69.7
808 Mw	**1.5**	1.8	55.9	46.1	**1.7**	1.9	68.5	58.6	**2.1**	2.4	79.2	69.1
3.2 Mw	**2.3**	2.7	55.7	45.8	**1.7**	2.2	67.6	57.4	**2.0**	2.5	77.6	67.0
Avg	**2.2**		55.5	46.3	**1.9**		68.3	58.9	**2.1**		79.0	69.2

finiteness, imposing an upper bound to the numbers of distinct n-grams, attained for very large *corpora*. Figure 2b shows, for given k and n, a non-monotonic evolution of $W_d(k, C; L, n)$ predictions for growing *corpora*: although it is expected $W(k, C; L, n)$ to grow with C for small and moderate *corpora*, when C becomes very large, there is a more pronounced slowdown of the $D(k, C; L, n)$ growth rate, which, compared to the $D(k+1, C; L, n)$ growth rate, leads to the predicted $W(k, C; L, n)$ decrease. Similar behaviour is predicted for higher n-gram sizes.

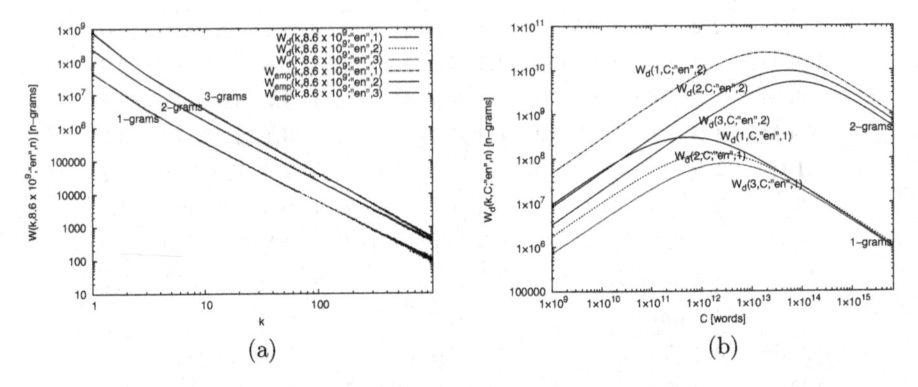

Fig. 2. (a) Empirical curves $W_{emp}(k, C; L, n)$ *versus* n-gram frequency (k) and their estimations by the $W_d(k, C; L, n)$ model, for different n-gram sizes; (b) Evolution of $W_d(k, C; L, n)$ predictions *versus* very large *corpus* sizes (C) in log-log; examples for English 1-grams and 2-grams and $k \in \{1, 2, 3\}$

5 Conclusions

This paper presents a new model, which applies, in a uniform way, to different natural languages (L) and n-gram sizes ($n \geq 1$), for predicting, in a *corpus* of size C, the cumulative numbers of distinct n-grams, $D(k, C; L, n)$, with frequencies greater or equal to a given value k, with $k \geq 1$. This is used to calculate the sizes of the equal-frequency (k) n-gram groups, $W(k, C; L, n)$, for the low-frequency n-grams, as a function of the *corpus* size. The low-frequency n-grams are often ignored by other studies, in particular for wide ranges of large *corpora*, although the majority of the relevant n-grams are of low frequency. Also, to the best of our knowledge, we have not found any other fully principled model for predicting the cumulative numbers of distinct n-grams with frequencies greater or equal to a given value k, with $k \geq 1$, and the sizes of equal-frequency groups, as a function of *corpus* sizes, applying to n-grams ($n \geq 1$), for the low frequency n-grams and large *corpora*. The model assumes that, at each temporal epoch, each language has a finite n-gram vocabulary. It is based on a principled approach that models the evolution of $D(k, C; L, n)$, for growing *corpora* sizes, based on two influences: a cumulative form of preferential attachment and the finiteness of the n-gram language vocabulary $V(L, n)$. The estimation of the model parameters, lying on their independence on C, showed to be consistent with the results in all test *corpora*. The model was validated for English and French, with n-gram sizes from 1 to 6, in a wide range of test *corpora* sizes from few Mega to few Giga words, with un-truncated frequency data. It achieved stable relative errors around 3%, over all the ranges considered for k, n and C, in both languages. This low relative error magnitude suggests the model robustness to deal with very large scale applications. The model has the potential to apply to other languages, as no specific language morphosyntactic constraints are imposed.

Acknowledgment. This work is supported by NOVA LINCS (UIDB/04516/2020) with the financial support of FCT.IP.

References

1. Albert, R., Barabási, A.L.: Statistical mechanics of complex networks. Rev. Mod. Phys. **74**(1), 47–97 (2002)
2. Bacaër, N.: Verhulst and the logistic equation (1838). In: A Short History of Mathematical Population Dynamics, pp. 35–39. Springer, London (2011). https://doi.org/10.1007/978-0-85729-115-8_6
3. Balasubrahmanyan, V.K., Naranan, S.: Algorithmic information, complexity and Zipf law. Glottometrics **4**, 1–26 (2002)
4. Bass, F.M.: A new product growth for model consumer durables. Manage. Sci. **15**(5), 215–227 (1969)
5. Bernhardsson, S., da Rocha, L.E.C., Minnhagen, P.: Size dependent word frequencies and translational invariance of books. CoRR abs/0906.0716 (2009)
6. Booth, A.D.: A "law" of occurrences for words of low frequency. Inf. Control **10**, 386–393 (1967)
7. Brants, T., Popat, A.C., Xu, P., Och, F.J., Dean, J.: Large language models in machine translation. In: Joint Conference on EMNLP - CoNLL, pp. 858–867. ACL (2007)
8. Buck, C., Heafield, K., van Ooyen, B.: N-gram counts and language models from the Common Crawl. In: LREC'14. European Language Resources Association (2014)
9. Egghe, L.: Untangling Herdan's law and Heaps' law: mathematical and informetric arguments. J. Am. Soc. Inf. Sci. Technol. **58**(5), 702–709 (2007)
10. Goncalves, C., Silva, J.F., Cunha, J.C.: n-gram cache performance in statistical extraction of relevant terms in large *corpora*. In: Rodrigues, J.M.F., et al. (eds.) ICCS 2019. LNCS, vol. 11537, pp. 75–88. Springer, Cham (2019). https://doi.org/10.1007/978-3-030-22741-8_6
11. Lü, L., Zhang, Z.K., Zhou, T.: Deviation of Zipf and Heaps laws in human languages with limited dictionary sizes. Sci. Rep. **3**, 1082 (2013). https://doi.org/10.1038/srep01082
12. Mandelbrot, B.: On the theory of word frequencies and on related Markovian models of discourse. Struct. Lang. Math. Aspects **12**, 190–219 (1953)
13. Newman, M.: Power laws, Pareto distributions and Zipf law. Contemp. Phys. **46**(5), 323–351 (2005)
14. Price, D.S.: A general theory of bibliometric and other cumulative advantage processes. J. Am. Soc. Inf. Sci. **27**(5), 292–306 (1976)

15. Silva, J.F., Cunha, J.C.: An empirical model for n-gram frequency distribution in large corpora. In: Lauw, H.W., Wong, R.C.-W., Ntoulas, A., Lim, E.-P., Ng, S.-K., Pan, S.J. (eds.) PAKDD 2020. LNCS (LNAI), vol. 12085, pp. 840–851. Springer, Cham (2020). https://doi.org/10.1007/978-3-030-47436-2_63
16. Silva, J.F., Cunha, J.C.: A model for predicting n-gram frequency distribution in large *corpora*. In: Paszynski, M., Kranzlmüller, D., Krzhizhanovskaya, V.V., Dongarra, J.J., Sloot, P.M.A. (eds.) ICCS 2021. LNCS, vol. 12742, pp. 699–706. Springer, Cham (2021). https://doi.org/10.1007/978-3-030-77961-0_55
17. Silva, J.F., Gonçalves, C., Cunha, J.C.: A theoretical model for n-gram distribution in big data corpora. In: 2016 IEEE International Conference on Big Data, pp. 134–141 (2016)
18. Simon, H.: On a class of skew distribution functions. Biometrika **42**(3/4), 425–440 (1955)
19. Zipf, G.K.: Human Behavior and the Principle of Least-Effort. Addison-Wesley, Cambridge (1949)

Meta-Reinforcement Learning Algorithm Based on Reward and Dynamic Inference

Jinhao Chen[1], Chunhong Zhang[2], and Zheng Hu[1(✉)]

[1] State Key Laboratory Of Networking and Switching Technology, Beijing University of Posts and Telecommunications, Beijing 100088, China
huzheng@bupt.edu.cn

[2] Key Laboratory of Universal Wireless Communications, Ministry of Education, Beijing University of Posts and Telecommunications, Beijing 100088, China

Abstract. Meta-Reinforcement Learning aims to rapidly address unseen tasks that share similar structures. However, the agent heavily relies on a large amount of experience during the meta-training phase, presenting a formidable challenge in achieving high sample efficiency. Current methods typically adapt to novel tasks within the Meta-Reinforcement Learning framework through task inference. Unfortunately, these approaches still exhibit limitations when faced with high-complexity task space. In this paper, we propose a Meta-Reinforcement Learning method based on reward and dynamic inference. We introduce independent reward and dynamic inference encoders, which sample specific context information to capture the deep-level features of task goals and dynamics. By reducing task inference space, agent effectively learns the shared structures across tasks and acquires a profound understanding of the task differences. We illustrate the performance degradation caused by the high task inference complexity and demonstrate that our method outperforms previous algorithms in terms of sample efficiency.

Keywords: Meta-Reinforcement Learning · Variational Inference · Hidden Feature

1 Introduction

The advancement of deep Reinforcement Learning (RL) has fueled multiple high-profile successes in solving problems associated with sequential decision-making [1,2]. However, conventional RL methods encounter a common challenge requiring millions of interactions with the environment to train a task-specific agent [3,4] and exhibit suboptimal performance when adapting to novel tasks. In response to this challenge, the field of machine learning has emerged Meta-Reinforcement Learning (Meta-RL) [5,6], which leverage extensive historical experiences to learn shared structural knowledge. This paradigm of learning-to-learn equips agents to efficiently utilize their accumulated experiences [7].

The conventional method in context-based Meta-RL methods involves the training of a task encoder responsible for mapping trajectory transition sequences

D.-N. Yang et al. (Eds.): PAKDD 2024, LNAI 14647, pp. 223–234, 2024.
https://doi.org/10.1007/978-981-97-2259-4_17

with a latent task variable [8,9]. The computation of differences between trajectories to obtain distinct latent variables does not provide a profound knowledge of tasks [10]. When an agent can discern shared structures related to task goals and dynamics, it becomes adept at leveraging past experiences to efficiently solve tasks [11]. Our method focuses on endowing the agent with the ability to differentiate tasks.

The task distribution consists of reward and dynamic distributions in the setting of Meta-RL, when two distinct tasks share an identical reward function, the rewards obtained by the agent that pertain solely to the current state remain consistent, despite differences in the entire trajectory sequences. Similarly, if tasks share the dynamics function, the trajectory may diverge, but the dynamics transition sequences related to the current state and selected actions remain consistent. Minor discrepancies within sampled trajectories can lead to significant variations in task inference, thereby impeding the effective utilization of historical experience and necessitating more samples for learning [12]. We attribute the heightened complexity issue to the typically large inference space in the conventional task inference process. When the task space contains more capturable deep shared experiential structures, traditional single-task encoders face challenges in effectively utilization. Additionally, the two hidden features, reward and dynamic, require distinct information. Therefore, the inclusion of all trajectories as input to the task inference encoder may introduce additional noise, potentially diminishing sample efficiency.

In summary, to further enhance the sample efficiency of Meta-Reinforcement Learning algorithms, we propose a novel context-based Meta-Reinforcement Learning algorithm that addresses the issue of high complexity in the traditional methods by reducing the size of the inference space through inferring deep hidden shared representations of task goals and dynamics, and we effectively reduce inference uncertainty by leveraging specific minimally relevant context experiences. We employ a trick, introducing a reward inference encoder for approximating task goals inference. Our primary contributions include:

1. We introduce a novel Meta-Reinforcement Learning algorithm that captures deep-level features of reward and dynamic, thus enhancing sample efficiency.
2. We demonstrate that our method exhibits superior sample efficiency performance compared to other advanced baselines in common benchmarks.
3. We design a task distribution scenario where task goals and dynamics are configured as Cartesian product combinations to validate the performance degradation caused by the complexity of task inference.

2 Background

2.1 Meta-Reinforcement Learning

In the standard setting of Meta-RL, we possess a task distribution $p(M)$ over Markov Decision Processes (MDPs) from which we sample during meta-training. A task, denoted by an MDP $M_i \sim p(M)$, is defined by a tuple

$M_i = (S, A, R_i, T_i, \gamma)$ with S a set of states, A a set of actions, $R_i(s)$ a reward function, $T_i(s_{t+1}|s_t, a_t)$ a dynamic transition function, and γ a discount factor elucidating the degree of future rewards discounting. Across tasks, the reward and dynamic transition functions vary but share some structure.

Meta-RL can be categorized into two types: gradient-based methods [13–17], which extend policy gradient methods and learn from aggregated experience using policy gradients [18]. Noteworthy examples include Model-Agnostic Meta-Learning (MAML) [19]. The second category is context-based methods [20–23], which we will introduce in Sect. 2.2.

2.2 Context-Based Meta-Reinforcement Learning

Context-based Meta-Reinforcement Learning leverages context information to infer latent task space and learn a universal policy, enhancing decision-making capabilities during meta-testing and adaptability to novel tasks [11,24,25]. The context information is typically derived from transition trajectory data specific to each individual task. During the meta-testing phase, adaptation can be viewed as a special case of RL in a Partially Observable Markov Decision Process (POMDP) [26], where the task serves as the unobserved part of the state. When the agent possesses complete information regarding both latent variable z and state s, it can learn a universal policy $\pi_\theta(s, z)$, which is equivalent to addressing a standard MDP problem, leading to an attainable optimal policy.

Pearl [9] stands as a high-sample-efficiency context-based Meta-RL algorithm. It separates task inference from agent execution, thereby enabling off-policy training and significantly improving the inherent inefficiencies associated with sample utilization during meta-training. VariBAD [27] utilizes latent variables for task inference to construct predictive models for rewards and dynamics, enabling tractable approximate Bayes-optimal exploration. However, traditional methods face a challenge of high complexity in task inference when using a single task encoder for extracting task information. We discuss the challenge in Sect. 3.

2.3 Parametric Task Distributions

In the setting of Meta-RL, a fundamental precondition for enabling the agent to generalize effectively through meta-training is the presence of shared structural knowledge across tasks within the task distribution $p(M)$. Yu et al. [28] categorized task distributions into parametric and non-parametric task distributions. Within parametric task distributions, the specific tasks M_i differ only in the reward function parameters θ_{R_i} and dynamic function parameters θ_{T_i}. Formally, the parameterized task distribution $p(M)$ is denoted as below.

$$p(M) = \{M_i\}_i = \{\theta_{R_i}, \theta_{T_i}\}_i \tag{1}$$

Agent can learn a universal policy by capitalizing on these shared parametric structures. Our approach is based on parametric task distributions, where both reward and dynamic latent variables as task-specific features, exerting influence over the execution of each specific task.

3 Problem Statement

In Context-based Meta-Reinforcement Learning, task inference functions by assigning a abstract task label M_i to each trajectory within a task [29]. This process provides additional information for execution policy when tackling tasks. However, the efficacy of this information is limited when the task distribution is a combination of task goals and dynamics. The rationale behind this challenge is that the abstract encoder captures task latent variables that are not intrinsically related to the specific features of the current task, but rather pertain solely to the transition trajectory. In fact, the encoder learns low-dimensional feature representations of tasks, which may exhibit significant discrepancies even among trajectories of the same task, making it challenging to learn effective shared structures inherent to tasks.

To elucidate the relation between higher complexity problems and task latent variables, we assume a straightforward case wherein the task distribution $p(M)$, reward distribution $p(R)$, and dynamics distribution $p(T)$ are all discrete. In such an ideal setting, the task distribution space equates to the Cartesian product of the reward distribution space and the dynamic distribution space [30], denoted as below.

$$p(M) = p(T) \times p(R) = \{(T_i, R_i)|T_i \in T \wedge R_i \in R\} \quad (2)$$

For effective task inference, the encoder must accurately capture the deviations in agent executed trajectories stemming from the combined impact of both rewards and dynamics. The task inference space of conventional methods is related to the size of Cartesian product. The left side of the following equation represents the task inference space of conventional methods, while the right side represents the individual rewards and dynamics inference space.

$$|p(M)| = |p(T)| \cdot |p(R)| \Longrightarrow |p(T)| + |p(R)| \quad (3)$$

Consequently, the complexity of traditional task inference approaches tends to be higher. Even when certain features of two tasks are similar, these approaches maintain higher uncertainty concerning the distinctions present in the sampled trajectories. This necessitates the acquisition of additional samples for improve task inference and the effective utilization of historical experience.

4 Method

Task latent variables present task-specific features, including task goals, reward function, and dynamic. In view of the intricacies associated with comprehending task goals during agent training and the rewards garnered within the task environment, reflecting the level of goal attainment. Task goals can, to a certain extent, be inferred through these reward signals. We therefore reframe the task latent variable inference process into the inference of two distinct features: reward functions and dynamic functions.

4.1 Reward and Dynamics Inference

In this section, we introduce the concepts of reward inference and dynamic inference in our approach. The task distribution $p(M)$ encompasses both reward distribution $p(R)$ and dynamics distribution $p(T)$, and the task inference process essentially entails the inference of both the reward and the dynamic. Each individual task M_i can be conceived as sampling a reward parameter θ_{R_i} from the reward distribution $p(R)$ and a dynamic parameters θ_{T_i} from the dynamic distribution $p(T)$. Disparities in rewards and dynamics across different tasks present the hidden features of tasks. For every task $M = (S, A, R, T, \gamma)$ within the task distribution $p(M)$, the inference of the reward function $R(s)$ and the dynamic function $T = P(s'|s, a)$ can be independently deconstructed into two parts, as follows.

$$p(M) = p(T, R) = p(T)p(R) \quad (4)$$

Acquiring accurate representations within the reward and dynamic distributions enables the agent to reconstruct the task distribution and effectively capture the distinctions and characteristics across tasks, thereby improving the efficiency in adapting. However, accurately computing posterior probabilities proves to be challenging. We infer the uncertainty of reward and dynamic distributions using latent variables. Inferencing the reward function serves to equip the agent with a cognition of the shared reward signals intrinsic to the current task, enabling rapid policy adjustments accordingly. Similarly, when the agent infers the dynamics function, it can learn about the shared physical attributes of the environment, effectively utilizing historical experience to accelerate adaptation.

4.2 Meta-Reinforcement Learning Algorithm Based on Reward and Dynamics Inference Encoders

We introduce the framework of our method in this section, based on Reward and Dynamic Inference Meta-Reinforcement Learning Algorithm (RDIMRL). Within RDIMRL, we design two key components, the reward inference encoder and the dynamic inference encoder, integrated into context-based Meta-Reinforcement Learning framework. For the current unknown Markov Decision Process (MDP), our objective is for these two encoders to capture deep-level representations of reward and dynamic through meta-learning. The framework is shown in Fig. 1. The reward inference encoder $q_{\varphi_r}(z_r|c_r)$ and the dynamic inference encoder $q_{\varphi_p}(z_p|c_p)$ infers the current MDP's reward z_r and dynamic z_p. Here, c_r encompasses the context information $\{(s_i', r_i)\}_i$ necessary for reward inference, and c_p represents the context information $\{(s_i, a_i, s_i')\}_i$ for dynamic inference.

Both the Pearl and VariBAD methods integrate context information into the task inference, encompassing all historical trajectories, including state S, action selection A, next state S', and reward R. However, as the information requisite for inferring reward features and dynamic features differs, the inclusive incorporation of all types of information as input introduces unnecessary noise, potentially impacting the learning process of the deep-level task feature inference encoder. Specifically, in the case of reward inference, environmental

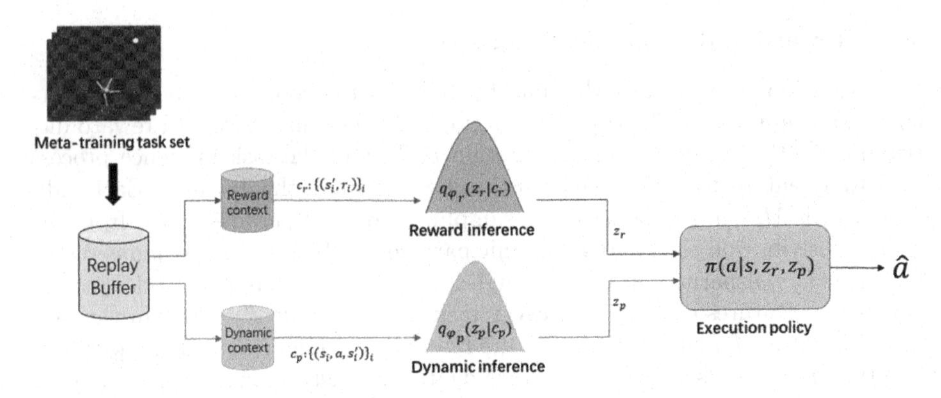

Fig. 1. RDIMRL framework: The replay buffer collects trajectories of states, actions, and rewards, which are then divided into reward-inference context information c_r and dynamic-inference context information c_p. These are passed to two separate inference encoders for processing, producing posterior embeddings z_r and z_p. The universal policy $\pi_\theta(a|s, z_r, z_p)$ conditions on the posteriors to select actions in the environment.

dynamic transition information unrelated to reward may potentially interfere with training the reward inference encoder. Similarly, additional task goals disparities information may also potentially impede the dynamic model's learn.

To address this issue, we encode only task-relevant context information based on prior knowledge. Both encoders within our approach exclusively receive information relevant to their respective inferences, cr and cp, avoiding the introduction of extraneous noise. We use a variational inference approach [31,32] to approximate the belief over the reward and dynamic. The two latent variables acquired from the encoders, z_r and z_p, serve as conditional inputs for the execution policy $\pi_\theta(a|s, z_r, z_p)$. RDIMRL builds on the Soft Actor-Critic algorithm (SAC) [33]. We define the training loss functions for the actor and critic as follows.

$$\mathcal{L}_{\text{actor}} = \mathbb{E}_{s\sim\mathcal{B}, a\sim\pi_\theta, z_r\sim q_{\varphi_{z_r}}, z_p\sim q_{\varphi_{z_p}}} \left[\mathrm{D}_{KL} \left(\pi_\theta(a|s, \overline{z}_r, \overline{z}_p) \parallel \exp\left(Q_\theta(s, a, \overline{z}_r, \overline{z}_p) \right) / Z_\theta(s) \right) \right] \tag{5}$$

$$\mathcal{L}_{\text{critic}} = \mathbb{E}_{(s,a,r,s')\sim\mathcal{B}, z_r\sim q_{\varphi_{z_r}}, z_p\sim q_{\varphi_{z_p}}} \left[Q_\theta(s, a, z_r, z_p) - (r + \overline{V}(s', \overline{z}_r, \overline{z}_p)) \right]^2 \tag{6}$$

The KL divergence term, an information bottleneck in the variational approximation [32], imposes constraints on the mutual information between z_r, z_p, c_r, and c_p. We aim for the inference encoders to focus on minimal, sufficient, and relevant information in the samples, avoiding the modeling of irrelevant dependencies.

During the meta-training, we train two inference encoders q_{ϕ_r} and q_{ϕ_p}. These encoders are used to estimate the posterior distributions of reward z_r and dynamic z_p. Furthermore, we learn a universal policy conditioned on z_r and z_p. In the adaptation, the agent can infer the reward z_r and dynamic z_p of the current task based on recent historical experiences in the novel task, and the posterior distributions of z_r and z_p are updated based on the sampled context information.

5 Experiment

We show the performance of our algorithm RDIMRL in terms of sample efficiency in this part. We compare the results to prior Meta-Reinforcement Learning methods by employing on four MuJoCo continuous control tasks [34]. The baseline algorithms for comparison encompassed eight advanced methods, as shown in the Table 1. In Sect. 5.1, we evaluate our algorithm's performance in MuJoCo environments, which are commonly used in the Meta-RL. In Sect. 5.2, we set the task space as the Cartesian product combination of reward distribution and dynamic distribution to investigate the higher task inference complexity issue in previous methods. We demonstrate the sample efficiency performance of our algorithm in this complex combined task space.

Table 1. Baseline Meta-Reinforcement Learning Algorithms

Method	Category	Method	Category
MAML (2017)	Gradient-Based	E-MAML (2019)	Gradient-Based
ProMP (2022)	Gradient-Based	RL^2 (2016)	Context-Based
Pearl (2019)	Context-Based	TrMRL (2022)	Context-Based
RoML (2023)	Context-Based	VariBAD (2019)	Context-Based

5.1 Common MuJoCo Environments

Our primary objective is to validate whether our algorithm RDIMRL offers sample efficiency advantages when compared to other algorithms on four environments focused around robotic locomotion in this experiment. These locomotion task families only require adaptation across goals while maintaining the same dynamic damping coefficients. For example, the simulated agent needs to achieve different velocities or orientations. The complexity of the task space in such cases depends on the uncertainty of the reward distribution. The experimental results are shown in Fig. 2.

Across all environments, our algorithm RDIMRL outperforms all previous algorithms. Both RoML and gradient-based methods exhibit suboptimal sample efficiency, Pearl stands out as the best-performing algorithm among the baselines. Our algorithm demonstrates approximately twice the sample efficiency and superior asymptotic performance compared to Pearl in most cases. However, as shown in Fig. 2b and c, our algorithm only achieves performance comparable to Pearl in the cheetah environment. We attribute this to the design of the task space, which maintains a consistent dynamic across these environments. The

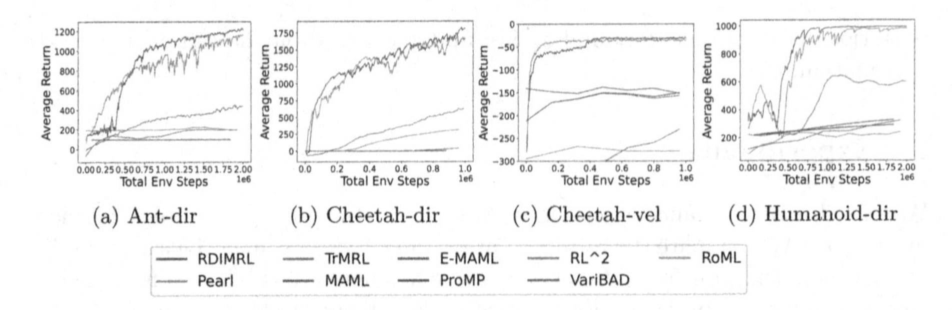

Fig. 2. Illustration of the test task performance and samples collected during meta-training in four continuous control MuJoCo environments, where the task space is configured with varying task goals.

strength of our algorithm lies in its ability to effectively learn when the task space contains more capturable shared knowledge. In Sect. 5.2, we further illustrate how our algorithm effectively enhances performance when the task space is more complex. We visualize the effectiveness of the z_p and z_r inference encoders, as shown in Fig. 3. We observe that our algorithm RDIMRL, after training on 500,000 samples, is able to recognize hidden setting of the dynamics parameter as a constant when adapting to new tasks. Additionally, it makes distinct reward inferences based on different task goals. This demonstrates that RDIMRL can capture deep-seated task features beyond merely as a task label.

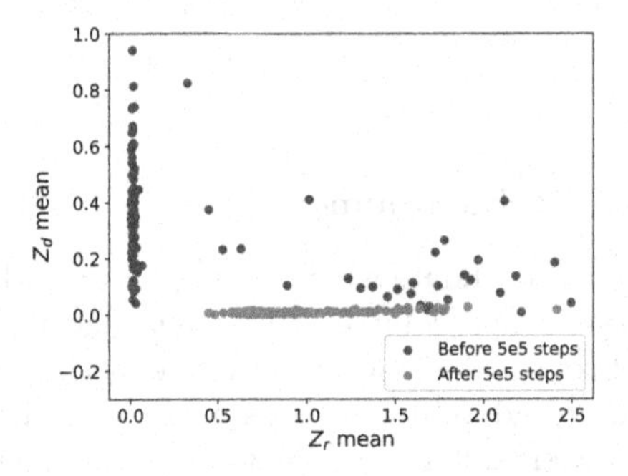

Fig. 3. Visualization of what the z_p and z_r inference encoders learn during meta-training. Shown is the inferred value they take after meta-training for a new task. Each dot represents one randomly sampled task from the task distribution.

5.2 Cartesian Product Combinations of Tasks with Different Goals and Dynamics

To gain insight into the issue of high task space complexity, we systematically arranged various combinations of task goals and dynamics functions, resulting in the total number of tasks being the Cartesian product of task goals and dynamics distributions. We maintain the same number of task goals in each environment as in experiment 5.1 and adjust the damping coefficient of the agent's MuJoCo simulation model to make the dynamics variable for each task. This adjustment impacts the transition probabilities when executing the same action in the same state. According to the specific counts of task goals and dynamics in each environment, we estimate the task inference space, reward inference space, and dynamic inference space as shown in Table 2.

Table 2. The size of inference space. The size of inference space for different task space configurations in four MuJoCo environments.

Environment	Goal inference space	Dynamic inference space	Task inference space
Ant-dir	10	4	40
Cheetah-dir	10	4	40
Cheetah-vel	80	8	640
Humanoid-dir	30	5	150

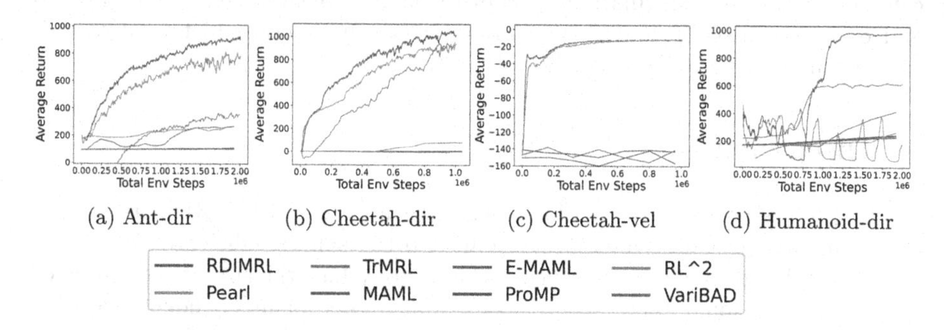

Fig. 4. Illustration of the performance of RDIMRL in a task space defined by the Cartesian product of task goals and dynamics distributions. RDIMRL achieves performance comparable to the best-performing prior method Pearl with only 5e5 samples, equivalent to Pearl's performance with 1e6 samples.

Among all baseline algorithms, our algorithm RDIMRL exhibits the highest performance in the modified task environments, as shown in Fig. 4. In Cheetah-dir environment where our algorithm demonstrates nearly twice the sample efficiency and a twenty percent improvement in asymptotic performance after the task distribution modifications designed to include different dynamics. We find Pearl fails to adapt and resulting in low-return in Humanoid-dir environment. We attribute this to the increased complexity of the task space.

The performance degradation due to the increased complexity of task inference is shown in Fig. 5. Our method also experiences a performance drop as the

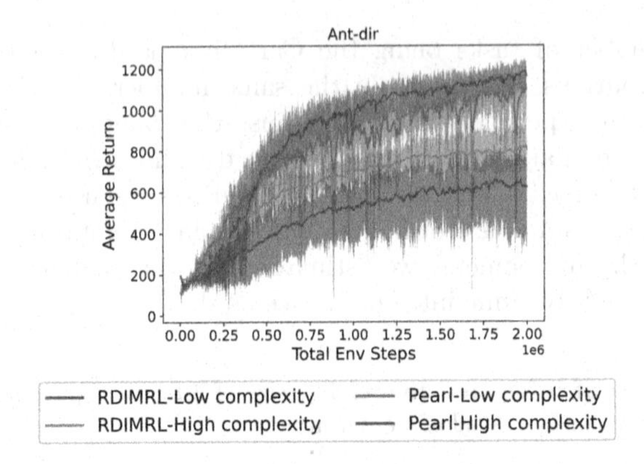

Fig. 5. We analyze the performance of meta-Reinforcement Learning methods as the size of the inference space increases. Two methods experience performance degradation in high complexity scenarios.

size of the inference space increases, but it is less affected compared to other methods, demonstrating greater adaptability. Our algorithm RDIMRL show superior sample efficiency in these experiments. We believe that if the agent can utilize context information to capture latent variables related to reward and dynamics distributions, it can more effectively leverage historical experience when adapting to novel tasks.

6 Discussion

We propose a novel Meta-Reinforcement Learning algorithm RDIMRL based on reward and dynamic inference to enhance the sample efficiency. We meta-train reward and dynamic inference encoders to reduce the size of the inference space in conventional inference processes. This allows for independent inference and learning of shared structures related to task-specific reward features and dynamic features within the Meta-RL framework. By capitalizing on these shared structures and past experiences, the agent extracts valuable insights from previously solved tasks and applies them effectively to tackle novel tasks, even in scenarios with higher task space complexity. Furthermore, by collecting specific context information, we reduce the potential noise that could be introduced by redundant information input. This enables the task feature inference encoder to more effectively distinguish differences and shared attributes across tasks, thereby further enhancing sample efficiency. We demonstrate the superior performance of our algorithm RDIMRL in common benchmarks, surpassing other Meta-Reinforcement Learning algorithms.

Acknowledgments. This work was supported by Beijing University of Posts and Telecommunications China Mobile Research Institute Joint Innovation Center.

References

1. Bellemare, M.G., et al.: Autonomous navigation of stratospheric balloons using reinforcement learning. Nature **588**(7836), 77–82. https://doi.org/10.1038/s41586-020-2939-8. https://www.nature.com/articles/s41586-020-2939-8
2. Miki, T., Lee, J., Hwangbo, J., Wellhausen, L., Koltun, V., Hutter, M.: Learning robust perceptive locomotion for quadrupedal robots in the wild. Sci. Robot. **7**(62), eabk2822. https://doi.org/10.1126/scirobotics.abk2822. https://www.science.org/doi/full/10.1126/scirobotics.abk2822
3. Lake, B.M., Ullman, T.D., Tenenbaum, J.B., Gershman, S.J.: Building machines that learn and think like people. Behav. Brain Sci. **40**, e253 (2017). https://doi.org/10.1017/S0140525X16001837
4. Peng, M., Zhu, B., Jiao, J.: Linear representation meta-reinforcement learning for instant adaptation. arXiv arXiv:2101.04750v1 (2021)
5. Beck, J., et al.: A survey of meta-reinforcement learning. arXiv arXiv:2301.08028 (2023). https://doi.org/10.48550/arXiv.2301.08028
6. Imagawa, T., Hiraoka, T., Tsuruoka, Y.: Off-policy meta-reinforcement learning with belief-based task inference. IEEE Access **10**, 49494–49507. https://doi.org/10.1109/ACCESS.2022.3170582. https://ieeexplore.ieee.org/abstract/document/9763505
7. Wang, J.X., et al.: Learning to reinforcement learn. arXiv arXiv:1611.05763 (2017)
8. Melo, L.C.: Transformers are meta-reinforcement learners. arXiv arXiv:2206.06614 (2022)
9. Rakelly, K., Zhou, A., Quillen, D., Finn, C., Levine, S.: Efficient off-policy meta-reinforcement learning via probabilistic context variables, p. 10 (2019)
10. Jiang, P., Song, S., Huang, G.: Exploration with task information for meta reinforcement learning. IEEE Trans. Neural Netw. Learn. Syst. 34(8), 4033–4046 (2023). https://doi.org/10.1109/TNNLS.2021.3121432. https://ieeexplore.ieee.org/document/9604770/
11. Humplik, J., Galashov, A., Hasenclever, L., Ortega, P.A., Teh, Y.W., Heess, N.: Meta reinforcement learning as task inference. arXiv arXiv:1905.06424 (2019)
12. Han, X., Wu, F.: Meta reinforcement learning with successor feature based context. arXiv arXiv:2207.14723 (2022)
13. Gupta, A., Mendonca, R., Liu, Y., Abbeel, P., Levine, S.: Meta-reinforcement learning of structured exploration strategies. In: Advances in Neural Information Processing Systems, vol. 31. Curran Associates, Inc. (2018). https://proceedings.neurips.cc/paper/2018/hash/4de754248c196c85ee4fbdcee89179bd-Abstract.html
14. Stadie, B.C., et al.: Some considerations on learning to explore via meta-reinforcement learning. arXiv arXiv:1803.01118 (2018)
15. Rothfuss, J., Lee, D., Clavera, I., Asfour, T., Abbeel, P.: ProMP: proximal meta-policy search (2018). https://doi.org/10.48550/arXiv.1810.06784. http://arxiv.org/abs/1810.06784
16. Zintgraf, L., Shiarli, K., Kurin, V., Hofmann, K., Whiteson, S.: Fast context adaptation via meta-learning. In: Proceedings of the 36th International Conference on Machine Learning, pp. 7693–7702. PMLR (2018). ISSN 2640-3498. https://proceedings.mlr.press/v97/zintgraf19a.html
17. Vuorio, R., Beck, J., Farquhar, G., Foerster, J., Whiteson, S.: No dice: an investigation of the bias- variance tradeoff in meta-gradients (2022)

18. Mendonca, R., Gupta, A., Kralev, R., Abbeel, P., Levine, S., Finn, C.: Guided meta-policy search. In: Advances in Neural Information Processing Systems, vol. 32. Curran Associates, Inc. (2019). https://proceedings.neurips.cc/paper/2019/hash/d324a0cc02881779dcda44a675fdcaaa-Abstract.html
19. Finn, C., Abbeel, P., Levine, S.: Model-agnostic meta-learning for fast adaptation of deep networks, p. 10 (2017)
20. Korshunova, I., Degrave, J., Dambre, J., Gretton, A., Huszár, F.: Exchangeable models in meta reinforcement learning (2020)
21. Raileanu, R., Goldstein, M., Szlam, A., Fergus, R.: Fast adaptation via policy-dynamics value functions (2020). https://doi.org/10.48550/arXiv.2007.02879. http://arxiv.org/abs/2007.02879
22. He, J.Z.Y., Raghunathan, A., Brown, D.S., Erickson, Z., Dragan, A.D.: Learning representations that enable generalization in assistive tasks (2022). https://doi.org/10.48550/arXiv.2212.03175. https://arxiv.org/abs/2212.03175v1
23. Beck, J., Jackson, M.T., Vuorio, R., Whiteson, S.: Hypernetworks in meta-reinforcement learning (2022). https://doi.org/10.48550/arXiv.2210.11348. https://arxiv.org/abs/2210.11348v1
24. Duan, Y., Schulman, J., Chen, X., Bartlett, P.L., Sutskever, I., Abbeel, P.: RL^2: fast reinforcement learning via slow reinforcement learning. arXiv arXiv:1611.02779 (2017)
25. Greenberg, I., Mannor, S., Chechik, G., Meirom, E.: Train hard, fight easy: robust meta reinforcement learning (2023). https://doi.org/10.48550/arXiv.2301.11147. http://arxiv.org/abs/2301.11147
26. Kaelbling, L.P., Littman, M.L., Cassandra, A.R.: Planning and acting in partially observable stochastic domains. Artif. Intell. **101**(1), 99–134 (1998). https://doi.org/10.1016/S0004-3702(98)00023-X. https://www.sciencedirect.com/science/article/pii/S000437029800023X
27. Zintgraf, L., et al.: VariBAD: a very good method for Bayes-adaptive deep RL via meta-learning (2020). https://doi.org/10.48550/arXiv.1910.08348. https://arxiv.org/abs/1910.08348v2
28. Yu, T., et al.: Meta-world: a benchmark and evaluation for multi-task and meta reinforcement learning, p. 17 (2021)
29. Yang, R., Xu, H., Wu, Y., Wang, X.: Multi-task reinforcement learning with soft modularization. arXiv arXiv:2003.13661 (2020)
30. Li, L., Huang, Y., Chen, M., Luo, S., Luo, D., Huang, J.: Provably improved context-based offline meta-RL with attention and contrastive learning, p. 21 (2021)
31. Kingma, D.P., Welling, M.: Auto-encoding variational bayes (2022). https://doi.org/10.48550/arXiv.1312.6114. http://arxiv.org/abs/1312.6114
32. Alemi, A.A., Fischer, I., Dillon, J.V., Murphy, K.: Deep variational information bottleneck (2019). https://doi.org/10.48550/arXiv.1612.00410. http://arxiv.org/abs/1612.00410
33. Haarnoja, T., Zhou, A., Abbeel, P., Levine, S.: Soft actor-critic: off-policy maximum entropy deep reinforcement learning with a stochastic actor (2018). https://doi.org/10.48550/arXiv.1801.01290. http://arxiv.org/abs/1801.01290
34. Todorov, E., Erez, T., Tassa, Y.: MuJoCo: a physics engine for model-based control. In: 2012 IEEE/RSJ International Conference on Intelligent Robots and Systems, pp. 5026–5033 (2012). ISSN 2153-0866. https://doi.org/10.1109/IROS.2012.6386109. https://ieeexplore.ieee.org/abstract/document/6386109

Security and Privacy

SecureBoost+: Large Scale and High-Performance Vertical Federated Gradient Boosting Decision Tree

Tao Fan[1,2(✉)], Weijing Chen[2], Guoqiang Ma[2], Yan Kang[2], Lixin Fan[2], and Qiang Yang[1,2]

[1] Department of Computer Science and Engineering, HKUST, Hong Kong, China
{tfanac,qyang}@cse.ust.hk
[2] WeBank, Shenzhen, China
{weijingchen,zotrseeewma,yangkang,lixinfan}@webank.com

Abstract. Gradient boosting decision tree (GBDT) is an ensemble machine learning algorithm that is widely used in industry. Due to the problem of data isolation and the requirement of privacy, many works try to use vertical federated learning to train machine learning models collaboratively between different data owners. SecureBoost is one of the most popular vertical federated learning algorithms for GBDT. However, to achieve privacy preservation, SecureBoost involves complex training procedures and time-consuming cryptography operations. This causes SecureBoost to be slow to train and does not scale to large-scale data. In this work, we propose SecureBoost+, a large-scale and high-performance vertical federated gradient boosting decision tree framework. SecureBoost+ is secure in the semi-honest model, which is the same as SecureBoost. SecureBoost+ can be scaled up to tens of millions of data samples faster than SecureBoost. SecureBoost+ achieves high performance through several novel optimizations for SecureBoost, including ciphertext operation optimization and the introduction of new training mechanisms. The experimental results show that SecureBoost+ is 6–35x faster than SecureBoost but with the same accuracy and can be scaled up to tens of millions of data samples and thousands of feature dimensions.

Keywords: Vertical Federated Learning · GBDT · SecureBoost

1 Introduction

In a real-world scenario, data is stored in different organizations, forming a phenomenon known as data isolation. Federated learning (FL) [17,18,23], a distributed collaborative machine learning paradigm, is a promising approach to deal with the data isolation challenge. It enables local models to benefit from all parties while keeping local data private [11]. In addition, combined with various cryptography techniques, local data privacy can be effectively protected. Federated learning can be generally divided into vertical federated learning, horizontal

D.-N. Yang et al. (Eds.): PAKDD 2024, LNAI 14647, pp. 237–249, 2024.
https://doi.org/10.1007/978-981-97-2259-4_18

federated learning, and federated transfer learning. Vertical federated learning on logistic regression [3,9,22], GBDT [5], neural network [10,12], transfer learning [16], etc., have been previously studied.

Gradient boosting decision tree (GBDT) is an ensemble machine learning algorithm that is widely used in industry [4,13] due to its good performance and easy interpretation. For example, GBDT and its variants are usually used in recommendation [13,20], fraud detection [1], and click-through rate prediction [21] tasks with large-scale data. Due to the advantage of GBDT, many works try to apply vertical federated learning to the GBDT model [5,8,14]. SecureBoost [5] is one of the most popular vertical federated learning algorithms for GBDT. However, as far as we know, some previous studies have focused more on the security of algorithms and less on the gap between research and industrial applications. Some works are only tested on public experimental data sets instead of million-scale or high-dimensional data, which are common in real-world scenarios, and some works are tested on relatively larger data sets but can not be completed in a reasonable time. According to our practical experience, SecureBoost's performance is still unsatisfactory.

In this work, we propose SecureBoost+, a large-scale and high-performance *vertical* federated gradient boosting decision tree framework. It is developed based on the work SecureBoost [5], which has been integrated into FATE [15]. The main contributions of SecureBoost+ are:

- We propose a ciphertext operation optimization framework based on vertical federated gradient boosting decision tree methods, designed for Homomorphic Encryption (HE) [19] encryption schema.
- We propose two novel training mechanisms for SecureBoost+: Mix Tree mode and Layered Tree mode. The novel training mechanisms can significantly reduce the number of interactions and communication costs between parties.
- We investigate the performance of SecureBoost+ on a large-instance and a high-dimension dataset. Experimental results demonstrate that SecureBoost+ is 6–35x faster than SecureBoost but with the same accuracy and can be scaled up to tens of millions of data samples and thousands of features.

2 Preliminaries

2.1 Gradient Boosting Decision Tree

GBDT is an ensemble machine learning algorithm that is widely used in industry [4,13]. Let f be a decision tree function, the prediction for an instance is given by the sum of all K decision tree is obtained by:

$$\hat{y}_i = \sum_{k=1}^{K} f_k(\mathbf{x}_i) \tag{1}$$

When training an ensemble tree model, we add a new decision tree f_t at iteration t to minimize the following second-order approximation loss:

$$L^{(t)} = \sum_{i \in \mathbf{I}} \left[l(y_i, \hat{y}_i^{(t-1)}) + g_i f_t(\mathbf{x}_i) + \frac{1}{2} h_i f_t^2(\mathbf{x}_i) \right] + \Omega(f_t) \tag{2}$$

where $\Omega(f_t)$ is the regularization term and g_i, h_i are the first and second derivatives of $l(y_i, \hat{y}_i^{(t-1)})$ with respect to $\hat{y}_i$.

We rewrite Eq. 2 in a leaf-weight format [4]:

$$L^{(t)} = \sum_{j \in \mathbf{T}} \sum_{i \in \mathbf{I}_j} \left[l(y_i, \hat{y}_i^{(t-1)}) + g_i w_j + \frac{1}{2} h_i w_j^2 \right] + \frac{\lambda}{2} w_j^2 \tag{3}$$

By setting the second-order approximation function above, we derive the split gain function for splitting a node:

$$gain = \frac{1}{2} \left[\frac{(\Sigma_{i \in I_L} g_i)^2}{\Sigma_{i \in I_L} h_i + \lambda} + \frac{(\Sigma_{i \in I_R} g_i)^2}{\Sigma_{i \in I_R} h_i + \lambda} - \frac{(\Sigma_{i \in I} g_i)^2}{\Sigma_{i \in I} h_i + \lambda} \right] \tag{4}$$

where I_L, I_R, and I are the instances space of the left, right, and parent nodes, respectively. The leaf output of an arbitrary leaf j put is given by:

$$w_j = -\frac{\Sigma_{i \in I_j} g_j}{\Sigma_{i \in I_j} h_j + \lambda} \tag{5}$$

2.2 Paillier Homomorphic Encryption

The Paillier Homomorphic Encryption (PHE) [19] schema is an additive homomorphic cryptosystem, and it is the key component of many privacy-preserving ML algorithms [2,5,23]. The core properties of PHE are homomorphic addition and scalar multiplication:

$$[[x_1]] \oplus [[x_2]] = [[x_1 + x_2]] \tag{6}$$

$$x_1 \otimes [[x_2]] = [[x_1 \times x_2]] \tag{7}$$

where $[[.]]$ denotes the encryption operator.

2.3 SecureBoost

In SecureBoost [5], all parties conduct a privacy-preserving instance intersection at the beginning. After the intersection, we get $\left\{ \mathbf{X}^k \in \mathbb{R}^{n \times d_k} \right\}_{k=1}^{m}$ as the final aligned data matrices distributed on each party. Each party holds a data matrix $\mathbf{X}^{d_k}$ with n instances and d_k features. For ease of description, we use $\mathbf{X}$ to denote a federated instance whose features are distributed on multiple parties and use $\mathbf{I} = \{1, ...n\}$ to represent the instance space. $\mathbf{F}^k \in \{f_1, f_2, ..., f_{d_k}\}$ is the feature set on the k-th party. Each party holds its unique features, for any two feature sets F^i and F^j from i-th party and j-th party, $\mathbf{F}^i \cap \mathbf{F}^j = \emptyset$. We define two party roles: **Active Party** and **Passive Party**. The active party is the data provider who holds both a data matrix and the class label $\mathbf{y} \in \mathbb{R}^n$. The passive party is the data provider who only holds a data matrix and does not hold the class label.

The goal of SecureBoost is to train an ensemble gradient boosting tree model on the federated dataset $\left\{\mathbf{X}^k \in \mathbb{R}^{n \times d_k}\right\}_{k=1}^{m}$ and label $\mathbf{y}$. SecureBoost uses a histogram-based split finding strategy similar to XGBoost [4]. Therefore, before constructing the tree, every party uses the quantile binning method to transform their feature values into bin indices.

To ensure that the label information is not leaked through from g_i and h_i, the active party needs to apply homomorphic encryption on g_i and h_i and then send them to the passive party. With the additive property of homomorphic encryption on g_i and h_i, the passive party can calculate the ciphertext histogram and then construct split-info. To protect the feature information of the passive party, the passive party marks split-info with unique IDs, randomly shuffles this split-info, and then sends it to the active party. The global optimal split node using features of all parties can be calculated collaboratively through the above process. The active party can calculate the local split information and then receive the encrypted, split information from the passive party and decrypt it. The active party is able to find the optimal split node without leaking labels and knowing information about the passive party features. The party that holds the optimal split information will be responsible for splitting and assigning current node instances to child nodes. The result of instances assignment will be synchronized to all parties. Starting at the root node, We repeat the split finding layer-wise until the convergence condition is reached.

2.4 Performance Bottlenecks Analysis for SecureBoost

The most expensive parts of histogram-based gradient boosting algorithms are the histogram building and split node finding [4,13]. Due to homomorphic encryption operations, these two parts are particularly expensive to compute in SecureBoost. From the review of SecureBoost, we have the following observations:

- **Encryption and Addition Operation Cost:** Histogram calculation of the passive party will be especially time-consuming because it is calculated on the homomorphic ciphertext. Homomorphic addition to ciphertext usually involves multiplication and modular operations on large integers and is much more expensive than plaintext addition [19].
- **Decryption Operation Cost:** Since we know that the gradient and hessian are encrypted, we have to decrypt a batch of encrypted split information when we split the node. Decryption is expensive to train on large amounts of data.
- **Communication Cost:** At the beginning of building each tree, we need to synchronize the encryption gradient and hessian to the passive parties. At the

same time, a batch of encrypted split information needs to be sent to calculate the split node of each layer of the tree. This entails heavy communication overhead.

3 Proposed SecureBoost+ Framework

In this section, we propose our solution to improve the performance of SecureBoost in the following directions: Ciphertext Operation Optimization and Training Mechanism Optimization.

3.1 Ciphertext Operation Optimization

Rough Estimate of Ciphertext Operation Cost: Suppose we train a binary-classification task, we have n_i samples, n_f features, every feature has n_b bins, the tree depth is d, and then there are $n_n = 2^d$ nodes. We have a rough cost estimate for building a decision tree in SecureBoost:

- The ciphertext computation cost: ciphertext histogram computation + bin cumsum:
$$cost_{comp} = 2 \times n_i \times d + 2 \times n_n \times n_f \times n_b \tag{8}$$
- The encryption and decryption cost: $g\&h$ encryption + split-info decryption:
$$cost_{ende} = 2 \times n_i + 2 \times n_b \times n_f \times n_n \tag{9}$$
- The communication cost: encrypted $g\&h$ + split-infos batches:
$$cost_{comm} = 2 \times n_i + 2 \times n_b \times n_f \times n_n \tag{10}$$

GH Packing: We first introduce GH Packing. From SecureBoost, we notice that gradient and hessian have the same operations in the process of histogram construction and split-finding: the gradient and hessian of a sample will be encrypted at the same time, and the encrypted gradient and hessian will go to the same bucket, but be added to different indices. The gradient and hessian of a split-info will also be decrypted together.

The Paillier [19] is a homomorphic encryption algorithm used in SecureBoost. During training, a key length of 1024 or 2048 is usually used. When 1024 key length is used, the upper bound on the positive integer allowed in encryption is usually 1023 bits in length. This upper bound is much larger than the gradient and hessian at fixed points. As a result, a lot of plaintext space is wasted in the encryption process. Inspired by recently proposed gradient quantization and gradient batching techniques [24], we adopt a GH(gradient: g and hessian: h) Packing method in our work. We bundle g and h using bit arithmetic: We move g to the left by a certain number of bits and then add h to it.

For example, in a binary-classification task, the range of g is $[-1, 1]$, $g_{max} = 1$ and the range of h is $[0, 1]$, $h_{max} = 1$. Considering that g might be negative, we can offset g to ensure that it is a positive number. In the classification case, the offset number g_{off} for g will be 1. In the split finding step, we can remove the offset number when recovering g, h.

After offsetting g, we adopt a fixed-point encoding strategy to encode g and h, which is shown in Eq. 12, where r is usually 53. Then, we can pack g and h into one large positive number. To avoid overflowing aggregation results during histogram calculation, we need to reserve more bits for g and h. Assuming we have n_i instances, we need to make sure that no cumulative gradients and hessians in the histogram will be larger than g_{cmax} and h_{cmax} in Eq. 11. Therefore, we assign b_g bits for g and b_h bits for h in Eq. 11:

$$\begin{aligned} g_{cmax} &= n_i \times \lfloor (g_{max} + g_{off}) \times 2^r \rfloor \\ h_{cmax} &= n_i \times \lfloor h_{max} \times 2^r \rfloor \\ b_g &= BitLength(g_{cmax}) \\ b_h &= BitLength(h_{cmax}) \\ b_{gh} &= b_g + b_h \end{aligned} \tag{11}$$

Finally, we calculate a large positive number gh and homomorphic encryption ciphertext $[[gh]]$ as follow:

$$\begin{aligned} g &= \lfloor (g + g_{off}) \times 2^r \rfloor \\ h &= \lfloor h \times 2^r \rfloor \\ gh &= g << b_h + h \\ [[gh]] &= HE(gh) \end{aligned} \tag{12}$$

Figure 1 shows details of the GH packing process. The active party synchronizes $[[gh]]$ ciphertext to the passive party. The passive party only needs to calculate one ciphertext per sample when constructing the histogram, while in SecureBoost, it needs to calculate two ciphertexts per sample. From the GH Packing optimization, the whole ciphertext computation cost is **reduced by half.**

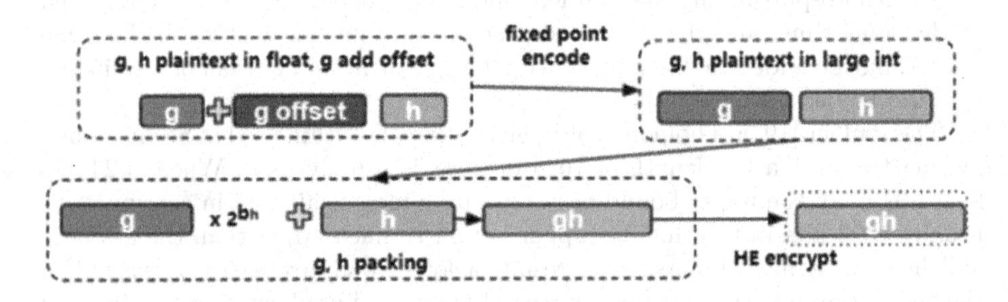

Fig. 1. The Process of GH Packing

Ciphertext Compression: Although the GH packing algorithm uses a lot of plaintext space, the plaintext space is not fully utilized. Remind that we usually have a 1023 bit-length integer as the plaintext upper bound when using a 1024-bit key length in the Paillier [19]. Assuming instances number n is $1,000,000$, r is 53, we will assign 74 bits for g and 73 bits for h, according to Eq. 11. The b_{gh} is 147, which is still less than 1023.

To fully use the plaintext space, we can perform ciphertext compression on the passive parties. This technique utilizes the property of our built-in HE algorithm, which is that an addition operation or a scalar multiplication operation is less costly than a decryption operation. We use addition and multiplication operations to compress several encrypted numbers in split-info into one encrypted number and make sure that the plaintext space is fully utilized. The idea of this method is the same as the GH Packing method: we multiply a ciphertext C by $2^{b_{gh}}$, shifting its plaintext by b_{gh} bits to the left. Then, we can add another ciphertext to it, compressing the two ciphertexts into one. We can repeat this process until no more bits are left in the plaintext space to compress, as shown in Eq. 13. Once the active party receives an encrypted number, it only needs to decrypt one encrypted number to recover several gh statistical results and reconstruct split-info.

$$\begin{aligned} C_{cum0} &= C_0 \\ C_{cum1} &= C_{cum0} \times 2^{b_{gh}} + C_1 \\ &.... \\ C_{cumk} &= C_{cumk-1} \times 2^{b_{gh}} + C_{k-1} \end{aligned} \tag{13}$$

In the case above, given a 1023 bit-length plaintext space and b_{gh} of 147, We are able to compress $\lfloor 1023/147 \rfloor = 6$ split points into 1. By using ciphertext compression, decryption operations and decryption time can be reduced by up to 6 times. On the other hand, the communication cost of encrypted split-info is also reduced by up to 6 times. Figure 2 illustrates the process of ciphertext compression.

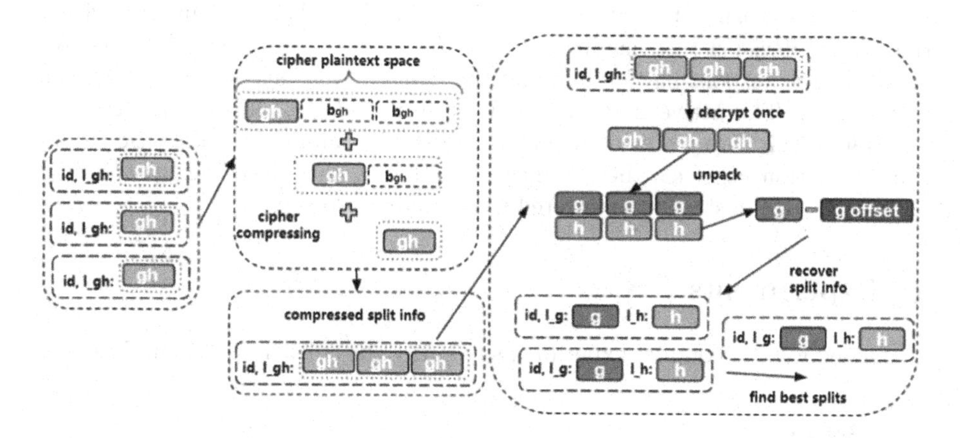

Fig. 2. The Process of Ciphertext Compression.

Ciphertext Histogram Subtraction: Histogram subtraction is a classic optimization technique in the histogram-based tree algorithms. When splitting nodes, samples are distributed to either the left or right nodes. The parent histogram is obtained for each feature by adding the set of histograms of the left and right children nodes. To take advantage of this property, in SecureBoost+, both the active and passive parties cache the histogram of the previous layer. When growing from the current layer, the active and passive parties first calculate the histogram of nodes with fewer samples and finally obtain the histogram of sibling nodes by histogram subtraction with the previous layer cache information. Since the ciphertext histogram calculation for passive parties is always the largest overhead, the ciphertext histogram subtraction will greatly speed up the calculation, and at least half of the ciphertext calculation is reduced.

3.2 Training Mechanism Optimization

At the training mechanism level, two optimization methods are proposed: mixed tree mode and layered tree mode. These two methods can significantly reduce the number of interactions and communication costs between parties.

Mix Tree Mode: In the Mix Tree Mode, every party will build several trees using their local features. When the active party builds a decision tree, the active party builds the tree locally. When the tree is built in the passive party, the relevant passive party receives the encrypted g and h, assisted by the active party, to find the best-split point. The structure of the tree and split point is retained on the passive party, while the leaf weight is retained on the active party. This mode is suitable for vertical federated learning under data balancing. In this way, the number of interactions and communication costs can be significantly reduced, thus speeding up training. Figure 3(a) demonstrates the mix tree training process.

Layered Tree Mode: In the Layered Tree Mode, the active and the passive party are responsible for different layers in the tree when building a decision tree. Every decision tree has in total $d = d_{active} + d_{passive}$ layers. The passive party will be responsible for building the first $d_{passive}$ layers, assisted by the active party. The active party will build the next d_{active} layers. All trees will be built in this 'layered' manner. In this way, the number of interactions and communication costs can be significantly reduced, thus speeding up training. Figure 3(b) demonstrates the layered tree training process.

4 Experiments

In this section, we conduct experiments to verify the efficiency of SecureBoost+.

4.1 Setup

Baseline Algorithm: SecureBoost and XGBoost.

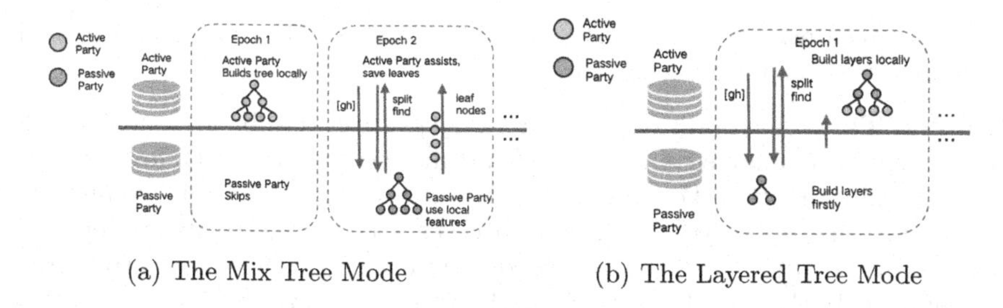

(a) The Mix Tree Mode

(b) The Layered Tree Mode

Fig. 3. Two proposed Training Mechanism.

Environment: Our experiments are conducted on two machines, which are active and passive parties, respectively. Each machine has 16 cores and 32GB RAM. We run our experiments in a 1GBps LAN network setup.

Parameters: We set $tree\_depth = 5$, $max\_bin\_num = 32$ and $learning\_rate = 0.3$. For encryption parameters, we use *Paillier* as the encryption schema with $key\_length = 1024$. For the efficiency of conducting experiments, our baseline methods will run $tree\_num = 25$ trees. In our framework SecureBoost+, we use a goss subsample and set the $top\_rate = 0.2$ and the $other\_rate = 0.1$. In the Layered Tree Mode, set the $d\_active = 2$ and $d\_passive = 3$, and in the Mix Tree Mode, every party is responsible for building $tree\_per\_party = 1$ tree.

Datasets: We evaluate our framework on four open datasets (*Give-credit*[1], *Susy* and *Higgs* [4,7], *Epsilon* [6]) to test model performance and training speed. The detail of the datasets is listed in Table 1. The data sets contain a large number of instances with high-dimension features. We vertically and equally divide every data set into two to make data sets for the active and passive parties.

Table 1. Data Set Details

datasets	# instance	# features	# active feature	# passive features	# labels
Give credit	150,000	10	5	5	2
Susy	5,000,000	18	4	14	2
Higgs	11,000,000	28	13	15	2
Epsilon	400,000	2000	1000	1000	2

4.2 Ciphertext Operation Optimization Evaluation

In this section, we will conduct experiments to compare SecureBoost+ with the Baseline algorithms in the setting in which ciphertext operation optimization is

[1] https://github.com/FederatedAI/FATE/tree/master/examples/data.

applied in SecureBoost+. SecureBoost+_Ciphertext is defined as SecureBoost+ with ciphertext operation optimization.

Training Time: We only consider the time spent on tree building and ignore the time spent on non-tree building, such as time spent on data I/O, feature engineering, and evaluation. For both SecureBoost+ and SecreuBoost, we build 25 trees and calculate the average time spent on each tree. The results are reported in Fig. 4. We can see that the proposed SecureBoost+ obviously outperforms SecureBoost: In the four datasets, Secureboost+ are **6.63x, 6.03x, 7.33x, 22x** faster than SecureBoost. From the experimental results, we can find that with the increase in the number of instances and feature dimensions, SecureBoost+ brings more advantages.

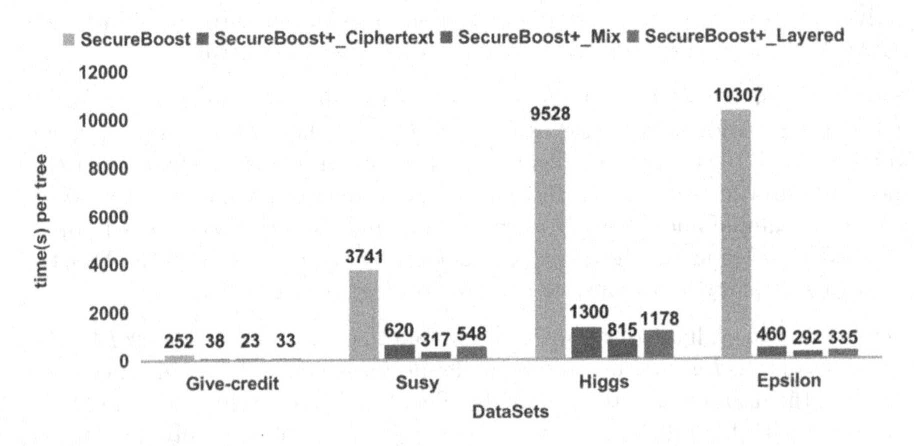

Fig. 4. Comparison of Tree Building Time in Binary Classification

Model Performance: The results are shown in Table 2. From the results, we can observe that SecureBoost+ performs just as well as XGBoost and SecureBoost in this setting.

4.3 Training Mechanism Optimization Evaluation

In this section, we will conduct experiments to compare SecureBoost+ with the Baseline algorithms in the setting where training mechanism optimizations are applied in SecureBoost+. In our experiment, when we enable Mix Tree Mode or Layered Tree Mode in SecureBoost+, we use ciphertext operation optimization in SecureBoost+ by default. SecureBoost+_Mix is defined as SecureBoot+ with the mix tree mode, while SecureBoost+_Layered is defined as SecureBoot+ with layered tree mode.

Training Time: Figure 4 reports the average tree-building time spent on each tree. It demonstrates that SecureBoost+ with the mix tree mode or layered tree mode builds a tree faster than SecureBoost+ with ciphertext operation optimization only and also faster than SecureBoost. This is because it skips global split-finding and inter-party communications. In the four datasets, SecureBoost+_Mix are **1.65x, 1.96x, 1.60x, 1.58x** faster than SecureBoost+_Ciphertext, while SecureBoost+_Mix are **10.96x, 11.80x, 11.69x, 35.30x** faster than SecureBoost. In the four datasets, SecureBoost+_Layered are **1.15x, 1.13x, 1.10x, 1.37x** faster than SecureBoost+_Ciphertext., while SecureBoost+_Layered are **7.64x, 6.83x, 8.09x, 30.77x** faster than SecureBoost.

Model Performance: The results are shown in Table 2. SecureBoost+_Mix and SecureBoost+_Layered have slight performance loss compared to XGBoost and SecureBoost+_Ciphertext, but the losses are very small. According to our experimental analysis, the two modes sometimes require a few more trees to catch up with the performances of SecureBoost+_Ciphertext. But from the result of *Epsilon* data set, we believe that the two training mechanisms can obviously accelerate training when the data set contains high dimensional features and features are balanced and distributed.

Table 2. Comparison of Model Performance in Binary Classification: AUC Metrics

Dataset	XGBoost	SecureBoost	SecureBoost+_Ciphertext	SecureBoost+ _Mix	SecureBoost +_Layered
Give-credit	0.872	0.874	0.873	0.87	0.871
Susy	0.864	0.873	0.873	0.869	0.87
Higgs	0.808	0.806	0.8	0.795	0.796
Epsilon	0.897	0.897	0.894	0.894	0.894

5 Conclusion

In this work, we proposed SecureBoost+, a large-scale and high-performance vertical federated gradient boosting decision tree framework. Based on Secureboost, we propose several optimization schemes, including Ciphertext Operation Optimization and Training Mechanism Optimization. The experimental results show that the speed of SecureBoost+ is significantly faster than that of SecureBoost, and the accuracy of SecureBoost+ is comparable to that of SecureBoost and XGBoost. With the increase in the number of instances and feature dimensions, SecureBoost+ will have more obvious advantages.

References

1. Cao, S., Yang, X., Chen, C., Zhou, J., Li, X., Qi, Y.: Titant: online real-time transaction fraud detection in ant financial. arXiv preprint arXiv:1906.07407 (2019)
2. Chai, D., Wang, L., Chen, K., Yang, Q.: Secure federated matrix factorization. IEEE Intell. Syst. (2020)

3. Chen, C., et al.: When homomorphic encryption marries secret sharing: secure large-scale sparse logistic regression and applications in risk control. In: Proceedings of the 27th ACM SIGKDD Conference on Knowledge Discovery and Data Mining, pp. 2652–2662 (2021)
4. Chen, T., Guestrin, C.: XGBoost: a scalable tree boosting system. In: Proceedings of the 22nd ACM SIGKDD International Conference on Knowledge Discovery and Data Mining, pp. 785–794 (2016)
5. Cheng, K., et al.: SecureBoost: a lossless federated learning framework. IEEE Intell. Syst. **36**(6), 87–98 (2021)
6. Dorogush, A.V., Ershov, V., Gulin, A.: CatBoost: gradient boosting with categorical features support. arXiv preprint arXiv:1810.11363 (2018)
7. Fu, F., Jiang, J., Shao, Y., Cui, B.: An experimental evaluation of large scale GBDT systems. arXiv preprint arXiv:1907.01882 (2019)
8. Fu, F., et al.: VF2Boost: very fast vertical federated gradient boosting for cross-enterprise learning. In: Proceedings of the 2021 International Conference on Management of Data, pp. 563–576 (2021)
9. Hardy, S., et al.: Private federated learning on vertically partitioned data via entity resolution and additively homomorphic encryption. arXiv preprint arXiv:1711.10677 (2017)
10. He, Y., et al.: A hybrid self-supervised learning framework for vertical federated learning. arXiv preprint arXiv:2208.08934 (2022)
11. Kairouz, P., et al.: Advances and open problems in federated learning. Found. Trends® Mach. Learn. **14**(1–2), 1–210 (2021)
12. Kang, Y., He, Y., Luo, J., Fan, T., Liu, Y., Yang, Q.: Privacy-preserving federated adversarial domain adaptation over feature groups for interpretability. IEEE Trans. Big Data (2022)
13. Ke, G., et al.: LightGBM: a highly efficient gradient boosting decision tree. In: Advances in Neural Information Processing Systems, vol. 30, pp. 3146–3154 (2017)
14. Li, Q., Wen, Z., He, B.: Practical federated gradient boosting decision trees. In: Proceedings of the AAAI Conference on Artificial Intelligence, pp. 4642–4649 (2020)
15. Liu, Y., Fan, T., Chen, T., Xu, Q., Yang, Q.: Fate: an industrial grade platform for collaborative learning with data protection. J. Mach. Learn. Res. **22**(226), 1–6 (2021). http://jmlr.org/papers/v22/20-815.html
16. Liu, Y., Kang, Y., Xing, C., Chen, T., Yang, Q.: A secure federated transfer learning framework. IEEE Intell. Syst. **35**(4), 70–82 (2020)
17. Liu, Y., et al.: Vertical federated learning: concepts, advances and challenges. arXiv preprint arXiv:2211.12814 (2022)
18. McMahan, B., Moore, E., Ramage, D., Hampson, S., Arcas, B.A.: Communication-efficient learning of deep networks from decentralized data. In: Artificial Intelligence and Statistics, pp. 1273–1282. PMLR (2017)
19. Paillier, P.: Public-key cryptosystems based on composite degree residuosity classes. In: Stern, J. (ed.) EUROCRYPT 1999. LNCS, vol. 1592, pp. 223–238. Springer, Heidelberg (1999). https://doi.org/10.1007/3-540-48910-X_16
20. Shahbazi, Z., Byun, Y.C.: Product recommendation based on content-based filtering using XGBoost classifier. Int. J. Adv. Sci. Technol **29**, 6979–6988 (2019)
21. Wang, X., He, X., Feng, F., Nie, L., Chua, T.S.: Tem: tree-enhanced embedding model for explainable recommendation. In: Proceedings of the 2018 World Wide Web Conference, pp. 1543–1552 (2018)
22. Yang, K., Fan, T., Chen, T., Shi, Y., Yang, Q.: A quasi-newton method based vertical federated learning framework for logistic regression. arXiv preprint arXiv:1912.00513 (2019)

23. Yang, Q., Liu, Y., Chen, T., Tong, Y.: Federated machine learning: concept and applications. ACM Trans. Intell. Syst. Technol. (TIST) **10**(2), 1–19 (2019)
24. Zhang, C., Li, S., Xia, J., Wang, W., Yan, F., Liu, Y.: BatchCrypt: efficient homomorphic encryption for cross-silo federated learning. In: 2020 {USENIX} Annual Technical Conference ({USENIX}{ATC} 20), pp. 493–506 (2020)

Construct a Secure CNN Against Gradient Inversion Attack

Yu-Hsin Liu, Yu-Chun Shen, Hsi-Wen Chen, and Ming-Syan Chen(✉)

Department of Electrical Engineering, National Taiwan University, Taipei, Taiwan
{yhliu,ycshen,hwchen}@arbor.ee.ntu.edu.tw, mschen@ntu.edu.tw

Abstract. Federated learning enables collaborative model training across multiple clients without sharing raw data, adhering to privacy regulations, which involves clients sending model updates (gradients) to a central server, where they are aggregated to improve a global model. Despite its benefits, federated learning faces threats from gradient inversion attacks, which can reconstruct private data from gradients. Traditional defenses, including cryptography, differential privacy, and perturbation techniques, offer protection but may suffer from a reduction in computational efficiency and model performance. Thus, in this paper, we introduce *Secure Convolutional Neural Networks (SecCNN)*, a novel approach embedding an upsampling layer into CNNs to inherently defend against gradient inversion attacks. SecCNN leverages Rank Analysis for enhanced security without sacrificing model accuracy or incurring significant computational costs. Our results demonstrate SecCNN's effectiveness in securing federated learning against privacy breaches, thereby building trust among participants and advancing secure collaborative learning.

Keywords: Gradient Inversion Attack · Privacy Preserving

1 Introduction

To enable better data privacy and ownership, federated learning [5,28] aims to orchestrate multiple clients in a distributed environment to collaboratively train a neural network model at a central server without moving their data to the central server. Instead of sending the actual data, each client computes a model update, i.e., a gradient by its local data using the latest copy of the global model, and then sends the gradient to the central server. Then, the server aggregates these updates (typically by averaging) to construct a global model and sends the new model parameters to all clients. Enabling clients to engage in training without directly disclosing their data aligns more effectively with data privacy regulations like the California Consumer Privacy Act (CCPA) [8] and the General Data Protection Regulation (GDPR) [27]. This alignment is crucial for sensitive applications, especially within industries such as healthcare [26] and finance [22].

Despite the successful implementation of federated learning in practical AI applications, various studies have underscored the continued presence of

D.-N. Yang et al. (Eds.): PAKDD 2024, LNAI 14647, pp. 250–261, 2024.
https://doi.org/10.1007/978-981-97-2259-4_19

unresolved issues and vulnerabilities, notably the *Gradient Inversion Attack* [9,13,21,40]. Generally, the gradient inversion attack aims to minimize the difference between estimated and actual gradients by generating dummy gradients. The attacker can feed random data-label pairs into the global model at the server. By treating the gradient difference as an error, the recovery process becomes an iterative optimization problem. Upon convergence, private data is fully reconstructed. This attack enables an immoral service to recover a client's private data by extracting gradients [13]. A surveillance attacker can also intercept communication between the clients and the server, potentially accessing other clients' private data [9].

Several strategies have been proposed to defend against the gradient inversion attack, including cryptography-based [1,4], differential privacy-based [33,35], and perturbation-based techniques [6,14]. Their common goal is to prevent privacy breaches resulting from gradient sharing. Nevertheless, these methods aim to establish secure gradient communication, potentially incurring additional computational costs [15] and resulting in a certain level of performance degradation [23]. Therefore, in this paper, we aim to prevent the gradient inversion attack from a different perspective: *Can we devise a model that offers fundamental security?*

e introduce a straightforward yet powerful CNN module called *Secure Convolutional Neural Networks (SecCNN)* to mitigate potential threats. Inspired by Rank Analysis [39], we observe that adding an upsampling layer to the CNN model provides inherent capability to defend against gradient inversion attacks. Experimental results demonstrate the effectiveness of SecCNN in mitigating the impact of gradient inversion attacks while preserving model accuracy. Deploying SecCNN bolsters security and nurtures trust between clients and servers. This proactive approach addresses concerns about potentially compromising sensitive information due to malicious server actions using gradient inversion attacks.

Our main contributions are summarized as follows:

- We present a novel building block in CNN, namely *Secure Convolutional Neural Networks (SecCNN)*, which can withstand gradient inversion attacks without compromising model performance.
- SecCNN offers a safer model in federated learning, fostering trust between clients and the server, reducing suspicion of malicious intent.
- Experimental results demonstrate SecCNN's effectiveness in mitigating the impact of gradient inversion attacks while maintaining model accuracy.

2 Preliminary

2.1 Federated Learning

Here, we briefly introduce the standard federated learning algorithm [25]. Every client i samples a mini-batch (x_i, y_i) from its dataset to compute the gradients as follows.

$$\nabla\theta_i = \frac{\partial L(\theta(x_i), y_i)}{\partial \theta}, \tag{1}$$

where the model θ and loss L are shared by default for synchronized distributed optimization. Following FedAvg [24], the gradients are averaged across the N servers and then used to update the weights as follows.

$$\nabla\theta = \frac{1}{N}\sum_{i=1}^{N} \nabla\theta_i \text{ , and } \theta' = \theta - \eta\nabla\theta. \tag{2}$$

The objective of a gradient inversion attack is to utilize gradients $\nabla\theta_i$ received from another client i to potentially infer or steal client i's training data (x_i, y_i).

2.2 Gradient Inversion Attack

Following DLG [40], the attacker begins by randomly initializing a dummy input, denoted as x', along with a labeled output, y', to retrieve data from gradients. Subsequently, the attacker inputs these dummy data into models and acquires corresponding dummy gradients as follows.

$$\nabla\theta' = \frac{\partial l(\theta(x'), y')}{\partial\theta}. \tag{3}$$

By minimizing the discrepancy between dummy and actual gradients, the dummy data will more closely resemble the actual training data, thus allowing an attacker to reconstruct the original training data, which can be expressed as follows.

$$\arg\min_{x',y'} \|\nabla\theta - \nabla\theta'\|^2. \tag{4}$$

The objective can be optimized using L-BFGS [2] to obtain the actual data x and label y. Directly optimizing the above objective is suboptimal for more complex architectures, especially those with arbitrary parameter vectors. While the magnitude of a parameter gradient only reflects the training state, the high-dimensional gradient direction contains valuable information. Therefore, Geiping et al. [7] propose a new objective function.

$$\arg\min_{x\in[0,1]^n} 1 - \frac{\langle\nabla\theta', \nabla\theta\rangle}{\|\nabla\theta'\|\|\nabla\theta\|} + \alpha\text{TV}(x), \tag{5}$$

where TV represents total variation, serving as a regularization method to ensure the input distribution x is realistic. Optimizing Eq. 5 enables the attacker to find images that provoke similar model prediction changes as the ground truth, diverging from solely aligning with observed gradients. This approach parallels minimizing the Euclidean cost function with normalized gradient vectors, as seen in Eq. 4.

2.3 Recursive Gradient Attack on Privacy (R-GAP)

In this paper, we utilize the *Recursive Gradient Attack on Privacy (R-GAP)* [39] as our attack method. This approach introduces a closed-form, recursive procedure for recovering data from gradients in deep neural networks. Unlike the previously mentioned optimization-based attacks [20,37], R-GAP is not susceptible

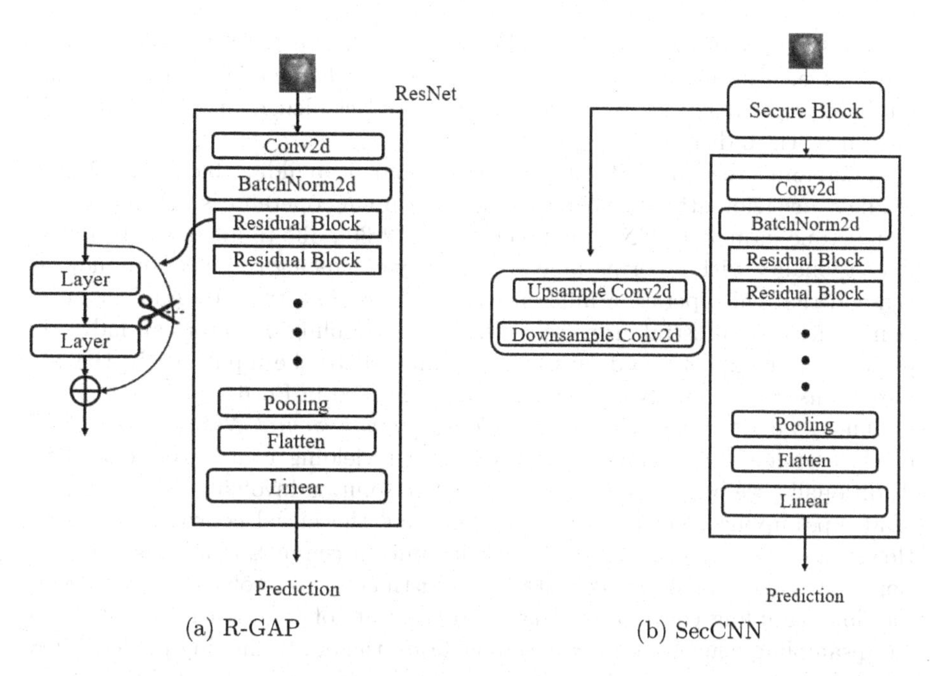

Fig. 1. The comparison of R-GAP and SecCNN to defense against gradient inversion attack

to local optima and operates orders of magnitude faster due to its deterministic runtime. This efficiency is achieved by establishing specific conditions where optimization attacks fail. Specifically, R-GAP uses the *Rank Analysis* technique to identify which network architectures allow for full recovery and which result in noisy recovery due to rank deficiency. The fundamental constraints necessary for executing a successful gradient inversion attack are formally outlined as follows.

$$|W_i| + |z_i| > |x_i|, \tag{6}$$

where $|x_i|$ denotes the number of input entries at the i-th layer, $|z_i|$ represents the number of output entries, and $|W_i|$ signifies the number of weights. Essentially, if the rank of the projection matrix and its associated output is lower than that of the input, a lossless transformation through the model cannot be achieved, resulting in noisy reconstruction outcomes.

3 Secure Convolutional Neural Networks

To effectively prevent gradient inversion attacks, R-GAP employs ResNet as its threat model, disrupting skip connections in the essential residual block to defend against such attacks while maintaining accuracy. This method is straightforward because the skip connection ensures that Eq. 6 always holds, leading to a perfect input reconstruction. However, several limitations arise with this cut-based defense strategy. Firstly, its applicability is confined to ResNet-based architectures, given the absence of skip connections in other model types. Secondly, multiple residual blocks in ResNet models pose challenges in pinpointing the critical block to disrupt.

Therefore, we aim to develop a model-agnostic building block to mitigate gradient inversion attacks. While the down-sampling operator is recognized as a cornerstone in modern CNN architectures [2,12,30], due to its ability to reduce dimensionality with fewer parameters, thereby enhancing training efficiency, we contend that this operation inherently introduces robustness to gradient inversion attacks. As illustrated in Eq. 6, the down-sampling operator can halve the projection weight $|W_i|$ and output $|z_i|$ compared to the input $|x_i|$ in the i-th layer, consistently satisfying the constraint specified in Eq. 6.

Thus, we introduce the *Secure Convolutional Neural Network (SecCNN)* block, seamlessly adaptable to any CNN and yielding comparable outcomes. Additionally, we propose a simple yet effective solution involving the integration of an upsampling convolutional layer following the initial convolutional layer. However, a drawback arises due to differing output channels in the first convolutional layer of each model, necessitating adjustments. To resolve this, we relocate the downsampling convolutional layer to the start of the model and introduce an upsampling convolutional layer to facilitate the downsampling process effectively. Figure 1 illustrates the difference between SecCNN and R-GAP.

4 Experiment

In our study, we evaluate the effectiveness of our SecCNN block in mitigating gradient inversion attacks. We employ the CIFAR-100 dataset [18] for training, which comprises $50,000$ training images and $10,000$ testing images distributed across 100 classes, each with dimensions of 32×32 pixels. Additionally, we select ResNet-18 [11] and LeNet-5 [19] as our base models while utilizing R-GAP [39] as the gradient inversion attack method. The learning rate is set to 0.01, and we use SGD as the optimizer. It's worth noting that we resize the images to dimensions of 224×224 and apply data augmentation techniques to ensure compatibility with the input requirements of ResNet.

4.1 Quantitative Results

In evaluating the trade-off between defensive effectiveness and model performance against gradient inversion attacks, we analyze the quality of reconstructed images using the Peak Signal-to-Noise Ratio (PSNR) and assess model performance through the top-1 accuracy score. PSNR [39] is a widely recognized metric for evaluating the fidelity of reconstructed or compressed images compared to their original, uncompressed counterparts. Lower PSNR values in the context of gradient inversion attacks indicate enhanced protection, deterring potential attackers from successfully reconstructing data.

Table 1. Quantitative Results.

	PSNR avg	PSNR std	PSNR Max	PSNR Min	Top-1 Accuracy
ResNet-18	11.6411	3.0229	18.2063	4.0333	0.7320
ResNet-18 + Gaussian noise ($\sigma^2 = 0.1$)	**8.0947**	1.0531	**10.1889**	5.4480	0.2580
ResNet-18 + Gaussian noise ($\sigma^2 = 0.01$)	8.0959	1.0609	10.2260	5.4058	0.6220
ResNet-18 + Gaussian noise ($\sigma^2 = 0.001$)	8.9805	1.3379	11.2437	4.8213	0.7224
ResNet-18 + Gaussian noise ($\sigma^2 = 0.0001$)	11.1793	2.8875	16.5728	**3.8044**	0.7273
ResNet-18 + Secure Block (ours)	9.0941	1.1912	11.2700	4.4268	**0.7325**

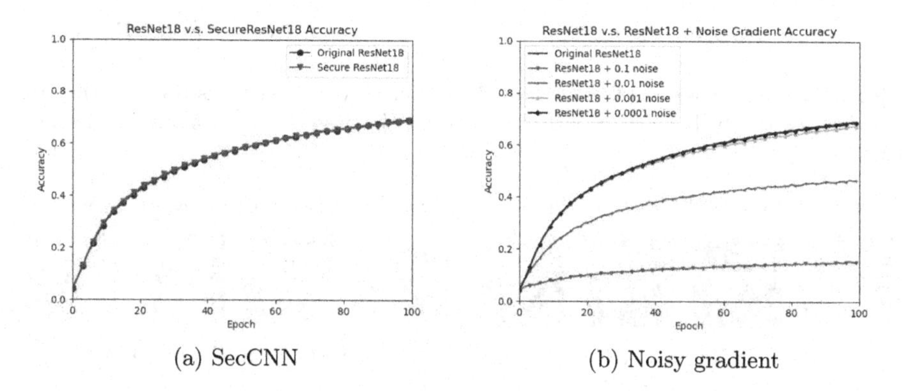

(a) SecCNN (b) Noisy gradient

Fig. 2. The learning curve of SecCNN and pertubation-based defense method.

We evaluate the performance of our SecCNN compared to perturbation-based defense methods [6,14] by introducing Gaussian noise into the gradient at varying levels. As shown in Table 1, ResNet-18 achieves the highest PSNR, indicating limited capability to defend against gradient inversion attacks as the input image is perfectly reconstructed. As the level of Gaussian noise increases, the PSNR progressively deteriorates, suggesting improved defense efficacy. However, adding Gaussian noise with variance 0.1 notably decreases the PSNR, leading to a significant drop in top-1 accuracy to 0.258, rendering it an inadequate image classifier. Similarly, noise with a level of 0.01 results in a decline to 0.622. Conversely, more minor variances in Gaussian noise, such as 0.001 or 0.0001, lead to only minor accuracy reductions, accompanied by relatively high PSNR values compared to our SecCNN.

In contrast, our SecCNN maintains accuracy and demonstrates slight improvement, with a lesser impact on accuracy compared to noisy gradient techniques. This analysis underscores the effectiveness of our approach in preserving model accuracy while integrating defense mechanisms against gradient inversion attacks. Figure 2 also illustrates that SecCNN has minimal impact on the training process, resulting in comparable model accuracy to the original model.

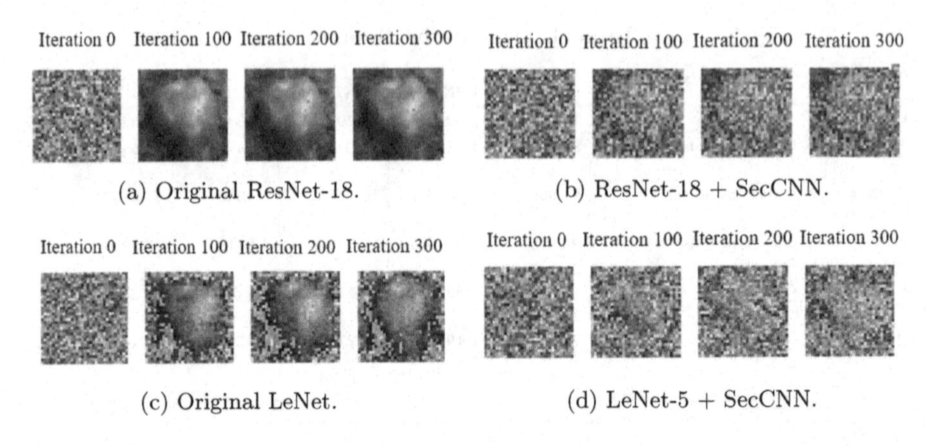

(a) Original ResNet-18.

(b) ResNet-18 + SecCNN.

(c) Original LeNet.

(d) LeNet-5 + SecCNN.

Fig. 3. The images reconstructed through gradient inversion attacks from (a) the original ResNet-18, (b) ResNet-18 with SecCNN, (c) the original LeNet-5, and (d) LeNet-5 with SecCNN.

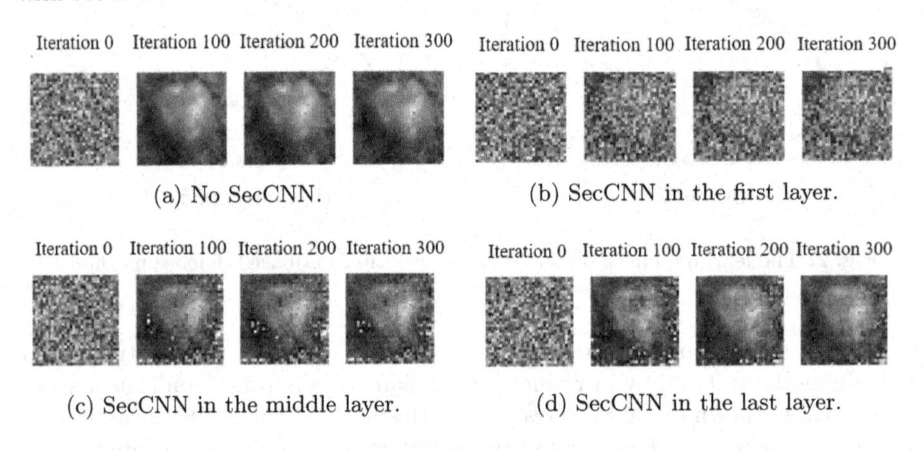

(a) No SecCNN.

(b) SecCNN in the first layer.

(c) SecCNN in the middle layer.

(d) SecCNN in the last layer.

Fig. 4. The effect of the position of SecCNN.

In contrast, increasing noise levels significantly degrades the perturbation-based model's performance.

4.2 Quantitative Results

Figure 3 illustrates the reconstructed images from the gradient inversion attacks on ResNet-18 and LeNet-5 and the SecCNN-defended models. The attack on the original ResNet-18 displays recognizable contours and colors corresponding to the original image, and conversely, incorporating SecCNN results in reconstructed images that lack similar recognizable features, indicating an improvement in defense. Although LeNet-5 is inherently more robust against gradient inversion attacks due to its absence of residual connections like ResNet to allevi-

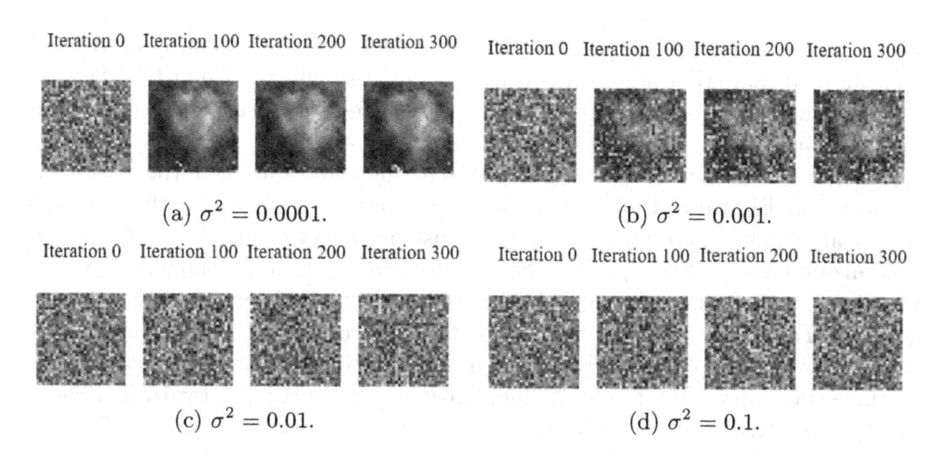

Fig. 5. The images reconstructed through gradient inversion attacks by varying different level noise in gradient.

ate the gradient vanishing problem, our SecCNN still significantly disrupts the reconstructed image compared to the original one.

Figure 4 examines the effectiveness of placing the SecCNN module at different positions within the network architecture. Positioning the SecCNN at the beginning of the model provides robust defense against gradient inversion attacks. However, relocating SecCNN to the middle (between layer 2 and layer 3) or towards the end (just before the final average pooling layer) results in the successful reconstruction of the image, indicating these placements are ineffective at thwarting gradient inversion attacks. The optimal protection is afforded when the SecCNN is situated at the model's inception. This is because the gradients of the first layer have the most direct correlation with the input data, making them critical for the reconstruction process. Thus, positioning the secure block here impacts the model's resilience to such attacks.

Figure 5 also provides visualizations of the reconstructed image from the perturbation-based defense method. The decimal values preceding the noise indicate the scaling factor applied to the gradients. Remarkably, adding 0.0001 Gaussian noise to the gradients makes it easy for the server to reconstruct the original data from the gradients. However, as 0.001 noise is added, the effectiveness improves, making it more difficult to recover the original data. Furthermore, increasing the noise level makes retrieving the original data nearly impossible.

5 Related Work

Pioneered by Google [17,24], Federated learning offers a privacy-preserving method for distributed model training across devices without exchanging raw data. McMahan et al. [24] introduce the FedAvg algorithm, enabling clients to contribute to a global model by sharing updates with a central server, which aggregates them. Then, Zhao et al. [38] extend the setting to support heterogeneous and non-IID data, highlighting the adverse effects of non-IID data on model accuracy and suggesting mitigating these effects by incorporating a shared global dataset.

However, a series of recent studies have shown that even the gradients uploaded in FL carry the risk of privacy leakage [1], known as the gradient inversion attack, where attackers can reconstruct network inputs by iteratively minimizing error to find optimal layer inputs. Phong et al. [1] show that inputs to a perceptron could be recovered using the gradient of the weight over the bias gradient. Fan et al. [6] adapt this method for convolutional layers by reformatting them into fully connected layers using stacked filters. Zhu and Blaschko [39] advance this approach for convolutional layers by integrating forward and backward propagation to frame the recovery process as solving a system of linear equations, explicitly targeting the image data of the first convolutional layer.

To defend against gradient inversion attacks, cryptographic methods offer robust protection for individual gradients, maintaining their usefulness. Lia et al. [20] and Bonawitz et al. [3,20] employ Multi-party computation to enable secure model updates, keeping each gradient's information private. Additionally, Kim et al. [16] and Zhang et al. [34] apply Homomorphic Encryption (HE) to execute operations within the encrypted domain, ensuring that gradients remain secure against decryption attempts, even amidst man-in-the-middle (MITM) attacks. However, adopting these cryptographic solutions requires changes to the traditional training setup and significantly increases the computational load, bandwidth needs, and storage capacity [36].

Another line of research involves gradient perturbation to ensure data privacy by applying some noise to the gradient to make the input image un reconstructible. Truex et al. [31] utilize Gaussian noise addition within the differential privacy (DP) framework to safeguard gradients. He et al. [10] delve into the influence of iterative training on privacy, establishing a correlation between privacy preservation and model generalization. Sun et al. [29] underscore the significance of the data representation layer in defending against gradient inversion attacks, advocating for targeted gradient perturbation within this layer. Additionally, Wei et al. [32] propose a method that dynamically adjusts noise levels, enhancing resilience to GradInv attacks while minimizing the impact on model accuracy. However, these approaches typically involve a discernible trade-off between model performance and the effectiveness of protection [39].

6 Limitation and Conclusion

In this paper, we introduce a new CNN component called Secure Convolutional Neural Networks (SecCNN), designed as an innovative safeguard against gradient inversion attacks. SecCNN employs Rank Analysis to bolster security without compromising accuracy or inflating computational demands. Experimental results manifest that SecCNN effectively enhances the security of federated learning, fostering trust among stakeholders and facilitating secure collaboration. Nevertheless, the impact of SecCNN varies across different models. For instance, models like VGG16, inherently more resistant to such attacks, may see slight improvement from incorporating SecCNN. Conversely, simpler models such as LeNet benefit greatly from SecCNN's defensive capabilities against gradient inversion attacks, though this comes with the trade-off of longer training times. Therefore, incorporating SecCNN calls for a comprehensive assessment of the trade-off between improved security and the training cost.

Acknowledgement. This work was supported in part by the National Science and Technology Council, Taiwan, under grant NSTC-112-2223-E-002-015, and by the Ministry of Education, Taiwan, under grant MOE 112L9009.

References

1. Aono, Y., Hayashi, T., Wang, L., Moriai, S., et al.: Privacy-preserving deep learning via additively homomorphic encryption. IEEE Trans. Inf. Forensics Secur. **13**(5), 1333–1345 (2017)
2. Berahas, A.S., Nocedal, J., Takác, M.: A multi-batch l-bfgs method for machine learning. In: Advances in Neural Information Processing Systems, vol. 29 (2016)
3. Bonawitz, K., et al.: Towards federated learning at scale: system design. Proc. Mach. Learn. Syst. **1**, 374–388 (2019)
4. Bonawitz, K., et al.: Practical secure aggregation for privacy-preserving machine learning. In: Proceedings of the 2017 ACM SIGSAC Conference on Computer and Communications Security, pp. 1175–1191 (2017)
5. Chilimbi, T., Suzue, Y., Apacible, J., Kalyanaraman, K.: Project Adam: building an efficient and scalable deep learning training system. In: 11th {USENIX} Symposium on Operating Systems Design and Implementation ({OSDI} 14), pp. 571–582 (2014)
6. Fan, L., et al.: Rethinking privacy preserving deep learning: how to evaluate and thwart privacy attacks. In: Federated Learning: Privacy and Incentive, pp. 32–50 (2020)
7. Geiping, J., Bauermeister, H., Dröge, H., Moeller, M.: Inverting gradients - how easy is it to break privacy in federated learning? In: Larochelle, H., Ranzato, M., Hadsell, R., Balcan, M., Lin, H. (eds.) Advances in Neural Information Processing Systems, vol. 33, pp. 16937–16947. Curran Associates, Inc. (2020)
8. Harding, E.L., Vanto, J.J., Clark, R., Hannah Ji, L., Ainsworth, S.C.: Understanding the scope and impact of the California consumer privacy act of 2018. J. Data Protect. Privacy **2**(3), 234–253 (2019)
9. Hatamizadeh, A., et al.: Gradvit: gradient inversion of vision transformers. In: Proceedings of the IEEE/CVF Conference on Computer Vision and Pattern Recognition, pp. 10021–10030 (2022)
10. He, F., Wang, B., Tao, D.: Tighter generalization bounds for iterative differentially private learning algorithms. In: Uncertainty in Artificial Intelligence, pp. 802–812. PMLR (2021)
11. He, K., Zhang, X., Ren, S., Sun, J.: Deep residual learning for image recognition. In: Proceedings of the IEEE Conference on Computer Vision and Pattern Recognition, pp. 770–778 (2016)
12. Howard, A., et al.: Searching for mobilenetv3. In: Proceedings of the IEEE/CVF International Conference on Computer Vision, pp. 1314–1324 (2019)

13. Huang, Y., Gupta, S., Song, Z., Li, K., Arora, S.: Evaluating gradient inversion attacks and defenses in federated learning. In: Advances in Neural Information Processing Systems, vol. 34, pp. 7232–7241 (2021)
14. Huang, Y., Song, Z., Li, K., Arora, S.: Instahide: instance-hiding schemes for private distributed learning. In: International Conference on Machine Learning, pp. 4507–4518. PMLR (2020)
15. Huang, Z., Wang, Y., Mitra, S., Dullerud, G.E.: On the cost of differential privacy in distributed control systems. In: Proceedings of the 3rd International Conference on High Confidence Networked Systems, pp. 105–114 (2014)
16. Kim, J., Koo, D., Kim, Y., Yoon, H., Shin, J., Kim, S.: Efficient privacy-preserving matrix factorization for recommendation via fully homomorphic encryption. ACM Trans. Privacy Secur. (TOPS) **21**(4), 1–30 (2018)
17. Konečný, J., McMahan, H.B., Yu, F.X., Richtárik, P., Suresh, A.T., Bacon, D.: Federated learning: strategies for improving communication efficiency. arXiv preprint arXiv:1610.05492 (2016)
18. Krizhevsky, A., Hinton, G., et al.: Learning multiple layers of features from tiny images (2009)
19. LeCun, Y., Bottou, L., Bengio, Y., Haffner, P.: Gradient-based learning applied to document recognition. Proc. IEEE **86**(11), 2278–2324 (1998)
20. Lia, D., Togan, M.: Privacy-preserving machine learning using federated learning and secure aggregation. In: 2020 12th International Conference on Electronics, Computers and Artificial Intelligence (ECAI), pp. 1–6. IEEE (2020)
21. Liang, H., Li, Y., Zhang, C., Liu, X., Zhu, L.: Egia: an external gradient inversion attack in federated learning. IEEE Trans. Inf. Forensics Secur. (2023)
22. Long, G., Tan, Y., Jiang, J., Zhang, C.: Federated learning for open banking. In: Yang, Q., Fan, L., Yu, H. (eds.) Federated Learning. LNCS (LNAI), vol. 12500, pp. 240–254. Springer, Cham (2020). https://doi.org/10.1007/978-3-030-63076-8_17
23. Mangold, P., Perrot, M., Bellet, A., Tommasi, M.: Differential privacy has bounded impact on fairness in classification. In: International Conference on Machine Learning, pp. 23681–23705. PMLR (2023)
24. McMahan, B., Moore, E., Ramage, D., Hampson, S., Arcas, B.A.: Communication-efficient learning of deep networks from decentralized data. In: Artificial Intelligence and Statistics, pp. 1273–1282. PMLR (2017)
25. McMahan, H.B., Moore, E., Ramage, D., Arcas, B.A.: Federated learning of deep networks using model averaging, **2**, 2. arXiv preprint arXiv:1602.05629 (2016)
26. Pfitzner, B., Steckhan, N., Arnrich, B.: Federated learning in a medical context: a systematic literature review. ACM Trans. Internet Technology (TOIT) **21**(2), 1–31 (2021)
27. Regulation, G.D.P.: General data protection regulation (GDPR). Intersoft Consulting, Accessed in October **24**(1) (2018)
28. Shokri, R., Shmatikov, V.: Privacy-preserving deep learning. In: Proceedings of the 22nd ACM SIGSAC Conference on Computer and Communications Security, pp. 1310–1321 (2015)
29. Sun, J., Li, A., Wang, B., Yang, H., Li, H., Chen, Y.: Provable defense against privacy leakage in federated learning from representation perspective. arXiv preprint arXiv:2012.06043 (2020)
30. Tan, M., Le, Q.: Efficientnet: rethinking model scaling for convolutional neural networks. In: International Conference on Machine Learning, pp. 6105–6114. PMLR (2019)

31. Truex, S., Liu, L., Chow, K.H., Gursoy, M.E., Wei, W.: LDP-FED: federated learning with local differential privacy. In: Proceedings of the Third ACM International Workshop on Edge Systems, Analytics and Networking, pp. 61–66 (2020)
32. Wei, W., Liu, L., Wu, Y., Su, G., Iyengar, A.: Gradient-leakage resilient federated learning. In: 2021 IEEE 41st International Conference on Distributed Computing Systems (ICDCS), pp. 797–807. IEEE (2021)
33. Ye, D., Shen, S., Zhu, T., Liu, B., Zhou, W.: One parameter defense-defending against data inference attacks via differential privacy. IEEE Trans. Inf. Forensics Secur. **17**, 1466–1480 (2022)
34. Zhang, C., Li, S., Xia, J., Wang, W., Yan, F., Liu, Y.: {BatchCrypt}: efficient homomorphic encryption for {Cross-Silo} federated learning. In: 2020 USENIX Annual Technical Conference (USENIX ATC 20), pp. 493–506 (2020)
35. Zhang, Q., Ma, J., Xiao, Y., Lou, J., Xiong, L.: Broadening differential privacy for deep learning against model inversion attacks. In: 2020 IEEE International Conference on Big Data (Big Data), pp. 1061–1070. IEEE (2020)
36. Zhang, R., Guo, S., Wang, J., Xie, X., Tao, D.: A survey on gradient inversion: attacks, defenses and future directions. arXiv preprint arXiv:2206.07284 (2022)
37. Zhao, B., Mopuri, K.R., Bilen, H.: IDLG: improved deep leakage from gradients. arXiv preprint arXiv:2001.02610 (2020)
38. Zhao, Y., Li, M., Lai, L., Suda, N., Civin, D., Chandra, V.: Federated learning with non-IID data. arXiv preprint arXiv:1806.00582 (2018)
39. Zhu, J., Blaschko, M.B.: R-{gap}: Recursive gradient attack on privacy. In: International Conference on Learning Representations (2021). https://openreview.net/forum?id=RSU17UoKfJF
40. Zhu, L., Liu, Z., Han, S.: Deep leakage from gradients. In: Advances in Neural Information Processing Systems, vol. 32 (2019)

Backdoor Attack Against One-Class Sequential Anomaly Detection Models

He Cheng and Shuhan Yuan(✉)

Utah State University, 84322 Logan, UT, USA
{he.cheng,shuhan.yuan}@usu.edu

Abstract. Deep anomaly detection on sequential data has garnered significant attention due to the wide application scenarios. However, deep learning-based models face a critical security threat - their vulnerability to backdoor attacks. In this paper, we explore compromising deep sequential anomaly detection models by proposing a novel backdoor attack strategy. The attack approach comprises two primary steps, trigger generation and backdoor injection. Trigger generation is to derive imperceptible triggers by crafting perturbed samples from the benign normal data, of which the perturbed samples are still normal. The backdoor injection is to properly inject the backdoor triggers to comprise the model only for the samples with triggers. The experimental results demonstrate the effectiveness of our proposed attack strategy by injecting backdoors on two well-established one-class anomaly detection models.

Keywords: Backdoor Attack · Anomaly Detection · Sequential Data

1 Introduction

Deep learning models have been widely used for sequential anomaly detection [7,13,16]. However, deep learning models are also vulnerable to various attacks, such as backdoor attacks. When compromised by a backdoor attack, a deep learning model behaves normally with benign samples but activates backdoors upon the appearance of triggers, resulting in mispredictions. Due to the various applications of deep sequential anomaly detection models, it is crucial to explore backdoor attacks against these models. If backdoor triggers are injected into deep sequential anomaly detection models, it presents a substantial security concern.

In the context of anomaly detection, the points of interest are anomalies. Therefore, we focus on conducting backdoor attacks to make the anomaly detection model predict abnormal sequences with triggers as normal. Meanwhile, in this work, we focus on attacking the distanced-based one-class sequential anomaly detection models, such as Deep SVDD [13] and OC4Seq [16], which detect anomalies based on their distances to the center of normal samples. It is not straightforward to conduct backdoor attacks against sequential anomaly detection models. First, it is challenging to craft invisible triggers for sequential data. The naive dirty-label attack strategy, which injects some abnormal

D.-N. Yang et al. (Eds.): PAKDD 2024, LNAI 14647, pp. 262–274, 2024.
https://doi.org/10.1007/978-981-97-2259-4_20

samples into the training dataset and marks them as normal, is not practical because the abnormal samples could be filtered out by rule-based inspection. Second, as the abnormal sequences are not available during the training phase, how to ensure the infected models label the abnormal sequences with triggers as normal is challenging.

To address the above challenges, we develop an attack strategy consisting of two key components: trigger generation and backdoor injection. In the trigger generation phase, we craft perturbed samples with specific sequential patterns that can be learned by the anomaly detection models as triggers. Importantly, these perturbed sequences exclude anomalies, rendering them inconspicuous and difficult to discern. In the backdoor injection phase, we extend the Deep SVDD-based objective function with two new learning objectives. The goal is to make the perturbed samples close to their benign counterparts as well as the center of normal samples. In this way, when the attacker conducts the backdoor attacks by leveraging the triggers in perturbed samples, the infected models can have a high chance of labeling the backdoored sequences as normal.

Our contributions can be summarized as follows: 1) we propose a novel backdoor attack framework for distance-based one-class sequential anomaly detection models; 2) to achieve an imperceptible attack, both trigger generation and backdoor injection steps do not involve any anomalies; 3) we apply the developed attack methodology to established anomaly detection models, and our experimental results demonstrate the effectiveness of the proposed approach.

2 Related Work

Numerous studies have investigated the vulnerability of machine learning models to backdoor attacks. BadNets [6] introduced the first backdoor attack by poisoning the training dataset. It randomly selected benign training samples and replaced them with poisoned samples, subsequently assigning target labels to the poisoned samples. However, these visible triggers are easily observable. To enhance the imperceptibility of backdoor attacks, invisible backdoor attacks have been proposed in both image and text domains [2,3,10,12,14,15,17]. For instance, BppAttack [15] employs image quantization and dithering techniques to generate imperceptible triggers, utilizing contrastive adversarial training to enable victim models to accurately learn the triggers. To attack the text classification model, invisible triggers can be concealed within specific syntactic templates [11]. However, the study on backdoor attacks against anomaly detection models is still very limited in the literature.

3 Preliminaries

3.1 Deep One-Class Sequential Anomaly Detection

Denote a sequence consisting of K entries as $\mathbf{x} = [e_1, \dots, e_k, \dots, e_K]$, where e_k indicates the k-th entry in $\mathbf{x}$. Deep one-class anomaly detection models usually assume the availability of a set of normal sequences, $\mathcal{X} = \{\mathbf{x}_1, \mathbf{x}_2, \ldots, \mathbf{x}_n\}$, and further detect abnormal sequences that deviate from normal samples.

Deep SVDD [13] aims to learn a model $f_\theta : \mathcal{X} \rightarrow \mathcal{R}$ parameterized by θ that can enclose the normal samples into a hypersphere and minimize the volume of the hypersphere, where $\mathcal{R} = \{\mathbf{r}_1, \mathbf{r}_2, \ldots, \mathbf{r}_n\}$ indicates the representations of samples. The training objective of Deep SVDD is to make the normal sample representations close to the center of the hypersphere $\mathbf{c} = \text{Mean}(\mathcal{R})$, defined as:

$$\mathcal{L}_{SVDD} = \min_{\theta} \frac{1}{N} \sum_{n=1}^{N} ||f_\theta(\mathbf{x}_n) - \mathbf{c}||_2^2 + \lambda||\theta||_F^2. \tag{1}$$

When applying Deep SVDD for sequential anomaly detection, an LSTM or GRU is commonly adopted as the instance of f_θ, and the representation $\mathbf{r}_n$ can be derived as the last hidden state of LSTM or GRU. After training, any sequences with distances to $\mathbf{c}$ greater than a threshold τ can be labeled as abnormal.

Recently, OC4Seq [16] is proposed to extend the vanilla Deep SVDD model into a hierarchical structure for sequential anomaly detection. Besides using the objective function defined in Eq. 1 to learn the representations of whole sequences, OC4Seq further assumes that subsequence information can enhance anomaly detection abilities. Therefore, given a sequence, OC4Seq utilizes the sliding window technique to create subsequences and aims to make the representations of subsequences close to the center of subsequences. Formally, the objective function of OC4Seq can be defined as

$$\mathcal{L}_{OC4Seq} = \mathcal{L}_{SVDD} + \eta \cdot \mathcal{L}_{local}, \tag{2}$$

where η represents a hyperparameter used to control the contribution from the local level; $\mathcal{L}_{local}$ can be formulated as $\mathcal{L}_{local} = \min_{\theta_l} \frac{1}{N} \sum_{n=1}^{N} \sum_{s=1}^{S} ||f_{\theta_l}(\mathbf{x}_n^s) - \mathbf{c}_l||_2^2 + \lambda||\theta_l||_F^2$, where $\mathbf{x}_n^s$ indicates the s-th subsequence derived from $\mathbf{x}_n$; $\mathbf{c}_l$ is the center of the hypersphere corresponding to subsequences in the latent space and θ_l is the parameters of another sequential model.

3.2 Mutual Information Maximization

Mutual information is widely used to quantitatively measure the relationship between random variables. Assuming that X and Y are two variables, their mutual information $\mathcal{I}(X;Y)$ can be expressed using the Kullback-Leibler (KL) divergence [9] as $\mathcal{I}(X;Y) = D_{KL}(\mathbb{J}||\mathbb{M})$, where $\mathbb{J} = P_{(X,Y)}(x,y)$ represents the joint probability distribution function of X and Y, and $\mathbb{M} = P_X(x)P_Y(y)$ denotes the product of the marginal probability distribution functions of X and Y.

The goal of many machine learning tasks is to maximize the mutual information $\mathcal{I}(X;Y)$. Mutual Information Neural Estimation (MINE) [1] employs the Donsker-Varadhan representation of the KL-divergence [5] to estimate the lower-bound of MI as:

$$\mathcal{I}(X;Y) = D_{KL}(\mathbb{J}||\mathbb{M}) \geq \mathbb{E}_{\mathbb{J}}[M(x,y)] - \log\mathbb{E}_{\mathbb{M}}[e^{M(x,y)}], \tag{3}$$

where $M : \mathcal{X} \times \mathcal{Y} \to \mathbb{R}$ is a discriminator function. To find the M^* that can maximize $\mathcal{I}(X;Y)$, MINE uses a neural network with parameters ω to model M, so maximizing the value of $\mathcal{I}(X;Y)$ can be achieved by optimizing M_ω. Following MINE, Deep InfoMax (DIM) [8] finds it unnecessary to use the KL-based formulation to maximize $\mathcal{I}(X;Y)$. An alternative, Jensen-Shannon mutual information estimator [8], is proposed to estimate $I(X;Y)$ as follows:

$$\hat{\mathcal{I}}^{(JSD)}_{\omega,\phi}(X;E_\phi(X)) := \mathbb{E}_{\mathbb{P}}[-sp(-M_{\omega,\phi}(x;E_\phi(x)))] - \mathbb{E}_{\mathbb{P}\times\tilde{\mathbb{P}}}[sp(M_{\omega,\phi}(x';E_\phi(x)))], \tag{4}$$

where $sp(z) = \log(1 + e^z)$ represents the softplus function, $\mathbb{P}$ is the distribution of X, $E_\phi : \mathcal{X} \to \mathcal{Y}$ is a differentiable parametric function, and x' is a sample from $\tilde{\mathbb{P}} = \mathbb{P}$.

4 Methodology

4.1 Threat Model

In this paper, we consider the victim to be a user aiming to build an anomaly detection application but cannot afford expensive computation resources. The attacker is a malicious computation service provider who takes the user's training samples and requirements to generate a model. Additionally, the user has a private validation dataset to validate the performance of the received model.

Attacker's Goal: The attacker's goal is to provide an infected model with the following properties. *Utility*: The infected model should perform well on benign data. Specifically, the infected model should be capable of detecting abnormal data without triggers. *Effectiveness*: Any anomaly containing the triggers should be classified as normal. *Stealthiness*: Any perturbations to the clean training data for the backdoor injection must be minimal to evade detection by data auditing applications or human observers.

Attacker's Capabilities: Following the assumption from the existing approaches [3,4,10,15,17], we assume that the attacker has the control of the training dataset and training process but cannot access the private validation dataset.

4.2 The Proposed Attack

We propose a novel backdoor attack approach, consisting of trigger generation and backdoor injection, to compromise the classical one-class sequential anomaly detection models, Deep SVDD and OC4Seq. As illustrated in Fig. 1, our approach initiates by randomly selecting a subset of samples to create perturbed samples containing a trigger. Subsequently, these perturbed samples are drifted toward specified locations in the latent space. In the attack phase, the abnormal samples containing the trigger can be located in the shown hypersphere, enabling them to be misclassified as normal samples. We first use the vanilla Deep SVDD model as the victim model to illustrate our attacking approach and then extend to attacking OC4Seq.

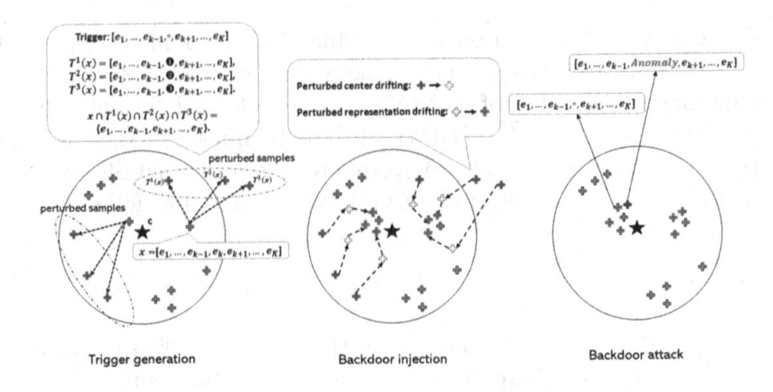

Fig. 1. Backdoor attacks against one-class anomaly detection models.

4.2.1 Trigger Generation

Trigger generation aims to generate perturbed samples with imperceptible triggers to the training dataset without incorporating any detectable anomalies, either by human observers or detection tools.

To this end, we first randomly select a small subset of samples from $\mathcal{X}$ to create a base set $\mathcal{X}'$. For each sequence $\mathbf{x} \in \mathcal{X}'$, we generate a set of perturbed samples $\mathcal{P}$ by replacing the entry at the k-th position with t different normal entries, denoted as $\mathcal{P} = \{T^1(\mathbf{x}), T^2(\mathbf{x}), \ldots, T^t(\mathbf{x})\}$, where $T^t(\mathbf{x})$ indicates the t-th perturbed sample derived from $\mathbf{x}$. As a result, the perturbed samples in $\mathcal{P}$ and the original sample $\mathbf{x}$ are only different at the k-th position, i.e., $\mathbf{x} \cap T^1(\mathbf{x}) \cap T^2(\mathbf{x}) \cap \cdots \cap T^t(\mathbf{x}) = \{e_1, \ldots, e_{k-1}, e_{k+1}, \ldots, e_K\}$.

By doing so, we can treat the subsequence $\{e_1, \ldots, e_{k-1}, e_{k+1}, \ldots, e_K\}$ as a trigger pattern. The backdoor attacks can be conducted by injecting an abnormal entry at the k-th position. Because for a sample $\mathbf{x} \in \mathcal{X}'$, we generate a large number of normal sequences with the only difference at the k-th position, the model would pay more attention to the subsequence $\{e_1, \ldots, e_{k-1}, e_{k+1}, \ldots, e_K\}$ instead of the specific entry at the k-th position. By weakening the attention of the anomaly detection model at the k-th position, we can inject an abnormal entry at this position and make the abnormal entry evade detection. In short, the k-th position is a placeholder for the potential abnormal entry when conducting attacks.

In real-world scenarios, successful attacks often require a series of coordinated actions rather than a single action. Therefore, instead of replacing one entry at the k-th position, for each perturbed sequence, we choose m entries as placeholders and replace them with some randomly chosen normal entries. Similarly, the unchanged subsequence can be considered as a trigger pattern. In this way, in the attacking phase, the attacker can inject at most m abnormal entries. Note that the attacker can either randomly or continuously choose m entries to derive a perturbed sequence. Meanwhile, in practice, to conduct the perturbation, the attacker can first gather some normal tokens from the training dataset to form a candidate set based on the domain knowledge and then randomly pick one from the candidate set for replacement.

Finally, a perturbed dataset $\mathcal{X}_p$ is crafted by combining all perturbed samples derived from $\mathcal{X}'$, i.e., $\mathcal{X}_p = \bigcup_{\mathbf{x}_j \in \mathcal{X}'} \mathcal{P}_j$, where $\mathcal{P}_j$ is the set of perturbed samples of $\mathbf{x}_j$ in $\mathcal{X}'$. Subsequently, we employ the combined dataset $\mathcal{X}_c \cup \mathcal{X}_p$ as the updated training dataset to train the deep anomaly detection model, where $\mathcal{X}_c = \mathcal{X} \setminus \mathcal{X}'$.

4.2.2 Backdoor Injection

The trigger generation aims to derive the undetectable triggers. However, the success of evade detection is not guaranteed by injecting abnormal entries into the triggers. To further achieve evade detection of backdoored samples, we propose two learning objectives, perturbed sequence center drifting and perturbed sequence representation drifting. The perturbed sequence center drifting aims to ensure the center of perturbed sequences in $\mathcal{X}_p$ close to the center of benign sequences in $\mathcal{X}_c$. The perturbed sequence representation drifting makes the perturbed sequences indistinguishable from their benign counterparts in the latent space.

Perturbed Sequence Center Drifting. Deep SVDD detects the anomalies based on their distances to the normal center $\mathbf{c}$. Therefore, it is reasonable to assume that attaching abnormal entries to a normal sample extremely close to $\mathbf{c}$ may push the sample away from $\mathbf{c}$ but still remain within the boundary of the hypersphere and be classified as normal. Conversely, if a normal sample is not close to $\mathbf{c}$ and is already near the hypersphere boundary, attaching abnormal entries can easily push it outside the hypersphere. Therefore, to make anomalies evade detection, a potential strategy is to attach abnormal entries to samples that are extremely close to $\mathbf{c}$. Because the backdoored samples are generated from perturbed samples in $\mathcal{X}_p$, we propose a learning objective that drifts the center of perturbed samples towards $\mathbf{c}$ in the latent space. Specifically, we compute a new center $\mathbf{c}_p$ by averaging the representations of the perturbed samples in the latent space, i.e., $\mathbf{c}_p = \frac{1}{|\mathcal{X}_p|} \sum_{\mathbf{x}_j \in \mathcal{X}_p} f_\theta(\mathbf{x}_j)$. Subsequently, the objective is to align $\mathbf{c}_p$ with $\mathbf{c}$ in the latent space, defined as:

$$\mathcal{L}_c = ||\mathbf{c}_p - \mathbf{c}||_2. \tag{5}$$

Note that $\mathbf{c}$ is derived from the benign sample set $\mathcal{X}_c$.

By minimizing the distance between $\mathbf{c}_p$ and $\mathbf{c}$, the perturbed samples can become close to $\mathbf{c}$. As a result, when conducting attacks, filling up m placeholders in a perturbed sample with abnormal entries still has a high chance of keeping the sequence inside the hypersphere boundary.

Perturbed Sequence Representation Drifting. Besides making the center of perturbed sequences in $\mathcal{X}_p$ close to the center of benign normal sequences, we further aim to ensure the distribution of perturbed sequences in $\mathcal{X}_p$ similar to the corresponding original ones. That said, if the perturbed sequences and the original ones are similar in latent space, we can further improve the chance that after putting abnormal entries in perturbed sequences, the abnormal sequences can still be similar to the benign counterpart in the latent space. To achieve this goal, we propose to maximize the mutual information between the representations of perturbed samples in $\mathcal{X}_p$ and their original versions.

For any $\mathbf{x}_p \in \mathcal{X}_p$, let $f_\theta(\mathbf{x}_p)$ denote its latent space representation. Similarly, for its benign counterpart $\mathbf{x} \in \mathcal{X}'$, $f_\theta(\mathbf{x})$ represents the corresponding latent space representation. We update Eq. 4 as follows:

$$\hat{\mathcal{I}}^{(JSD)}_{\omega,\theta}(f_\theta(\mathcal{X}); f_\theta(\mathcal{X}_p)) := \mathbb{E}_{\mathbb{P}}[-sp(-M_{\omega,\theta}(f_\theta(\mathbf{x}), f_\theta(\mathbf{x}_p)))] - \mathbb{E}_{\mathbb{P}\times\tilde{\mathbb{P}}}[sp(M_{\omega,\theta}(f_\theta(\mathbf{x}'), f_\theta(\mathbf{x}_p)))], \tag{6}$$

where $\mathbb{P}$ is the distributions of benign samples in $\mathcal{X}$ and $\mathbf{x}'$ is a sample from the distribution $\tilde{\mathbb{P}} = \mathbb{P}$. $M_{\omega,\theta}$ is a deep neural network and defined as:

$$M_{\omega,\theta} = C_\omega \circ H(f_\theta(\mathbf{x}), f_\theta(\mathbf{x}_p)), \tag{7}$$

where H is a function that computes the square of element-wise difference of the representations between perturbed samples and their benign versions and C_ω is a fully connected neural network. Therefore, for all perturbed samples in $\mathcal{X}_p$, the learning objective is to maximize the mutual information between $f_\theta(\mathbf{x})$ and $f_\theta(\mathbf{x}_p)$:

$$\mathcal{L}_r = \frac{1}{|\mathcal{X}_p|} \sum_{\mathbf{x}_p \in \mathcal{X}_p} \hat{\mathcal{I}}^{(JSD)}_{\omega,\theta}(f_\theta(\mathbf{x}); f_\theta(\mathbf{x}_p)). \tag{8}$$

To train the infected Deep SVDD model, the new objective function is defined as:

$$\mathcal{L'}_{SVDD} = \mathcal{L}_{SVDD} + \alpha \cdot \mathcal{L}_c - \beta \cdot \mathcal{L}_r, \tag{9}$$

where α and β balance the proposed backdoor objectives.

Extend the Proposed Approach to Attack OC4Seq. As OC4Seq detects sequential anomalies from both local and global levels, we extend the above attacking strategies by further applying the perturbed sequence center and representation drifting at local levels, defined as follows:

$$\mathcal{L'}_{OC4Seq} = \mathcal{L'}_{SVDD} + \eta \cdot \mathcal{L'}_{local}, \quad \mathcal{L'}_{local} = \mathcal{L}_{local} + \alpha \cdot \mathcal{L}_{c_l} - \beta \cdot \mathcal{L}_{r_l}, \tag{10}$$

where

$$\mathcal{L}_{c_l} = ||\mathbf{c}_{p_l} - \mathbf{c}_l||_2, \quad \mathcal{L}_{r_l} = \frac{1}{|\mathcal{X}_p|} \sum_{\mathbf{x}_p \in \mathcal{X}_p} \sum_{s=1}^{S} \hat{\mathcal{I}}^{(JSD)}_{\omega,\theta}(f_\theta(\mathbf{x}^s); f_\theta(\mathbf{x}^s_p)). \tag{11}$$

In Eq. 11, $\mathbf{c}_{p_l}$ is the mean representations of perturbed subsequences and $\mathbf{x}^s$ is the s-th subsequence derived from $\mathbf{x}$.

4.3 Post-deployment Attack

After deploying the infected model, the attacker can attach abnormal entries at m placeholders into a sequence in $\mathcal{X}_p$. This poisoned sequence can activate the backdoor in the infected model, leading to the model erroneously classifying this sequence as normal.

5 Experiments

5.1 Experimental Setup

5.1.1 Datasets

We evaluate the proposed attack against the anomaly detection models on two datasets [18], BlueGene/L (BGL) and Thunderbird, which are commonly used for evaluating sequential anomaly detection. We set all the sequences with a fixed length of 40. Table 1 shows the statistics of the datasets. In the training phase, 1/10 of training sequences are perturbed sequences. For each dataset, we create a benign test set to evaluate the infected model for anomaly detection and a poisoned test set to check whether the infected model can predict abnormal sequences with triggers as normal.

To derive the perturbed dataset $\mathcal{X}_p$, in the trigger generation phase, we randomly select 50 sequences to create $\mathcal{X}'$ and generate 200 perturbed sequences for each sequence in $\mathcal{X}'$, leading to 10,000 perturbed sequences. Meanwhile, we choose $M = 6$ entries as placeholders so that the maximum number of abnormal entries that can be injected during the attacking phase is 6. The poisoned sequences are generated by replacing the placeholders with abnormal entries.

Table 1. Statistics of training and evaluate datasets.

Dataset		BGL	Thunderbird
Training	Benign	90,000	90,000
	Perturbed	10,000	10,000
Benign Test Set	Normal	5,000	5,000
	Abnormal	500	500
Poisoned Test Set	Abnormal	10,000	10,000

5.1.2 Evaluatin Metric

We adopt the following metric the evaluate the effectiveness of the proposed attack approach. 1) **Benign performance (BP)** is to evaluate the performance of infected models on benign datasets, including precision, recall, and F-1 score as evaluation metrics. 2) **Attack success rate (ASR)** is defined as the fraction of poisoned samples identified as normal by the infected models when injecting real abnormal entries into $\mathcal{X}_p$.

5.1.3 Baseline

As there is no backdoor attack approach against sequential anomaly detection models in the literature, we compare the performance of the infected model with a benign model that is trained on the benign training set.

5.1.4 Implementation Details

We set $\eta = 1$ and hyperparameters $\alpha = 0.5$ and $\beta = 0.5$. We represent log entries in BGL and Thunderbird as embedding vectors with a size of 100 and use a single-layer LSTM model with a hidden size of 256 to learn sequence representations. We use a small validation set to get the threshold τ for anomaly detection. The code is available online[1].

5.2 Experimental Results

5.2.1 Performance of Infected Models on Benign Data for Anomaly Detection

We first compare the performance of benign models and infected models for anomaly detection on benign datasets. The results are presented in Table 2. We observe that both Deep SVDD and OC4Seq infected models can maintain performance close to that of the benign ones, demonstrating the effectiveness of infected models for anomaly detection. The fluctuation in BP between benign and infected models may be attributed to changes in hypersphere boundaries. The incorporation of perturbed sequences could slightly shift the original distribution of benign data, leading to the derivation of different hyperspheres compared to a benign setting.

Table 2. Benign and infected models for anomaly detection on benign datasets.

Model	Dataset	Metrics	Benign	Infected
DeepSVDD	BGL	Precision	93.47	96.12
		Recall	94.40	94.20
		F-1 score	93.93	95.15
	Thunderbird	Precision	95.59	95.60
		Recall	95.40	95.60
		F-1 score	95.50	95.60
OC4Seq	BGL	Precision	91.80	98.94
		Recall	94.00	93.20
		F-1 score	92.89	95.98
	Thunderbird	Precision	82.47	91.45
		Recall	82.80	70.60
		F-1 score	82.63	79.68

[1] https://github.com/Serendipity618/BA-OCAD.

5.2.2 Performance of Infected Models on Poisoned Data for Evade Detection

We then evaluate the effectiveness of the proposed backdoor attack by injecting varying numbers of abnormal entries. To achieve this, we inject m abnormal entries into sequences in $\mathcal{X}'$, with m ranging from 1 to 6. Table 3 presents the results of both benign and infected models.

Table 3. Attack success rate on poisoned datasets with various abnormal entries.

Model	Dataset		m = 1	m = 2	m = 3	m = 4	m = 5	m = 6
DeepSVDD	BGL	Benign	98.47	98.79	96.67	91.92	75.83	40.86
		Infected	98.58	98.32	97.98	97.29	94.29	86.95
	Thunderbird	Benign	99.94	97.15	86.51	64.22	39.04	14.20
		Infected	100.00	100.00	99.94	99.03	94.91	86.98
OC4Seq	BGL	Benign	100.00	100.00	98.00	98.00	98.00	32.65
		Infected	100.00	99.99	99.77	98.89	95.02	82.41
	Thunderbird	Benign	80.17	41.82	5.05	0.34	0.00	0.00
		Infected	94.67	91.87	84.57	66.84	35.49	11.48

We observe that when only one abnormal entry is injected into sequences, the benign models can also achieve high ASR. This finding aligns with our earlier assumption that injecting anomalies into sequences could push them from the center but possibly still within the hypersphere. However, when m is large, the ASR for benign models dramatically decreases. In contrast, infected models maintain a high ASR even when multiple abnormal entries are injected.

5.2.3 Sensitivity Analysis

Hyperparameter α. We analyze the impact of the parameter α by varying its values from 0 to 1.0 in increments of 0.2. As illustrated in Fig. 2, increasing the values of α generally leads to a slight increase in ASR. For both models, ASR

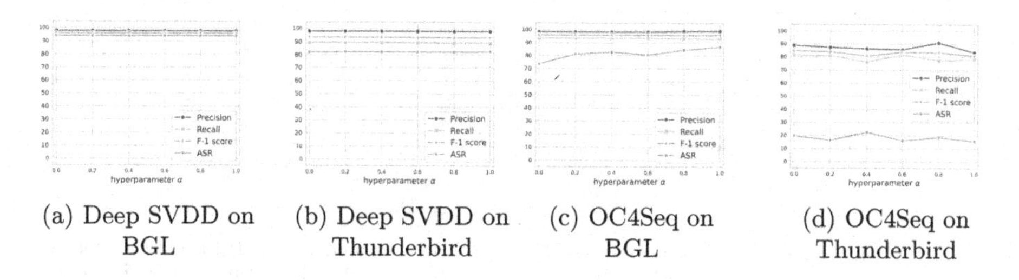

(a) Deep SVDD on BGL (b) Deep SVDD on Thunderbird (c) OC4Seq on BGL (d) OC4Seq on Thunderbird

Fig. 2. Results of backdoor attack for various hyperparameter α.

consistently stabilizes at a higher level with different values of α. It is noticeable that when $\alpha = 0$, this case differs from a benign setting, as the training dataset contains perturbed samples and meanwhile, the new learning objective function still includes the perturbed sequence representation drifting term.

Hyperparameter β. We also investigate the impact of the parameter β by varying its value from 0 to 1.0 in increments of 0.2. The results are presented in Fig. 3. It is noticeable that with the increase of β, the ASR generally keeps rising and then stabilizes at a high level. For OC4Seq on Thunderbird, the ASR starts to decrease when $\beta > 0.6$.

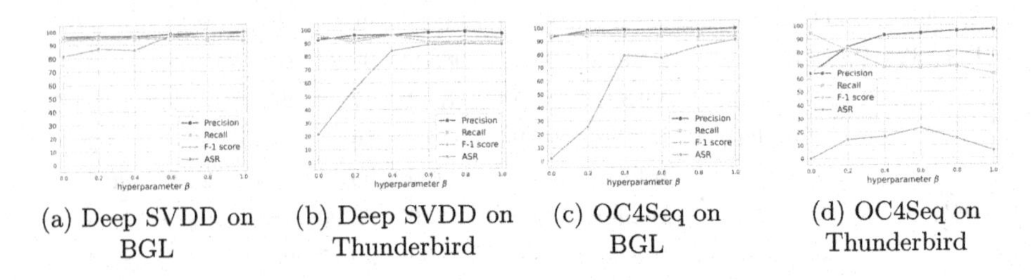

(a) Deep SVDD on BGL (b) Deep SVDD on Thunderbird (c) OC4Seq on BGL (d) OC4Seq on Thunderbird

Fig. 3. Results of backdoor attack for various hyperparameter β.

5.2.4 Visualization

We further visualize representations of benign, perturbed, and poisoned sequences for the infected Deep SVDD model in the BGL dataset. We randomly selected 5000 benign sequences to create their corresponding perturbed and poisoned sequences. The results are shown in Fig. 4.

Figure 4a shows that the representations of perturbed and poisoned sequences derived by the benign model are far from the benign center. Meanwhile, Figs. 4b and 4c reveal that by employing the proposed perturbed sequence center drifting or perturbed sequence representation drifting, we can either move perturbed sequences closer to the benign center or establish a correlation between perturbed and benign sequences, but cannot achieve both simultaneously.

Figure 4d illustrates that using both training objectives, the infected model brings perturbed sequences close to the benign center and all the poisoned samples are also close to the perturbed counterpart, making them challenging to

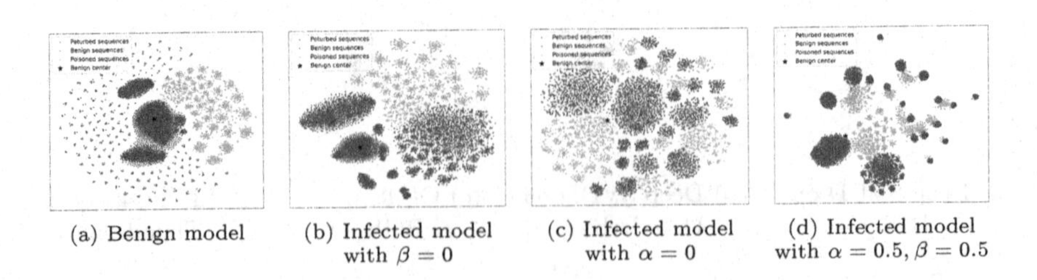

(a) Benign model (b) Infected model with $\beta = 0$ (c) Infected model with $\alpha = 0$ (d) Infected model with $\alpha = 0.5, \beta = 0.5$

Fig. 4. Visualization of benign, perturbed, and poisoned sequences.

detect. Figures 4b, 4c, and 4d also demonstrate that poisoned sequences are consistently associated with their corresponding perturbed sequences, providing evidence that the models ignore the placeholders and focus on the patterns outlined in our proposed trigger generation.

6 Conclusions

In this paper, we have developed a novel backdoor attack framework against one-class anomaly detection models on sequential data, enabling anomalies to evade detection. Our framework comprises two essential components, trigger generation and backdoor injection. Trigger generation is to derive imperceptible backdoor triggers from the normal sequences, while the backdoor injection is to inject backdoor patterns to infected models during the training phase by developing two learning objectives. After deployment, the attacker can conceal abnormal entries within the sequences, which enables anomalies to evade detection by the infected model. Our experiments on one-class anomaly detection models demonstrate the effectiveness of our proposed backdoor attack strategy. In the future, we plan to study how to effectively defend the backdoor attack against the sequential anomaly detection models.

Acknowledgement. This work was supported in part by NSF 2103829.

References

1. Belghazi, M.I., et al.: Mutual information neural estimation. In: International Conference on Machine Learning, pp. 531–540. PMLR (2018)
2. Chen, X., Liu, C., Li, B., Lu, K., Song, D.: Targeted backdoor attacks on deep learning systems using data poisoning. arXiv preprint arXiv:1712.05526 (2017)
3. Doan, K., Lao, Y., Zhao, W., Li, P.: Lira: learnable, imperceptible and robust backdoor attacks. In: ICCV (2021)
4. Doan, K.D., Lao, Y., Li, P.: Marksman backdoor: backdoor attacks with arbitrary target class. In: NeurIPS (2022)
5. Donsker, M.D., Varadhan, S.S.: Asymptotic evaluation of certain Markov process expectations for large time. iv. Commun. Pure Appl. Math. **36**(2), 183–212 (1983)
6. Gu, T., Liu, K., Dolan-Gavitt, B., Garg, S.: Badnets: evaluating backdooring attacks on deep neural networks. IEEE Access **7**, 47230–47244 (2019)
7. Guo, H., Yuan, S., Wu, X.: LogBERT: log anomaly detection via BERT. In: 2021 International Joint Conference on Neural Networks (IJCNN), pp. 1–8. IEEE (2021)
8. Hjelm, R.D., et al.: Learning deep representations by mutual information estimation and maximization. In: ICLR (2019)
9. Joyce, J.M.: Kullback-leibler divergence. In: Dubitzky, W., Wolkenhauer, O., Cho, K.H., Yokota, H. (eds.) Encyclopedia of Systems Biology, pp. 720–722. Springer, New York (2011). https://doi.org/10.1007/978-1-4419-9863-7_100751
10. Nguyen, T.A., Tran, A.T.: Wanet - imperceptible warping-based backdoor attack. In: International Conference on Learning Representations (2021)
11. Qi, F., et al.: Hidden killer: invisible textual backdoor attacks with syntactic trigger. arXiv preprint arXiv:2105.12400 (2021)

12. Qi, F., Yao, Y., Xu, S., Liu, Z., Sun, M.: Turn the combination lock: learnable textual backdoor attacks via word substitution. arXiv preprint arXiv:2106.06361 (2021)
13. Ruff, L., et al.: Deep one-class classification. In: International Conference on Machine Learning, pp. 4393–4402. PMLR (2018)
14. Wang, T., Yao, Y., Xu, F., An, S., Tong, H., Wang, T.: An invisible black-box backdoor attack through frequency domain. In: Avidan, S., Brostow, G., Cissé, M., Farinella, G.M., Hassner, T. (eds.) ECCV 2022. LNCS, vol. 13673, pp. 396–416. Springer, Cham (2022). https://doi.org/10.1007/978-3-031-19778-9_23
15. Wang, Z., Zhai, J., Ma, S.: Bppattack: stealthy and efficient trojan attacks against deep neural networks via image quantization and contrastive adversarial learning. In: CVPR (2022)
16. Wang, Z., Chen, Z., Ni, J., Liu, H., Chen, H., Tang, J.: Multi-scale one-class recurrent neural networks for discrete event sequence anomaly detection. In: ACM SIGKDD (2021)
17. Zhong, N., Qian, Z., Zhang, X.: Imperceptible backdoor attack: from input space to feature representation. In: IJCAI (2022)
18. Oliner, A., Stearley, J.: What supercomputers say: a study of five system logs. In: IEEE/IFIP International Conference on Dependable Systems and Networks (2007)

Semi-supervised and Unsupervised Learning

DALLMi: Domain Adaption for LLM-Based Multi-label Classifier

Miruna Bețianu[1], Abele Mălan[1,2(✉)], Marco Aldinucci[3], Robert Birke[3], and Lydia Chen[1,2]

[1] Delft University of Technology, Delft, Netherlands
m.betianu@student.tudelft.nl
[2] University of Neuchâtel, Neuchâtel, Switzerland
{abele.malan,yiyu.chen}@unine.ch
[3] University of Turin, Torino, Italy
{marco.aldinucci,robert.birke}@unito.it

Abstract. Large language models (LLMs) increasingly serve as the backbone for classifying text associated with distinct domains and simultaneously several labels (classes). When encountering domain shifts, e.g., classifier of movie reviews from IMDb to Rotten Tomatoes, adapting such an LLM-based multi-label classifier is challenging due to incomplete label sets at the target domain and daunting training overhead. The existing domain adaptation methods address either image multi-label classifiers or text binary classifiers. In this paper, we design DALLMi, **D**omain **A**daptation **L**arge **L**anguage **M**odel **i**nterpolator, a first-of-its-kind semi-supervised domain adaptation method for text data models based on LLMs, specifically BERT. The core of DALLMi is the novel variation loss and MixUp regularization, which jointly leverage the limited positively labeled and large quantity of unlabeled text and, importantly, their interpolation from the BERT word embeddings. DALLMi also introduces a label-balanced sampling strategy to overcome the imbalance between labeled and unlabeled data. We evaluate DALLMi against the partial-supervised and unsupervised approach on three datasets under different scenarios of label availability for the target domain. Our results show that DALLMi achieves higher mAP than unsupervised and partially-supervised approaches by 19.9% and 52.2%, respectively.

Keywords: Large language models · Domain adaptation · Multi-label classification · Semi-supervised learning

1 Introduction

Text classification is a fundamental task in Natural Language Processing with a diverse range of applications, including sentiment analysis [14], spam detection [6], and document categorization [11]. Large Language Models (LLMs) are the state-of-the-art backbone for such classification tasks [22]. By pre-training on large and heterogeneous corpora, they achieve good baseline performance across

D.-N. Yang et al. (Eds.): PAKDD 2024, LNAI 14647, pp. 277–289, 2024.
https://doi.org/10.1007/978-981-97-2259-4_21

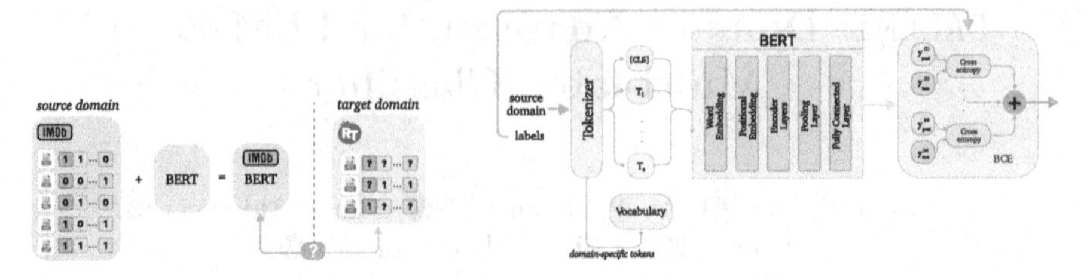

Fig. 1. Adapting *BERT* from IMDb to Rotten Tomatoes.

Fig. 2. *BERT* multi-label classification flow.

various domains, even in cases where multiple labels[1] may be present simultaneously. When encountering a new domain with a sufficient amount of labeled data, fine-tuning LLMs can further increase the accuracy of classification tasks.

In real life scenarios, there is only limited labeled data in many target domains, making regular fine-tuning ineffective and requiring a domain shift from source domain. Figure 1 outlines an example of such a challenge in classifying text into multiple labels/classes. Assume we have an LLM trained to predict the genres of a movie from an IMDb dataset, as the source domain. Each movie is associated with multiple genres/labels. We then have to perform the same task on a new dataset from Rotten Tomatoes, as the target domain, which contains incomplete labels. The annotation of the target domain is a time-consuming and resource-intensive task, nonetheless unfeasible. The research challenge is how to utilize the knowledge of the source domain and the limited labels for the target domain to enhance the performance of LLM-based multi-label classifiers.

Existing text classification adaptation research [21] focuses on the multi-class case, where each sample corresponds to only one label from the available set, like in the ubiquitous sentiment analysis task or the even simpler binary one. The multi-label scenario, where samples correspond to any number of labels, poses an increased challenge, further increased by possibly underrepresented labels in the available data. Most prior multi-label domain adaptation efforts lie within the modality of image data [15,20]. Moreover, only few unsupervised approaches exist to finetune LLMs [17,18], which are usually limited to single label classification.

This paper introduces `DALLMi`, a novel semi-supervised technique for LLM adaptation to different textual domains. We combine supervised fine-tuning on a source text domain with semi-supervised fine-tuning on a partially labeled text target domain. Specifically, we focus on scenarios where there are limited positive samples for each label, and the remaining samples are unlabeled, as shown in Fig. 1. `DALLMi` introduces a variational loss that leverages labeled and unlabeled information to maximize the knowledge extracted from all samples. To further boost its classification ability, `DALLMi` augments the target dataset with synthetic samples generated by mixing labeled and unlabeled ones. We extensively

[1] Each label represents a possible class.

evaluate `DALLMi` under various settings for multiple datasets, including ablations on relevant features. To recap, we make the following contributions:

- We design `DALLMi`, the first semi-supervised LLM domain adaptation framework for multi-label text classification.
- We design a novel variational loss for the model to learn from both labeled and unlabeled parts of target instances.
- We interpolate representations of labeled and unlabeled samples into new synthetic ones to regularize our variational loss.
- `DALLMi` outperforms partial fine-tunning and unsupervised approaches by 19.9% and 52.2%, respectively.
- We make the code available at https://github.com/mirunabetianu/DALLMi.

2 Language Model and Domain Adaptation

BERT for Multi-label Classification. Figure 2 shows a high-level overview of the architecture of BERT's multi-label classifier, which assumes a full set of labels. The input sentence is initially passed through the tokenizer to obtain tokens, paddings, and truncation to prepare it for BERT. This process may also include the addition of new tokens to BERT's vocabulary. The encoded sentence proceeds through BERT's forward method, which incorporates embedding and encoding layers, to obtain the internal hidden representations. A pooling layer decreases the hidden representation's dimension, from which a classification head (usually a single fully connected layer) obtains the final logits. An activation function may be applied to determine predicted labels based on a threshold. The default training approach uses multiple binary cross entropy losses to treat the multi-label classification as several binary problems.

Prior Art Domain Adaptation. Domain adaptation addresses the dataset shift [16] between the source and target domains. It can be split into four main categories: fully supervised, semi-supervised, weakly supervised, and fully unsupervised. Fully supervised methods rely heavily on many target labels, aiming to generalize the model to both domains [13]. When the target domain labels are limited, the problem becomes semi-supervised. Common approaches combine source and domain data to improve model training [27]. Weakly supervised methods aim to adapt the model by considering the uncertainty and limitations of the target domain [7]. Unsupervised domain adaptation aims to adapt the model without any information about the target labels, approaches including batch normalization tuning [24], feature alignment [8], or whole network training [12]. Most methods operate within image domains. Nonetheless, some techniques exist for LLMs, with fine-tuning being the most common adaptation method [2,9,10]. Other methods expand the LLM's vocabulary, enhancing the corpus with specific domain tokens [19], or its architecture, adding new adapter layers [5]. In terms of classification adaptation, the majority of LLM-based approaches are designed for sentiment analysis, using adversarial adaptation with knowledge distillation [17] or supervised fine-tuning [18]. Our method, however, targets general classification workloads and employs semi-supervised fine-tuning.

3 DALLMi

Adapting a fine-tuned LLM to partially labeled target data is a challenging task, primarily due to the need for a substantial amount of labels to ensure that the model gives reliable predictions. To address the challenge posed by the scarcity of the target domain, we introduce a new semi-supervised fine-tuning method based on Positive Unlabeled (PU) learning and data augmentation. Our method takes inspiration from VPU [4], estimating separate losses for positive and unlabeled samples and combining positive-unlabeled pairs to generate artificial samples to compensate for the limited number of labels.

Formally, we aim to learn a multi-label classifier, $\Phi : X^t \rightarrow Y$, where $X^t \in \mathbb{R}^d$ is the input data from the target domain, and $Y = [y_1 \dots y_L], \forall y_l \in [1, 0]$ is the value vector for each label from 1 to L. Given some label l, $y_l = 1$ denotes the presence of a positive example, while $y_l = 0$ denotes the absence of an annotation. For each possible label $l \in \{0 \dots L\}$, we separate X^t into two sets: the unlabeled set $\mathcal{U}_N^l$ with samples s_u and the positive set $\mathcal{P}_N^l$ with samples s_p.

Primer on VPU. Positive unlabeled learning [1] focuses on the scenarios where labeled samples are always a subset of the positive ones, and unlabeled ones may be positive or negative. The goal of the trained model is to distinguish between positive and negative instances, as in the classic fully labeled case [1]. Rooted in the variational principle, VPU [4] uses a novel variation loss function to approximate the ideal **binary** classifier without the class prior, given unlabeled and positive samples. The variational loss function aims to quantify the empirical difference between the predicted outcomes for unlabeled and positive samples. In addressing the lack of positive samples, the VPU algorithm introduces a consistency regularisation term that incorporates the principles of MixUp, mitigating the problem of overfitting and increasing the robustness of the model. The MixUp term uses data augmentation techniques to generate synthetic samples by interpolating between positive and unlabeled instances. In addition, the algorithm quantifies the consistency between the model's predictions and the predictions made on these interpolated samples by the mean squared logarithmic error. We note that PU-MLC considers the problem of expanding VPU for multi-label classification [25], but in the image analysis domain.

Overview of DALLMi. To jointly leverage both positive labeled and unlabeled data, DALLMi fine-tunes the LLM-based multi-label classifier in a semi-supervised manner. We extend the design idea of VPU, combining the variation loss per label and MixUp regularization based on data interpolation, for a multi-label classifier. In contrast to image data or tabular data, such an interpolation is not straightforward for text data, which is represented as tokens in the LLM. We thus need to design a novel loss function and interpolation scheme to compute the MixUp, and then train the multi-label classifier and backbone BERT.

DALLMi consists of three key novelties, namely label-balanced sampling, variational loss and MixUp regularization. Figure 3 shows the primary flow of DALLMi

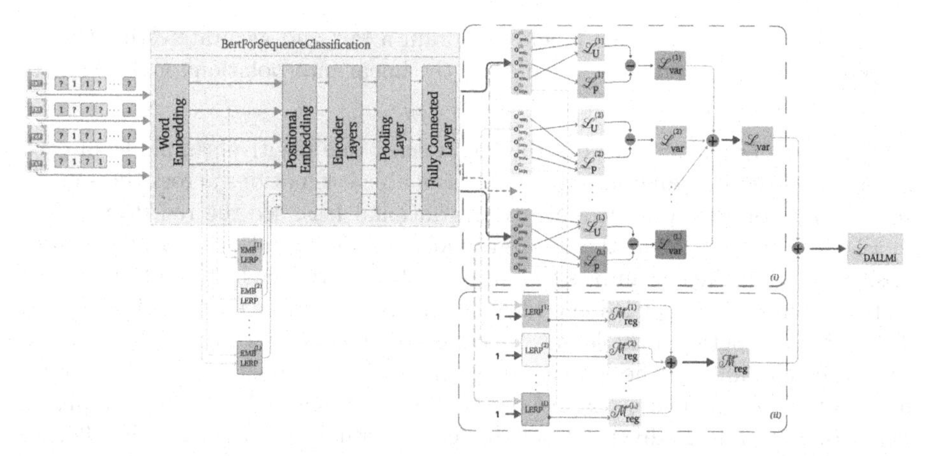

Fig. 3. `DALLMi` flow: Unlabeled (?) and positive (1) samples from the target domain are fed through BERT, generating label-specific output logits. The logits are used to compute partial per-label variational losses for unlabeled and positive samples (dashed box (i)). The MixUp regularization combines per label linear interpolations (LERPs) applied to both inputs and outputs (dashed box (ii)).

for computing the proposed variational loss and MixUp regularization for an example set of 4 samples. The target domain is pre-processed to ensure compatibility with the BERT classifier (blue shaded box). After the initial word embeddings layer, a distinctive input interpolation mechanism is introduced for each label. Here, unlabeled and positive embeddings are paired to generate new artificial embeddings via a linear interpolation block (LERP). After interpolation, the original and newly generated embeddings pass through the remaining layers of the BERT classifier. The final layer features a classification head represented as a single fully connected layer outputting L logit scores. These logit scores are used to compute a specialised multi-label variational loss, $\mathcal{L}_{var}$, which sums up all per-label variation losses (dashed block (i)). The MixUp regularization $\mathcal{M}_{reg}$ sums up per label combinations of input interpolations and output interpolations, which are determined by linearly interpolating the logits derived from the input interpolation and the true output (dashed block (ii)).

Label-Balanced Sampling. In multi-label classification, where labels are representable as a sparse matrix, the challenge is creating label-balanced batches, particularly as the label count increases. Since, for any label there are often more unlabeled samples than positive ones, we need to overcome this disparity by ideally boosting the number of labeled samples from each label. Therefore, the first step in `DALLMi` involves an efficient and balanced sampler that ensures at least one positive sample from each label exists in every batch.

To achieve such balance, we introduce a cycle sampler that iterates through every label and retrieves one positive sample, repeating the process as many as needed until to reach the desired batch size. By ensuring the presence of at least one positive sample for each label, we obtain a fair and accurate estimation of the variational loss and allow calculating the linear interpolation for the MixUp.

Variational Loss per Label. Having at least one positive sample from each label allows the loss function to make reasonable estimations across the sample set. In this context, we introduce the variational loss; see the red dashed box in Fig. 3. For each label, the loss, grounded in the variational principle, uses positive and unlabeled samples to minimize the divergence between the positive distributions of the modeled and ideal classifiers as a proxy for the original task. The difference between the log of the expected sigmoids for unlabeled data and the expected value of the log sigmoids for labeled data serves as the approximation for the classifier's positive distribution. Interestingly, through empirical analysis, our findings diverge from this conventional approach. Specifically, we observe that replacing the log operations with the norm yields significant performance improvements. This norm-based approach is particularly advantageous when working within a specific domain. In addition, using the logarithm can exacerbate errors in certain situations, especially when dealing with labels with very low probabilities. This problem becomes more pronounced as the width of the datasets increases. In both cases, aggregating the values across all possible labels gives the final variational loss:

$$\mathcal{L}^{l}_{var}(\Phi) = \frac{1}{|\mathcal{U}^{l}_{N}|} \sum_{s_u \in \mathcal{U}^{l}_{N}} \sigma(\Phi(s_u)) - \frac{1}{|\mathcal{P}^{l}_{N}|} \sum_{s_p \in \mathcal{P}^{l}_{N}} ||\sigma(\Phi(s_p))||, \tag{1}$$

where $\sigma(\cdot)$ is the sigmoid function, and $||\cdot||$ represents the norm of a vector.

Although the effectiveness of variational loss is proven [4,25], it has limitations, notably the need for a substantial number of positive samples in the dataset. To mitigate this limitation, we use data augmentation.

MixUp Regularization. We compute the MixUp [23,26] term based on synthetically augmented data. The primary purpose of the MixUp is to generate additional samples by interpolating between positive and unlabeled samples, thus compensating for the lack of positive labels. The interpolations are weighted combinations using the Beta distribution to facilitate a smooth transition. In addition to effectively expand the dataset size, the MixUps quantify the consistency between the predicted and interpolated outputs.

LLM inputs consist of sequences of tokens that are not necessarily closer in value if they are closer in meaning. Therefore, simple token-level interpolation could result in problematic combinations that may hinder the model's learning process. For the interpolation to be possible, we need to delve into the BERT model layers to identify which internal representations of the text data would be suitable [23]. Figure 4 illustrates three possible locations where to extract internal representations to perform MixUps and generate artificial samples. The first location is after the tokens pass the embedding layer of BERT. We extract

Algorithm 1. `DALLMi` – Loss Calculation

Input: Label-balanced sampled data batch B, BERT model Φ, hyperparameters $\alpha/\beta/\lambda$
Output: Loss value for samples in B

1: **for** each label l **do**
2: Select unlabeled samples U_N^l and positive samples P_N^l from batch B
3: Calculate variational loss by Eq. 1
4: Select one pair of unlabeled and positive inputs, s_u and s_p
5: Sample $\mu \sim Beta(\alpha, \beta)$
6: e_u, e_p = Φ.embeddings(s_u), Φ.embeddings(s_p)
7: Compute interpolation by Eq. 4
8: Compute artificial model prediction by Eq. 5
9: Compute MixUp term by Eq. 2
10: **end for**
11: Compute the overall loss $\mathcal{L}_{\text{final}}(\Phi) = \sum_{l=1}^{L} \mathcal{L}_{var}^l(\Phi) + \lambda \mathcal{M}_{reg}^l(\Phi)$
12: Backpropagate $\mathcal{L}_{\text{final}}(\Phi)$

the word embeddings for the chosen unlabeled and positive sentences and perform interpolation using the Beta distribution (see Eq. 2). The resulting new embedding undergoes the subsequent layers of BERT to generate predictions for this new sample. Similarly, the other two locations extract, interpolate and feed back the embeddings at the deeper encoding and pooling layers of the model. Our empirical evaluation of the all interpolation locations favours the word embedding variant. We then compute the MixUp via a norm-based mean squared error to determine the consistency between the predicted outcome for unlabeled samples and of interpolated one:

$$\mathcal{M}_{reg}^l(\Phi) = (||\sigma(\tilde{\Phi})|| - ||\sigma(\Phi(\tilde{e}))||)^2, \tag{2}$$

$$\text{where } \mu \overset{iid}{\sim} Beta(\alpha, \beta) \tag{3}$$

$$\tilde{e} = \mu \cdot e_p + (1 - \mu) \cdot e_u \tag{4}$$

$$\tilde{\Phi} = \mu \cdot 1 + (1 - \mu) \cdot \Phi(s_u) \tag{5}$$

where e_p and e_u represent the embedding corresponding to a positive s_p and unlabeled s_u sample, respectively; the hyperparameters α and β control the shape of the Beta distribution.

Algorithm 1 summarises the overall computation of `DALLMi` loss. Following the steps displayed in Fig. 3, a balanced batch is selected and used for the calculation of per-label variational loss. Additionally, the algorithm creates input and output interpolations to compute per-label MixUp values. The last step of the algorithm combines the variational loss and MixUp regularization terms weighted via the λ hyperparameter. The gradients of the final loss are computed using backpropagation, which are then used to update the BERT classifier parameters and weights, gradually guiding the model towards better performance.

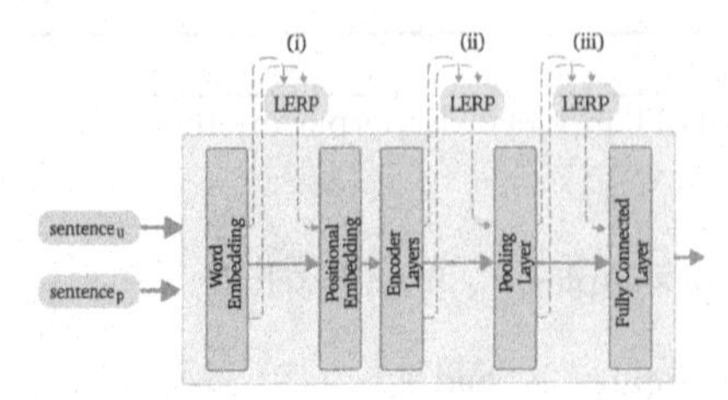

Fig. 4. Examples of possible MixUp strategies by linear interpolation in LLM hidden representations: (i) word embedding, (ii) encoding, (iii) sentence embedding.

Table 1. Evaluation datasets source problem domains & properties.

Dataset	Source/Target	#Samples	#Labels
PubMed	Male/Female	5757	14
arXiv	Old/New	7000	14
Movies	Wiki/IMDb	1000	20

4 Experiments

We assess the effectiveness of `DALLMi` against different LLM fine-tuning methods on three multi-label datasets: *PubMed*, *ArXiv*, and *Movies*.

Evaluation Setup. We use three datasets from different textual categories: healthcare, academia, and entertainment. *PubMed* is a medical dataset consisting of research articles from the PubMed[2] repository. The articles' subheadings denote the source and target domains, namely *female* and *male* patients. The labels represent different biological categories. *ArXiv* provides a collection of research paper abstracts, where labels represent subjects into which to categorize the abstract. Here, the target and source domains are old and new articles scrapped from the ArXiv repository[3]. The *Movies* dataset contains a collection of movie summaries that may belong to different genres. The source and target domains for the movie overviews are Wikipedia and IMDb, respectively. Table 1 shows an overview of the datasets' properties. Each dataset from both domains is split inot a 80% training and 20% testing using random partitioning. To simulate missing labels we discard 50%, 70%, and 90% of the labels for each sample.

Training Details. We use pretrained BERT *bert-base-uncased*[4]. Specifically, we employ the *BertForSequenceClassification* model and fine-tune it using the *Trainer* class for easier reproducibility, efficient GPU memory utilization, and simplified workflow. All runs use batch size equal to 64, learning rate equal to 5e-5, and 12 training epochs. We set the λ, α, β to 1, 0.3, 0.3. We run each experiment thrice and report the average mAP score [3] on the target domain. Furthermore, we do not freeze BERT's layers during the fine-tuning.

Baselines. We compare `DALLMi` with supervised fine-tuning, no fine-tuning, unsupervised domain adaptation method AAD [18], and an asymptotic upper

[2] https://pubmed.ncbi.nlm.nih.gov.
[3] https://arxiv.org.
[4] https://huggingface.co/bert-base-uncased.

Table 2. mAP scores under different ratios of available labels per column for: target domain data of source model, regular fine-tuning (supervised), AAD [18] (unsupervised), and `DALLMi` (semi-supervised).

Dataset	Method & Ratio Available Labels								
	None	Fine-Tuning				AAD	`DALLMi`		
	-	100%	50%	30%	10%	-	50%	30%	10%
PubMed	57.1	$62.0_{.28}$	$50.6_{0.40}$	$45.3_{.36}$	$42.5_{.05}$	$53.7_{.40}$	$58.9_{.48}$	$58.2_{0.38}$	$52.2_{.57}$
arXiv	21.8	$25.0_{.24}$	$18.5_{.12}$	$15.7_{.45}$	$13.7_{.19}$	$21.5_{.03}$	$24.5_{.32}$	$23.2_{.60}$	$20.5_{.48}$
Movies	20.7	$24.7_{.39}$	$15.8_{.55}$	$13.2_{.09}$	$12.6_{.00}$	$17.6_{.12}$	$26.5_{.21}$	$26.2_{.48}$	$26.0_{.36}$

bound using fully supervised fine-tuning without removing any labels. For AAD, we follow the authors' work, freezing BERT's layers and training for three epochs.

Empirical Results. Table 2 reports the mAP results of `DALLMi` compared to the baseline methods for different ratios of discarded labels per column. We observe that `DALLMi` consistently outperforms the supervised fine-tuning technique among all the dataset and label ratio combinations. Our method outperforms the unsupervised AAD in almost all cases, except for scenarios with only 10% label availability, where it remains competitive. Consistent with expectations, the supervised fine-tuning, having all labels, outperforms our proposed algorithm. However, in the case of the Movies dataset, our method achieves marginally higher performance than fully supervised fine-tuning. The variation can be explained by the dataset's low sample count and the wide nature of the dataset. We argue that our approach's effectiveness is due to its incorporation of multiple interpolations and the generation of artificial samples, which contribute to higher results in such data settings. Figure 5 presents an overview of how the mAP score varies over the 12 epochs in the case of 50% label removal for each dataset within a fixed seed. Our method's performance fits between that of the fully supervised fine-tuning with and without label removal. Here, in Fig. 5c, we can better observe the phenomenon from Table 2 in Movies. Our method achieved slightly better performance than supervised fine-tuning without label removal.

Ablation Studies. To establish the effectiveness of `DALLMi`, we conduct three ablation studies. First, we analyze the impact of varying interpolation methods. Second, we compare our norm-based approach to the original log-based VPU. Third, we evaluate the efficiency of our sub-sampling technique.

Table 3(i) illustrates the differences in mAP scores resulting from the different interpolation methods, as opposed to the base word embedding interpolation. The reference mAP scores for the comparative analysis are those from Table 2. In particular, the encoding and sentence embeddings both show analogous scores,

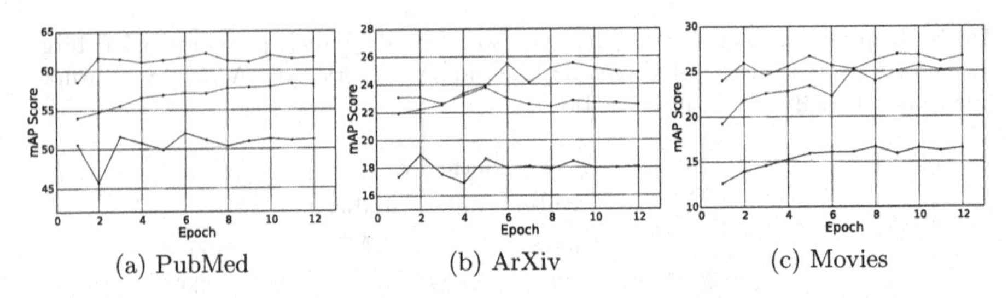

(a) PubMed (b) ArXiv (c) Movies

Fig. 5. mAP scores/epoch for: supervised fine-tuning w/ 50% labels (blue), `DALLMi` w/ 50% labels (green), and supervised fine-tuning w/ 100% labels (red). (Color figure online)

Table 3. Relative changes in mAP score for `DALLMi`'s Table 2 results when changing (i) the word-level MixUp to encoding/sentence-level and (ii) the Norm-based loss to the Log-based formulation (negative percentage means score decrease).

Dataset	Ablation & Ratio Available Labels					
	(i) Encoding/Sentence MixUp			(ii) Log-based Loss		
	50%	30%	10%	50%	30%	10%
PubMed	−0.018%	−0.167%	−0.937%	−20.94%	−21.80%	−16.73%
arXiv	−0.497%	−0.455%	−1.234%	−24.94%	−23.34%	−18.16%
Movies	+0.547%	−0.966%	−0.935%	−39.76%	−44.47%	−46.75%

having slightly lower performance compared to the word embeddings in the PubMed and ArXiv datasets. In contrast, using encoding or sentence-based interpolations in the Movies dataset results in a marginal mAP score increase. This finding highlights the adaptability of the interpolation mechanism, indicating its potential for customization to the specifics of individual datasets. The Movies dataset constitutes a unique case study, where our method is validated under conditions characterized by a limited number of samples in a wide dataset.

Table 3(ii) shows the changes in mAP scores obtained by our method using a log-based variational approach. Exclusively, these scores display a decline in performance across all three datasets and even fall below those achieved by supervised fine-tuning with label removal. We argue that the complexities imposed by the internal representations of textual data make log-based methods overly strict with respect to deviations from learning, leading to a pronounced amplification of errors and, consequently, to poorer performance. In contrast, our proposed method consistently outperforms the baselines, with competitive results in the few cases where it does. This contrast highlights the robustness and validity of our approach over logarithmic variational strategies.

Table 4 illustrates the differences in training time and mAP score associated with different batch sampling strategies. In the first version of the `DALLMi`, we use a standard unweighted sampler, where the batches have a higher risk of being imbalanced, leading to a poorer performance, even if the training time

Table 4. Training time and mAP score w/ 50% labels for: unweighted sampler, and nested batching. DALLMi achieves training time similar to the unweighted sampler with performance on the level of nested batching.

Dataset	Training time (s)		mAP Score	
	Unweighted Sample	Nested Batch	Unweighted Sample	Nested Batch
PubMed	1793.45	2607.62	58.13	58.86
arXiv	1553.97	2796.19	22.92	23.61
Movie Genres	369.54	631.78	25.36	25.22

is low. In the second version, we use nested batching to overcome the imbalance, pairing a large batch of unlabeled samples with smaller batches associated with positive samples for each specific label. Despite showing comparable results to our suggested method, nested batching significantly increases training time. In our final approach, we use a cycle sampler whereby each batch contains at least one positive sample from every label. This adapted sampling technique maintains quality predictions while substantially reducing the training duration.

5 Conclusion

To tackle the domain shift in classifying unlabeled text data, we propose a semi-supervised domain adaptation approach, DALLMi, specifically for LLM-based multi-label classifiers. By incorporating a novel variational loss, MixUp regularizer, and label-balanced sampling, DALLMi effectively exploits the limited positively labeled and abundant unlabeled text data during domain adaptation. We evaluate DALLMi against partial-supervised and unsupervised methods across three datasets, accounting for various label availability scenarios, demonstrating its efficiency. Importantly, our method achieves higher mAP scores, exceeding unsupervised and partial-supervised approaches by 19.9% and 52.2%, respectively. Our results showcase DALLMi's effectiveness and resilience of in enhancing multi-label classification tasks in the presence of domain shifts and scarce labels.

Acknowledgements. This work has been supported by the Spoke "FutureHPC & BigData" of the ICSC - Centro Nazionale di Ricerca in "High Performance Computing, Big Data and Quantum Computing", funded by EU - NextGenerationEU and the EuPilot project funded by EuroHPC JU under G.A. 101034126.

References

1. Bekker, J., Davis, J.: Learning from positive and unlabeled data: a survey. Mach. Learn. **109**(4), 719–760 (2020)
2. Buonocore, T.M., Crema, C., Redolfi, A., Bellazzi, R., Parimbelli, E.: Localizing in-domain adaptation of transformer-based biomedical language models. J. Biomed. Informatics **144**, 104431 (2023)

3. Cartucho, J., Ventura, R., Veloso, M.: Robust object recognition through symbiotic deep learning in mobile robots. In: IEEE/RSJ IROS, pp. 2336–2341 (2018)
4. Chen, H., Liu, F., Wang, Y., Zhao, L., Wu, H.: A variational approach for learning from positive and unlabeled data. In: NeurIPS, vol. 33, pp. 14844–14854 (2020)
5. Chronopoulou, A., Peters, M.E., Dodge, J.: Efficient hierarchical domain adaptation for pretrained language models. In: NAACL, pp. 1336–1351 (2022)
6. Crawford, M., Khoshgoftaar, T.M., Prusa, J.D., Richter, A.N., Najada, H.A.: Survey of review spam detection using machine learning techniques. J. Big Data **2**, 23 (2015)
7. Deng, A., Wu, Y., Zhang, P., Lu, Z., Li, W., Su, Z.: A weakly supervised framework for real-world point cloud classification. Comput. Graph. **102**, 78–88 (2022)
8. Eastwood, C., Mason, I., Williams, C.K.I., Schölkopf, B.: Source-free adaptation to measurement shift via bottom-up feature restoration. In: ICLR (2022)
9. Grangier, D., Iter, D.: The trade-offs of domain adaptation for neural language models. In: ACL, pp. 3802–3813 (2022)
10. Guo, Y., Rennard, V., Xypolopoulos, C., Vazirgiannis, M.: Bertweetfr: domain adaptation of pre-trained language models for French tweets. In: W-NUT, pp. 445–450 (2021)
11. Lee, L.H., Wan, C.H., Rajkumar, R., Isa, D.: An enhanced support vector machine classification framework by using euclidean distance function for text document categorization. Appl. Intell. **37**(1), 80–99 (2012)
12. Liu, H., Long, M., Wang, J., Wang, Y.: Learning to adapt to evolving domains. In: NeurIPS, vol. 33, pp. 22338–22348 (2020)
13. Motiian, S., Piccirilli, M., Adjeroh, D.A., Doretto, G.: Unified deep supervised domain adaptation and generalization. In: IEEE ICCV, pp. 5716–5726 (2017)
14. Nasukawa, T., Yi, J.: Sentiment analysis: capturing favorability using natural language processing. In: K-CAP, pp. 70–77 (2003)
15. Pham, D.D., Koesnadi, S.M., Dovletov, G., Pauli, J.: Unsupervised adversarial domain adaptation for multi-label classification of chest x-ray. In: IEEE ISBI, pp. 1236–1240 (2021)
16. Quinonero-Candela, J., Sugiyama, M., Schwaighofer, A., Lawrence, N.D.: Dataset Shift in Machine Learning. MIT Press (2022)
17. Rietzler, A., Stabinger, S., Opitz, P., Engl, S.: Adapt or get left behind: domain adaptation through BERT language model finetuning for aspect-target sentiment classification. In: LREC, pp. 4933–4941 (2020)
18. Ryu, M., Lee, G., Lee, K.: Knowledge distillation for BERT unsupervised domain adaptation. Knowl. Inf. Syst. **64**(11), 3113–3128 (2022)
19. Sachidananda, V., Kessler, J.S., Lai, Y.: Efficient domain adaptation of language models via adaptive tokenization. In: SustaiNLP@EMNLP, pp. 155–165 (2021)
20. Singh, I.P., Ghorbel, E., Kacem, A., Rathinam, A., Aouada, D.: Discriminator-free unsupervised domain adaptation for multi-label image classification. In: Proceedings of the IEEE/CVF Winter Conference on Applications of Computer Vision (2023)
21. Singhal, P., Walambe, R., Ramanna, S., Kotecha, K.: Domain adaptation: challenges, methods, datasets, and applications. IEEE Access **11**, 6973–7020 (2023)
22. Sun, X., et al.: Text classification via large language models. In: EMNLP 2023 Findings (2023). https://aclanthology.org/2023.findings-emnlp.603/
23. Verma, V., et al.: Manifold Mixup: better representations by interpolating hidden states. In: ICML, vol. 97, pp. 6438–6447 (2019)
24. Wang, D., Shelhamer, E., Liu, S., Olshausen, B.A., Darrell, T.: Tent: fully test-time adaptation by entropy minimization. In: ICLR (2021)

25. Yuan, Z., Zhang, K., Huang, T.: Positive label is all you need for multi-label classification. arXiv preprint arXiv:2306.16016 (2023)
26. Zhang, H., Cissé, M., Dauphin, Y.N., Lopez-Paz, D.: Mixup: beyond empirical risk minimization. In: ICLR (2018)
27. Zhang, Y., Zhang, H., Deng, B., Li, S., Jia, K., Zhang, L.: Semi-supervised models are strong unsupervised domain adaptation learners. arXiv preprint arXiv:2106.00417 (2021)

Contrastive Learning for Unsupervised Sentence Embedding with False Negative Calibration

Chi-Min Chiu(✉), Ying-Jia Lin, and Hung-Yu Kao

Department of Computer Science and Information Engineering, National Cheng Kung University, Tainan, Taiwan
a868111817@gmail.com, hykao@mail.ncku.edu.tw

Abstract. Contrastive Learning, a transformative approach to the embedding of unsupervised sentences, fundamentally works to amplify similarity within positive samples and suppress it amongst negative ones. However, an obscure issue associated with Contrastive Learning is the occurrence of False Negatives, which treat similar samples as negative samples that will hurt the semantics of the sentence embedding. To address it, we propose a framework called **FNC** (False Negative Calibration) to alleviate the influence of false negatives. Our approach has two strategies to amplify the effect, i.e. false negative elimination and reuse. Specifically, in the training process, our method eliminates false negatives by clustering and comparing the semantic similarity. Next, we reuse those eliminated false negatives to reconstruct new positive pairs to boost contrastive learning performance. Our experiments on seven semantic textual similarity tasks demonstrate that our approach is more effective than competitive baselines.

Keywords: Sentence embedding · Contrastive learning · False negative

1 Introduction

Effective sentence representation plays a crucial role in various tasks of natural language processing (NLP) [14], such as sentiment analysis, question answering, text classification, and machine translation. In recent years, pre-trained language models like BERT [10] and RoBERTa [16] have emerged as powerful tools that facilitate the learning of sentence representations.

Supervised methods, such as SBERT [21], have a tendency to produce superior sentence representations. However, their effectiveness heavily relies on extensive fine-tuning using copious amounts of labeled data. Recently, unsupervised contrastive learning frameworks without any labels have drawn closer to similar semantics and removed dissimilar ones. Contrastive learning encompasses a variety of data augmentation aimed at constructing positive pairs that are more similar to each other than negative pairs. Prominent examples of these augmentations include back-translation, word deletion, cut-off, and dropout.

D.-N. Yang et al. (Eds.): PAKDD 2024, LNAI 14647, pp. 290–301, 2024.
https://doi.org/10.1007/978-981-97-2259-4_22

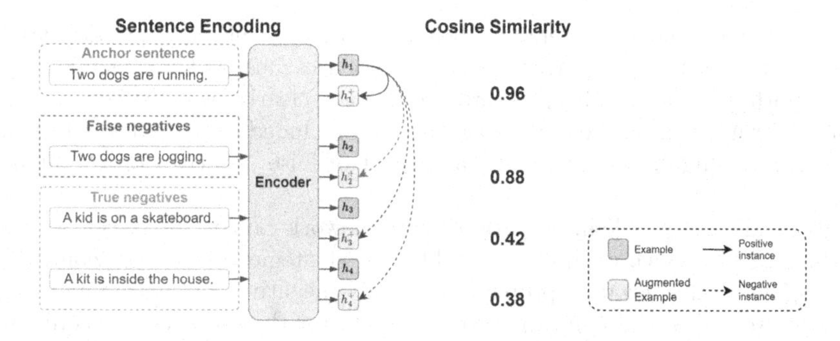

Fig. 1. This diagram illustrates the false negative problem in SimCSE framework. SimCSE is a method for generating sentence embeddings. An 'anchor sentence' is processed by an encoder to produce an embedding ($h1$), which is then augmented by applying dropout to create a 'positive pair' (h_1^+). Other sentences in the same batch are treated as 'negative pairs'. Within the negative pairs, highly similar sentences are deemed false negatives, as highlighted by the red box in the accompanying figure. (Color figure online)

SimCSE [11] proposed a dropout-based contrastive learning framework that applies the standard dropout twice to obtain two different embeddings as "positive pairs". SimCSE considers other sentences within the same mini-batch as "negative pairs." Despite its seemingly simple nature, this approach not only surpasses other training objectives, such as predicting the next sentences [17] and more complex data augmentation (such as word deletion and replacement), by a considerable margin but also achieves comparable performance to previous supervised methods.

Nevertheless, despite the effectiveness of in-batch negatives, there is a problem known as **false negative** which pairs negatives with indeed semantic close sentences. As illustrated in Fig. 1, the anchor sentence is 'Two dogs are running'. However, within the in-batch negatives, there exists a sentence 'Two dogs are jogging', which exhibits a high degree of similarity to the anchor. Consequently, this sentence is categorized as a false negative sentence. It will affect the performance of sentence representation to push away similar sentences. A false negative problem would occur in the SimCSE framework because they sample negatives from training data at random.

Table 1. The Impact of False Negative problem on SimCSE framework. **FNs**: False Negatives

Models	STS12	STS13	STS14	STS15	STS16	STS-B	SICK-R	Avg.
SimCSE-with-FNs	44.92	52.68	45.48	63.39	55.43	53.54	60.16	53.66
SimCSE-without-FNs	**60.95**	**73.08**	**65.64**	**78.51**	**72.33**	**71.78**	**69.36**	**70.24**

To investigate the adverse effects stemming from false negatives, we conducted experiments using natural language inference (NLI) [23] dataset. In this particular experiment, sentences labeled as "entailment" in the NLI dataset were treated as false negatives, as these sentences bear a highly similarity to the hypothesis. The results presented in Table 1 clearly demonstrate a significant performance gap when false negatives are included compared to when they are excluded, highlighting the critical impact of false negatives on the model's efficacy.

To tackle this problem, we propose a framework called False Negative Calibration (FNC), which comprises two different strategies to improve contrastive learning with false negative problems, i.e., false negative elimination and reuse. Initially, we utilize the K-means method to cluster the sentence embeddings in the same mini-batch. After clustering, we remove the sentence that belongs to the identical cluster of the anchor sentence from the contrastive loss, called false negative elimination. Then we reuse those false negatives to construct a positive pair to enhance the effect of contrastive learning, called false negative reuse.

For our FNC framework, our goal is to address the problem of false negatives in unsupervised contrastive learning methods. Consequently, our framework can generally be applied to other existing approaches. For SimCSE, we simply integrate our framework without modifying any settings and our framework boosts performance from 76.25% to 77.19% in BERT_{base}.

Our contributions are summarized as follows:

(1) We present an innovative framework FNC for unsupervised learning of sentence representations. It leverages clustering to mitigate the impact of false negatives and utilizes their reuse to enhance performance.
(2) Our framework seamlessly integrates with existing methods, effectively enhancing performance without the need for redundant tuning or adjustments.
(3) Experimental results from seven semantic textual similarity tasks demonstrate the efficacy of our framework.

2 Related Work

In this section, we review the related works of contrastive learning and introduce SimCSE in detail.

2.1 Contrastive Learning

Contrastive learning is a popular and effective unsupervised learning framework that originated in computer vision [12]. The main concept is to bring the features of images in the same category closer while increasing the distance between features from different categories. In contrastive learning, Data augmentation plays a vital role such as random cropping and image rotation [7]. In NLP, CLEAR [24] is a framework in that word and phrase deletion, phrase order switching, and synonym substitution are served as data augmentation. ConSERT [26] uses Adversarial Attack, Token Shuffling, and Cutoff as data augmentation. SimCSE [11] uses dropout as data augmentation to create positive examples for representation learning.

2.2 SimCSE

Assume a set of paired sentences $\{x_i, x_i^+\}_{i=1}^m$, where x_i and x_i^+ are semantically related and will form positive pairs. SimCSE uses the identical sentence to pass the same encoder twice by applying the standard dropout to obtain two different embeddings. Let h_i and h_i^+ denote the representations of x_i and x_i^+ with a mini-batch of N pairs, the contrastive learning objective is formulated as follows:

$$l_{\text{i}} = -\log \frac{e^{\text{sim}(h_i, h_i^+)/\tau}}{\sum_{j=1}^{N} e^{\text{sim}(h_i, h_j^+)/\tau}} \tag{1}$$

where τ is a temperature hyperparameter and $sim(h_i, h_i')$ is cosine similarity function.

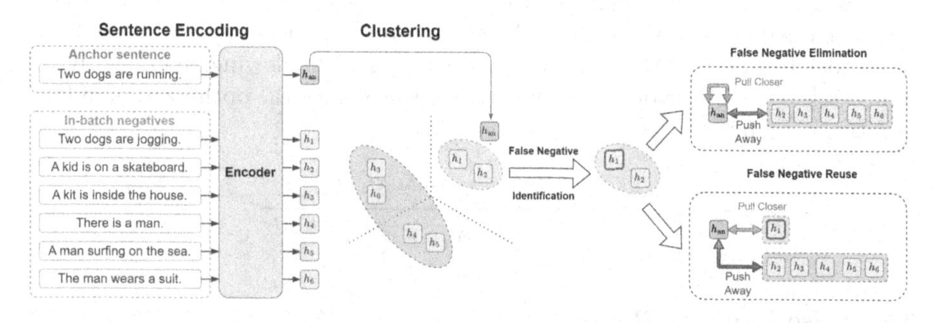

Fig. 2. Illustration of FNC. Two approaches to managing false negatives are depicted: elimination and reuse. In elimination, the model aims to pull semantically similar sentences (like h_1) closer to the anchor while pushing dissimilar ones away. Conversely, in reuse, the model reclassifies h_1 as a positive example, pulling it even closer to h_{an} and maintaining the distance from the other negatives.

3 Method

The core idea of our proposed framework False Negative Calibration (FNC) is to reduce the influence of false negatives and reuse them to enhance the effect of contrastive learning. Our proposed approach has two strategies: false negative elimination and reuse, aimed at amplifying the effect of contrastive learning. An illustration of FNC is shown in Fig. 2.

In the beginning, we employ k-means clustering on in-batch samples to group them into k clusters. The k-th cluster centroid is denoted as c_k and represents the k-th cluster group. In false negative elimination, we treat the negative samples in the same cluster c_k as false negatives and remove them on contrastive loss. Then, in false negative reuse, we can utilize the eliminated false negatives to create additional positive pairs, thereby enhancing performance. Finally, our framework combines those two strategies for contrastive learning. The subsequent section will provide a detailed introduction to these two components.

3.1 False Negative Elimination

The most straightforward method for suppressing the influence of false negatives is to exclude them from the contrastive comparison. In this strategy, We consider negative samples within the same cluster c_k as false negative candidates. Additionally, in order to effectively identify false negatives, we conduct a cosine similarity comparison among the candidates. Any false negative candidate with a semantic similarity higher than the threshold ϕ will be discarded. The weight α defined as follows:

$$\alpha_{h_j^+}[h_j^+ \in c_i] = \begin{cases} 0 & , \mathrm{sim}(h_i, h_j^+) \geq \phi \\ 1 & , \mathrm{sim}(h_i, h_j^+) < \phi \end{cases} \tag{2}$$

When negative h_j^+ belongs to the same cluster c_i as anchor sample h_i and the cosine similarity exceeds the threshold value ϕ, the weight will be set to 0 in order to mitigate the impact of false negatives. Following the objective of SimCSE [11], we incorporate a weight factor to enhance the optimization of false negative elimination.

$$l_{\mathrm{elim}} = -\log \frac{e^{\mathrm{sim}(h_i, h_i^+)/\tau}}{\sum_{j=1}^{N} \alpha_{h_j^+} \times e^{\mathrm{sim}(h_i, h_j^+)/\tau}} \tag{3}$$

3.2 False Negative Reuse

Although removing false negatives mitigates the negative impact of contrasting with them, it disregards the valuable information provided by true positives. Hence, we reuse the false negatives as true positives to improve contrastive learning.

After clustering negatives and excluding the false negative, we reconstruct the positive pairs by reusing the false negatives as true positives. F_i is the false negative set of sentence i. This leads to the subsequent expression of the contrastive loss:

$$l_{\mathrm{reuse}} = -\frac{1}{|F_i|} \left(\sum_{\hat{h} \in F_i} \log \frac{e^{\mathrm{sim}(h_i, \hat{h})/\tau}}{\sum_{j=1}^{N} e^{\mathrm{sim}(h_i, h_j^+)/\tau}} \right) \tag{4}$$

Finally, we optimize these two strategies by combining their losses with a weighting coefficient λ:

$$l = l_{\mathrm{elim}} + \lambda \cdot l_{\mathrm{reuse}} \tag{5}$$

Table 2. The performance on STS tasks (Spearman's correlation) for different sentence embedding models. ♣: results from [21] ♠: result from [11] ♢:result from [15] ♡:result from [22] ♡1:result from [13]

Models	STS12	STS13	STS14	STS15	STS16	STS-B	SICK-R	Avg.
GloVe embeddings(avg.)♣	55.14	70.66	59.73	68.25	63.66	58.02	53.76	61.32
$\text{BERT}_{\text{base}}$(first-last avg.)	39.70	59.38	49.67	66.03	66.19	53.87	62.06	56.70
$\text{BERT}_{\text{base}}$-flow♢	58.40	67.10	60.85	75.16	71.22	68.66	64.47	66.55
$\text{BERT}_{\text{base}}$-whitening♡	57.83	66.90	60.90	75.08	71.31	68.24	63.73	66.28
SimCSE-$\text{BERT}_{\text{base}}$ ♠	68.40	82.41	74.38	80.91	78.56	76.85	**72.23**	76.25
FNC-$\text{BERT}_{\text{base}}$	**69.03**	**82.67**	**75.36**	**83.77**	**78.68**	**79.39**	71.44	**77.19**
$\text{RoBERTa}_{\text{base}}$(first-last avg.)	40.88	58.74	49.07	65.63	61.48	58.55	61.63	56.57
$\text{RoBERTa}_{\text{base}}$-whitening♡	46.99	63.24	57.23	71.36	68.99	61.36	62.91	61.73
SimCSE-$\text{RoBERTa}_{\text{base}}$ ♠	**70.16**	81.77	73.24	81.36	80.65	80.22	68.56	76.57
FNC-$\text{RoBERTa}_{\text{base}}$	69.17	**82.68**	**74.43**	**82.58**	**81.61**	**81.14**	**69.52**	**77.30**
$\text{PromptBERT}_{\text{base}}$ ♡1	**71.56**	84.58	76.98	84.47	**80.60**	81.60	69.87	78.54
FNC-$\text{PromptBERT}_{\text{base}}$	71.29	**84.84**	**77.00**	**84.58**	80.29	**81.89**	**70.17**	**78.58**
$\text{DiffCSE}_{\text{base}}$(reproduce)	69.67	82.89	74.31	**83.10**	**80.53**	**80.38**	71.84	77.53
FNC-$\text{DiffCSE}_{\text{base}}$	**69.84**	**83.58**	**75.55**	83.01	80.08	79.71	**72.81**	**77.80**

4 Experiment and Analysis

4.1 Setup

In our experiment, we follow the setting of unsupervised SimCSE [11] and PromptBERT [13]. We use 1 million randomly sampled sentences from English Wikipedia for training. The capability of sentence embeddings is measured by the task of semantic textual similarity(STS). We conduct our experiments on 7 STS tasks: STS 2012-2016 [1–5], STS Benchmark [6] and SICKRelatedness [18].

4.2 Training Details

We start from pre-trained checkpoints of BERT (uncased) or RoBERTa (cased) and take the [CLS] representation as the sentence embedding. We train our models for one epoch using the Adam optimizer with the batch size = 64 and the temperature $\tau = 0.05$. Moreover, following SimCSE, we evaluate the model every 125 training steps on the development set of STS-B and keep the best checkpoint for the final evaluation on test sets.

4.3 Main Result

We evaluate our framework on baseline models consisting of average GloVe embeddings [19], as well as averaged first and last layer BERT embeddings. Post-processing techniques such as BERT-flow [24] and BERT-whitening [26] are also considered. Furthermore, we compare our framework to SimCSE [11].

Table 2 presents the results of our experiments on 7 STS tasks. FNC outperforms SimCSE, increasing the average Spearman's correlation from 76.25% to 77.19%.

PromptBERT [13] introduces a new contrastive learning technique to improve sentence representation. It addresses issues found in BERT's approach by using prompts to create embeddings and a denoising template for training. We combined FNC with PromptBERT, resulting in a slight increase in the average STS performance from 78.54% to 78.58%, indicating an incremental yet positive enhancement through the integration of our method with PromptBERT.

DiffCSE [9] introduces an unsupervised approach for teaching computer sentence meaning by comparing original and slightly altered sentences, yielding superior results in semantic textual similarity tasks. We integrated our method, FNC, with DiffCSE, enhancing the average performance on STS tasks from 77.53% to 77.80%. The above improvement demonstrates that our approach can effectively combine with existing methods.

4.4 Short Text Clustering

Following [28] work, we also conduct an experiment for short text clustering on 8 datasets, SearchSnippets (SS) [20], StackOverflow (SO) [25], Biomedical (Bio) [25], AgNews (AG) [29], Tweet [27], and GoogleNews (G-T, G-S, G-TS) [27]. We apply K-means to the representations generated by each model and report the clustering accuracy as the evaluation metric. Show in Table 3, FNC improves the performance of two models, SimCSE-BERT$_{base}$ and SimCSE-RoBERTa$_{base}$.

Table 3. Clustering accuracy reported on eight short text clustering datasets.

Models	AG	Bio	Go-S	G-T	G-TS	SS	SO	Tweet	Avg.
SimCSE-BERT$_{\text{base}}$	74.46	35.64	59.01	57.92	64.18	67.09	50.78	54.71	57.97
FNC-BERT$_{\text{base}}$	**75.16**	**37.72**	**59.64**	**58.99**	**65.93**	**69.87**	**70.75**	**55.78**	**61.73**
SimCSE-RoBERTa$_{\text{base}}$	**69.71**	**37.35**	60.89	**57.66**	65.05	46.90	69.00	**51.89**	57.31
FNC-RoBERTa$_{\text{base}}$	63.39	37.26	**61.53**	56.53	**65.72**	**59.18**	**70.64**	50.04	**58.04**

4.5 Ablation Study

Our framework mainly consists of two parts, false negative elimination and reuse. To investigate the contribution of those components, we conduct an ablation study on 7 STS tasks, comparing average Spearman's correlation. Ablation study is provided in Table 4. The results show that both of the FNC strategies are helpful for the STS tasks.

We extend our ablation study to analyze the impact of our FNC framework on short text clustering tasks, using the average as a metric. The results are presented in Table 5. It is evident that the full implementation of FNC with BERT-base achieves the highest performance at 61.73. When individual components are removed, we observe a decrease in performance: without elimination,

Table 4. Ablation of our approach on the test set of seven STS tasks.

Model	$\text{BERT}_{\text{base}}$	$\text{RoBERTa}_{\text{base}}$
FNC	**77.19**	**77.30**
w/o Elimination	76.49(−0.70)	76.64(−0.66)
w/o Reuse	76.56(−0.63)	76.78(−0.52)
w/o Cluster	76.83(−0.36)	76.88(−0.42)
$\text{SimCSE-BERT}_{\text{base}}$	76.25	76.57

the performance drops by 1.99 points; without reuse, it decreases by 1.01 points; and without clustering, it falls by 3.08 points. This indicates that all components contribute to the framework's effectiveness, with clustering having the most significant impact. In comparison, the baseline SimCSE-BERT-base model scores 57.97, further illustrating the enhancements brought about by our FNC approach.

Table 5. Ablation of our approach on the test set of seven STS tasks.

Model	Short Text Cluster-avg.
$\text{FNC-BERT}_{\text{base}}$	**61.73**
w/o Elimination	59.74(−1.99)
w/o Reuse	60.72(−1.01)
w/o Cluster	58.65(−3.08)
$\text{SimCSE-BERT}_{\text{base}}$	57.97

A comprehensive analysis of both tables suggests that our FNC strategies not only contribute positively to the STS tasks but also to short text clustering tasks, with each component playing a crucial role in the overall performance. The clustering component, in particular, shows a substantial influence in both domains, underscoring its importance in our approach. The improvements over the baseline SimCSE models in both tables reaffirm the robustness and versatility of our FNC framework across different NLP tasks.

4.6 Comparison with Other False Negative Solutions

DCLR. Debiased Contrastive Learning of unsupervised sentence Representations (DCLR) [30] framework offers a solution by introducing an instance weighting mechanism to mitigate the impact of improper negatives and by generating noise-based negatives, which ensures a more uniform representation space.

In the provided comparative analysis, we evaluate the efficacy of the FNC approach against DCLR, using BERT and RoBERTa models. Figure 3 illustrates that FNC, with its reuse component, outperforms the version without

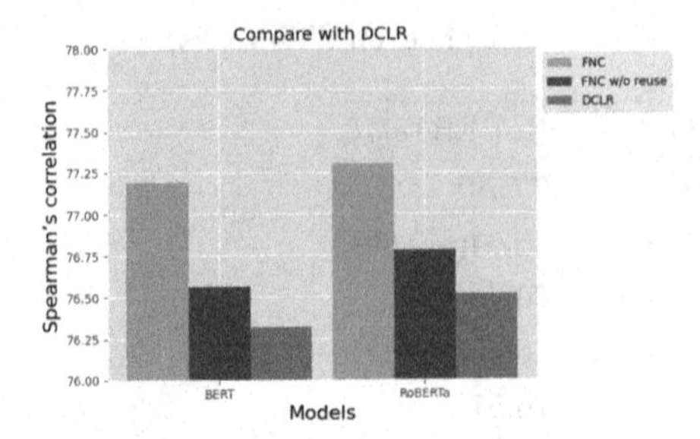

Fig. 3. Comparison with DCLR.

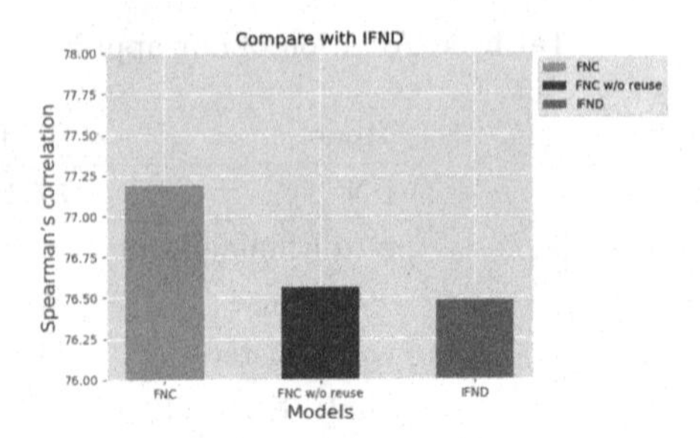

Fig. 4. Comparison with IFND.

reuse and DCLR for BERT, highlighting the importance of the reuse component in sentence representation. For RoBERTa, while FNC still leads, DCLR shows competitive performance, indicating model-specific advantages with debiasing strategies. This suggests a nuanced interplay between model architecture and contrastive learning techniques, meriting further investigation into targeted optimizations for unsupervised sentence representation.

IFND. In the computer vision (CV) area, self-supervised learning, particularly through contrastive learning, has been a breakthrough in vision tasks. It distinguishes individual images but often overlooks their semantic connections, pushing away similar images known as "false negatives." This issue is more pronounced in large datasets with diverse concepts. INCREMENTAL FALSE NEGATIVE DETECTION (IFND) [8] research presents a new framework that progressively identifies and removes these false negatives as the model learns and the embedding space refines. Strategies for eliminating false negatives lead to superior performance across various benchmarks, even with limited resources.

In Fig. 4, there are FNC, its variant without the reuse feature (FNC w/o reuse), and the Incremental False Negative Detection (IFND) approach. FNC leads with the highest correlation, indicating its effectiveness in generating sentence representations. Removing the reuse feature from FNC results in a performance drop, highlighting the feature's contribution to the model's success. This comparison underscores the importance of the reuse component in achieving superior sentence embeddings within the contrastive learning framework.

5 Conclusion

In this paper, we propose FNC, an innovative framework aimed at addressing the issue of false negatives in unsupervised contrastive learning for sentence representation. We employ clustering to group sentence embeddings, enabling us to identify false negatives. By eliminating these false negatives, we minimize their impact. Additionally, we repurpose the eliminated false negatives to create positive pairs, further enhancing performance. Moreover, our framework seamlessly integrates into unsupervised contrastive learning methods, requiring no unnecessary tuning or adjustments. To validate the effectiveness of FNC, we conduct experiments on STS tasks. We aim to bridge gap between supervised and unsupervised methods.

Acknowledgements. This work was funded in part by Qualcomm through a Taiwan University Research Collaboration Project NAT-487842.

References

1. Agirre, E., et al.: SemEval-2015 task 2: semantic textual similarity, English, Spanish and pilot on interpretability. In: Proceedings of the 9th International Workshop on Semantic Evaluation (SemEval 2015), pp. 252–263. Association for Computational Linguistics, Denver, Colorado (2015). https://doi.org/10.18653/v1/S15-2045, https://aclanthology.org/S15-2045
2. Agirre, E., et al.: SemEval-2014 task 10: multilingual semantic textual similarity. In: Proceedings of the 8th International Workshop on Semantic Evaluation (SemEval 2014), pp. 81–91. Association for Computational Linguistics, Dublin, Ireland (2014). https://doi.org/10.3115/v1/S14-2010, https://aclanthology.org/S14-2010
3. Agirre, E., et al.: SemEval-2016 task 1: semantic textual similarity, monolingual and cross-lingual evaluation. In: Proceedings of the 10th International Workshop on Semantic Evaluation (SemEval-2016), pp. 497–511. Association for Computational Linguistics, San Diego, California (2016). https://doi.org/10.18653/v1/S16-1081, https://aclanthology.org/S16-1081
4. Agirre, E., Cer, D., Diab, M., Gonzalez-Agirre, A.: SemEval-2012 task 6: a pilot on semantic textual similarity. In: *SEM 2012: The First Joint Conference on Lexical and Computational Semantics – Volume 1: Proceedings of the Main Conference and the Shared Task, and Volume 2: Proceedings of the Sixth International Workshop on Semantic Evaluation (SemEval 2012), pp. 385–393. Association for Computational Linguistics, Montréal, Canada (2012). https://aclanthology.org/S12-1051
5. Agirre, E., Cer, D., Diab, M., Gonzalez-Agirre, A., Guo, W.: *SEM 2013 shared task: semantic textual similarity. In: Second Joint Conference on Lexical and Computational Semantics (*SEM), Volume 1: Proceedings of the Main Conference and the Shared Task: Semantic Textual Similarity, pp. 32–43. Association for Computational Linguistics, Atlanta, Georgia, USA (2013). https://aclanthology.org/S13-1004
6. Cer, D., Diab, M., Agirre, E., Lopez-Gazpio, I., Specia, L.: SemEval-2017 task 1: semantic textual similarity multilingual and crosslingual focused evaluation. In: Proceedings of the 11th International Workshop on Semantic Evaluation (SemEval-2017), pp. 1–14. Association for Computational Linguistics, Vancouver, Canada (2017). https://doi.org/10.18653/v1/S17-2001, https://aclanthology.org/S17-2001
7. Chen, T., Kornblith, S., Norouzi, M., Hinton, G.: A simple framework for contrastive learning of visual representations. In: III, H.D., Singh, A. (eds.) Proceedings of the 37th International Conference on Machine Learning. Proceedings of Machine Learning Research, vol. 119, pp. 1597–1607. PMLR (2020)
8. Chen, T.S., Hung, W.C., Tseng, H.Y., Chien, S.Y., Yang, M.H.: Incremental false negative detection for contrastive learning. arXiv preprint arXiv:2106.03719 (2021)

9. Chuang, Y.S., et al.: DiffCSE: difference-based contrastive learning for sentence embeddings. In: Annual Conference of the North American Chapter of the Association for Computational Linguistics (NAACL) (2022)
10. Devlin, J., Chang, M.W., Lee, K., Toutanova, K.: BERT: pre-training of deep bidirectional transformers for language understanding. In: Proceedings of the 2019 Conference of the North American Chapter of the Association for Computational Linguistics: Human Language Technologies, Volume 1 (Long and Short Papers), pp. 4171–4186. Association for Computational Linguistics, Minneapolis, Minnesota (2019). https://doi.org/10.18653/v1/N19-1423, https://aclanthology.org/N19-1423
11. Gao, T., Yao, X., Chen, D.: SimCSE: simple contrastive learning of sentence embeddings. In: Empirical Methods in Natural Language Processing (EMNLP) (2021)
12. Hadsell, R., Chopra, S., LeCun, Y.: Dimensionality reduction by learning an invariant mapping. In: 2006 IEEE Computer Society Conference on Computer Vision and Pattern Recognition (CVPR 2006), vol. 2, pp. 1735–1742 (2006). https://doi.org/10.1109/CVPR.2006.100
13. Jiang, T., et al.: PromptBERT: improving BERT sentence embeddings with prompts. In: Proceedings of the 2022 Conference on Empirical Methods in Natural Language Processing, pp. 8826–8837. Association for Computational Linguistics, Abu Dhabi, United Arab Emirates (2022). https://aclanthology.org/2022.emnlp-main.603
14. Kiros, R., et al.: Skip-thought vectors. In: Cortes, C., Lawrence, N., Lee, D., Sugiyama, M., Garnett, R. (eds.) Advances in Neural Information Processing Systems, vol. 28. Curran Associates, Inc. (2015). https://proceedings.neurips.cc/paper_files/paper/2015/file/f442d33fa06832082290ad8544a8da27-Paper.pdf
15. Li, B., Zhou, H., He, J., Wang, M., Yang, Y., Li, L.: On the sentence embeddings from pre-trained language models. In: Proceedings of the 2020 Conference on Empirical Methods in Natural Language Processing (EMNLP), pp. 9119–9130. Association for Computational Linguistics (2020). https://doi.org/10.18653/v1/2020.emnlp-main.733, https://aclanthology.org/2020.emnlp-main.733
16. Liu, Y., et al.: Roberta: a robustly optimized BERT pretraining approach. arXiv preprint arXiv:1907.11692 (2019)
17. Logeswaran, L., Lee, H.: An efficient framework for learning sentence representations. In: International Conference on Learning Representations (2018). https://openreview.net/forum?id=rJvJXZb0W
18. Marelli, M., Menini, S., Baroni, M., Bentivogli, L., Bernardi, R., Zamparelli, R.: A SICK cure for the evaluation of compositional distributional semantic models. In: Proceedings of the Ninth International Conference on Language Resources and Evaluation (LREC 2014), pp. 216–223. European Language Resources Association (ELRA), Reykjavik, Iceland (2014). http://www.lrec-conf.org/proceedings/lrec2014/pdf/363_Paper.pdf
19. Pennington, J., Socher, R., Manning, C.: GloVe: global vectors for word representation. In: Proceedings of the 2014 Conference on Empirical Methods in Natural Language Processing (EMNLP), pp. 1532–1543. Association for Computational Linguistics, Doha, Qatar (2014). https://doi.org/10.3115/v1/D14-1162, https://aclanthology.org/D14-1162
20. Phan, X.H., Nguyen, L.M., Horiguchi, S.: Learning to classify short and sparse text & web with hidden topics from large-scale data collections. In: Proceedings of the 17th international conference on World Wide Web, pp. 91–100 (2008)

21. Reimers, N., Gurevych, I.: Sentence-BERT: sentence embeddings using Siamese BERT-networks. In: Proceedings of the 2019 Conference on Empirical Methods in Natural Language Processing and the 9th International Joint Conference on Natural Language Processing (EMNLP-IJCNLP), pp. 3982–3992. Association for Computational Linguistics, Hong Kong, China (2019). https://doi.org/10.18653/v1/D19-1410, https://aclanthology.org/D19-1410
22. Su, J., Cao, J., Liu, W., Ou, Y.: Whitening sentence representations for better semantics and faster retrieval. arXiv preprint arXiv:2103.15316 (2021)
23. Williams, A., Nangia, N., Bowman, S.: A broad-coverage challenge corpus for sentence understanding through inference. In: Walker, M., Ji, H., Stent, A. (eds.) Proceedings of the 2018 Conference of the North American Chapter of the Association for Computational Linguistics: Human Language Technologies, Volume 1 (Long Papers), pp. 1112–1122. Association for Computational Linguistics, New Orleans, Louisiana (2018). https://doi.org/10.18653/v1/N18-1101, https://aclanthology.org/N18-1101
24. Wu, Z., Wang, S., Gu, J., Khabsa, M., Sun, F., Ma, H.: Clear: contrastive learning for sentence representation. arXiv preprint arXiv:2012.15466 (2020)
25. Xu, J., et al.: Self-taught convolutional neural networks for short text clustering. Neural Networks **88**, 22–31 (2017). https://doi.org/10.1016/j.neunet.2016.12.008, https://www.sciencedirect.com/science/article/pii/S0893608016301976
26. Yan, Y., Li, R., Wang, S., Zhang, F., Wu, W., Xu, W.: ConSERT: a contrastive framework for self-supervised sentence representation transfer. In: Proceedings of the 59th Annual Meeting of the Association for Computational Linguistics and the 11th International Joint Conference on Natural Language Processing (Volume 1: Long Papers), pp. 5065–5075. Association for Computational Linguistics (2021). https://doi.org/10.18653/v1/2021.acl-long.393, https://aclanthology.org/2021.acl-long.393
27. Yin, J., Wang, J.: A model-based approach for text clustering with outlier detection. In: 2016 IEEE 32nd International Conference on Data Engineering (ICDE), pp. 625–636 (2016). https://doi.org/10.1109/ICDE.2016.7498276
28. Zhang, D., et al.: Pairwise supervised contrastive learning of sentence representations. In: Proceedings of the 2021 Conference on Empirical Methods in Natural Language Processing, pp. 5786–5798. Association for Computational Linguistics, Online and Punta Cana, Dominican Republic (2021). https://doi.org/10.18653/v1/2021.emnlp-main.467, https://aclanthology.org/2021.emnlp-main.467
29. Zhang, X., LeCun, Y.: Text understanding from scratch. arXiv preprint arXiv:1502.01710 (2015)
30. Zhou, K., Zhang, B., Zhao, W.X., Wen, J.R.: Debiased contrastive learning of unsupervised sentence representations. arXiv preprint arXiv:2205.00656 (2022)

Recovering Population Dynamics from a Single Point Cloud Snapshot

Yuki Wakai(✉), Koh Takeuchi, and Hisashi Kashima

Kyoto University, Kyoto, Japan
yuki_wakai@ml.ist.i.kyoto-u.ac.jp, {takeuchi,kashima}@i.kyoto-u.ac.jp

Abstract. Discovering population dynamics from point cloud data has experienced increased popularity in various applications, including GPS behavior prediction, multi-target tracking, and single cell analysis. Existing methods require data in multiple time periods. However, to address privacy concerns and observational restrictions, our method estimates trajectories solely from a single snapshot without time series information or features other than coordinates. We propose a model that recovers vector fields by solving an optimal transport problem and introducing the smoothness of point movements as regularization terms. Experiments with point cloud data generated from typical vector fields show that our method can accurately recover the original vector fields and predict the trajectories at arbitrary coordinates from just one point cloud snapshot.

Keywords: trajectory prediction · optimal transport · vector fields

1 Introduction

Recent advancements in location monitoring technologies like GPS, stereo cameras, and 3D laser imaging have led to the ubiquity of various point cloud data types. These point cloud data have been utilized in practical applications in many domains including behavior prediction using GPS data [9,10], multi-target tracking [13], density observation in meteorology [7], observation of dynamics among wild animal population [14], and single cell analysis [3,16]. For example, if we can estimate population dynamics from point cloud data characterizing the GPS locations of automobiles, bicycles, and pedestrians, we can attempt to predict future congestion locations in cities and prevent unnecessary transportation durations.

When predicting trajectories of points in point cloud data, a common method is to track each point over multiple time periods to recover population dynamics models from continuously acquired data. However, this method is difficult to apply when the same point cloud cannot be monitored along a time series owing to observational constraints or privacy issues. For example, in single cell analysis, measuring cell features requires destroying the cells. Therefore, the sampled cells cannot be continuously monitored after the analysis and the entire cell

D.-N. Yang et al. (Eds.): PAKDD 2024, LNAI 14647, pp. 302–315, 2024.
https://doi.org/10.1007/978-981-97-2259-4_23

population does not remain the same before and after the observation. Moreover, because privacy preservation has become significantly important recently, data features that can be tied to personal information must be handled carefully. Consequently, there may be cases where time series information or point-to-point matching between each point is intentionally hidden owing to recent concerns regarding individual privacy issues.

To address this issue, we propose a novel approach called Single-Shot Neural Vector Field (SNVF) that recovers the underlying population dynamics models from a single-time point cloud snapshot. Accordingly, we introduce a pseudo-trajectory recovering problem regarding points observed from only coordinates data without time series information and auxiliary point cloud features. We make two assumptions to deal with such an extreme case. The first assumption is that the neighbors of each point represent its coordinates in the recent past or near future, and each point will move to its neighbors at the next time step. This assumption allows us to track the movement of each point in the pseudo-time series. The second assumption is that the population dynamics model, which we consider as a vector field that determines point trajectories, is smooth. This means that the trajectory vectors of each point and its neighbors are similar. Based on these two assumptions, we formulate the pseudo-tracking problem as an optimal transport problem and introduce regularization terms. In our experiments, we used point cloud data artificially generated from typical vector fields and evaluated the results using ground truth vector fields. The results showed that SNVF achieves high accuracy in reconstructing trajectories and vector fields from a single-time point cloud snapshot.

The contributions of our research are threefold. First, while existing studies have focused on the case where snapshots of point clouds can be obtained at multiple time points or where each point has various features other than coordinates, this research tackles a new problem where these assumptions do not hold. Second, the paper proposes a novel algorithm, SNVF, specifically designed for predicting point trajectories from a single snapshot, where only coordinate data without time series are accessible. Lastly, this research defines an evaluation metric for the methods that predict the trajectory of each point in point cloud data, evaluates comparative approaches quantitatively, and demonstrates the effectiveness of SNVF for the problem, namely, recovering population dynamics from a single point cloud snapshot.

2 Related Work

Previous research focus on the case where the same point cloud can be tracked continuously over multiple times but point-to-point correspondence is not available. Bunne et. al. [2] used a neural network model called JKONET, to predict point cloud configurations from point cloud in the multiple unaligned time series. Our study focuses on predicting trajectories based solely on a single-time snapshot without time series data.

Single cell analysis handles the case where each point in a single-time point cloud snapshot has rich features besides coordinate information. In single-cell analysis, sampled cells are destroyed during the analysis process, making it impossible to track the same cell population before and after the observation. Additionally, tagging cells for continuous tracking incurs tremendous cost and effort. To address this, trajectory inference methods [15] are employed using dimensionality reduction techniques or constructing graphs with cells as the vertices. Therefore, in single cell analysis, time series information is not usually available but each point has rich features besides its coordinates. In contrast, our study focuses on the problem where no data other than the coordinates is accessible.

Previous studies have utilized neural networks to learn vector fields for various applications, such as temporal super-resolution [8], deformable image registration [6], and continuous completion of vector fields [12]. However, these methods require the ground truth vector fields. In contrast, our study aims to predict the movement of each point from a single-time point cloud snapshot without the ground truth vector field data.

3 Problem Setting

We aim to predict the underlying population dynamics model that determines how each point will move at the next time step given the current coordinates of a point set in a D-dimensional Euclidean space $\mathbb{R}^D$. The input of our problem is the set of coordinates of a point cloud in $\mathbb{R}^D$. We denote the number of points by N and coordinates of point i by $\boldsymbol{x}_i \in \mathbb{R}^D$ $(i = 1, \ldots, N)$. The output is a vector field $\boldsymbol{V} : \mathbb{R}^D \rightarrow \mathbb{R}^D$ that determines the velocity vector at arbitrary coordinates in $\mathbb{R}^D$; for simplicity, we assume discrete time in this study, that is, the point at x_t at time t moves to $x_t + f(x_t)$ in the next time step $t+1$. We denote a vector from point i cloud $\boldsymbol{x}_i$ to point j cloud $\boldsymbol{x}_j$ by $\boldsymbol{v}_{ij} = \boldsymbol{x}_j - \boldsymbol{x}_i$.

An example problem is presented in Fig. 1, where the task is to recover the trajectory that each point should follow using a single point cloud snapshot. Let us consider that Fig. 1(a) represents the coordinate data of a set of point clouds at a specific time. We show the ground truth vector field of this example in Fig. 1(b). In Fig. 1(b), the gray scale of each arrow represents the magnitude of the trajectory vector, while the angle of the arrow indicates the trajectory direction.

Existing studies have dealt with cases where the same set of points can be tracked continuously over multiple time periods, where the correspondence between points cannot be obtained [2], or where a pseudo time series is generated based on the features of each point [3]. However, these studies assume that the same set of points can be continuously tracked over multiple times, or that each point has features other than coordinates. Because these assumptions do not often hold in real applications, we in this study supposes a more challenging case where only coordinates of each point are available, and we can observe a snapshot of the entire point cloud.

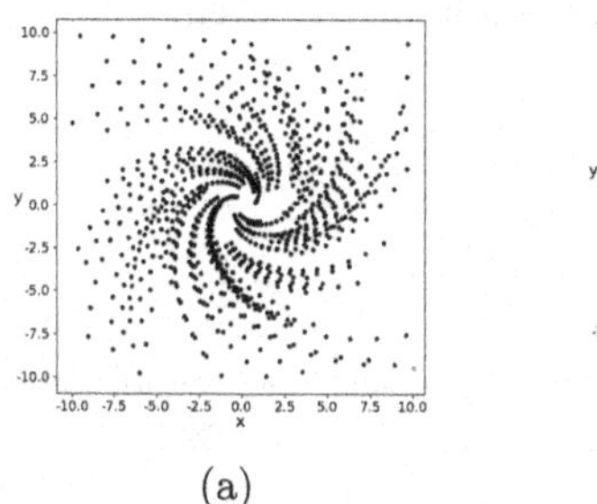

(a)

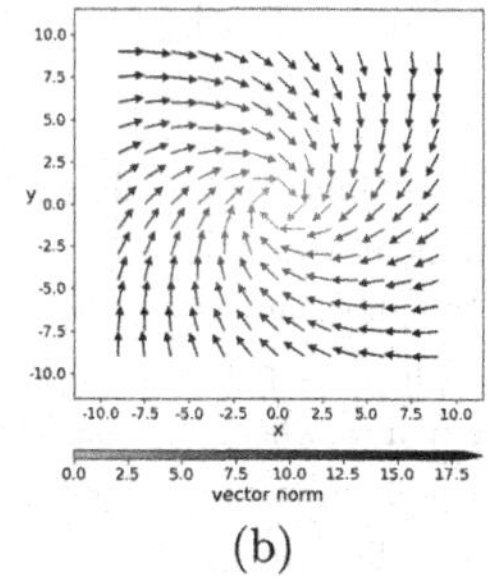

(b)

Fig. 1. An example of our problem setting. The goal is to predict (b) population dynamics (a vector field) from (a) a single-time point cloud snapshot.

4 Proposed Method

We address the problem of estimating the underlying vector field that determines the movement of each point from a single snapshot of a point cloud. We introduce a regression model $f : \mathbb{R}^D \to \mathbb{R}^D$ to predict the continuous vector field for any arbitrary points in $\mathbb{R}^D$. We denote the output for $\boldsymbol{x}_i$ as $\boldsymbol{f}_i = f(\boldsymbol{x}_i) \in \mathbb{R}^D$. In order to learn such a model from a single snapshot, we introduce several assumptions and define an objective function based on an optimal transport problem. In addition, we introduce several smoothness constraints on the trajectory vectors as regularization terms. Because the vector fields tend to contain complex dynamics over the input field, we employ a neural network as f to deal with the possibility of non-linearity of vector fields.

As we show in Fig. 1, to recover a velocity vector field from a single snapshot of point clouds is significantly challenging. Because estimating a vector field from a single shot of point cloud is nearly unfeasible, we make two assumptions in our problem setting so that this problem is tractable. These assumptions can be regarded as too strict to satisfy in the real-world datasets. However, since no existing study has addressed to estimate the vector field from a snapshot, we start from a simple problem setting in this study.

Assumption 1. *Each point's neighbors represent its coordinates in the recent past or near future, and each point will move to its neighbors at the next time step.*

We consider that this assumption makes it possible to track the movement of each point in the pseudo-time series.

Assumption 2. *The vector field is smooth in the neighborhood space, that is, nearby point clouds have similar trajectory vectors.*

These assumptions help estimate the trajectory vector at each point, considering those of the neighboring points. We formulate the aforementioned two assumptions as constraints in the objective function for the model f.

Table 1. Notation in our research

Symbol	Description
$\boldsymbol{x}_i$	Coordinates of point i in a D-dimensional Euclidean space $\mathbb{R}^D$
$\boldsymbol{v}_{ij}$	Vector from point i to point j ($= \boldsymbol{x}_j - \boldsymbol{x}_i$)
K	Number of neighboring points used for inference
N_i	Set of K neighboring points of point i
P	Probability matrix where p_{ij} represents the probability of transition from point i to point j
f	Regression model that approximates the underlying vector field
θ	Parameters of the model f
$\boldsymbol{f}_i$	Trajectory vector at $\boldsymbol{x}_i$ (the coordinate of point i) estimated by the regression model f

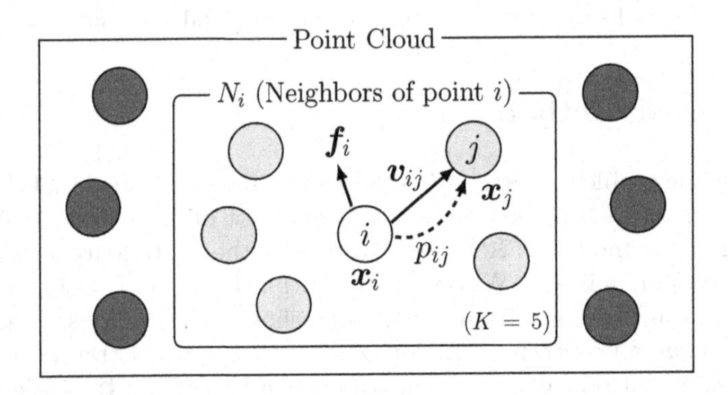

Fig. 2. Notation in our research

4.1 Notation

The notation in our research is described in Table 1 and Fig. 2. We introduce a latent transition probability p_{ij} that represents the possibility of a movement from point i to its neighbor point j based on Assumption 1. We suppose that each point cannot transit towards itself, which means $p_{ii} = 0$. We denote transition probabilities for pairs of point clouds as P, which satisfy a doubly stochastic matrix condition

$$\sum_{i=1}^{N} p_{ij} = 1 \ (\forall j,\ 1 \leq j \leq N),\ \sum_{j=1}^{N} p_{ij} = 1 \ (\forall i,\ 1 \leq i \leq N). \qquad (1)$$

We denote the number of neighbors of each point as K, and the set of K neighboring points of point i as N_i, where $|N_i| = K$. Note that N_i excludes point i itself.

4.2 Optimal Transport for Recovering the Vector Field

We consider that the locations of neighboring points in N_i represent the possible coordinates of i-th point in the near past or future, and that each point tends to move towards its surrounding neighbors after one time step based on Assumption 1. In other words, if the probability p_{ij}, which indicates the movement possibility from point i to point j, is larger than that of the other neighbors $p_{ij'}$ ($\forall j' \in N_i \backslash j$), the predicted trajectory vector $\boldsymbol{f}_i$ should be more similar to $\boldsymbol{v}_{ij}$ compared to $\boldsymbol{v}_{ij'}$ ($j' \in N_i \backslash j$).

However, because transition probabilities P are not available in a single snapshot, we employ an optimal transport problem [5] to consider both the latent probability p_{ij} and similarity between $\boldsymbol{f}_i$ and $\boldsymbol{v}_{ij}$. Optimal transport is a problem to find a transportation plan P that minimizes the cost of moving objects from one point set to another point set. We define an optimal transport for recovering the vector field as follows:

$$\sum_{i=1}^{N} \sum_{j \in N_i} p_{ij} \|\boldsymbol{f}_i - \boldsymbol{v}_{ij}\|_2^2 + \lambda_1 \sum_{i=1}^{N} \|\boldsymbol{f}_i\|_2^2, \tag{2}$$

where $\|\boldsymbol{f}_i\|_2^2$ is an ℓ_2 regularization term and $\lambda_1 \geq 0$ is its weight. We can estimate the latent transition probability P by solving this. Once we estimate P by solving the optimal transport problem, this equation can be regarded as a sum of the ℓ_2 loss function for $\boldsymbol{f}_i$ weighted by the transition probabilities p_{ij}.

Therefore, we employ an unbalanced optimal transport [4] that relaxes a doubly stochastic condition Eq. (1). Now, the problem is written as

$$\sum_{i=1}^{N} \sum_{j \in N_i} p_{ij} \|\boldsymbol{f}_i - \boldsymbol{v}_{ij}\|_2^2 + \tau D(P\mathbf{1} \| \mathbf{1}) + \tau D(P^T \mathbf{1} \| \mathbf{1}) + \lambda_1 \sum_{i=1}^{N} \|\boldsymbol{f}_i\|_2^2, \tag{3}$$

where $D(\boldsymbol{a} \| \boldsymbol{b})$ is a KL divergence between vector $\boldsymbol{a}$ and $\boldsymbol{b}$ with the hyper parameter τ.

4.3 Vector and Acceleration Smoothing

We assume that the vector field is smooth in Assumption 2. Intuitively, we suppose that the trajectory vector of each point has similar values to those of its neighbors. Here, we formulate this assumption as a constraint between trajectory vectors $\boldsymbol{f}_i$ ($i = 1, \ldots, N$) predicted by the model f. We extend the idea of trend filtering [11,17] to recover the velocity vector field from a point cloud.

We introduce a constraint imposing smoothness of the velocity vectors in terms of acceleration and formulate it as a sum of differences between each point i and its neighboring points in N_i:

$$\sum_{i=1}^{N} \sum_{j \in N_i} \|\boldsymbol{f}_i - \boldsymbol{f}_j\|_2^2. \tag{4}$$

Minimizing this term ensures that the magnitudes and orientations of the trajectory vectors of point i and that of its neighboring point j are close to each other.

The closer $\boldsymbol{f}_i$ and $\boldsymbol{f}_j$ are in terms of magnitude and orientation, the smaller is the value of $\|\boldsymbol{f}_i - \boldsymbol{f}_j\|_2^2$, which ensures that the velocity function is smooth. Because trajectory vector $\boldsymbol{f}$ represents how each point moves in one time step, it can be interpreted as the velocity of each point at the time when the point cloud snapshot is taken. Therefore, this regularization term is henceforth referred to as a "velocity smoothing term".

We have just introduced velocity smoothing to enforce the smoothness of a vector field. Pushing this idea further, we impose a higher order of smoothness on a vector field, namely, acceleration smoothing. Just as velocity smoothness is approximated by the norm of the difference between two adjacent velocity vectors, the acceleration smoothness is approximated by the norm of the "difference in differences" of three adjacent velocity vectors as:

$$\sum_{i=1}^{N} \sum_{j \in N_i} \sum_{k \in N_j} \|(\boldsymbol{f}_i - \boldsymbol{f}_j) - (\boldsymbol{f}_j + \boldsymbol{f}_k)\|_2^2 = \sum_{i=1}^{N} \sum_{j \in N_i} \sum_{k \in N_j} \|\boldsymbol{f}_i - 2\boldsymbol{f}_j + \boldsymbol{f}_k\|_2^2. \quad (5)$$

A vector $\boldsymbol{f}_i$ represents the velocity at the time the point cloud snapshot was taken; therefore, the difference between the two consecutive vectors can be interpreted as acceleration. Similarly, the difference of differences of vectors can be regarded as jerk. Therefore, this regularization term is henceforth referred to as an "acceleration smoothing term".

4.4 Objective Function and Algorithm

Our overall objective function for our proposed method is defined using Eq. (3), Eq. (4), and Eq. (5). We define a loss for SNVF as follows:

$$\begin{aligned} &\sum_{i=1}^{N} \sum_{j \in N_i} p_{ij} \|\boldsymbol{f}_i - \boldsymbol{v}_{ij}\|_2^2 + \tau D(P\mathbf{1}\|\mathbf{1}) + \tau D(P^T\mathbf{1}\|\mathbf{1}) + \lambda_1 \sum_{i=1}^{N} \|\boldsymbol{f}_i\|_2^2 \\ &+ \lambda_2 \sum_{i=1}^{N} \sum_{j \in N_i} \|\boldsymbol{f}_i - \boldsymbol{f}_j\|_2^2 + \lambda_3 \sum_{i=1}^{N} \sum_{j \in N_i} \sum_{k \in N_j} \|\boldsymbol{f}_i - 2\boldsymbol{f}_j + \boldsymbol{f}_k\|_2^2, \end{aligned} \quad (6)$$

where $\lambda_2, \lambda_3 > 0$ are regularization coefficients. Note that, we estimate the vector field by training a neural vector field f_θ whose parameter is denoted as θ.

Because the objective function Eq. (6) is non-convex, no straightforward way to minimize this function exists. To address this problem, we use an alternating optimization method that estimates the transition probability matrix P and parameter θ of neural vector field f. Because P is a probability matrix that satisfies a doubly stochastic condition expressed as Eq. (1), we can minimize the objective function expressed in Eq. (6) with respect to P by solving an optimal transport problem. Therefore, once θ is fixed, we solve the unbalanced optimal transport problem [4] using the Sinkhorn algorithm [5]. On the other hand, the objective function Eq. (6) is a quadratic function with respect to $\{\boldsymbol{f}_i\}_i$. Once P is fixed, gradient-based optimization procedures, such as stochastic gradient descent, are used to estimate the parameter θ of the neural vector field f. The whole algorithm of alternating optimization is described in Algorithm 1.

Algorithm 1. The algorithm of alternating optimization

1: $\boldsymbol{f} \leftarrow \mathbf{0}$
2: cost_matrix $\leftarrow \infty$
3: prob_matrix $\leftarrow \boldsymbol{O}$
4: **for** $i = 0, \ldots, N-1$ **do** ▷ N is the number of points
5: **for all** $j \in N_i$ **do** ▷ if point j is included in N_i
6: prob_matrix$[i][j] \leftarrow 1.0/K$ ▷ K is the number of neighborhoods
7: **end for**
8: **end for**
9:
10: **for** $k = 1, 2, \ldots$ **do**
11: $\boldsymbol{f} \leftarrow$ calculate_optimal_vectors(prob_matrix)
12: ▷ solve optimization problem with respect to $\boldsymbol{f}$
13: **for** $i = 0, \ldots, N-1$ **do**
14: **for all** $j \in N_i$ **do**
15: cost_matrix$[i][j] \leftarrow \|\boldsymbol{f}_i - \boldsymbol{v}_{ij}\|_2^2$
16: ▷ update cost matrix used in unbalanced optimal transport problem
17: **end for**
18: **end for**
19: prob_matrix $\leftarrow$ calculate_optimal_transport(cost_matrix)
20: ▷ solve unbalanced optimal transport problem with respect to P
21: **end for**

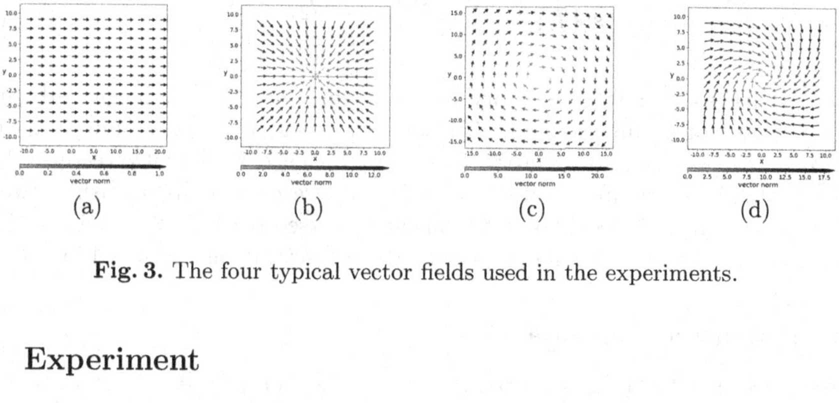

(a) (b) (c) (d)

Fig. 3. The four typical vector fields used in the experiments.

5 Experiment

5.1 Datasets

To verify the accuracy of the trajectory vectors estimated by the proposed method, we generated point cloud datasets using the following four typical vector fields: a uniform vector field $\boldsymbol{V}(x, y) = (1, 0)$, an irrotational vector field $\boldsymbol{V}(x, y) = (-x, -y)$, an incompressible vector field $\boldsymbol{V}(x, y) = (y, -x)$, and a whirlpool vector field $\boldsymbol{V}(x, y) = (-x + y, -x - y)$. Figure 3 depicts the four vector fields. Following Aris [1], we refer to the four point cloud datasets as uniform, irrotational, incompressible, and whirlpool, respectively.

We used following procedures to generate datasets. Let N_0 denote the number of initial points and T denote the number of iterations. Starting from the initial point $\boldsymbol{x}^{(0)} \in \mathbb{R}^D$, which is uniformly randomly chosen, the subsequent points $\boldsymbol{x}^{(t)}$

are obtained by $\boldsymbol{x}^{(t)} = \boldsymbol{x}^{(t-1)} + \boldsymbol{V}(\boldsymbol{x}^{(t-1)})\Delta t$ $(t = 1, \ldots, T-1)$. Consequently, all subsequent points are added to the dataset. This procedure is repeated N_0 times, which results in $N = N_0 T$ points in the entire point cloud.

5.2 Comparison Methods

In our experiments, we compared six different estimation methods. The first baseline method estimates a trajectory vector by simply averaging the position vectors over the K nearby points, which is expressed as $\boldsymbol{f}_i = \frac{1}{K} \sum_{j \in N_i} \boldsymbol{v}_{ij}$. We employ another baseline that considers only the optimal transport problem expressed in Eq. (3). We consider two variants: $\lambda_1 = 1$ (with L2 regularization) and $\lambda_1 = 0$ (with no regularization). In addition, double smoothing is employed, which further includes two regularization terms: the velocity smoothing term in Eq. (4) and acceleration smoothing term in Eq. (5).

5.3 Evaluation Metrics

The accuracy of the estimated trajectory vectors was measured based on the score s, which is defined as follows:

$$s(\{\boldsymbol{f}_i\}_i) = \left| \frac{1}{N} \sum_{i=1}^{N} \frac{\langle \boldsymbol{f}_i, \boldsymbol{f}'_i \rangle}{\|\boldsymbol{f}_i\| \cdot \|\boldsymbol{f}'_i\|} \right|, \tag{7}$$

where $\boldsymbol{f}_i$ is the estimated trajectory vector at point i, and $\boldsymbol{f}'_i$ is the ground truth vector. The scoring function Eq. (7) is defined as the averaged cosine similarity between the estimated trajectory vectors and the ground truth vectors. The reason why we calculate absolute value is that we cannot distinguish between solutions in which all vector orientations are reversed because only a single-time point cloud snapshot is available in our problem setting, as discussed in Sect. 3.

5.4 Parameter Configuration

In our experiments, the ground truth vector fields $\boldsymbol{V}(x, y) \in \mathbb{R}^2 \rightarrow \mathbb{R}^2$ are simple linear functions of x and y as described in Sect. 5.1, we employed a three-layer perceptron model whose hidden dimension is 100 as a regression model $f : \mathbb{R}^2 \rightarrow \mathbb{R}^2$. This regression model serves as an approximation of the ground truth vector field $\boldsymbol{V}(x, y) \in \mathbb{R}^2 \rightarrow \mathbb{R}^2$. We employed ReLU function as an activation function of our model. We implemented our proposed model with PyTorch and trained parameters with stochastic gradient descent. In the objective function described in Sect. 5.2, the regularization coefficients were set as follows: $\lambda_1 = \lambda_2 = \lambda_3 = 1$ and $\mu_1 = \mu_2 = 1$. Also, we set the number of neighbors K to 5. Although there is room for tuning these hyperparameters, we set arbitrary parameters as above to avoid overfitting to the point cloud data used in the experiments.

In the data generation procedure described in Sect. 5.1, the total number of points N was set to $N = 1,000$. The initial number of points N_0 was set to

Table 2. The accuracy of six different methods in the evaluation metric in Eq. (7).

methods	vector field types			
	uniform	irrotational	incompressible	whirlpool
average of position vectors to neighboring points	0.014	0.242	0.088	0.165
L2 regularization ($\lambda = 0$)	0.019	0.264	0.084	0.143
L2 regularization ($\lambda = 1$)	0.019	0.264	0.084	0.143
velocity smoothing	0.057	0.567	0.311	0.480
accleration smoothing	0.123	0.908	0.390	0.682
double smoothing	**0.133**	**0.920**	**0.397**	**0.695**

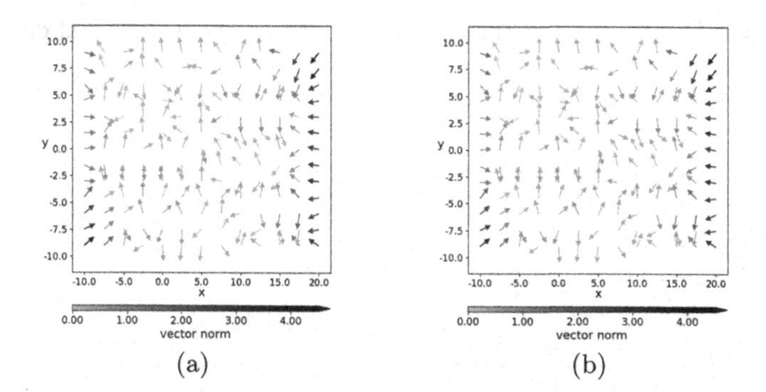

Fig. 4. Visualization of the results obtained on the uniform data.

$N_0 = 100$ and the number of steps T was set to $T = 10$. The (x, y) coordinates of the initial point were defined within the range $-10 \leq x, y \leq 10$. The time step parameter Δt was set to $\Delta t = 1.0$ for the uniform vector field and $\Delta t = 0.1$ for the others.

5.5 Result

We compared six different methods described in Sect. 5.2 on the four types of vector field data explained in Sect. 5.1. The evaluation was based on the evaluation metric expressed in Eq. (7) in Sect. 5.3.

Table 2 summarizes the accuracy of the trajectory vectors predicted in the alternating optimization process with the Algorithm 1. Figures 4, 5, 6, and 7 visualize the vector fields estimated by SNVF for uniform, irrotational, incompressible, and whirlpool, respectively. The magnitudes of the trajectory vectors are represented by a grey scale. Each figure is composed of two parts. In each figure, the left part (a) shows the result when predicting velocity vectors by averaging the position vectors over the neighboring points. The right part (b) shows the result when SNVF is employed in the prediction procedures.

Out of the six methods compared in Table 2, the double smoothing method expressed in Eq. (6) achieved the best prediction performance on all types of

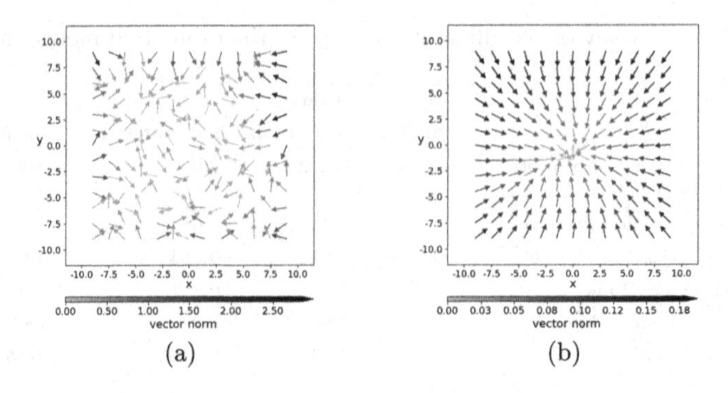

Fig. 5. Visualization of the results obtained on the irrotational data.

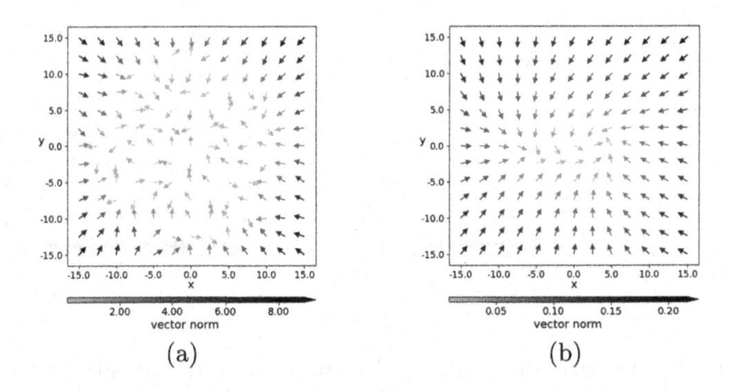

Fig. 6. Visualization of the results obtained on the incompressible data.

data. Both the velocity and acceleration smoothing terms significantly contribute to the performance improvements. The remarkable improvement in the evaluation metric is ascribed to the regularization, which facilitates the alignment of the trajectory vector $\boldsymbol{f}$ among adjacent points in terms of the vectors' magnitude and orientation. Remarkably, the estimated vector fields of the irrotational and whirlpool data using SNVF (see Figs. 5 and 7) demonstrate similar norms and directions of the trajectory vectors in the neighborhood. This qualitative confirmation further strengthens the proximity of the estimation results to the accurate vector field.

As reported in Table 2, the uniform data exhibits significantly lower evaluation values across all methods compared to the other data types. This disparity arises from the estimated trajectory vectors for the uniform data, which comprise both positive and negative vectors along the x-axis. Conversely, the correct vectors exhibit solely positive orientation along the x-axis, as illustrated in Fig. 3(a). In the evaluation metrics described in Eq. (7), the absolute value is calculated after averaging the cosine similarity at each point. Consequently, the uniform data are observed to yield significantly lower evaluation values than those obtained with other data types.

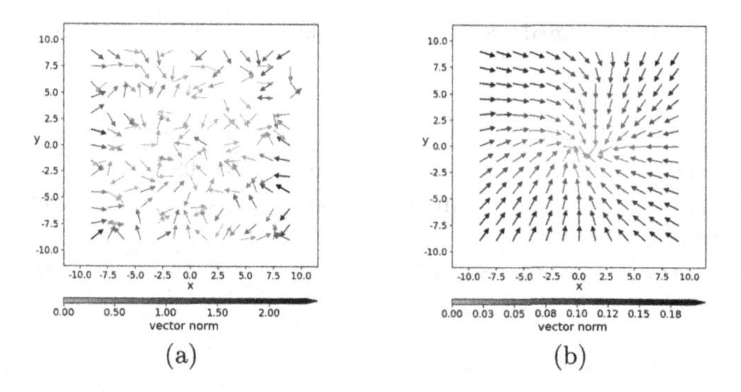

Fig. 7. Visualization of the results obtained on the whirlpool data.

5.6 Ablation Study

We conducted an ablation study to examine the contribution of each term in the objective function Eq. (6). We used RMSE as the evaluation metric:

$$\mathrm{RMSE}(\{\boldsymbol{f}_i\}_i) = \sqrt{\frac{1}{N}\sum_{i=1}^{N}\|\boldsymbol{f}_i - \boldsymbol{f}'_i\|_2^2}.$$

RMSE considers both the magnitude and direction of the estimated velocity vectors; therefore it serves as a comprehensive evaluation metric.

We showed the results in Table 3 and observed that velocity and acceleration regularization terms have different impacts on the evaluation metrics. For instance, in the case of irrotational data, the velocity regularization term works effectively. This is because the correct vectors exhibit movements oriented along a specific direction. The velocity regularization term plays an important role in incorporating the magnitude difference in neighboring vectors. Therefore, the velocity regularization term handles the difference of magnitude in the overall objective function and significantly improves the evaluation metrics. As for the case of incompressible data or whirlpool data, the correct vectors involve rotational movements around the origin. Therefore, the acceleration regularization term plays an important role in incorporating the direction changes of velocity vectors among the neighboring points; thereby, in the overall optimization, contributing to improved estimation accuracy.

5.7 Discussion and Limitation

This study deals with the problem of estimating the underlying vector field from a single point cloud snapshot and obtaining the trajectory vectors at individual points. Because the trajectory of each point is estimated from a single-time point cloud snapshot, it is not possible to determine whether each point moves in the direction of the correct vector or in the exact opposite direction.

Table 3. Results of ablation study.

Objective Function	Vector Field Type			
	uniform	irrotational	incompressible	whirlpool
Loss function	1.0152	0.4719	0.8608	0.1047
Loss function + L2 regularization ($\lambda = 1$)	1.0119	0.5299	0.8491	0.0711
Loss function + velocity smoothing	1.0207	0.4459	0.8650	0.0805
Loss function + accleration smoothing	0.9995	0.4785	0.8464	0.0610
Loss function + velocity smoothing + accleration smoothing	1.0144	0.4797	0.8459	0.0520

There remains a dataset expansion in future work. In the problem setting of this study, we assumed that the trajectory vector of each point is similar to the trajectory vectors of its neighbors and varies smoothly. In the experiment, point cloud data was generated based on this assumption; therefore the point cloud data used in the experiment did not include more complicated cases such as intersecting point sequences generated from two initial points. In addition, it is necessary to verify whether the proposed method is robust to noisy data considering its application to real-world data.

6 Conclusion

This research proposes a novel method to recover population dynamics from a single-time point cloud snapshot, addressing the cases where tracking along time series is not possible due to observational constraints or privacy issues. Our approach pseudo-tracks each point's movement by assuming that its neighbors represent past and future coordinates and that the underlying vector field is smooth.

In this study, the problem of estimating the underlying vector field is formulated as an optimization problem in which the objective function incorporates an optimal transport problem and introduces the smoothness of the velocity and acceleration of trajectories as regularization terms.

In the experiments, point cloud data were artificially generated from four typical vector fields, and the trajectories of each point estimated by the proposed method were evaluated by comparing them with the actual vectors in the original vector fields. The experiments showed that the velocity and acceleration smoothing terms enable estimation that considers the dynamics of the neighborhood; accordingly, it became possible to estimate the correct vector fields with high accuracy from a single point cloud snapshot.

Acknowledgment. This study is partially supported by JSPS KAKENHI JP21H-05299 and 20H04244.

References

1. Aris, R.: Vectors, Tensors and the Basic Equations of Fluid Mechanics (1989)
2. Bunne, C., Papaxanthos, L., Krause, A., Cuturi, M.: Proximal optimal transport modeling of population dynamics. In: AISTATS (2022)
3. Campbell, K.R., Yau, C.: Uncovering pseudotemporal trajectories with covariates from single cell and bulk expression data, **9** (2018)
4. Chizat, L., Peyré, G., Schmitzer, B., Vialard, F.-X.: Scaling algorithms for unbalanced optimal transport problems. Math. Comput. **87** (2018)
5. Cuturi, M.: Sinkhorn distances: lightspeed computation of optimal transport. In: Advances in Neural Information Processing Systems, vol. 26 (2013)
6. de Vos, B.D., et al.: End-to-end unsupervised deformable image registration with a convolutional neural network. In: Deep Learning in Medical Image Analysis and Multimodal Learning for Clinical Decision Support (2017)
7. Fisher, M., Nocedal, J., Trémolet, Y., Stephen, J.W.: Data assimilation in weather forecasting: a case study in PDE-constrained optimization. Optim. Eng. **10** (2009)
8. Han, J., Wang, C.: TSR-VFD: generating temporal super-resolution for unsteady vector field data. Comput. Graph. **103** (2022)
9. Huang, X., et al.: Grab-posisi: an extensive real-life GPS trajectory dataset in southeast Asia. In: SIGSPATIAL (2019)
10. Huang, Y., Weng, Y., Yu, S., Chen, X.: Diffusion convolutional recurrent neural network with rank influence learning for traffic forecasting. In: TrustCom/BigDataSE (2019)
11. Kim, S.-J., Koh, K., Boyd, S., Gorinevsky, D.: l_1 trend filtering. SIAM Rev. **51** (2009)
12. Kuroe, Y., Mitsui, M., Kawakami, H., Mori, T.: A learning method for vector field approximation by neural networks. In: IEEE International Joint Conference on Neural Networks Proceedings. IEEE World Congress on Computational Intelligence (1998)
13. Luo, W., Xing, J., Milan, A., Zhang, X., Liu, W., Kim, T.-K.: Multiple object tracking: a literature review. Artif. Intell. (2020)
14. Maeda, T., et al.: Aerial drone observations identified a multilevel society in feral horses. Sci. Rep. **11** (2021)
15. Saelens, W., Cannoodt, R., Todorov, H., Saeys, Y.: A comparison of single-cell trajectory inference methods. Nat. Biotechnol. **37** (2019)
16. Schiebinger, G., et al.: Optimal-transport analysis of single-cell gene expression identifies developmental trajectories in reprogramming. Cell **176** (2019)
17. Wang, Y.-X., Sharpnack, J., Smola, A.J., Tibshirani, R.J.: Trend filtering on graphs. J. Mach. Learn. Res. **17** (2016)

SAWTab: Smoothed Adaptive Weighting for Tabular Data in Semi-supervised Learning

Morteza Mohammady Gharasuie[1], Fengjiao Wang[2(✉)], Omar Sharif[1], and Ravi Mukkamala[1]

[1] Old Dominion University, Norfolk, VA 23529, USA
mmoha014@odu.edu, osharif@odu.edu, mukka@cs.odu.edu
[2] University of Utah, Salt Lake City, UT 84112, USA
u6053554@utah.edu

Abstract. Self-supervised and Semi-supervised learning (SSL) on tabular data is an understudied topic. Despite some attempts, there are two major challenges: 1. Imbalanced nature in the tabular dataset; 2. The one-hot encoding used in these methods becomes less efficient for high-cardinality categorical features. To cope with the challenges, we propose SAWTab which uses a target encoding method, Conditional Probability Representation (CPR), for efficient representation in the input space of categorical features. We improve this representation by incorporating the unlabeled samples through pseudo-labels. Furthermore, we propose a Smooth Adaptive Weighting mechanism in the target encoding to mitigate the issue of noisy and biased pseudo-labels. Experimental results on various datasets and comparisons with existing frameworks show that SAWTab yields best test accuracy on all datasets. We find that pseudo-labels can help improve the input space representation in the SSL setting, which enhances the generalization of the learning algorithm.

Keywords: Semi-supervised learning · Feature representation · Pseudo-label · Tabular domain · adaptive weighting

1 Introduction

The progress of **S**emi-**S**upervised **L**earning (SSL) in the tabular domain is limited compared to the image and language domains. Usually, tabular data consists of diverse features, discrete categorical and continuous numerical data, each characterized by its unique data type and distribution; causing the learning from such data a challenging task.

Existing approaches [1,2] heavily rely on one-hot encoding, which is not a best encoding for big datasets with high-cardinality categorical features. The one-hot encoding of high-cardinality categorical features introduces challenges such as the curse of dimensionality, sparse representation, and substantial resource requirements. Common solutions such as hashing technique [3] and denser encodings like target encoding attempt to mitigate these issues. The target encoding is

D.-N. Yang et al. (Eds.): PAKDD 2024, LNAI 14647, pp. 316–328, 2024.
https://doi.org/10.1007/978-981-97-2259-4_24

not thoroughly studied in semi-supervised learning (SSL) settings, as it relies on target labels unavailable for the unlabeled set. In this paper, we explore a variation of target encoding, Conditional Probability Representation (CPR), for SSL setting and propose enhancing CPR with pseudo-labels from unlabeled samples.

On the other side, existing SSL frameworks for tabular data overlook the challenge of class imbalance, introducing bias toward majority classes. When the model is trained on imbalanced labeled set, it overpredict the majority classes and underpredict the minory classes. Subsequent training intensifies this bias. While methods like pseudo-label alignment [4] and resampling techniques [5] address this issue, our proposed approach takes a unique perspective. We aim to learn unbiased and robust feature representation in the input space, offering a novel solution for class-imbalance challenges in tabular data.

We propose SAWTab technique to address the challenge of using target encoding (CPR) for imbalanced tabular datasets in SSL settings. The proposed technique progressively updates CPR for categorical data using both labels and pseudo-labels. We present a novel weighting scheme for updating CPR with two objectives: 1) mitigate class bias introduced by pseudo-labels; 2) distinguish the trustworthiness of labels and pseudo-labels. The proposed technique initially trains the model on a small labeled set. Then, it generates pseudo-labels on unlabeled set and updates CPR using both labeled and unlabeled sets, which improves the model's generalization and performance.

The main contribution of this paper can be summarized as follows:

- We present *SAWTab* that enhances categorical feature representation by utilizing pseudo-labels, smoothed adaptive weighting, and progressive feature upgrading.
- The proposed technique is integrated into an existing semi-supervised learning algorithms, called *VIME* [1] and enhance its performance.
- The proposed technique shows improvement in comprehensive experiments with datasets containing high-cardinality categorical features. It outperforms existing frameworks on three tabular datasets.

2 Related Works

2.1 Semi-supervised Learning on Tabular Data

Real-world data distributions are often imbalanced and more challenging in SSL settings. There are some efforts in the image domain to solve the issues related to imbalance datasets from angle of debiasing the pseudo-labels [6], resampling [7], and adaptive thresholding [8].

Despite imbalance nature of tabular datasets, current SSL frameworks on tabular data have not paid attention to imbalance issue. *VIME* [1] introduces a novel tabular data augmentation scheme, pretext task, and mask vector estimation to improve the representation learning of the autoencoder on several tabular datasets in SSL setting. SubTab [2] learns the representation through an autoencoder that gets multiple subset of features as input and reconstructs the whole input features. Contrastive learning is used in SCARF [9] that learns the representation from corrupted random features of input using contrastive learning. Moreover, Contrastive Mixup [10] trains an autoencoder using reconstructing input from mixed-up latent representation of input samples that have the same label or pseudo-label. Finally, the pre-trained encoders are used in a self-supervised or semi-supervised setting on the downstream task. This paper proposes an approach to learn robust representation for tabular data while mitigating the bias induced from imbalanced nature of the data.

2.2 Representation of Categorical Data

Representing tabular data for machine learning models can be challenging because the results are highly dependent on the quality of the data and preprocessing steps. One-hot and target-encodings are common preprocessing approaches to handle non-numerical data. One-hot is unsuitable for high-cardinality categorical features in large datasets due to high-dimensional sparse representation and computational problems [11]. In Contrast, there are some works that try to tackle the problem of high-cardinality features. Cerdar P. Varoquax G. [11] propose min-hash encoding and Gamma Poisson factorization. They also propose similarity encoding [12] to encode dirty, non-curated categorical data. Also, Slakey A. et al. [13] proposed a CBM encoding approach to represent the categorical features in low dimensions.

3 Methodology

We propose *SAWTab* to apply CPR as a form of target encoding for categorical data in a SSL setting on tabular data. It improves input representation by updating the Conditional Probability Representation (CPR) for categorical data, incorporating both labels and pseudo-labels from labeled and unlabeled sets. A weighting scheme distinguishes label and pseudo-label trustworthiness, improving generalization and model performance by incorporating more data and features from the unlabeled set.

In the following, we introduce notations, the definition of conditional probability representation, and the progressive CPR Upgrade with smooth adaptive weighting in *SAWTab*. *VIME* serves as the base model for demonstration and validation, and the framework is adaptable to other base architectures.

3.1 Preliminaries

To present the proposed technique, we first formally define the semi-supervised learning problem. Consider a dataset $D = (X, Y)$ with N samples. In this dataset, there is a small subset of labeled samples $D^L = \{(X_n^L, Y_n)\}_{n=1}^{N_L}$ and a large subset of the unlabeled samples $D^U = \{(X_n^U)\}_{n=1}^{N_U}$, where $N = N_U + N_L$. We consider the setting where $N_U >> N_L$. In the labeled dataset D^L, for pair (X_n^L, Y_n^L), X_n^L denotes the features for n^{th} data sample, while Y_n^L denotes the class label for n^{th} sample, and $Y_n \in \{0, 1, ..., C\}$. Furthermore, for the feature vector X_n^L we denote that $X_n^L = \{X_{n,0}^L, X_{n,1}^L, ..., X_{n,M}^L\}$, where M indicates the number of features in the dataset. We denote m^{th} feature column as $X_{\cdot,m}^T$, assume feature column $X_{\cdot,m}^T$ is a categorical feature column with cardinality K_m. Finally, $X_{n,m}$ indicates the categorical value for n^{th} sample and m^{th} feature.

3.2 Smoothed Adaptive Weighting

Smoothed Adaptive Weighting (*SAW*) addresses predictive biases [14] in models trained on imbalanced datasets. *SAW* introduces smoothed adaptive weights to the loss function, which improves the model performance by assigning higher weights (w_k) to less frequent classes (minority) and lower weights to more frequent classes (majority). This encourages the model to prioritize challenging classes, with the weight w_k inversely proportional to the number of samples for each class, i.e.,

$$w_k \propto \frac{1}{E_k}, E_k = (\frac{1 - \beta^{n_k}}{1 - \beta}), \beta = \frac{N - 1}{N}, N = \frac{\sum n_k}{C} \tag{1}$$

where C is number of classes in the dataset and n_k is number of predicted samples as class $k \in \{1, 2, ..., C\}$. If $n_k = 0$, n_k is adjusted by $n_k = max(n_k, 1)$. β determines the degree of flattening (smoothness). When $\beta = 0$, the weights are uniform. As β approaches 1, E_k converges to n_k. $E_k \to n_k$ corresponds to the weights by strictly inverse class frequency.

3.3 Conditional Probability Representation with Weighting Schema

Conditional probability representation (CPR) is a form of target encoding used to encodes categorical values within tabular data. It is a more compact representation than one-hot encoding, and outperforms one-hot encoding in some datasets [15,16]. Generally, different encoding methods have varying performance across datasets. Given the notation in the previous section, now we can define the CPR representation as following:

$$X_{n,m} \equiv [\frac{N_{n,m,1}}{N_{n,m}}, \frac{N_{n,m,2}}{N_{n,m}}, ..., \frac{N_{n,m,C}}{N_{n,m}}] \tag{2}$$

where $X_{n,m}$ is the categorical value belongs to m^{th} feature in n^{th} sample, and the vector is the corresponding CPR representation. This vector has the same number of dimensions as the number of classes. In the equation, $N_{n,m}$ is the number of observations of categorical value $X_{n,m}$ in $X_{\cdot,m}^T$, while $N_{n,m,c}$ is among those observations, how many times the corresponding label is same with c.

We propose to utilize SAW for upgrading the CPR representation rather than weighting the loss. The proposed weighting schema increases the frequency of minority classes (under-predicted) and reduces the frequency of majority classes (over-predicted) in pseudo-labels. Therefore, incorporating the weights into the

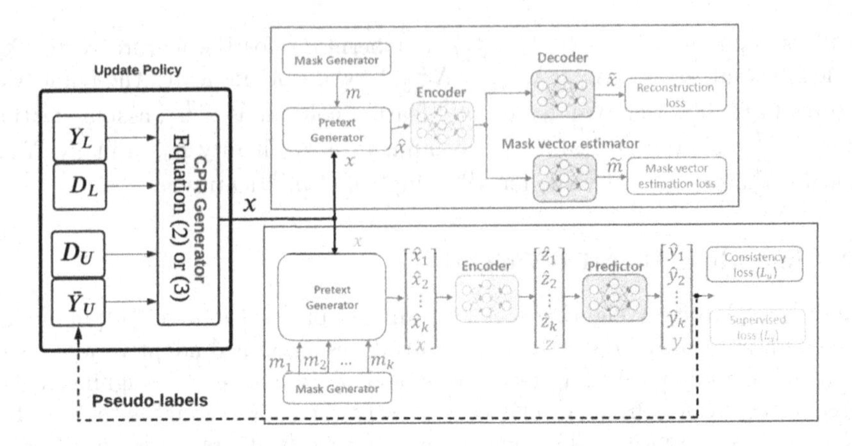

Fig. 1. *VIME* Framework with *SAWTab* Additions: Green and blue boxes represent the training autoencoder and predictor, respectively, while the non-shadowed area illustrates the newly proposed *SAWTab* components. (Color figure online)

CPR representation mitigates the effect of the bias. To achieve this, we introduce equation (3) that has two types of weights: class-wise weights (w_c^u) for unlabeled data, which is measured by *SAW*, and a fixed weight (w_l) for labeled data. The w_c^u mitigate the effect of noisy pseudo-labels and the w_l on labeled data adjust the effectiveness of the labels versus pseudo-labels. Since pseudo-labels are noisy, and the weights can not entirely eliminate the impact of incorrect pseudo-labels, the w_l is always larger than w_c^u to put more trust on the labeled data.

$$X_{n,m} \equiv [\frac{w^l N^l_{n,m,1} + w^u_1 N^u_{n,m,1}}{w^l N^l_{n,m} + \sum_{k=1}^{C} w^u_k N^u_{n,m,k}}, \ldots, \frac{w^l N^l_{n,m,C} + w^u_C N^u_{n,m,C}}{w^l N^l_{n,m} + \sum_{k=1}^{C} w^u_k N^u_{n,m,k}}] \quad (3)$$

where $X_{n,m}$ is defined similarly to Eq. (2). $N^l_{m,n,c}$ and $N^u_{m,n,c}$ are similar to $N_{m,n,c}$ in Eq. (2) but l and u show the number of observations in labeled and unlabeled sets with the corresponding labels or pseudo-labels. The weight of categorical values with a pseudo-label of c is represented by w_c^u. The categorical values with labels have a fixed weight, regardless of the class, represented by w_l.

3.4 Progressive Feature Upgrading and SAWTab

We assert that updating representations with pseudo-labels can enhance learning. *VIME*, a self/semi-supervised framework designed for tabular data, is employed to generate these pseudo-labels. As shown in Fig. 1, it learns the representation by optimizing an autoencoder that reconstructs a binary mask and original input from corrupted version of input. The pre-trained encoder of the autoencoder and pre-text generator is then applied in self-supervised or semi-supervised settings for downstream tasks, utilizing both supervised and unsupervised (consistency) losses. *SAWTab* incorporates the pseudo-labels generated by *VIME* for unlabeled samples to update and enhance CPR representation in equation (3) after measuring weights based on pseudo-labels. It iteratively refines the CPR by training the autoencoder and predictor, generating pseudo-labels, upgrading CPR and restarting the training process. Algorithm 1 provides the pseudo-code for *SAWTab*.

Algorithm 1. Pseudo-code for SAWTab technique

$W_L \leftarrow$ fixed_number
CPR $\leftarrow$ generate_CPR_representation(W_L, labels)
for $k \leftarrow 1$ to N **do**
  autoencoder $\leftarrow$ train_autoencoder(CPR)
  predictor $\leftarrow$ train_predictor(autoencoder)
  pseudo_labels $\leftarrow$ generate_pseudo_labels(predictor)
  $W_U \leftarrow$ measure_weights(pseudo_labels)
  CPR $\leftarrow$ generate_CPR_representation(W_L, W_U, labels, pseudo_labels)
end for

Table 1. Dataset summary, highlights feature and value statistics in each dataset, with the majority being categorical. The max cardinality column displays the feature with the highest number of distinct values (maximum cardinality). The last column displays the distribution of classes in datasets. #CC, #NC stands for number of categorical and numerical columns respectively.

Dataset	#CC	#NC	#Samples	Max cardinality	Class Distribution
Avazu	26	0	1,000,000	313,002	[839781, 160219]
Traffic Violations	14	12	1,578,154	163,365	[745051, 254949]
Display Advertising Challenge	26	11	1,000,000	321,439	[700429, 75533, 887, 78197]

4 Experiments and Evaluation

We assess *SAWTab*'s effectiveness through experiments comparing it to other models, using CPR in both Self-supervised and Semi-supervised contexts on large datasets. The goal is to highlight capabilities and limitations of *SAWTab*, including its potential for applications, and to demonstrate the impact of smoothed adaptive weights in CPR versus unsupervised loss.

4.1 Tabular Dataset

We carried out experiments using three different datasets (Traffic Violations [17], Avazu [18], and Display Advertising Challenge [19]). These datasets comprise high-cardinality categorical features, and are also imbalanced. More information about datasets are mentioned in Table 1.

We used all samples in the Traffic Violations dataset and a subset of one million samples from both Avazu and Display Advertising Challenge datasets, which contain high-cardinality categorical features. The Display Advertising dataset has 11 numerical and 26 categorical features, whereas the Traffic Violations dataset includes a mix of DateTime, numerical, boolean, and categorical features, with Avazu being entirely categorical. We discarded DateTime features from Traffic Violations, normalized numerical features with scikit-learn's standard scaler, and transformed categorical features into numerical ones using CPR representation.

Table 2. The prediction accuracy of *SAWTab*, *VIME*, and *Progressive VIME* in self-supervised methods

Method	Avazu	Display Advertising	Traffic Violations
Supervised	82.79% (± 0.5001)	70.44% (± 1.027)	77.932% (± 0.297)
VIME:Self-Supervised	78.25% (± 3.2843)	74.31% (± 0.458)	75.96% (± 2.794)
PVIME:Self+U	76.82% (± 1.7614)	73.166% (± 1.041)	78.89% (± 0.316)
PVIME:Self+U&R	81.462%(± 1.113)	74.438%(± 0.861)	78.77% (± 0.36)
Scarf-Self	72.68%(± 18.244)	73.136%(± 3.69)	72.80%(± 4.324)
SubTab-Self	83.93%(± 0.06503)	71.85 (± 0.4243)%)	79.80%(± 0.0716)
SAWTab-Self	**85.019%(± 0.00075)**	**74.88%(±0.042)**	**80.64%(± 0.2991)**

Table 3. The prediction accuracy of *SAWTab*, *VIME*, and *Progressive VIME* and *SAW* semi-supervised methods

Method	Avazu	Display Advertising	Traffic Violations
Supervised	82.79% (± 0.5001)	70.44% (± 1.027)	77.932% (± 0.297)
VIME:Semi-Supervised	84.10% (± 0.5618)	74.61% (± 0.414)	78.6% (± 0.47)
PVIME:Semi+U	80.92% (± 0.7146)	74.788% (± 0.368)	79.037% (± 0.283)
PVIME:Semi+U&R	84.34%(± 0.2401)	74.96%(± 0.226)	79.92%(± 0.504)
Contrastive Mixup	80.10%(± 4.879)	71.72%(± 1.376)	77.87%(± 0.703)
Scarf-Semi	72.30%(± 8.707)	74.62%(± 0.309)	68.39%(± 3.632)
SubTab-Semi	83.59%(± 0.1412)	69.77%(± 1.0878)	78.12%(± 0.1159)
SAW	84.61%(± 0.0511)	75.13%(± 0.0513)	79.29%(± 0.0030)
SAWTab-Semi	**85.0193%(± 0.00063)**	**75.54%(± 0.145)**	**80.90%(± 0.2076)**

4.2 Experiment Settings

The proposed *SAWTab* is evaluated on the mentioned datasets and compared to other frameworks [1,2,9,14,16]. Since transformer-based frameworks [20,21] utilize embedding layers to convert categorical data into numerical representations, and given that the categorical data is already available in a numerical format (CPR representation), we do not discuss and compare these frameworks in this paper. The proposed framework aims to assess performance with limited labeled data in self-supervised and semi-supervised settings. For the experiments, we allocated 80% of the data for training, with 10% of that labeled. The remaining 20% are used as the test set to assess model accuracy.

We compare *SAWTab* agains other self-supervised and semi-supervised frameworks (methods) on tabular data. The experiments underwent a two-phase training, starting with an autoencoder/encoder training on data by the corresponding method. The pre-trained encoder then is used to generate input for training the predictor on the downstream task through self-supervised or semi-supervised manners. The self-supervised manner uses pre-trained encoder's output on the labeled data to train the predictor, while semi-supervised manner employs the output of pre-trained encoder on the augmented versions of unlabeled data for consistency loss and original labeled data for supervised loss.

Some methods only report self-supervised performance. So, we add consistency loss and use the corresponding augmentation in their frameworks to train the predictor in a semi-supervised manner to have a correct comparison of semi-supervised learning of frameworks. For these frameworks, we maintain the same training protocol, hyperparameters, and consistent network architecture across tests. The hyperparamters for *VIME* and its variants including *SAWTab* is shown in Table 4.

In the following, we explain baseline methods outlined in Tables 2 and 3:

- **Supervised** method: trains a predictor (classifier) using original labeled samples without involving the pre-trained encoder.
- More explanation about **VIME**, **Progressive-VIME (PVIME)** and their variations including *Self-Superivsed*, *Semi-Supervised*, *Self-Supervised with update and refinement* (Self+U and Self+U&R), *Semi-Supervised with Update, and refinement* (Semi+U and semi+U&R) can be found in [16].
- **SAW** [14] method: We adapt *SAW* for use in *VIME:Semi–Supervised with update* to compare its performance in the tabular domain. We calculate class weights using all pseudo-labels. Then, the weights are integrated into the consistency loss during the training of the predictor.
- **Contrastive Mixup** [10], **Scarf-Self** [9], and **SubTab-Self** [2] methods are the original proposed frameworks without any change.
- **Scarf-Semi** and **SubTab-semi** methods are similar to *Scarf-Self* and *SubTab-Self*. They use the pre-trained encoder to train a predictor in semi-supervised manner, using augmentation and a consistency loss.

4.3 Results

Test Accuracy: The training results shown in Tables 2 and 3 indicate that the *SAWTab* approach outperforms other methods on all datasets. It demonstrates the crucial role that generating a better representation, by progressively updating

Table 4. Optimized hyper-parameters. (BS: Batch Size, act_fn: activation function, arch: netowrk architecture)

	Datasets → ↓ parameters	Avazu	Display Advertising	Traffic Violations
Supervised	arch	[44,100,100,2]	[65,100,100,2]	[104,100,100,4]
	act_fn	tanh	relu	tanh
	BS	100	64	128
Autoencoder	arch	[44,44,44]	[65,50,65]	[44,44,44]
	BS	256	256	256
	Epochs	15	20	15
	act_fn	relu	tanh	relu
	α	1.0	0.913	1.0
Self/Semi-Supervised	arch	[44,100,100,2]	[50,100,100,2]	[104,100,100,4]
	act_fn	tanh	relu	tanh
	β	0.75	0.349	1.0
Shared Parameters	p_k	0.4	0.7	0.3
	K	0.3	4.0	2.0
	w_l	2.9	6.23	2.0

the CPR and mitigating the bias, in the input space can play. *SCARF* underperforms the supervised method on the Avazu and Traffic Violations datasets, showing no improvement in self-supervised or semi-supervised learning using the augmentation in *SCARF* and its contrastive learning. However, it is not the case on the Display Advertising Challenge dataset. *SAW* only works in semi-supervised setting because it only works on the consistency loss. Table 3 shows that *SAW* outperforms other methods except *SAWTab*. Note, *SAW* is only investigated in *VIME: Semi-Supervised.* Both SAW and SAWTab imply that debiasing the input representation surpasses weighting the consistency loss.

By looking more at the tables, we can conclude that the augmentations proposed for tabular data exhibit varying performances across datasets and do not give rise to better generalization in all experiments including self-supervised and semi-supervised. Unlike other domains, it is not possible to combine these augmentations during model training. This is an indication of the need for more study on tabular data.

5 Ablation Study

Class-Wise Recall and Precision. Since datasets used for the experiments are all imbalanced datasets, we want to dive deep to investigate the performance (precision and recall) of various methods on both majority and minority classes. We used the traffic violations dataset, a severly imbalanced dataset with class distribution as follows: 45%, 4.8%, 0.05%, and 50.15% for class 0 to class 3 respectively. The experimental results in Fig. 2(a) and (b) shows that only *SAWTab* is able to predict the minority class (class 2) in the test set, indicating its superiority in handling imbalanced datasets. Both *SAWTab* and *SubTab* showed higher precision, with *SAWTab* performing particularly well for the minority class. While *SAWTab* and *SubTab* excelled in precision, they showed slightly lower recall for most classes, except the minority class in *SAWTab*. These findings highlight *SAWTab*'s superior performance, especially in accurately predicting samples from the minority class in severely imbalanced datasets.

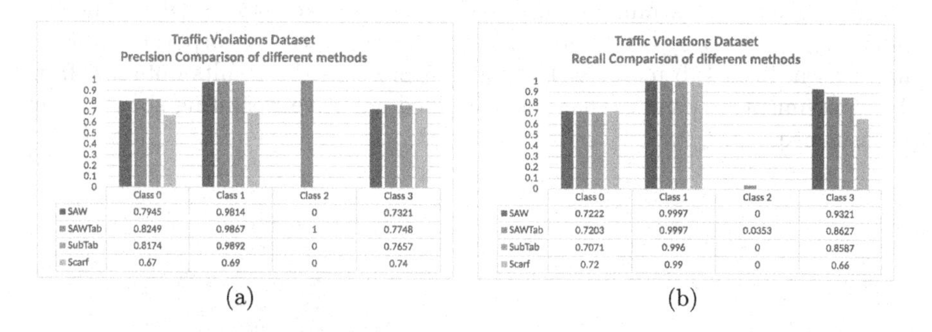

(a) (b)

Fig. 2. Comparing recall and precision performance across frameworks

Precision and Recall Analysis over Different Runs. This section demonstrates the per-class improvement of precision and recall through progressive updates of the CPR representation in successive runs on the traffic violations dataset. Figure 3(a) displays the precision and recall progress for the first class, both of which exhibit higher values in the later runs. In Fig. 3(b), for the second class, *SAWTab* initially has lower recall in the first run, but it improves in subsequent runs. Towards the end, the performance of *SAWTab* slightly surpasses that of *SAW*. As shown in Fig. 3(c) for class 2, *SAW* has higher precision but lower recall, whereas *SAWTab* exhibits the opposite behavior. With further progress in subsequent runs, *SAW* fails to predict any samples from class 2, while *SAWTab* achieves accurate predictions. Lastly, in Fig. 3(d), both precision and recall improve for *SAWTab* across the following runs, while in *SAW*, precision slightly decreases and recall improves. In conclusion, *SawTab* shows more robustness compared to *Saw* especially in minroity classes.

Table 5. Performance comparison of weighting methods. Incorporating two weighting methods in Eq. 3

Weighting Method	Self-Supervised	Semi-Supervised
Balanced Heuristic	79.47 (±0.54)	79.52 (±0.63)
Smoothed Adaptive Weighting	80.64 ±(0.2991)	80.90 (±0.2076)

Smoothed Adaptive Weighting Versus Other Weighting Methods. We conducted an experiment on the Traffic Violations Dataset to compare the performance of Smoothed adaptive weighting and another weighting method from scikit-learn[1]. To estimate class weights using the later method, we utilized the "compute_class_weight" function available in scikit-learn and named it *Balanced Heuristic.* This function calculates the class weights based on the formula: $n_{samples}/(n_{classes} * np.bincount(y))$. The weights are incorporated in Eq. 3 and used in SAWTab method. The Table 5 presents the results obtained from this experiment and shows that the smoothed adaptive weighting shows better performance.

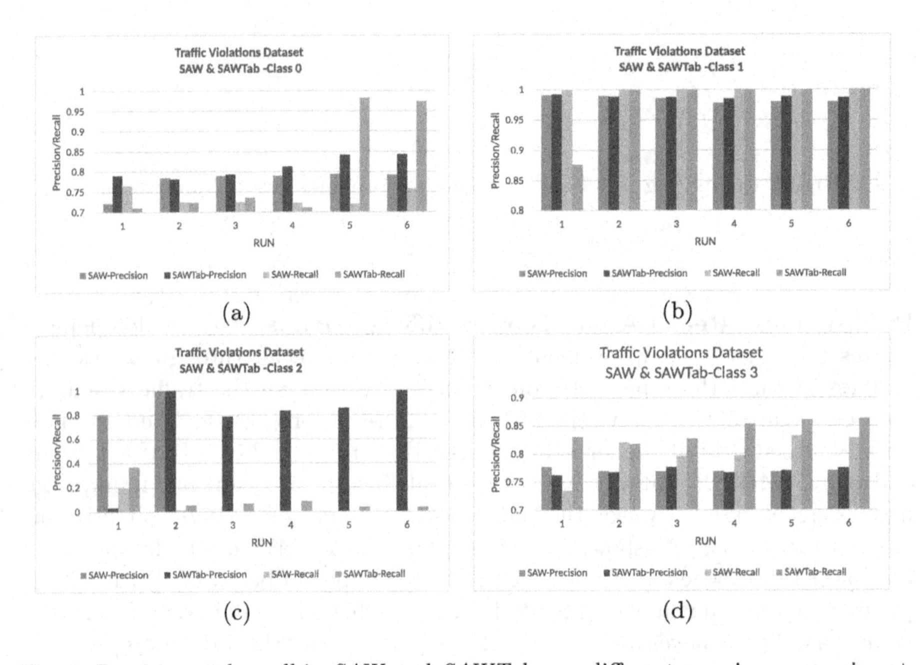

(a) (b) (c) (d)

Fig. 3. Precision and recall in *SAW* and *SAWTab* over different runs in an experiment

5.1 Discussion

Large tabular datasets with high-cardinality categorical features present unique challenges in SSL settings. One-hot encoding of these features inferiors to the target encoding because of sparse representation and computational limitations. However, target encoding in the SSL setting is underexplored and has its own challenges. Incorporating pseudo-labels becomes necessary in target encoding to handle numerous categorical values in the unlabeled data that do not exist in the labeled data.

[1] https://scikit-learn.org/stable/modules/generated/sklearn.utils.class_weight.compute_class_weight.html.

Limitations: For datasets that possess low-cardinality categorical features or where the categorical features lack significant predictive power for the target variable, the proposed approach is not well-suited. In this case, the target encoding does not have superiority over one-hot encoding, and upgrading the representation of CPR does not show improvement in the input space. This is because the categorical features may not add predictive power for the target variable. Upgrading the representation based on a target encoding method like CPR does not improve the predictive power of the machine learning algorithm.

6 Conclusion

The SAWTab method presents a semi-supervised solution for tabular data where categorical features are preprocessed with a target-based encoding technique. In semi-supervised learning, most samples are unlabeled, making it challenging to use all samples for the target-based encoding. We propose utilizing pseudo-labels of unlabeled samples for target-based encoding. We also propose using the smoothed adaptive weighting to upgrade the CPR representation progressively. Furthermore, the proposed approach is not contradictory to the existing semi-supervised learning approaches, but complementary to help produce robust representation. We hope with the awareness of many encoding tools for tabular domain data, it becomes easier to learn meaningful representations in tabular domain and use them in SSL setting.

References

1. Yoon, J., Zhang, Y., Jordon, J., van der Schaar, M.: Vime: extending the success of self-and semi-supervised learning to tabular domain. In: Advances in Neural Information Processing Systems, vol. 33 (2020)
2. Ucar, T., Hajiramezanali, E., Edwards, L.: Subtab: subsetting features of tabular data for self-supervised representation learning. In: Advances in Neural Information Processing Systems, vol. 34, pp. 18853–18865 (2021)
3. Weinberger, K., Dasgupta, A., Langford, J., Smola, A., Attenberg, J.: Feature hashing for large scale multitask learning. In: Proceedings of the 26th Annual International Conference on Machine Learning, pp. 1113–1120 (2009)
4. Kim, J., Hur, Y., Park, S., Yang, E., Hwang, S.J., Shin, J.: Distribution aligning refinery of pseudo-label for imbalanced semi-supervised learning. In: Advances in Neural Information Processing Systems, vol. 33, pp. 14567–14579 (2020)
5. Wei, C., Sohn, K., Mellina, C., Yuille, A., Yang, F.: Crest: a class-rebalancing self-training framework for imbalanced semi-supervised learning. In: Proceedings of the IEEE/CVF Conference on Computer Vision and Pattern Recognition, pp. 10857–10866 (2021)
6. Wang, X., Wu, Z., Lian, L., Yu, S.X.: Debiased learning from naturally imbalanced pseudo-labels for zero-shot and semi-supervised learning. CoRR, vol. abs/2201.01490 (2022). https://arxiv.org/abs/2201.01490
7. He, J., et al.: Rethinking re-sampling in imbalanced semi-supervised learning. CoRR, vol. abs/2106.00209 (2021). https://arxiv.org/abs/2106.00209

8. Guo, L.-Z., Li, Y.-F.: Class-imbalanced semi-supervised learning with adaptive thresholding. In: Proceedings of the 39th International Conference on Machine Learning, vol. 162, pp. 8082–8094. PMLR (2022). https://proceedings.mlr.press/v162/guo22e.html
9. Bahri, D., Jiang, H., Tay, Y., Metzler, D.: Scarf: self-supervised contrastive learning using random feature corruption. arXiv preprint arXiv:2106.15147 (2021)
10. Darabi, S., Fazeli, S., Pazoki, A., Sankararaman, S., Sarrafzadeh, M.: Contrastive mixup: self- and semi-supervised learning for tabular domain. arXiv:2108.12296 (2021)
11. Cerda, P., Varoquaux, G.: Encoding high-cardinality string categorical variables. IEEE Trans. Knowl. Data Eng. **34**, 1164–1176 (2020)
12. Cerda, P., Varoquaux, G., Kégl, B.: Similarity encoding for learning with dirty categorical variables. Mach. Learn. **107**(8), 1477–1494 (2018)
13. Slakey, A., Salas, D., Schamroth, Y.: Encoding categorical variables with conjugate bayesian models for wework lead scoring engine. arXiv preprint arXiv:1904.13001 (2019)
14. Lai, Z., Wang, C., Gunawan, H., Cheung, S.-C.S., Chuah, C.-N.: Smoothed adaptive weighting for imbalanced semi-supervised learning: improve reliability against unknown distribution data. In: Proceedings of the 39th International Conference on Machine Learning, vol. 162, pp. 11828–11843. PMLR (2022). https://proceedings.mlr.press/v162/lai22b.html
15. Verleysen, M., François, D.: The curse of dimensionality in data mining and time series prediction. In: Cabestany, J., Prieto, A., Sandoval, F. (eds.) IWANN 2005. LNCS, vol. 3512, pp. 758–770. Springer, Heidelberg (2005). https://doi.org/10.1007/11494669_93
16. Gharasuie, M.M., Wang, F.: Progressive feature upgrade in semi-supervised learning on tabular domain. In: 2022 IEEE International Conference on Knowledge Graph (ICKG), pp. 188–195 (2022)
17. Website, U.G.: Traffic violations dataset. https://catalog.data.gov/dataset/traffic-violations-56dda, non specified
18. Food, U., Administration, D.: CTR prediction contest (2015). https://www.kaggle.com/c/avazu-ctr-prediction
19. Criteo: Display advertising challenge (2015). https://www.kaggle.com/c/criteo-display-ad-challenge
20. Majmundar, K., Goyal, S., Netrapalli, P., Jain, P.: Met: masked encoding for tabular data. arXiv preprint arXiv:2206.08564 (2022)
21. Somepalli, G., Goldblum, M., Schwarzschild, A., Bruss, C.B., Goldstein, T.: Saint: improved neural networks for tabular data via row attention and contrastive pre-training. arXiv preprint arXiv:2106.01342 (2021)

Big Data

Improving Anti-money Laundering via Fourier-Based Contrastive Learning

Meihan Tong[1,2](✉), Shuai Wang[3], Xinyu Chen[1], and Jinsong Bei[1]

[1] China National Clearing Center, Beijing, China
[2] Tsinghua University, Beijing, China
tongmeihan@gmail.com
[3] Hashkey, Beijing, China
wangshuai@hashkey.com

Abstract. Anti-money laundering (AML) aims to detect money laundering from daily transactions, which is the key frontier of combating financial crimes. Previous deep-learning AML methods are not robust enough. To address the problem, we propose a novel Fourier-based contrastive learning model (FCLM) to improve AML. With contrastive learning, FCLM can maintain prediction consistency and be more robust in the face of data perturbations. Experiments on both the synthetic benchmark IBM2023 and the real-world benchmark show that FCLM outperforms seven state-of-the-art baselines, demonstrating the effectiveness of the proposed Fourier-based contrastive learning model.

Keywords: Anti-Money laundering · Contrastive Learning · Data Augmentation · Fourier Transformation

1 Introduction

Anti-money laundering aims to identify money laundering activities that conceal the source of criminal proceeds from massive transactions. Money laundering impairs the stability of the financial market, increases the operational risks of financial institutions, and has caused a great economic loss to the world. According to the report by Cybersecurity Ventures, the economic losses caused by global money laundering crimes will grow at an annual rate of 15% in the next two years, reaching 10.5$ trillion per year by 2025[1].

Due to the greater dangers of money laundering, anti-money laundering has received great attention from researchers around the world. Early AML methods are mainly rule-based methods, such as [4,26], which suffer from inflexible and high maintenance costs. Later, machine learning AML methods, such as SVM [28] and RF [1], became popular. However, these methods have low accuracy. Recently, deep learning AML methods, such as SkipGNN [33] and HAMLET

[1] https://s3.ca-central-1.amazonaws.com/esentire-dot-com-assets/assets/resourcefiles/2022-Official-Cybercrime-Report.pdf.

D.-N. Yang et al. (Eds.): PAKDD 2024, LNAI 14647, pp. 331–343, 2024.
https://doi.org/10.1007/978-981-97-2259-4_25

[29], have been proposed. However, these methods are not robust enough, that is by adding subtle and deliberate perturbations to the transactions, these deep-learning AML models will output erroneous results with a high confidence level.

We propose to leverage data augmentation and contrastive learning to handle the issue. The main idea is to generate the augmented view for each transaction and then leverage contrastive learning to pull the transaction closer to its augmented view to ensure that our model can maintain the consistency of discrimination results in the face of data perturbations, and avoid frequent misjudgments under minor disturbances. We utilize the Fourier data augmentation method to generate the augmented view. Compared to Gaussian noise [3], column sampling[2], mask token replacing [20], and window wrapping [25], Fourier data augmentation method maps the entire transaction features from the time domain to the frequency domain, resulting in a more differentiated augmented view and allowing our model capture a wider range of variations and different perspectives of the same sample.

In summary, we propose a Fourier-based contrastive learning model(FCLM) to improve anti-money laundering. FCLM uses the Transformer model as the backbone network. When detecting money laundering, FCLM first employs Fourier transformation to generate the augmented view of each transaction. Then, FCLM uses comparative pre-training to ensure the consistency of its predictions for original transactions and augmented views. Finally, FCLM distinguishes money laundering transactions from normal ones through the MLP classifier. The classifier takes both the original transaction and its augmentation view as input to synthesize both sides of information to improve money laundering detection.

We evaluate FCLM on a synthetic dataset *IBM2023*[3] and a real-world dataset. Experiments show that our proposed method consistently surpasses seven strong baselines on both datasets, demonstrating the superiority of the proposed Fourier-based contrastive learning model.

Our contributions can be summarized as follows:

- We propose a Fourier-based contrastive learning model (FCLM) to improve the robustness of money laundering detection.
- We proposed a Fourier-based data augmentation method, which transforms the transaction attributes from the time domain to the frequency domain to ensure the discrepancy of the augmented view, enabling our model to capture wider variations and become more robust in representation learning.
- Experiments on the synthetic dataset IBM2023 and the real-world dataset show that FCLM surpasses seven strong baselines. Detailed studies also show that the proposed Fourier data augmentation method surpasses three commonly used data augmentation algorithms.

---

[2] https://www.kaggle.com/code/bigironsphere/basic-data-augmentation-feature-reduction.

[3] https://www.kaggle.com/datasets/ealtman2019/ibm-transactions-for-anti-money-laundering-aml.

The following of the paper is organized as follows. Section 2 introduces commonly used anti-money laundering methods and related work on contrastive learning. Section 3 illustrates the overall architecture of our proposed Fourier-based contrastive learning model (FCLM) and details each module in the FCLM. Section 4 introduces the experimental datasets and hyper-parameters of FCLM, and presents extensive experimental results. Section 5 concludes the paper.

2 Related Work

2.1 Anti-Money Laundering(AML)

Anti-money laundering is a hot topic. Early AML models were mostly rules-based approaches [2,4,5]. They commonly use 1) a surge in transaction traffic in a short period, 2) the transaction amount exceeds a specified threshold for multiple consecutive days, etc. to detect money laundering. Rule-based methods are inflexible and can be easily broken by criminals through rule probing.

Other methods adopt machine learning to improve AML [13,19,27,30,35]. For instance, [28] proposes a set of abnormal behavior detection algorithms based on support vector machines (SVM). [1] employs the Light gradient Boosting Algorithm (LGBA) and XGBoost in distinguishing illegal money laundering exercises. These machine-learning AML methods have low accuracy, and their performance is easily affected by the quality of feature engineering.

Deep learning supervised AML methods [6,12,14,16,22,29,32] have gradually attracted the attention of scholars. CS-CNN [16] leverages cost-sensitive CNN and feature matrix for fraud detection. OCGTL [23] combines deep one-class classification with GNN for graph-level anomaly detection. ComGA [18] leverages a community-aware tailored GCN to handle AML. SkipGNN [33] utilizes a skip connection GNN network and Diga [15] employs the semi-supervised guided diffusion to identify money laundering activities. HAMLET [29] employs a hierarchical transformer to identify complex money laundering at both transaction and sequence levels. However, these methods are not robust enough and often produce inconsistent prediction results when faced with data disturbances. In this paper, we empower AML with contrastive learning to allow the proposed model to maintain prediction consistency under the perturbation of the Fourier augmented view to improve robustness in AML.

2.2 Contrastive Learning

Contrastive learning [9] is a self-supervised learning method that has been proven to be effective in a wide range of fields.

In anti-money laundering, we pay more attention to tabular data augmentation. Naive Gaussian [3] is a widely adopted tabular data augmentation method, which generates the augmented data by injecting Gaussian noise. Column sampling (see Footnote 2) adds noise to tabular data by replacing the feature with the value sampled from its overall distribution. Mask token replacement [20]

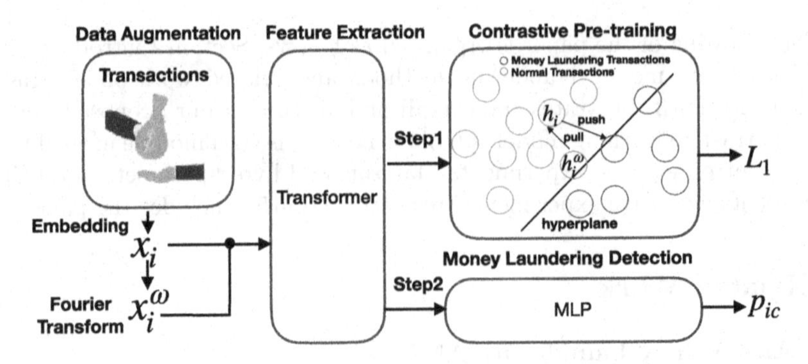

Fig. 1. Architecture of the proposed Fourier-based contrastive learning model(FCLM).

augments the tabular data by masking out the feature and replacing it with a variable-length [MASK] embedding. These methods perturb each feature individually, resulting in a small degree of diversity. Window wrapping [25] is a time-series data augmentation method that augments the time-series feature by speeding it up or slowing it down, which cannot be applied to category features in transactions, such as payee and payer. In this paper, we leverage Fourier to transform the whole features in the transactions from the time domain to the frequency domain to generate augmented views with high diversity.

3 Methdology

3.1 Task Definition

Formally, given the corpus $x_i = \{t_i, t_{i-1}, \dots, t_{i-m}|_{i=1}^{N}\}$, where t_i is the transaction to be classified, t_{i-m} is the historical transactions of t_i, N is the total number of transactions, the proposed model FCLM first minimizes $L_1 = log\frac{exp(h_i, h_i^{\omega})}{\sum_{k=1}^{K} exp(h_i, h_k)}$ to ensure prediction consistency in the face of data perturbations, where h_i and h_i^{ω} are the hidden representations of the original data x_i and its augmented view x_i^{ω}, K is the number of sampled negative examples. Then, FCLM maximizes the classification probability $p_{ic} = \frac{exp(o_{ic})}{\sum_{c=1}^{C} exp(o_{ic})}$ to detects money laundering transactions, where C is the category number, o_{ic} is the predicted probability that h_i belongs to the cth category.

3.2 Overall Architecture

Figure 1 illustrates the overall architecture of the proposed Fourier-based contrastive learning model(FCLM), which includes four modules: data augmentation module, feature encoding module, contrastive pre-training module, and money laundering detection module.

The data augmentation module is designed to generate augment views for transactions. The feature encoding module aims to encode the original transaction and its augmented view into hidden representations. The contrastive pre-training module aims to utilize contrastive learning to bring the hidden representation closer between the original transaction and its augmented view, and to push the distance between the original transaction and other transactions farther away. The money laundering detection module aims to distinguish money laundering transactions from normal ones.

3.3 Data Augmentation

In this section, we introduce the proposed Fourier data augmentation method.

We start with converting raw transactions into embeddings. We handle categorical and numeric attributes in raw transactions differently. For the categorical attribute, we randomly initialize an embedding for each class in the categorical attribute and then obtain the embedding of the categorical attribute based on the class index. For the numeric attribute, we first employ Z-score normalization for scaling. After that, we discretize the numeric attribute into equal-frequency buckets (each bucket has the same number of elements) and randomly initialize an embedding for each bucket in the numeric attributes to obtain the embedding of the numeric attribute.

After obtaining the embedding x_i of the transaction, we employ the Discrete Fourier Transform (DFT) to generate its augmented data. Formally, for each transaction x_i, the discrete Fourier transform can be formalized as:

$$F(\omega) = \sum_{i=0}^{N-1} e^{-i\frac{2\pi}{N}n\omega} x_i \tag{1}$$

where $\omega \in [0, N-1]$ is the angular frequency, e is the base of natural logarithms and i is the imaginary unit. Due to the large number of transactions, we employ the fast Fourier transform (FFT) algorithm [11] to solve DFT to improve computational efficiency.

In summary, after leveraging DFT, we get a complex number representing magnitude and phase in the frequency domain as the Fourier-augmented view of the transaction data.

3.4 Feature Encoding

In this section, we aim to build a feature encoder that can convert both original and Fourier-augmented transactions into hidden representations.

We employ Transformer [31] as our feature encoder to fuse features deeply. Transformer is a multi-layer deep learning architecture that aims to handle long-distance dependencies, which has achieved great success on a wide range of financial fraud tasks, such as credit card fraud detection [34] and insurance fraud detection [10]. The multi-head attention mechanism in each layer of the

Transformer can automatically adjust the attention weights during the training process, enabling the model to discover features that are important to anti-money laundering while ignoring irrelevant features.

We take the output of the last layer of the Transformer as the hidden representation h_i of the original transactions x_i, which is denoted as:

$$h_i = Encode(x_i) \tag{2}$$

In the same way, we obtain the hidden representation h_i^{ω} of the Fourier-augmented transaction x_i^{ω}.

3.5 Contrastive Pre-training

In this section, we perform comparative pre-training on the full amount of transaction data to optimize the feature encoder in Sect. 3.4.

Formally, the loss function of contrastive pre-training is defined as:

$$L_1(h_i, h_i^{\omega}, \theta_0) = \frac{1}{N}\sum_{i=1}^{N} log\frac{exp(h_i, h_i^{\omega})}{\sum_{k=1}^{K} exp(h_i, h_k)} \tag{3}$$

where N is the scale of the transaction data, h_i and h_i^{ω} refer to hidden representation of x_i (original transaction) and x_i^{ω} (Fourier augmented transaction) respectively, h_k refers to hidden representation of x_k ($x_k \neq x_i$), and K is the number of sampled examples.

Through comparative pre-training, we can make the hidden representation of the transaction closer to the hidden representation of its own augmented view and farther away from the hidden representation of other transactions, which can make our model more robust in the face of data perturbations.

3.6 Money Laundering Detection

The money laundering detection module is designed to distinguish normal transactions from money laundering transactions based on the optimized feature encoder.

Specifically, we first randomly undersample the normal transactions to rebalance the training data, and then, we build a multiple-layer perceptron classifier on the top of the optimized feature encoder to map the hidden representation into the classification space, which can be formally expressed as:

$$o_i = W(h_i + h_i^{\omega}) + b \tag{4}$$

where h_i and h_i^{ω} are the hidden representations of the original transaction and the Fourier augmented transaction respectively. The classifier takes both the original transaction and the Fourier-enhanced transaction as input to combine the time and frequency domain information when determining whether a transaction is a money laundering transaction.

We calculate the classification probability by:

$$p_{ic} = \frac{exp(o_{ic})}{\sum_{c=1}^{C} exp(o_{ic})} \tag{5}$$

where C is the number of categories. Finally, the training loss of the money laundering detection module is defined as:

$$L_2(h_i, \theta_1) = -\frac{1}{N}\sum_{i=1}^{N}\sum_{c=1}^{C} y_{ic} log(p_{ic}) \tag{6}$$

where p_{ic} is the predicted probability that h_i belongs to the cth category, and y_{ic} is the category label.

4 Experiment

4.1 Datasets

IBM2023 (see Footnote 3) is a large synthetic anti-money laundering dataset released by IBM Corporation. IBM2023 creates an entire virtual financial ecosystem by having artificial individuals, companies and banks transact with each other. IBM2023 contains six versions of the dataset, including HI-small, HI-medium, HI-large, LI-small, LI-medium, and LI-large. We adopt the HI-small version of the dataset for experimental training, which contains 515K bank accounts and 5M transaction data. The features in the transaction of IBM2023 include account, timestamp, amount, currency type, etc. Following the official instructions, we divide the datasets into 60%/20%/20% for training, validating, and testing respectively.

Unlike previous anti-money laundering models [29] that are only evaluated on synthetic datasets, we further evaluate our model on a real-world dataset. The real-world dataset is built from the China interbank clearing transaction system, which contains a total of 10k money laundering transactions and 10M normal transactions. To maintain data confidentiality, each transaction in this dataset is a summary of the transactions of both parties on the day. The transaction attributes include transaction time, payer, payer's bank, payer's account, payee, payee's bank, payee's account, the total transaction amount of the day, the number of night transactions, etc., a total of 19 features. We split the real-world dataset into 80%/10%/10% for training, validation, and testing respectively.

Following [7], we evaluate the performance of the proposed model through four metrics, including precision, recall, F1, and AUC. Due to data imbalance in AML scenarios, we use the macro average of these metrics as the final result.

4.2 Hyperparameters

Through grid search, we set the learning rate to 1e–4 in the pre-training stage and 2e–5 in the money laundering detection stage. The training step/batch size of the pre-training stage and the money laundering detection stage are 10,000/512 and 1,000/128 respectively. We adopt the 8-layer Transformer as our backbone network, with 5 heads of attention and 128 hidden representation dimensions. We select the checkpoint that performs best on the validation set as the final inference model and report the average result of 10 runs as the final results. We use Adam as the gradient descent optimizer. We use an 8-card A100 server with 80G GPU memory per card for training. The training time is about 10/0.5 h for the pre-training and the money laundering detection stage respectively.

Table 1. Overall Performance of FCLM(%).

Methods	IBM2023				Real-World			
	P	R	F1	AUC	P	R	F1	AUC
VAE	55.2	54.7	54.9	54.1	53.7	54.5	54.1	52.9
SVM	58.1	64.6	61.2	60.7	54.5	52.3	53.4	52.7
RF	56.8	60.3	58.5	64.7	54.7	56.1	55.4	58.5
CS-CNN	**85.6**	80.3	82.8	79.4	66.2	67.5	66.8	67.8
SkipGNN	72.4	80.6	76.3	78.1	66.3	71.6	68.8	70.2
Inspection-L	78.2	84.5	81.2	83.6	69.6	74.2	71.8	74.7
HAMLET	79.7	85.1	82.3	86.2	71.6	76.2	73.8	75.9
FCLM (ours)	82.5	**91.2**	**86.6**	**91.4**	**73.7**	**78.2**	**75.9**	**80.1**

4.3 Baselines

We compare our methods with seven AML baselines, including:

- VAE [8] uses a sparse autoencoder to learn the distribution of normal transactions and treat transactions that do not fit the distribution as money laundering transactions.
- SVM [21] proposes a support vector machine (SVM) classifier optimized with the random undersampling (RUS) technique to detect financial fraud.
- RF [24] is a traditional machine learning AML baseline, which leverages rich features and the random forest algorithm for money laundering detection.
- CS-CNN [16] leverages cost-sensitive CNN and feature matrix for fraud detection.
- SkipGNN [33] takes transactions as nodes and transaction flow as edges, and leverages skip connections GNN to detect AML at the graph level.
- Inspection-L [17] is a self-supervised graph neural network (GNN) framework based on Deep Graph Infomax (DGI) and Graph Isomorphism Network (GIN), with Random Forest (RF) to detect illicit transactions for anti-money laundering (AML).
- HAMLET [29] employs the hierarchical transformer to fuse features at both transaction and sequence levels to identify money laundering.

Table 1 presents the overall performance of the proposed Fourier-based contrastive learning model on IBM2023 and the real-world dataset. As shown in Table 1, our method outperforms seven state-of-the-art baselines, demonstrating the effectiveness of FCLM and the superiority of the Fourier data augmentation method.

FCLM (ours) outperforms VAE by 37.4% in AUC score on IBM2O23 and 27.4% on the real-world dataset. VAE is a semi-supervised AML method, which does not exploit money laundering transactions for training. While FCLM (ours) not only utilizes a large amount of normal transaction data for training, but also utilizes money laundering annotations as supervision signals, taking both sides into account, and thus has better performance.

Compared with non-transformer baselines SVM, RF, CS-CNN, SkipGNN, and Inspection-L, the transformer baselines HAMLET and FCLM(ours) perform better. This shows the strong representation capabilities of the transformer, which can effectively extract and integrate transaction features in anti-money laundering. Among the transformer's baseline HAMLET and FCLM (ours), FCLM (ours) performs better. This is because we add contrastive learning training to the basic transformer to allow our model to maintain prediction consistency in the face of data perturbations. With more stable prediction results, our model is more robust and therefore generalizes better on the unseen data.

Compared with Inpection-L which also uses contrastive learning, our model achieves better performance (83.6% VS 91.4% in IBM2023 and 74.7% VS 80.1% in the real-world dataset). One of the reasons is that Inspection-L augments data through graph corruption (randomly adding and deleting edges in the transaction graph), which may fabricate the occurrence of transactions and cannot guarantee that the augmented view and the original data still belong to the same class. As a result, wrong labels are mixed into the training, leading to a decrease in performance. Our method uses Fourier transform to obtain enhanced data, which can ensure the equivalence of the two and there is no risk that the augmented view and the original data do not belong to the same class.

4.4 Effectiveness of Data Augmentation

In this section, we hope to observe what happens when the proposed Fourier data augmentation method is replaced with other tabular data augmentation data. In addition to the non-augmentation baseline, we also adopt three baselines, including:

- Mask Replacing [20] augments the transaction data by randomly replacing the features in the transactions with the [MASK] token.
- Gaussian [3] augments the transaction by adding random Gaussian noise.
- Column sampling (see Footnote 2) adds noise to the transaction by replacing original features with sampled values from feature distributions.

For all baselines, we uniformly use the transformer as the classifier so that the baselines can be compared fairly with the proposed method.

Table 2. Performance on Different Data Augmentation Methods on IBM2023

Augmentation	AUC
Non-Augmentation	85.6
Mask Token Replacing	85.3
Gaussian	87.1
Column Sampling	87.3
Fourier(ours)	91.4

As shown in Table 2, Fourier (ours) outperforms the non-augmented baseline and the three data-augmented baselines by 5.8%, 6.1%, 4.3%, and 4.1% in AUC, achieving the best performance. The mask token replacing baseline is not as good as the non-augmentation baseline, which shows that data enhancement does not always improve AML and requires careful design. An in-depth analysis of the failure of the mask token replacement baseline shows that there is serious information lost in the process of replacing the original features with the [MASK] embedding, which makes the augmented view less informative in making money laundering decisions. The poor performance of the Gaussian method and column sampling method is because these two baselines perform data augmentation on each feature independently, resulting in smaller differences between the augmented view and the original data. In contrast, our method maps the entire transaction features from the time domain to the frequency domain. With more differentiated augmented views, our model can capture wider variations and different perspectives of the same sample, thereby becoming more robust in representation learning and better able to generalize to unseen data.

4.5 Future Direction

Two promising directions can be considered in future work. First, leverage more data sources for training. FCLM(ours) only incorporates transaction data in the training processes. Future work can consider using additional data information such as account data, portrait data, credit data, corporate financial report data, etc. to improve the monitoring performance of the AML model. Second, utilize the large language models. FCLM (ours) does not take advantage of large language models such as GPT4, PaLM-E, BLOOM, LLaMA, etc. Future work can consider enhancing the AML model with large language models to mine more complex gang-based money laundering behaviors and reduce the false positive rate of money laundering.

5 Conclusion

In this paper, we propose a Fourier-based contrastive learning model (FCLM) to improve the robustness of anti-money laundering. FCLM first leverages the

Fourier data augmentation method to transform the transaction attributes from the time to frequency domain to ensure the discrepancy of the augment views, and then utilize contrastive learning to refine the representation of transactions. Through contrastive learning, FCLM can maintain prediction consistency in the face of data perturbations and therefore has stronger robustness and generalization. Experiments on two benchmarks demonstrate the effectiveness of FCLM. Comparative experiments between different data augmentation methods further show the superiority of the proposed Fourier data augmentation method.

References

1. Ahmed, A.A.A.: Anti-money laundering recognition through the gradient boosting classifier. Acad. Accounting Fin. Stud. J. **25**(5), 1–11 (2021)
2. Ai, L.: Rule-based but risk-oriented approach for combating money laundering in Chinese financial sectors. J. Money Laundering Control **15**(2), 198–209 (2012)
3. Arslan, M., Guzel, M., Demirci, M., Ozdemir, S.: SMOTE and gaussian noise based sensor data augmentation. In: UBMK, pp. 1–5. IEEE (2019)
4. Bellomarini, L., Laurenza, E., Sallinger, E.: Rule-based anti-money laundering in financial intelligence units: experience and vision. RuleML+ RR **2644**(Suppl.), 133–144 (2020)
5. Butgereit, L.: Anti money laundering: rule-based methods to identify funnel accounts. In: 2021 Conference on Information Communications Technology and Society (ICTAS), pp. 21–26 (2021)
6. Chai, Z., et al.: Towards learning to discover money laundering sub-network in massive transaction network. In: Proceedings of the AAAI Conference on Artificial Intelligence (2023)
7. Charitou, C., Garcez, A.D., Dragicevic, S.: Semi-supervised GANs for fraud detection. In: IJCNN, pp. 1–8 (2020)
8. Chen, J., Shen, Y., Ali, R.: Credit card fraud detection using sparse autoencoder and generative adversarial network. In: IEMCON, pp. 1054–1059. IEEE (2018)
9. Chen, T., Kornblith, S., Norouzi, M., Hinton, G.E.: A simple framework for contrastive learning of visual representations. CoRR abs/2002.05709 (2020). https://arxiv.org/abs/2002.05709
10. Fursov, I., et al.: Sequence embeddings help detect insurance fraud. IEEE Access **10**, 32060–32074 (2022)
11. Heckbert, P.: Fourier transforms and the fast Fourier transform (FFT) algorithm. Comput. Graph. **2**(1995), 15–463 (1995)
12. Hu, B., Zhang, Z., Shi, C., Zhou, J., Li, X., Qi, Y.: Cash-out user detection based on attributed heterogeneous information network with a hierarchical attention mechanism. In: Proceedings of the AAAI Conference on Artificial Intelligence, vol. 33, pp. 946–953 (2019)
13. Kumar, A., Das, S., Tyagi, V., Shaw, R.N., Ghosh, A.: Analysis of classifier algorithms to detect Anti-money laundering. In: Bansal, J.C., Paprzycki, M., Bianchini, M., Das, S. (eds.) Computationally Intelligent Systems and their Applications. SCI, vol. 950, pp. 143–152. Springer, Singapore (2021). https://doi.org/10.1007/978-981-16-0407-2_11

14. Kute, D.V.: Explainable deep learning approach for detecting money laundering transactions in banking system. Ph. D. thesis (2022)
15. Li, X., Li, Y., Mo, X., Xiao, H., Shen, Y., Chen, L.: Diga: guided diffusion model for graph recovery in anti-money laundering. In: Proceedings of the 29th ACM SIGKDD Conference on Knowledge Discovery and Data Mining, pp. 4404–4413 (2023)
16. Liu, X., Zhang, X., Miao, Q.: A click fraud detection scheme based on cost-sensitive CNN and feature matrix. In: Tian, Y., Ma, T., Khan, M.K. (eds.) ICBDS 2019. CCIS, vol. 1210, pp. 65–79. Springer, Singapore (2020). https://doi.org/10.1007/978-981-15-7530-3_6
17. Lo, W.W., Kulatilleke, G.K., Sarhan, M., Layeghy, S., Portmann, M.: Inspection-l: self-supervised GNN node embeddings for money laundering detection in bitcoin. Appl. Intell. **53**, 19406–19417 (2023)
18. Luo, X., et al.: ComGA: community-aware attributed graph anomaly detection. In: Proceedings of the Fifteenth ACM International Conference on Web Search and Data Mining, pp. 657–665 (2022)
19. Misra, S., Thakur, S., Ghosh, M., Saha, S.K.: An autoencoder based model for detecting fraudulent credit card transaction. Procedia Comput. Sci. **167**, 254–262 (2020)
20. Onishi, S., Meguro, S.: Rethinking data augmentation for tabular data in deep learning. arXiv preprint arXiv:2305.10308 (2023)
21. Pambudi, B.N., Hidayah, I., Fauziati, S.: Improving money laundering detection using optimized support vector machine. In: 2019 International Seminar on Research of Information Technology and Intelligent Systems (ISRITI), pp. 273–278 (2019). https://doi.org/10.1109/ISRITI48646.2019.9034655
22. Pareja, A., et al.: EvolveGCN: evolving graph convolutional networks for dynamic graphs. In: Proceedings of the AAAI Conference on Artificial Intelligence, vol. 34, pp. 5363–5370 (2020)
23. Qiu, C., Kloft, M., Mandt, S., Rudolph, M.: Raising the bar in graph-level anomaly detection. arXiv preprint arXiv:2205.13845 (2022)
24. Raiter, O.: Applying supervised machine learning algorithms for fraud detection in anti-money laundering. J. Mod. Issues Bus. Res. **1**(1), 14–26 (2021)
25. Rashid, K.M., Louis, J.: Window-warping: a time series data augmentation of IMU data for construction equipment activity identification. In: ISARC. Proceedings of the International Symposium on Automation and Robotics in Construction, vol. 36, pp. 651–657. IAARC Publications (2019)
26. Ross, S., Hannan, M.: Money laundering regulation and risk-based decision-making. J. Money Laundering Control **10**(1), 106–115 (2007)
27. Sundarkumar, G.G., Ravi, V., Siddeshwar, V.: One-class support vector machine based undersampling: application to churn prediction and insurance fraud detection. In: 2015 IEEE International Conference on Computational Intelligence and Computing Research (ICCIC), pp. 1–7. IEEE (2015)
28. Tang, J., Yin, J.: Developing an intelligent data discriminating system of anti-money laundering based on SVM. In: 2005 International Conference on Machine Learning and Cybernetics, vol. 6, pp. 3453–3457. IEEE (2005)
29. Tatulli, M.P., Paladini, T., D'Onghia, M., Carminati, M., Zanero, S.: HAMLET: a transformer based approach for money laundering detection. In: Dolev, S., Gudes, E., Paillier, P. (eds.) International Symposium on Cyber Security, Cryptology, and Machine Learning, vol. 13914, pp. 234–250. Springer, Cham (2023). https://doi.org/10.1007/978-3-031-34671-2_17

30. Tundis, A., Nemalikanti, S., Mühlhäuser, M.: Fighting organized crime by automatically detecting money laundering-related financial transactions. In: Proceedings of the 16th International Conference on Availability, Reliability and Security, pp. 1–10 (2021)
31. Vaswani, A., et al.: Attention is all you need. CoRR abs/1706.03762 (2017). https://arxiv.org/abs/1706.03762
32. Wang, D., et al.: Temporal-aware graph neural network for credit risk prediction. In: Proceedings of the 2021 SIAM International Conference on Data Mining (SDM), pp. 702–710. SIAM (2021)
33. Weber, M., et al.: Anti-money laundering in bitcoin: experimenting with graph convolutional networks for financial forensics. arXiv preprint arXiv:1908.02591 (2019)
34. Yuan, M.: A transformer-based model integrated with feature selection for credit card fraud detection. In: 2022 7th International Conference on Machine Learning Technologies (ICMLT), pp. 185–190 (2022)
35. Zou, J., Zhang, J., Jiang, P.: Credit card fraud detection using autoencoder neural network. arXiv preprint arXiv:1908.11553 (2019)

A Novel SegNet Model for Crack Image Semantic Segmentation in Bridge Inspection

Rong Pang[1,2,3], Hao Tan[4], Yan Yang[1(✉)], Xun Xu[3], Nanqing Liu[1,3], and Peng Zhang[2]

[1] School of Computing and Artificial Intelligence, Southwest Jiaotong University, Chengdu, China
yyang@swjtu.edu.cn

[2] China Merchants Chongqing Road Engineering Inspection Center Co., Ltd., Chongqing, China

[3] The Institute for Infocomm Research (I2R), A-STAR, Singapore, Singapore
xux@i2r.a-star.edu.sg

[4] College of Mechanical and Vehicle Engineering, Chongqing University, Chongqing, China

Abstract. Cracks on bridge surfaces represent a significant defect that demands accurate and efficient inspection methods. However, current approaches for segmenting cracks suffer from low accuracy and slow detection speed, particularly when dealing with fine and small cracks that occupy only a few pixels. In this work, we propose a novel crack image semantic segmentation method based on an enhanced SegNet. The proposed approach addresses these challenges through three key innovations. First, we reduce the network depth to improve computational efficiency while maintaining accuracy. Furthermore, we employ ConvNeXt-V2 to effectively extract and fuse crack features, thereby improving segmentation performance. To handle pixel imbalance during loss calculation, we integrate the Dice coefficient into the original cross-entropy loss function. Experimental results demonstrate that our enhanced SegNet achieves remarkable improvements in mIoU for non-steel and steel crack segmentation tasks, reaching 82.37% and 77.26%, respectively. Our approach outperforms state-of-the-art methods in both inference speed and accuracy.

Keywords: image segmentation · bridge cracks · enhanced SegNet · deep learning

1 Introduction

Bridges are crucial components of transportation infrastructure that are constructed over bodies of water, valleys, or other obstacles [1]. The performance

Supported by the National Natural Science Foundation of China (No. 61976247) and Chongqing Traffic Science and Technology Project(No.CQJT2022ZC05).

D.-N. Yang et al. (Eds.): PAKDD 2024, LNAI 14647, pp. 344–355, 2024.
https://doi.org/10.1007/978-981-97-2259-4_26

of civil structures deteriorates due to factors such as structural deterioration, external loads, weather effects, poor workmanship and natural disasters [7]. As a significant defect in bridge infrastructure inspection, bridge cracks form due to the combined effects of inherent material properties and external factors. While it is not possible to completely eliminate cracks at their source, effective measures can be implemented to control and prevent their development during the early and middle stages. Presently, visual inspection and simple instrument measurements constitute the primary methods for detecting crack-related defects. Nonetheless, visual inspection proves to be a time-consuming and labor-intensive undertaking, leading to a notable occurrence of false positives and false negatives in identifying existing defects [11]. Therefore, manual crack detection is not suitable for large-scale inspections. In contrast, machine vision-based detection offers higher efficiency and, due to its non-contact nature, enables fast, safe, and reliable inspections without damaging the structural integrity of the facilities. Traditional machine vision methods have been widely applied in solving industrial problems, such as the histogram bimodal method, Otsu [14] method and the mean value method. They can be used for object detection, material contour measurement, distance measurement, and more. Detection and extraction of poorly contrasted linear structures in textured regions using the Markov random field model by [8]. [20] propose a model with a length criterion for scalable local image processing techniques and extract continuous textures with reference to the connectivity of the luminance and shape of the infiltrated regions. Although traditional vision techniques have made significant advancements, they often require extensive expert analysis and fine-tuning, making them unsuitable for handling complex problems. The integration of digital image processing techniques with deep learning has emerged as a new research direction in non-contact detection technology in recent years. Compared to manual inspection, in the task of crack detection, a large number of image data can be captured from the surface of engineering structures using unmanned aerial vehicles (UAVs) and climbing robots equipped with relevant devices, followed by the automatic detection of target defects using deep learning algorithms [13,15]. In crack detection tasks, deep learning can be applied for tasks such as classification, localization, segmentation, and even material performance assessment [10].

With the introduction of fully convolutional networks (FCNs) [12], which are semantic segmentation models based on deep learning methods, end-to-end convolutional neural networks have found increasing applications in semantic segmentation tasks. U-Net [16] has been used for medical image segmentation, improving detection performance with limited training data. DeeplabV3+ [5] introduced dilated convolutions and pyramid pooling, combined with a decoder, to address the issues of multi-scale objects and boundary information loss caused by repeated downsampling. Additionally, a self-attention-based semantic segmentation model, Swin U-Net [3], has been proposed for medical image segmentation, achieving a transition from local to global understanding. Among various image semantic segmentation methods, SegNet [2] stands out for the adoption of an encoder-decoder structure, highlighting the importance of the decoder in

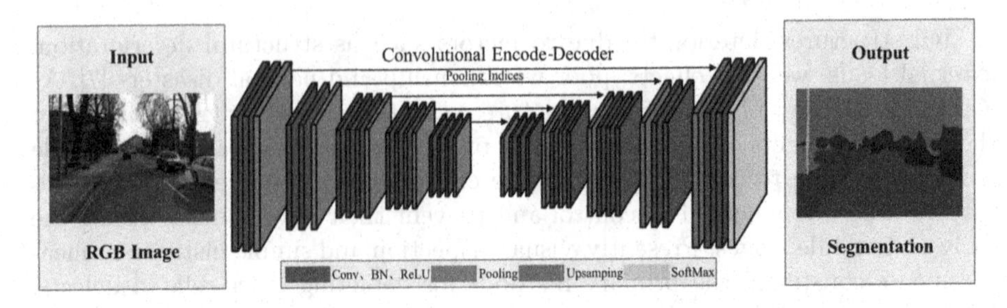

Fig. 1. An illustration of the architecture of the original SegNet.

the segmentation network. It demonstrates stable performance across different segmentation tasks. Compared to other methods, SegNet offers the following advantages: (1) It uses a symmetric encoder-decoder structure, which enables effective feature extraction and restoration of the image. (2) The dimension transformation of feature maps is achieved through upsampling, downsampling, and pooling indices, reducing computational complexity. (3) It has fewer training parameters, resulting in a faster inference speed of the network model. The network architecture of SegNet is illustrated in Fig. 1.

In the application of crack segmentation, [4] proposed a method that uses a feature pyramid backbone and a bottom-up feed-forward residual structure for limited crack datasets. [21,22] introduced a pyramid and hierarchical enhancement network for pavement crack detection. [6,17] implemented pixel-level segmentation of cracks based on SegNet, verifying the feasibility of SegNet in the field of crack segmentation. However, these methods have the following issues [19,23]: Firstly, these methods tend to focus on the relationship between network depth and global information during information propagation, which may not be suitable for small objects like cracks. Secondly, the ability of the models to extract crack features needs improvement. Lastly, the commonly used cross-entropy loss function in network training does not effectively capture the changes in crack pixels throughout the entire training process, as the number of crack pixels and background pixels in crack images is disproportionate.

In response to the aforementioned issues, we propose an improved crack image semantic segmentation method based on SegNet. The main improvements are as follows: Firstly, considering the complexity of the crack segmentation task, we reduce the network depth of the SegNet basic structure, which improves the accuracy and speed of the network to some extent. Secondly, we replace the original dual-layer convolution structure with ConvNeXtV2 [18] to enhance crack feature extraction by increasing contrast and selectivity among channels. Lastly, we incorporate the Dice coefficient-weighted cross-entropy loss function for loss calculation. Experimental results demonstrate that these improvements have a positive impact on the performance of the detection model.

2 Proposed Methods

2.1 Enhanced SegNet

The architecture of the enhanced SegNet network is illustrated in Fig. 2. It preserves the fundamental symmetric coding and decoding structure of SegNet while eliminating the six-layer convolution and its corresponding structure in the middle of the network. Both the encoder and decoder structures have four stages in terms of feature map size, while SegNet has five stages. In the encoder structure, the network depth is increased through convolutional structures, and the feature map size is reduced through pooling. The decoder structure is symmetric to the encoder structure, with the difference being the use of upsampling to restore the feature map size. Furthermore, in the fourth and fifth stages of both the encoder and decoder structures, the original dual-layer convolution structure of SegNet is replaced with the ConvNeXtV2 module, which enhances the feature representation of crack and further optimizes the number of network parameters. It is worth noting that pooling and upsampling does not require model learning. As shown in Fig. 3(a), the max pooling index table represents the positions of the maximum values obtained during the pooling process in the encoding stage. During the upsampling process in the decoding stage, the values are restored according to the corresponding positions in the index table. Compared to other methods of the same type, the use of pooled indexes in the decoding process on the one hand reduces the accuracy loss of the upsampling operation, on the other hand, it does not require additional parametric quantities for learning, but at the same time, it has the advantage of transpose convolution. In addition, the corresponding loss function is also used for the problem of a small percentage of cracked pixels in the training process of the network model. Then in the following sections, we will provide detailed explanations of each component.

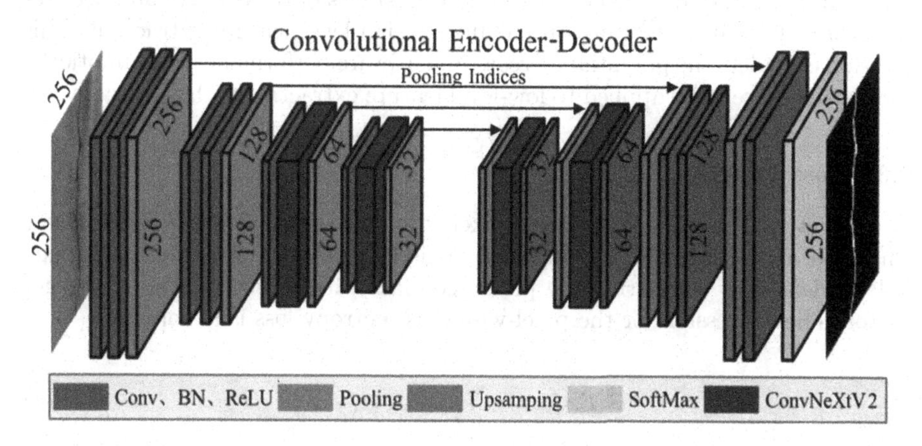

Fig. 2. An illustration of the architecture of our proposed enhanced SegNet.

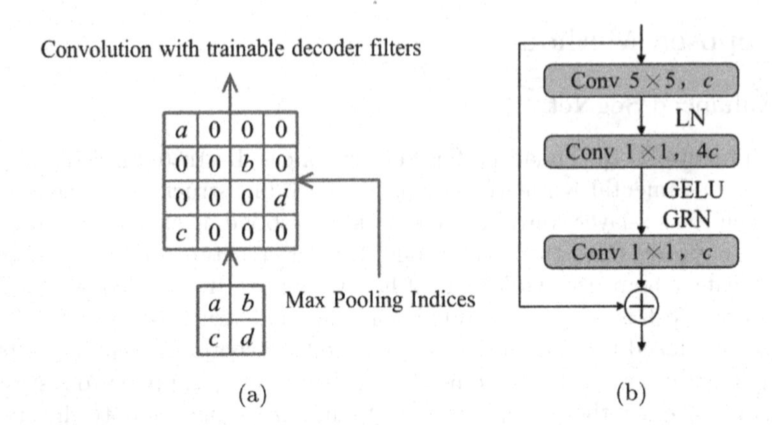

Fig. 3. (a)Upsampling process using pooling index. (b)Network structure of ConvNeXtV2 module.

2.2 ConvNeXtV2 Module

As depicted in Fig. 3 (b), the ConvNeXtV2 module, incorporating Global Response Normalization (GRN) and leveraging existing methodologies, forms a network structure. In this figure, the variable c represents the number of channels in the feature map. The larger convolutional kernels (e.g. 5×5 kernel size) enable the module to capture a broader receptive field. The GRN unit comprises three components: global feature aggregation, feature normalization, and feature calibration. The objective of the GRN unit is to enhance the contrast and selectivity of channels. By aggregating global features, normalizing them, and performing feature calibration, the GRN unit effectively improves the contrast and selectivity of channels in the feature map. It is worth noting that the module is specifically inserted into the fourth and fifth stages of the encoder and decoder structures, primarily focusing on enhancing high-level feature extraction. This decision considers the potential increase in hardware performance consumption if the module were to be applied to low-level feature extraction within the network.

2.3 Loss Function

In conventional segmentation tasks, pixel-wise cross-entropy loss (CE) is widely employed as the primary loss function. This loss function evaluates each pixel independently by comparing the predicted class with the corresponding target vector. The expression for the pixel-wise cross-entropy loss is as follows:

$$\mathcal{L}_{CE} = -\sum_{c=1}^{M} y_c \log(p_c) \tag{1}$$

where M represents the number of classes, y_c is a vector with elements that take only 0 and 1, and p_c represents the probability of the predicted sample belonging to each class. The cross-entropy loss function performs well in common semantic segmentation tasks. However, when dealing with crack images, there is often an imbalance between crack pixels and background pixels. Therefore, simply averaging the loss per pixel using the cross-entropy loss may not be suitable for addressing this imbalance. The Dice coefficient is a similarity measurement function commonly used to compute the similarity between two samples. It is often applied to assess the performance of segmentation tasks. The expression for the Dice coefficient is as follows:

$$\mathcal{L}_{Dice} = 1 - \frac{2|X \cap Y|}{|X| + |Y|} \tag{2}$$

where X represents the pixel labels, and Y represents the predicted pixel classes. In this paper, a weighted loss function is used that combines the Dice coefficient with the CE loss for loss calculation. The expression for the total loss $\mathcal{L}_{Total}$is as follows:

$$\mathcal{L}_{Total} = 1 + 0.6 \times \mathcal{L}_{CE} - 0.4 \times \mathcal{L}_{Dice} \tag{3}$$

3 Experiments

3.1 Experimental Setting and Datasets

The experiments are conducted using an AMD Ryzen7 5800H Processor with 16GB RAM, and an NVIDIA GeForce RTX 3060 Laptop with 6GB RAM GPU. The deep learning framework is Pytorch, and Adam with a momentum of 0.9 is chosen as the optimizer during training. the initial learning rate and minimum learning rate are set to 10-4 and 10-6, respectively, and the learning rate descent formula is cos. Due to memory constraints, the number of multi-threads and batch size is set to 4, and 50 epochs are used for training. The training process is analyzed using different sets of hyperparameters to select the model configuration for optimal validation. In order to evaluate the effectiveness of the method, the experiments used 2000 non-steel crack images and 2000 steel crack images from the Bridge Crack Library (BCL) [9], which is included in the Harvard database, for further validation. Both datasets are randomly divided into a training set, validation set, and test set according to the ratio of 8:1:1. All images in the experiments are 256 × 256 in size, RGB three channels, and the Masks are converted to single channel 8-bit grayscale maps of the same size. The experiments mainly compare the segmentation performance of this method on the non-steel crack dataset and also use the steel crack dataset for generalizability verification. In this study, the primary evaluation metrics used are the mean Intersection over Union (mIoU) and Frames Per Second (FPS), Accuracy, F1-score, and Mean Pixel Accuracy (mPA).

3.2 Ablation Experiments

To verify whether the optimization of the present method has a positive impact on the network performance, ablation experiments are performed on different network depths, adding the ConvNeXtV2 module and using the improved loss function on the non-steel crack dataset, respectively. The experiments are divided into five groups in total: the first group is the performance of SegNet; the second and third groups are the performance after cutting the network into four and three stages, respectively, with different gradient representations as shown in Sect. 1.1; the fourth group uses the ConvNeXtV2 module on top of the network in the second group, as shown in Fig. 2; the fifth group is the performance with the improved loss function; the sixth group is the pooling and upsampling operations with pooling indexes in the fifth group of networks are replaced by using convolution and deconvolution to up- and down-dimension the feature map dimensions.

From the experimental results in Table 1, it can be seen that the performance of SegNet under five to three stages is represented sequentially in the first three sets of experiments, and the stages represent the network depth in this paper, and it is obvious that the best performance is achieved at four stages, with the main performance index mIoU reaching 81.01%. The second and fourth sets of experiments, on the other hand, verify the improvement of the model performance after using the ConvNeXtV2 module. The fifth set of experiments verifies the performance improvement after using the loss function of this paper with the above improvements, and the final Accuracy is 98.93%, mIoU reaches 82.37%, F1-score is 89.43%, and mPA is 89.92, which are all significant improvements compared with the original SegNet. In addition, the last set of experiments shows a decrease in performance metrics compared to the fifth set of experiments, which illustrates the advantage of pooled indexing in crack semantic segmentation.

Table 1. Performance of different network structures

Methods	Accuracy	mIoU	F1-score	mPA
SegNet	0.9881	0.8010	0.8777	0.8643
Cut×4	0.9886	0.8101	0.8843	0.8771
Cut×3	0.9870	0.7800	0.8623	0.8358
ConvNeXtV2	0.9891	0.8121	0.8864	0.8658
TConv	0.9885	0.8148	0.8879	0.8957
Ours	**0.9893**	**0.8237**	**0.8943**	**0.8992**

3.3 Comparison with State-of-the-Art

Accuracy indicates the number of correctly predicted pixels as a percentage of all pixels, as shown in Fig. 4(a). The Accuracy metric of this method performs well on the non-steel dataset and the steel dataset, but the simple use of the Accuracy metric does not serve as a basis for evaluating the good or bad performance of the model because the cracked pixels account for a relatively small percentage of the cracked images.

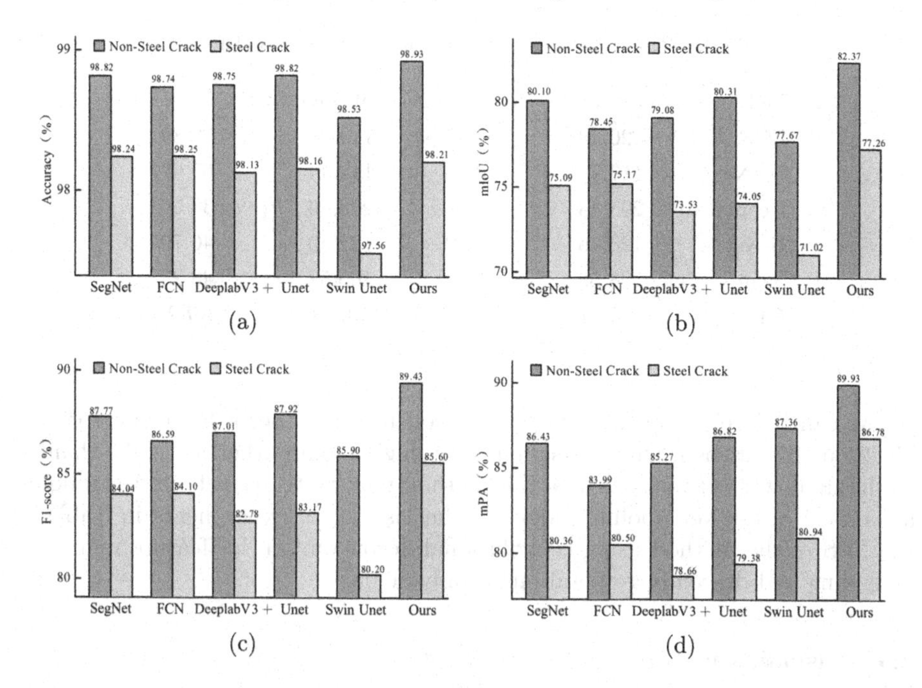

Fig. 4. (a)Comparison of Accuracy for different methods. (b)Comparison of mIoU for different methods. (c)Comparison of F1-score for different methods. (d)Comparison of mPA for different methods.

The mIoU comparisons of different methods are shown in Fig. 4(b). The mIoU metrics of this method perform well on the non-steel dataset and the steel dataset, with significant improvement relative to other methods, reaching 82.37% and 77.26% on the non-steel dataset and the steel dataset, respectively.

The F1-score and mPA comparisons of different methods are shown in Fig. 4(c) and 4(d). Obviously, the F1-score and mPA indexes of the method in this paper are significantly better than other methods both on the non-steel dataset and on the steel dataset, which reflects the superiority of this method.

Hardware bottlenecks usually become the main factor limiting functionality when performing crack segmentation method deployment. Table 2 mainly compares the number of parameters, floating point operation requirements, and stored model file size of the model of this paper's method with other segmentation methods. The number of model parameters in this paper is only 35.26M, the FLOPs is 17.29, and the stored file size is 33.68MB, which is significantly lower than other models.

Speed is an extremely important metric when the crack segmentation method is deployed, and a small and fast network has a great advantage in mobile deployments where performance is not high. Figure 4(b) mainly compares the difference in speed between this paper's method and other segmentation methods, where the test image is an RGB image of size 256×256.

Table 2. Performance of different network structures

Methods	Param (M)	FLOPs (G)	Storage (MB)	FPS
FCN	120.48	806.55	114.94	8.89
SegNet	117.78	40.08	112.44	61.67
Deeplabv3+	208.43	20.66	208.86	36.06
U-Net	128.36	50.70	122.50	46.79
Swin U-Net	108.58	76.90	105.52	43.23
Ours	**35.26**	**17.29**	**33.68**	**69.29**

As can be seen from Table 2, the method in this paper achieves an FPS of 69.29 on this testbed, which is significantly higher than other methods, thanks to the shallower network depth of this method relative to SegNet and the unique advantage of SegNet pooling index. In the last set of experiments in Table 1, the FPS of the method using convolutional deconvolution for feature map size ascending and descending operations is only 50.08.

3.4 Visualization

The visualization results of the ablation experiments are shown in Fig. 5, where the first column is the image to be detected, the second column is the real value of the crack mask, and the last six columns indicate the results of the actual detection of the corresponding item of the title, respectively. It can be seen that the detection results are superior and inferior basically match the performance index results, and the present method is optimal in terms of pixel continuity and accuracy, which is closest to the mask, while the third group of experimental effects is the worst due to the network retaining only three stages and lacking the acquisition of advanced features in the image.

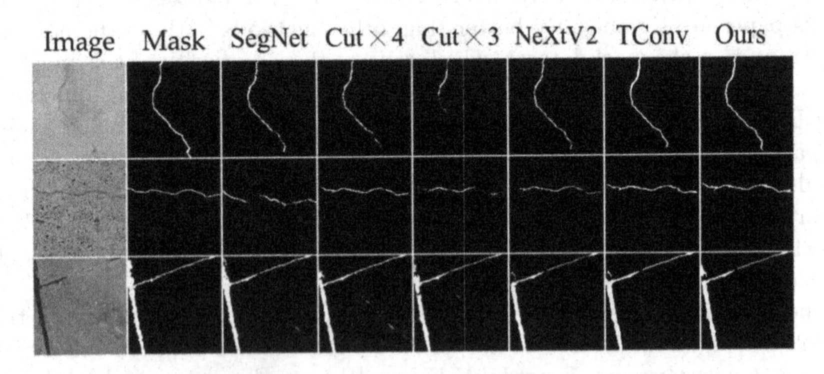

Fig. 5. The visualization of ablation experiments detection effects

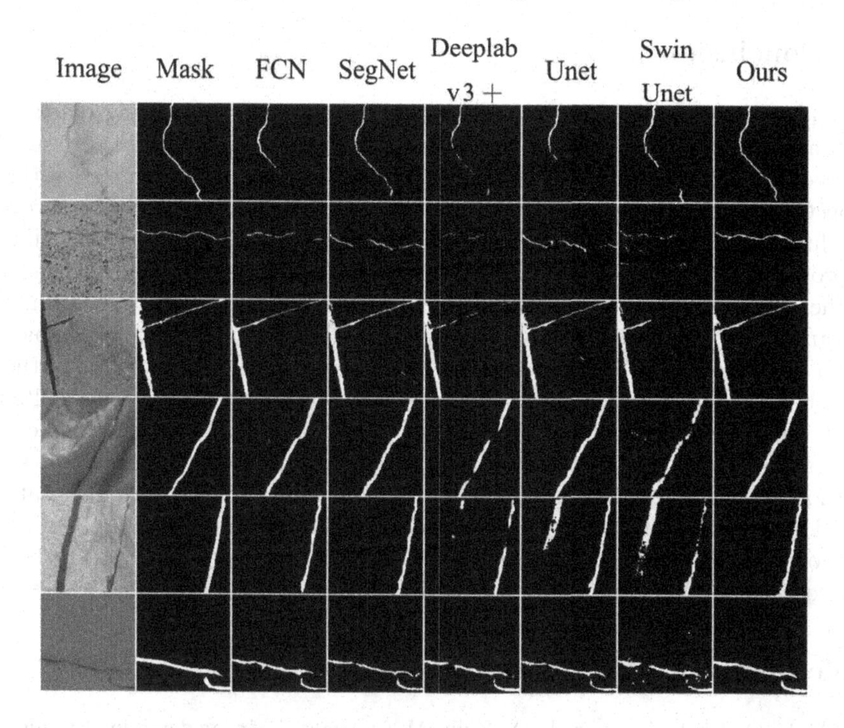

Fig. 6. The visualization of different methods detection effects

The visualization results of different methods are shown in Fig. 6. The first column is the image to be detected, the second column is the real value of the crack mask, and the last six columns indicate the actual detection results of different methods, respectively. The first three rows show the detection effect of cracks in non-steel structures, and the last three rows show the detection effect of cracks in steel structures.

As can be seen from the detection results in Fig. 6, Swin U-Net is less effective, and the method in this paper has the most complete and pure segmentation, while all other methods have the present images of orphaned pixels, segmentation discontinuities, and missegmentation. In particular, on the segmentation task of the fifth image, DeeplabV3+, U-Net, and Swin U-Net all have different degrees of missegmentation due to the presence of crack-like color interference in the image. In addition, a large number of discontinuous segmentation pixels and orphaned pixels within the image may have a small impact on the segmentation index, but in the actual segmentation task, these pixels can greatly affect the detector's judgment of crack type, size, and shape, which is undesirable, and this side reflects the superiority of the present method.

4 Conclusion

To address the problems of low accuracy and slow speed of existing crack segmentation methods, this paper proposes an improved SegNet and the research carried out for semantic segmentation of crack images mainly includes three aspects: reduced network depth, borrowed ConvNextV2 module and improved loss function in the training process. The experimental results show that all the improvements involved in this method have a positive impact on the performance of the crack segmentation model, and the final main performance index mIoU is significantly improved relative to SegNet, and also higher than other methods of the same type. In addition, from the actual segmentation results of these models on randomly selected crack image samples, the segmentation results of the present method are closest to the true values, and there are almost no cases of mis-segmentation and missed segmentation. At the same time, the present method outperforms other methods of the same type in terms of segmentation speed and is more suitable for deployment in real detectors. Therefore, the next step of work will be to deploy the model to perform actual detection tasks on devices such as wall-climbing robots and UAVs.

References

1. Abdel-Qader, I., Abudayyeh, O., Kelly, M.E.: Analysis of edge-detection techniques for crack identification in bridges. J. Comput. Civil Eng. **17**(4), 255–263 (2013)
2. Badrinarayanan, V., Kendall, A., Cipolla, R.: SegNet: a deep convolutional encoder-decoder architecture for image segmentation. IEEE Trans. Pattern Anal. Mach. Intell. **39**(12), 2481–2495 (2017)
3. Cao, H., et al.: Swin-Unet: unet-like pure transformer for medical image segmentation. In: European Conference on Computer Vision, pp. 205–218. Springer (2022). https://doi.org/10.1007/978-3-031-25066-8_9
4. Carr, T.A., Jenkins, M.D., Iglesias, M.I., Buggy, T., Morison, G.: Road crack detection using a single stage detector based deep neural network. In: 2018 IEEE Workshop on Environmental, Energy, and Structural Monitoring Systems (EESMS), pp. 1–5. IEEE (2018)
5. Chen, L.-C., Zhu, Y., Papandreou, G., Schroff, F., Adam, H.: Encoder-decoder with atrous separable convolution for semantic image segmentation. In: Proceedings of the European Conference on Computer Vision (ECCV), pp. 801–818 (2018)
6. Chen, T., et al.: Pavement crack detection and recognition using the architecture of SegNet. J. Ind. Inf. Integr. **18**, 100144 (2020)
7. Chong, W.-K., Low, S.-P.: Assessment of defects at construction and occupancy stages. J. Perform. Constr. Facil. **19**(4), 283–289 (2005)
8. Delagnes, P., Barba, D.: A markov random field for rectilinear structure extraction in pavement distress image analysis. In: Proceedings of International Conference on Image Processing, vol. 1, pp. 446–449. IEEE (1995)
9. Jin, T., Li, Z., Ding, Y., Ma, S., Ou, Y.: Bridge crack library. In: Harvard Dataverse (2021)
10. Jin, X., et al.: Development of nanomodified self-healing mortar and a u-net model based on semantic segmentation for crack detection and evaluation. Construct. Build. Mater. **365**, 129985 (2023)

11. Lei, B., Wang, N., Pengcheng, X., Song, G.: New crack detection method for bridge inspection using UAV incorporating image processing. J. Aerosp. Eng. **31**(5), 04018058 (2018)
12. Long, J., Shelhamer, E., Darrell, T.: Fully convolutional networks for semantic segmentation. In: Proceedings of the IEEE Conference on Computer Vision and Pattern Recognition, pp. 3431–3440 (2015)
13. Munawar, H.S., Hammad, A.W.A., Haddad, A., Pereira Soares, C.A., Waller, S.T.: Image-based crack detection methods: a review. Infrastructures **6**(8), 115 (2021)
14. Otsu, N.: A threshold selection method from gray-level histograms. IEEE Trans. Syst. Man Cybern. **9**(1), 62–66 (1979)
15. Qin, Y., Dong, S., Pang, R., Xia, Z., Zhou, Q., Yang, J.: Design and kinematic analysis of a wall-climbing robot for bridge appearance inspection. In: IOP Conference Series: Earth and Environmental Science, vol. 638, 012062. IOP Publishing (2021)
16. Ronneberger, O., Fischer, P., Brox, T.: U-Net: convolutional networks for biomedical image segmentation. In: Navab, N., Hornegger, J., Wells, W.M., Frangi, A.F. (eds.) MICCAI 2015. LNCS, vol. 9351, pp. 234–241. Springer, Cham (2015). https://doi.org/10.1007/978-3-319-24574-4_28
17. Song, C., et al.: Pixel-level crack detection in images using SegNet. In: Chamchong, R., Wong, K.W. (eds.) MIWAI 2019. LNCS (LNAI), vol. 11909, pp. 247–254. Springer, Cham (2019). https://doi.org/10.1007/978-3-030-33709-4_22
18. Woo, S., et al.: ConvNeXt V2: co-designing and scaling convnets with masked autoencoders. In: Proceedings of the IEEE/CVF Conference on Computer Vision and Pattern Recognition, pp. 16133–16142 (2023)
19. Xu, X., Nguyen, M.C., Yazici, Y., Lu, K., Min, H., Foo, C.-S.: Semi-supervised curvilinear structure segmentation. Semicurv. IEEE Trans. Image Process. **31**, 5109–5120 (2022)
20. Yamaguchi, T., Nakamura, S., Saegusa, R., Hashimoto, S.: Image-based crack detection for real concrete surfaces. IEEJ Trans. Electr. Electron. Eng. **3**(1), 128–135 (2008)
21. Yang, F., Zhang, L., Sijia, Yu., Prokhorov, D., Mei, X., Ling, H.: Feature pyramid and hierarchical boosting network for pavement crack detection. IEEE Trans. Intell. Transp. Syst. **21**(4), 1525–1535 (2019)
22. Zhang, J., Qian, S., Tan, C.: Automated bridge crack detection method based on lightweight vision models. Complex Intell. Syst. **9**(2), 1639–1652 (2023)
23. Zheng, Y., Gao, Y., Lu, S., Mosalam, K.M.: Multistage semisupervised active learning framework for crack identification, segmentation, and measurement of bridges. Comput.-Aided Civil Infrastruct. Eng. **37**(9), 1089–1108 (2022)

Graph-based Dynamic Preference Modeling for Personalized Recommendation

Jiaqi Wu, Yidan Xu, Bowen Zhang, Zekun Xu, and Bohan Li(✉)

Nanjing University of Aeronautics and Astronautics, Nanjing, China
bhli@nuaa.edu.cn

Abstract. Sequential Recommendation (SR) can predict possible future behaviors by considering the user's behavioral sequence. However, users' preferences constantly change in practice and are difficult to track. The existing methods only consider neighbouring items and neglect the impact of non-adjacent items on user choices. Therefore, how to build an accurate recommendation model is a complex challenge. We propose a novel Graph Neural Network (GNN) based model, Graph-based **D**ynamic **P**reference Modeling for **P**ersonalized **R**ecommendation (DPPR). In DPPR, the graph attention network (GAT) learns the features of long-term preference. The short-term graph computes items' dependencies on link propagation between items and attributes. It adjusts node features under the user's views. The module emphasizes skip features among entity nodes and incorporates time intervals of items to calculate the impact of non-adjacent items. Finally, we combine their representations to generate user preferences and aid decisions. The experimental results indicate that our model outperforms state-of-the-art methods on three public datasets.

Keywords: Sequential recommendation · Graph neural network · User preferences

1 Introduction

Recommender system uses advanced algorithms and technologies to analyze user behavioral data to understand user preferences and requirements. In early work, the recommender algorithms are generally based on collaborative filtering [13], Matrix factorization [7], and Markov chains [12]. However, these methods ignore the evolutionary trend of user preferences and cannot capture users' temporal behavior well. The current mainstream methods for recommendation are based on deep learning methods such as Graph Neural Network (GNN) [17,22,24], Graph Convolutional Network (GCN) [4,21,23], Attention [8,19,20]. These methods learn the dependencies between users and items and typically deal with complex recommendation tasks.

D.-N. Yang et al. (Eds.): PAKDD 2024, LNAI 14647, pp. 356–368, 2024.
https://doi.org/10.1007/978-981-97-2259-4_27

Although these methods are essential for the ability to improve recommendation accuracy, there are still two limitations in practical work: (1) The model only examines users' current interests and ignores the global features of sequence and the temporal features between items. (2) User interaction items are usually based not only on neighbors but might be affected by non-neighboring items. As shown in Fig. 1, clothing and bags are the most popular shopping types selected by users, which are their long-term preferences. Users have recently become interested in electronics, but the two shopping behavior types overlap, and there are different time intervals between other behaviors. Based on our analysis, the items users access are more connected to other entities through public attributes. In addition, users match a pair of shoes to a dress they buy and an electronic pen to a tablet. Therefore, the time interval between the entities in the link and the non-adjacent items also plays an auxiliary role in the user's decision.

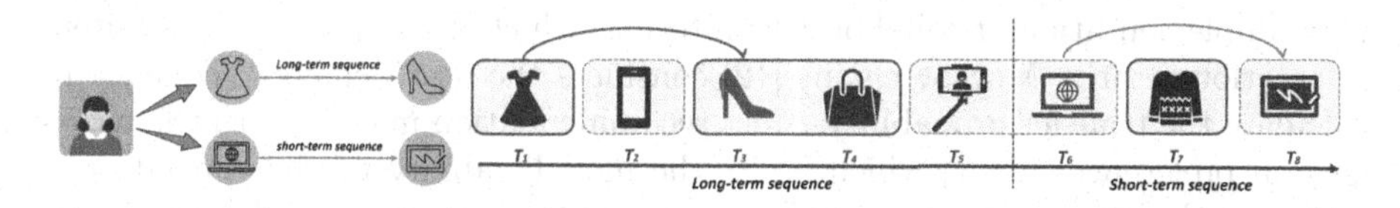

Fig. 1. An example of user sequence in E-commerce.

We propose DPPR model to solve the above problems and achieve personalized user sequence recommendations. Unlike the methods that only focused on the users' current sequence in the past, our model built modules for long-term and short-term sequence. We first encode the time position of the project for the long-term sequence. Next, we construct the global graph and use GAT to calculate the user's long-term preferences to assist the short-term preferences calculation. As for the short-term sequence, we also construct a graph, which takes users, items and attributes as inputs. It calculates the multi-hop neighbours according to the link propagation in the graph and introduces the user group perspective to supplement the item nodes. Besides, we also consider the influence of non-adjacent items and time intervals on user's decision. Finally, we adaptively integrate users' long-term and short-term preferences to get the final interest. DPPR effectively simulates user dynamic preferences change and improves the recommendation performance. We summarize the main contributions to this paper as follows:

- We construct a long-term and short-term sequential recommendation framework, which can effectively recommend the accuracy of tasks.
- The short-term module considers the path propagation of user-item-attributes in the interaction graph while adjusting the item embedding from the user's perspective, introducing time intervals, skipping features between different items in the modeling sequence, and obtaining current preferences.
- The long-term module encodes the position embedding with time intervals, which ensures the temporal orderliness of the items in the sequence. Then, global graph dependencies are constructed based on the graph network.

- To verify the effectiveness of DPPR, we conduct comprehensive experiments using three real-world data sets as the testing basis. Our model outperforms the state-of-the-art methods, as demonstrated by the results.

2 Related Work

2.1 Sequential Recommendation

Sequential recommendation tasks the user with predicting the next item that the user might be interested in based on the user's sequence. Traditional recommendation methods generally rely on displayed user behavioral data and item attribute information to recommend items most likely to match the user's preferences. For example, the collaborative filtering-based method [13] computes the similarity between users and items for recommendation. Matrix factorization [7] is an implementation of collaborative filtering, which decomposes the user-item interaction matrix. Markov chains [12] considers the user behavior sequence to calculate the transfer probability. Most recommendation models learn the user's general preferences alone, which limits the model's ability to recommend. The reasons include user sequence with long-term item dependencies and recent interaction changes.

2.2 Graph-Based Recommendation

Nowadays, various methods based on graph neural networks have been applied to recommendation system improvement. GNN allows the construction of a user-item interaction graph and then learning the nodes' embedding representation to improve the recommendation system accuracy. Liu et al. [9] build a hypergraph self-supervised learning model based on the Knowledge Graph (KG) for modeling complex relationships. Xia et al. [21] propose a dual channel hypergraph convolutional network for recommendation tasks and enhance hypergraph modeling. Wang et al. [18] construct the two graphs from the global and the current session view and fuse their representations after continuously updating the nodes in the graph. Chen et al. [1] propose a graph network-based temporal recommendation model named LESSR, which makes innovations in graph generation methods by aggregating the neighborhood node for learning. Pang et al. [11] construe a heterogeneous global graph to enhance user preferences representation with personalized recommendations utilizing interaction item transformations.

3 Problem Formulation

In our recommendation task, we utilize $U = \{u_i\}_{i=1}^{|U|}$ and $V = \{v_i\}_{i=1}^{|V|}$ to denote the set of users and items, where $|U|$ and $|V|$ represent the number of users and items, respectively. $\mathcal{S}_{(u_i)} = \{\mathcal{S}_{(u_i)_1}, \mathcal{S}_{(u_i)_2}, \dots, \mathcal{S}_{(u_i)_n}\}$ denotes the sequence of each user's interactions, where $\mathcal{S}_{(u_i)_j} = \left\{v_1^{\widehat{ij}}, v_2^{\widehat{ij}}, \dots, v_\ell^{\widehat{ij}}\right\}$ is listed in temporal sequential order. It is the $j-th$ session of user u_i and ℓ represents the session length. $\widehat{v_t^{ij}} \in V$ is the item which user u_i interacts with at time t. Furthermore, in order to improve the recommendation quality, we transform all the training sessions into a graph $\mathcal{G}_u = (\mathcal{E}_u, R_u)$ and use the graph to develop a study. The graph consists of the entity-relation-entity triplet $\{(h,r,t)|h \in \mathcal{E}_u, t \in \mathcal{E}_u, r \in R_u\}$, where $\mathcal{E}_u$ and R_u are the set of entities and relations of $\mathcal{G}_u$. For each user $u_i \in U$, we define the user's long-term sequence as $L^{(u_i)} = \left\{\mathcal{S}_{(u_i)_j}\right\}_{j=1}^{|T|}$, and the short term sequence as $S^{(u_i)} = \left\{\mathcal{S}_{(u_i)_j}\right\}_{j=|T|+1}^{n}$. Similarly, we transform the sessions into the long-term and the short-term graphs $\mathcal{G}_L$ and $\mathcal{G}_S$, respectively. Our task aims to recommend top-N items for users.

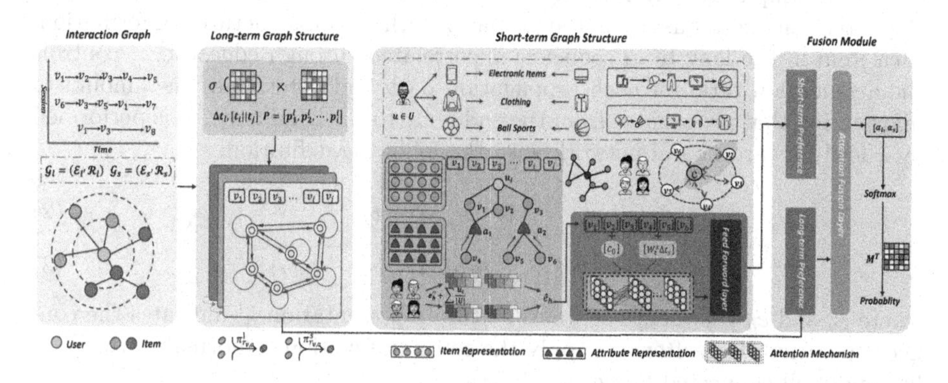

Fig. 2. The overall architecture of DPPR.

4 The Proposed Method

The overall workflow of DPPR is shown in Fig. 2. We describe the key components in the following sections from three aspects: long-term global graph, short-term global graph, preferences fusion and prediction.

4.1 Long-Term Graph Model

Time Embedding Layer. In the history sequence, some users interact frequently, with a large number of items they have interacted with in one day or one week, while other users may not. The layer first considers the temporal vectors for user interactions in the long-term sequence. The reason is to better describe the relationship between any two items. We define the interaction time t_i for item i and t_j for item j. A shorter time interval usually expresses tighter relationships between items compared to longer intervals. Here we only focus on the relative intervals $\Delta t_l = |t_i - t_j|$, while adding a learnable position embedding matrix $P = [p_1^l, p_2^l, \ldots, p_i^l]$ for items. Formally, we get the following item i position expression:

$$pt_i^l = \sigma(W_t^l[t_i \,||\, t_j] + \Delta t_l b_t^l)p_i^l, \tag{1}$$

where W_t^l and b_t^l are parameters of the long-term sequence. $\sigma(\cdot)$ is the sigmoid activation function.

Global Graph Construction. In a user-item graph, their interaction relationship reflects the degree of user preference, with node features containing abundant information about user interactions. To fully capture the long-term dependencies of the sessions, inspired by KGCN [16], we express the optimized long-term sequence input as a feature graph $\mathcal{G}_l = (\mathcal{E}_l, \mathcal{R}_l)$. The entity items' neighborhood field is represented as $\mathcal{R}_l^h = \{a|(v, r, a) \in \mathcal{D}_l\}$.

Given a candidate entity, we utilize the learnable linear transformation to convert the input vectors into high-level features and then derive weights with the attention mechanism. In the scenario with graph-structured information, each item in the long-term sequence is correlated through edges. $\pi_{r_{v,a}}^l$ controls the attenuation ratio for each propagation path and determines the amount of information that propagates from the end node to the head node. It is performed by a feed forward neural network with the following definition:

$$e_{r_{v,a}}^l = W_e^l(e_a||e_r^l||pt), \tag{2}$$

$$\pi_{r_{v,a}}^l = (W_{h1}^l e_v)^T tanh(W_{h2}^l e_{r_{v,a}}^l + b_l) \tag{3}$$

where e_v and e_a represent the items feature representation, $||$ indicates the connect operation. Hereafter, we apply the softmax function to normalize the coefficients of all connected triples as follows:

$$\pi_{r_{v,a}}^l = \frac{exp(\pi_{r_{v,a}}^l)}{\sum_{k \in \mathcal{N}_l} exp(\pi_{r_{v,k}}^l)} \tag{4}$$

where $\pi_{r_{v,a}}$ denotes the importance ascribed to node e_v concerning the feature of node e_a. It limits the amount of information propagated from the triplets.

The interaction items are computed by linear combinations of node features and with a sigmoid function to obtain the nodes' output vectors e_{ui}^l in the session, which ultimately make up the final output features E_u^l.

$$e_{uv}^l = ||_{l=0}^L \sigma(\sum_{a \in \mathcal{N}^l} \pi_{r_{v,a}}^l W_v^h e_a) \tag{5}$$

4.2 Short-Term Graph Model

Item-Attribute Graph Construction. GNN primarily focus on aggregating information from neighboring nodes to enhance the feature representation of nodes. [25] In the constructed short-term graph $\mathcal{G}_s = (\mathcal{E}_s, \mathcal{R}_s)$, we define two types of edges (i.e., user-item and attribute-item). The user-entity interaction items can infer user preferences, and the model recommends items with the same attributes for users to improve recommendation accuracy. A set of neighbors constituting k-hops is defined as $\mathcal{R}_s^h = \{g|(h, r, g) \in \mathcal{D}_s\}$.

The target user might have different preferences for various aspects of a particular interaction item. $\pi^s_{r_{h,g}}$ is introduced to describe the entity e_g's attention score specific to entity e_h [10]. Different neighbors of the target node can distinguish various contributions of the current node during information aggregation.

Based on the above illustration, the calculation process of target node information aggregation is as follows:

$$\pi^s_{r_{h,g}} = \frac{exp(ReLU(W^l_1 e_h \,||\, W^l_2 e_g \,||\, e^s_r))}{\sum_{k \in \mathcal{N}_s} exp(ReLU(W^l_1 e_h \,||\, W^l_2 e_k \,||\, e^s_r))}, \tag{6}$$

where $W^l_1, W^l_2 \in \mathbb{R}^{d \times d}$, e^s_r represents the embedding vector of the entity relationship r. Similar to the long-term graph, $\pi^s_{r_{h,g}}$ normalizes the neighborhood weights with the softmax function. After that, we calculate the linear combination of the given entity h to get its neighborhood embedding as follows:

$$e^{'}_h = \sum_{g \in \mathcal{N}_s} e_g W^s \pi^s_{r_{h,g}}, \tag{7}$$

where $e^{'}_h$ is the updated node feature, and e_g is the node embedding. By combining the multi-hop neighborhood features to supplement the node characteristics in the short-term graph [25].

$$e^{H+1}_h = W^s_h e^{'(H)}_h + b^s_h, \tag{8}$$

where H denotes the graph network's number of layers. We perform multilayer aggregation and use non-linear activation to yield the feature vector of the user interaction items $e^{''}_h$. To enhance the interaction between users and entities, following [14], we introduce a user-centric entity projection module. It enriches items features from multiple user views.

$$\vec{e}_h = tanh((e^{''}_h + \sum_{i \in |U|} \frac{u_i}{|U|})\hat{W}_1 + \hat{b}_1), \tag{9}$$

where $\hat{W}_1$ and $\hat{b}_1$ are parameters of the module. The final result is the item features $\vec{e}_h$ adjusted by the user projection.

Skip Features Mechanism. In the user behavioral sequence, neighboring items can influence interaction items, and the non-neighboring items can also affect the user's selection results. For example, the user bought some snacks and appliances. After a period of time, the user purchased shelves to place the snacks. There is no directly time link between the two items. Inspired by [3], we combine the skip features with the attention module while adding time intervals at session nodes to get the user's personalized features. Specifically, we introduce a virtual central node to which any two nodes in the graph can connect.

Firstly, each item node $n^l_{u_i}$ in the session graph aggregates its neighbor node $h^l_{u_i}$. Splicing context vectors and time vectors updates the neighbors. The item nodes aggregate neighborhood and global features as follows:

$$h_{ui}^{f} = MLP(\vec{e}_{hi-1} \,||\, \vec{e}_{hi+1} \,||\, W_t^s \Delta t_s), \tag{10}$$

$$n_{ui}^{f} = ReLU(W_n[h_{ui}^{f-1} \,||\, c^{f-1} \,||\, \vec{e}_{hi}]), \tag{11}$$

where Δt_s is the time interval between session nodes. W_* denotes the model's trainable linear projection matrix.

Secondly, the central node aggregates the item node with the previous state's node in the graph. For the given $c_u^{(f-1)} \in \mathbb{R}^d$, we update c_u^f based on the attention mechanism, which is proposed by [15]:

$$SDPA = softmax(\frac{QK^T}{\sqrt{d}})V \tag{12}$$

With the scaled dot-product attention (SDPA), the virtual central node could collect and consolidate information from all item nodes. Then, the module concats the previous state and user-projection item's embedding.

$$vc_u^{f-1} = ReLU(W_c[c_u^{f-1} \,||\, \vec{e}_h]), \tag{13}$$

$$c_u^f = SDPA(Q = vc_u^{f-1}, K = c_u^{f-1}, V = c_u^{f-1}), \tag{14}$$

where $W_c \in \mathbb{R}^{2d \times d}$. After F rounds of updates, similar to the central node, we utilize the attention mechanism and the feed forward neural network to obtain the final features of the short-term graph nodes e_u^s. It allows the extraction of short-term preferences E_u^s.

$$e_u^s = FFN(SDPA(Q = c_u^F, K = n_u^F, V = n_u^F)) \tag{15}$$

4.3 Preference Fusion and Prediction

In this section, DPPR fuses the user's long-term preference E_u^l and short-term preference E_u^s adaptively to generate the user's final preference P_u:

$$P_u = \alpha_l W_u^l e_u^l + \alpha_s W_u^s e_u^s + b_u, \tag{16}$$

$$[\alpha_l, \alpha_s] = softmax(W_\alpha^T [e_u^l, e_u^s]), \tag{17}$$

where $W_u^l, W_u^s, W_\alpha, b_u$ are parameters. α_L and α_s are the weights for long and short-term preferences. Based on the preference representation P_u and the initial embedding of candidate item M, we define the model output as the dot product of both as follows:

$$\hat{y}_i = softmax(p_u^T m_i), \tag{18}$$

where $\hat{y}_i$ is the recommended probability of the current candidate item. The model uses the cross-entropy loss as its loss function.

$$\mathcal{L}_{\mathcal{G}} = -\sum_{i=1}^{|V|} y_i log(\hat{y}_i) + (1 - y_i) log(1 - \hat{y}_i) \tag{19}$$

where $y_i \in \mathbb{R}^{|V|}$ represents the ground truth item's one-hot vector.

5 Experiments

In this section, we conduct experiments to evaluate the performance of our proposed method. Our experiments are based on three real datasets and aim to address three key questions (RQs):

- RQ1: How does the proposed DPPR perform in the sequential recommendation task compared with the most advanced baseline model?
- RQ2: Does the central node of the short-term module take effect?
- RQ3: What are the effects of various Hyper-parameter settings on DPPR?

5.1 Datasets

We employ Amazon-Beauty, Last-FM and Taobao to evaluate the recommendation effect of DPPR in the experiments. For each data set, we divide them into train/validation/test at a ratio of 70%/20%/10%. Table 1 shows the statistics of the data set. To improve the quality of them, we discard users and items less than five times.

Table 1. Statistics of the datasets.

Dataset	Amazon-Beauty	Last-FM	Taobao
#Users	3,805	21,752	86,709
#Items	15,462	43,694	30,536
#Interactions	59,865	139,095	384,150

5.2 Baselines

In this section, we introduce seven state-of-the-art recommendation methods and take them as the baselines to compare the performance of DPPR. There are the details of these models:

- POPRec: It is a simple recommendation ranking method that directly recommends the top-k items with the highest occurrence frequency in the datasets.
- FPMC [12]: It is based on matrix decomposition and Markov chains and is a hybrid personalized recommendation model.
- GRU4Rec [5]: It applies GRU to predict the next item in the user's sequence within an improved RNN in the recommender scenario.
- SASRec [6]: It uses sub-attention to capture the importance of different positions in the user's sequence and better understand their preference evolution.
- LESSR [1]: Its main task is to mine the temporal patterns in the sequence, using GNN to process user behavior efficiently.
- DRL-SRe [2]: It utilizes a time-slicing GNN method to model sequence from the global while capturing more detailed time.

- HG-GNN [11]: It constructs a heterogeneous global GNN model that employs a graph-enhanced encoder to get a better dynamic representation of users' preferences.
- DPPR: Our proposed model.

5.3 Experimental Setup

To evaluate the performance of models, we employ the widely used metrics: Hit Rate (HR@K) and Normalised Discounted Cumulative Gain (NDCG@K). We set K $=\{10, 20\}$. In the experiments, we keep the batch size constant at 256, the latent vector's dimension fixed at 64, and the hop number at 3 of the short-term graph. For the proposed DPPR, we utilize the Adam optimizer to optimize the model. The learning rate is set to be 0.001.

5.4 Overall Comparison (RQ1)

The empirical results of various models are shown in Table 2. We mark the best performance in bold and the second best in underlined.

Table 2. Performance comparisons of all models on datasets.

Dataset / Metric	Amazon-Beauty				Last-FM				Taobao			
	N=10	/	N=20		N=10	/	N=20		N=10	/	N=20	
Model	NDCG	HR	NDCG	HR	NDCG	HR	NDCG	HR	NDCG	HR	NDCG	HR
POPRec	0.0106	0.0311	0.0278	0.0432	0.0225	0.0419	0.0398	0.0617	0.0363	0.0564	0.0413	0.0791
FPMC	0.0325	0.0473	0.0397	0.0668	0.0409	0.0694	0.0589	0.0902	0.0421	0.0739	0.0673	0.1095
GRU4Rec	0.0577	0.0851	0.0794	0.0952	0.0631	0.1092	0.0807	0.1359	0.0833	0.1164	0.0934	0.1543
SASRec	0.0671	0.1007	0.0817	0.1215	0.0725	0.1287	0.0902	0.1623	0.0972	0.1503	0.1160	0.2056
GCE-GNN	0.0832	0.1128	0.0935	0.1426	0.0935	0.1321	0.1015	0.1855	0.1128	0.2025	0.1376	0.2391
LESSR	0.1073	0.1327	0.1118	0.1831	0.1026	0.1566	0.1235	0.1983	0.1371	0.2103	0.1592	0.2407
HG-GNN	0.1012	0.1669	0.1046	0.1702	0.1059	0.1924	0.1322	0.2270	0.1559	0.2247	0.1942	0.2752
Our model	**0.1091**	**0.1712**	**0.1149**	**0.1903**	**0.1105**	**0.2086**	**0.1392**	**0.2441**	**0.1621**	**0.2399**	**0.2047**	**0.2889**

We notice the following observations from Table 2. Traditional recommendation methods, such as POPRec and FPMC, rely solely on user feedback to make predictions. They cannot outperform other methods and cannot simulate the constantly changing interests of users, which is the reason behind this issue. GRU4Rec and SASRec consistently outperform POPRec and FPMC. As the amount of data increases, the performance ratios of these methods outperform traditional methods. It shows that users' dynamic preferences sometimes play a more significant role in decision-making. Furthermore, all the GNN-based recommendation methods outperform the other methods, where HG-GNN shows the most significant results in datasets. It accounts for the complex user-project interactions and item correlation. DPPR outperforms all other models on all datasets and evaluation metrics. It offers features combining session-based interactions with long-and short-term sequences.

5.5 Ablation Analysis (RQ2)

In short-term sequence modeling, We introduce a virtual central node to capture skipped user preferences features. This section considers whether the centre node is an experimental condition. The Short-NC model updates nodes sequentially based on the order of links in the session, excluding the central node. The results presented in Table 3 indicate the optimal performance of DPPR in both cases. It demonstrates the effectiveness of the central node. The central node acts as a hub that connects other nodes, allowing it to convey the global characteristics of sequence throughout the graph.

Table 3. Comparison of virtual central node processing methods.

Model \ Dataset / Metric	Amazon-Beauty		Last-FM		Taobao	
	Short-NC	DPPR	Short-NC	DPPR	Short-NC	DPPR
HR@10	0.1657	**0.1712**	0.2004	**0.2086**	0.2325	**0.2399**
HR@20	0.1852	**0.1903**	0.2320	**0.2441**	0.2744	**0.2889**
NDCG@10	0.1069	**0.1091**	0.1062	**0.1105**	0.1576	**0.1621**
NDCG@20	0.1095	**0.1149**	0.1313	**0.1392**	0.2022	**0.2047**

5.6 Hyper-parameter Analysis (RQ3)

Influence of Latent Dimensionality. Based on the observations from Fig. 3, it's evident that the size of the embedding dimension significantly impacts the model's performance. In general, the performance of all models improves as the value of d increases. Larger values of d may result in model overfitting.

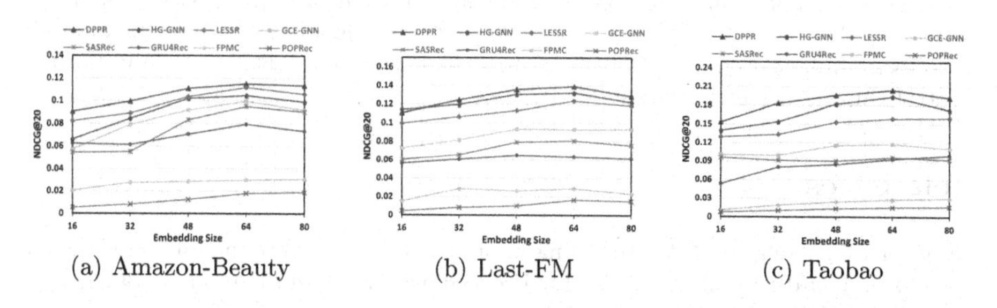

(a) Amazon-Beauty (b) Last-FM (c) Taobao

Fig. 3. Comparison model performance with different embedding settings.

Influence of Sequence Length. We limit the sequence length to improve the handling of sequence data. The results are shown in Fig. 4. We can observe that when length=50, HR@20 and NDCG@20 of DPPR in the Amazon-Beauty performs the best. It indicates that the sequence data at this length best reflects the user's preferences. However, the proper lengths for the three datasets are not the same. In conclusion, we suggest setting appropriate sequence lengths for different data sets to achieve better recommendation results.

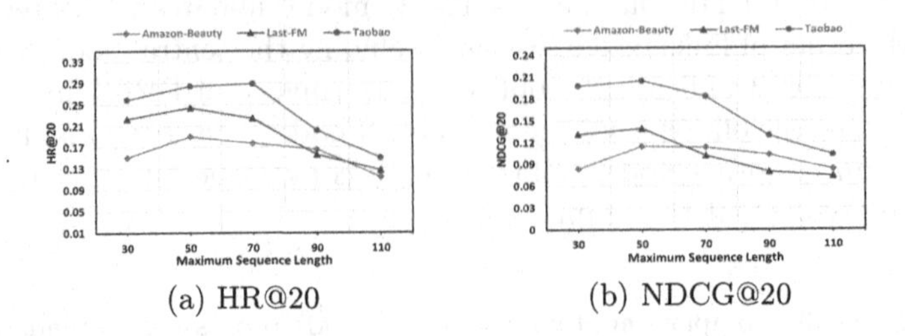

(a) HR@20 (b) NDCG@20

Fig. 4. Comparison model performance with different sequence length.

6 Conclusion

In this work, we proposed DPPR, a novel sequential recommendation model based on graph neural networks. DPPR builds a long-term graph which implements temporal location embedding and GAT. For the short-term graph, it considers the propagation properties of the user-item-attribute link and utilize GNN to process the data. In addition, the skip features and user projection method are introduced to enhance the short-term preferences features. Our model outperforms state-of-the-art models and is demonstrated through multiple experiments on three real-world datasets.

Acknowledgement. This work was supported in part by the "14th Five-Year Plan" Civil Aerospace Pre-Research Project of China under Grant D020101, the Natural Science Foundation of China under Grant No. 62302213, the Natural Science Foundation of Jiangsu Province under Grant No. BK20210280, Project of Hebei Key Laboratory of Software Engineering, No. 22567637H, and the Fundamental Research Funds for the Central Universities under Grant NS2022089.

References

1. Chen, T., Wong, R.C.W.: Handling information loss of graph neural networks for session-based recommendation. In: Proceedings of the 26th ACM SIGKDD, pp. 1172–1180 (2020)
2. Chen, Z., Zhang, W., Yan, J., Wang, G., Wang, J.: Learning dual dynamic representations on time-sliced user-item interaction graphs for sequential recommendation. In: Proceedings of the 30th ACM International Conference on Information & Knowledge Management, pp. 231–240 (2021)
3. Hao, J., Dun, Y., Zhao, G., Wu, Y., Qian, X.: Annular-graph attention model for personalized sequential recommendation. IEEE Trans. Multimedia **24**, 3381–3391 (2021)

4. He, X., Deng, K., Wang, X., Li, Y., Zhang, Y., Wang, M.: LightGCN: simplifying and powering graph convolution network for recommendation. In: Proceedings of the 43rd International ACM SIGIR, pp. 639–648 (2020)
5. Hidasi, B., Karatzoglou, A., Baltrunas, L., Tikk, D.: Session-based recommendations with recurrent neural networks. In: Proceedings of the International Conference on Learning Representations, pp. 1–10 (2016)
6. Kang, W.C., McAuley, J.: Self-attentive sequential recommendation. In: 2018 IEEE International Conference on Data Mining (ICDM). IEEE (2018)
7. Koren, Y., Bell, R., Volinsky, C.: Matrix factorization techniques for recommender systems. Computer **42**(8), 30–37 (2009)
8. Liu, Q., Zeng, Y., Mokhosi, R., Zhang, H.: STAMP: short-term attention/memory priority model for session-based recommendation. In: Proceedings of the 24th ACM SIGKDD International Conference on Knowledge Discovery & Data Mining, pp. 1831–1839 (2018)
9. Liu, Y., Xuan, H., Li, B., Wang, M., Chen, T., Yin, H.: Self-supervised dynamic hypergraph recommendation based on hyper-relational knowledge graph. In: Proceedings of the 32nd ACM International Conference on Information and Knowledge Management, pp. 1617–1626 (2023)
10. Liu, Y., Yang, S., Xu, Y., Miao, C., Wu, M., Zhang, J.: Contextualized graph attention network for recommendation with item knowledge graph. IEEE Trans. Knowl. Data Eng. **35**(1), 181–195 (2021)
11. Pang, Y., et al.: Heterogeneous global graph neural networks for personalized session-based recommendation. In: Proceedings of the Fifteenth ACM International Conference on Web Search and Data Mining, pp. 775–783 (2022)
12. Rendle, S., Freudenthaler, C., Schmidt-Thieme, L.: Factorizing personalized markov chains for next-basket recommendation. In: Proceedings of the 19th International Conference on World Wide Web, pp. 811–820 (2010)
13. Sarwar, B., Karypis, G., Konstan, J., Riedl, J.: Item-based collaborative filtering recommendation algorithms. In: Proceedings of the 10th International Conference on World Wide Web, pp. 285–295 (2001)
14. Tai, C.Y., Wu, M.R., Chu, Y.W., Chu, S.Y., Ku, L.W.: MVIN: learning multiview items for recommendation. In: Proceedings of the 43rd International ACM SIGIR Conference on Research and Development in Information Retrieval, pp. 99–108 (2020)
15. Vaswani, A., et al.: Attention is all you need. In: Advances in Neural Information Processing Systems, vol. 30 (2017)
16. Wang, H., Zhao, M., Xie, X., Li, W., Guo, M.: Knowledge graph convolutional networks for recommender systems. In: The World Wide Web Conference, pp. 3307–3313 (2019)
17. Wang, X., He, X., Cao, Y., Liu, M., Chua, T.S.: KGAT: knowledge graph attention network for recommendation. In: Proceedings of the 25th ACM SIGKDD International Conference on Knowledge Discovery & Data Mining, pp. 950–958 (2019)
18. Wang, Z., Wei, W., Cong, G., Li, X.L., Mao, X.L., Qiu, M.: Global context enhanced graph neural networks for session-based recommendation. In: Proceedings of the 43rd International ACM SIGIR Conference on Research and Development in Information Retrieval, pp. 169–178 (2020)
19. Wu, J., et al.: Time-aware preference recommendation based on behavior sequence. In: Asia-Pacific Web (APWeb) and Web-Age Information Management (WAIM) Joint International Conference on Web and Big Data. Springer (2023). https://doi.org/10.1007/s44196-023-00310-w

20. Wu, J., Zhang, Y., Li, Y., Zou, Y., Li, R., Zhang, Z.: SSTP: social and spatial-temporal aware next point-of-interest recommendation. Data Sci. Eng. **8**(4), 329–343 (2023)
21. Xia, X., Yin, H., Yu, J., Wang, Q., Cui, L., Zhang, X.: Self-supervised hypergraph convolutional networks for session-based recommendation. In: Proceedings of the AAAI Conference on Artificial Intelligence, vol. 35, pp. 4503–4511 (2021)
22. Xuan, H., Li, B.: Temporal-aware multi-behavior contrastive recommendation. In: International Conference on Database Systems for Advanced Applications, pp. 269–285. Springer (2023). https://doi.org/10.1007/978-3-031-30672-3_18
23. Xuan, H., Liu, Y., Li, B., Yin, H.: Knowledge enhancement for contrastive multi-behavior recommendation. In: Proceedings of the Sixteenth ACM International Conference on Web Search and Data Mining, pp. 195–203 (2023)
24. Yu, J., Yin, H., Xia, X., Chen, T., Cui, L., Nguyen, Q.V.H.: Are graph augmentations necessary? Simple graph contrastive learning for recommendation. In: Proceedings of the 45th International ACM SIGIR Conference on Research and Development in Information Retrieval, pp. 1294–1303 (2022)
25. Zhang, C., et al.: Multi-aspect enhanced graph neural networks for recommendation. Neural Netw. **157**, 90–102 (2023)

LEAF: A Less Expert Annotation Framework with Active Learning

Aishan Maoliniyazi[1,2], Chaohong Ma[1], Xiaofeng Meng[1](✉), and Yingtao Peng[1]

[1] Renmin University of China, Beijing 100872, China
{aishan,chaohma,xfmeng,pengyingtao}@ruc.edu.cn
[2] Xinjiang University of Finance and Economics, Urumqi 830012, China

Abstract. Many modern ML applications rely on large amounts of labeled data, which can be difficult and time-consuming to obtain. Active Learning (AL) is an advanced solution that addresses this problem. AL not only enables efficient training with limited data but also speeds up the labeling process and saves on labor costs. However, existing AL methods primarily focus on optimizing the query sampling strategy for single-task and fixed model scenarios, which is inefficient for real-world multi-task scenarios. In multi-task AL, multi-model hyperparameters optimization and multi-query strategies bring new challenges that require more labor. In this paper, we propose LEAF, a Less Expert Annotation Framework, to tackle those challenges and reduce the workload of both data experts and technical experts. In LEAF, we apply AutoML techniques to automatically optimize hyperparameters for multi-task and multi-model AL and design a heuristic adaptive query strategy for multi-query strategy in AL. Experimental results on three publicly available datasets show that our framework requires fewer iterations, less training time, and higher precision than conventional Active Learning frameworks. Additionally, we present a detailed case study that demonstrates the practical use and high quality of our proposed framework for real-world data annotation tasks.

Keywords: Data Annotation · Active Learning · AutoML

1 Introduction

Labeled data are essential for machine learning tasks. In recent years, modern applications have employed multiple machine learning tasks, and each task may rely on specific labeled data related to a separate data annotation task. For instance, in a general Dialog System, in order to comprehend the various questions posed by users, the system needs to be capable of recognizing and processing a variety of tasks including Named Entity Recognition (NER), Sentiment Analysis, Part Of Speech (POS), and more. Multiple machine learning tasks require different labeled data. For instance, NER requires the annotated

D.-N. Yang et al. (Eds.): PAKDD 2024, LNAI 14647, pp. 369–384, 2024.
https://doi.org/10.1007/978-981-97-2259-4_28

entities, and Sentiment Analysis needs to label the polarity. Although various efforts have been made to develop labeled resources, which are annotated by different annotation tasks, they are task-specific and carried out independently of each other. For example, UPenn annotation efforts for the Penn Treebank [1] WSJ segments include annotation of sentences with part-of-speech tags (POS), parse trees, and discourse annotation from the Penn Discourse Treebank [2].

However, as the processes of data annotation are both time-consuming and labor-consuming, we have to consider more advanced techniques to facilitate data annotation for multiple tasks in a system. The proposed Active Learning (AL) paradigm [3] is a more flexible solution to reducer the human labeling effort. In AL, it selects the more informative samples for annotation, which reduces the amount of data to be annotated, thus saving human efforts.

Nonetheless, most of the existing works in AL focus on the query strategy. For example, [4] is for the Sequence Tagging task, [5] is designed for a classification task of open-set data which unknown classes. [6] explored the use of multi-task active learning (MT-AL) for pre-trained Transformer-based NLP models. The multi-task scenarios bring new challenges in multi-type data annotation (i.e., different types of data in different tasks). On one hand, multi-tasks have different models, which need to be optimized separately. Although there have been studies on multi-task active learning, challenges still remain. In existing methods, the model is fixed and hyperparameters are manually selected and optimized. This is particularly inefficient and time-consuming in multi-task scenarios.

In this paper, we propose LEAF, a **L**ess **E**xpert **A**nnotation **F**ramework with Active Learning, to address the challenges of data annotations in multiple tasks. We also take the advanced AL solutions [6,7] to annotate data with less data expert labor. However, different from existing AL techniques, which target a single and specific task and do not consider automatic model optimization, our LEAF is designed to facilitate the general data annotation for multiple tasks and automatically optimize model hyperparameters. Unlike existing active learning methods that generally rely solely on manual annotation, our novel approach synergizes the mechanics of both manual and automatic annotation. LEAF reduces both the labor of data annotation and model optimization in active learning.

At a high level, LEAF has three key components. **First**, LEAF utilizes an adaptive query strategy method in the query procedure to query samples to be annotated. The adaptive method is used to adjust the query strategy according to the current task. **Second** is the annotation process, in which the queried samples by the query procedure are annotated by a human annotator or Large Language Models (to further speed up the annotation process), for the most confident samples are annotated with an automatical process without the need for humans, to reduce the labor of data experts. **Third**, LEAF integrates an Auto Active Learning Model (AALM), which is based on AutoML, to accelerate the optimization procedures for multiple models. The AALM with automatic hyperparameter optimization reduces the labor of model experts.

The three interrelated components make LEAF a more practical active learning framework. As shown in Sect. 4, we conduct extensive experiments on several open datasets and different tasks. The results indicate that LEAF is able to achieve high accuracy while reducing the costs of data annotation and model optimization. In summary, we make the following contributions:

- We propose LEAF, a Less Expert Annotation Framework with Active Learning, that reduces the cost of data annotation.
- We propose an Auto Active Learning Model (AALM) training approach, which is autoML-powered and is able to automatically optimize multiple models with less cost.
- We design an adaptive query strategy adjustment method, which adjusts the query strategy according to the current task and data.
- We utilized the LEAF framework to annotate a real-world corpus of academic literature for entity recognition tasks and subsequently published our findings.

The rest of this paper is outlined as follows. Related work is reviewed in Sect. 2. We then present the framework of LEAF in Sect. 3. We show the experiments in Sect. 4 and give a result and discussion in Sect. 5. Section 6 concludes the paper.

2 Related Work

2.1 Large Language Models for Data Annotation

In machine learning, data annotation involves attaching meaningful labels to raw data such as images, text files, and videos. While data annotation is critical for ensuring the quality and reliability of machine learning models, especially in modern applications with abundant unlabeled data, it is often a time-consuming and labor-intensive task due to the difficulty and expense of obtaining labels. A large language model is trained on a vast amount of data, giving it a remarkable ability to deeply understand text, such as ChatGPT[1]. Recent studies have explored large language models for data annotation. For example, [8] demonstrated that ChatGPT outperforms crowd-workers for the annotation tasks for detecting relevance, topics, and frame. [9–11] apply ChatGPT to zero-shot text annotation, identification, and information extraction.

2.2 Active Learning

Active learning aims to reduce the amount of labeled data required for machine learning tasks by selecting the most informative samples for annotation, leading to greater accuracy using a smaller number of labeled samples than passive learning. There are several studies on Active learning, such as [7] proposes a sampling protocol for the multi-task scenario, and [12] designs an approach to balance the uncertain sampling scores between different tasks in a single model. However, those works primarily focus on the sampling strategy, without considering the multi-model automatic optimization.

[1] www.openai.com.

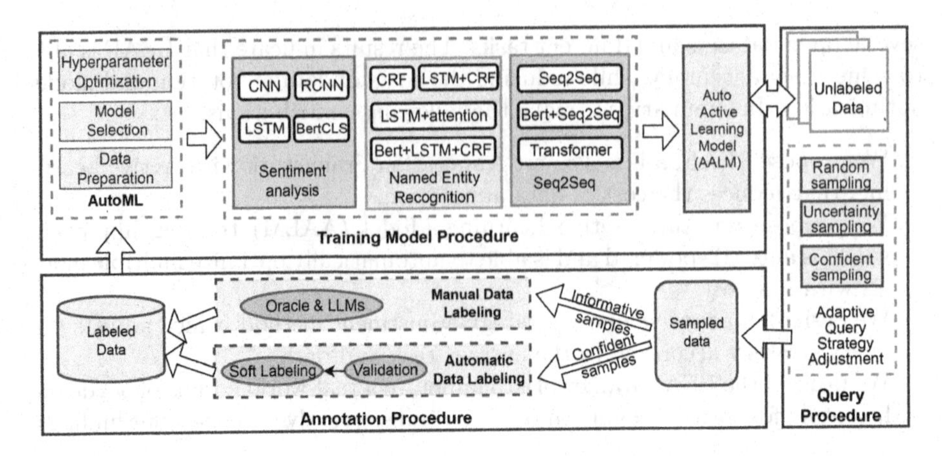

Fig. 1. The LEAF architecture.

2.3 Automatic Machine Learning

Automatic Machine Learning (AutoML) is a technique that automates the machine learning life cycle, including data preparation, feature engineering, model generation, and evaluation, reducing the workload of technical experts and enhancing model robustness. Parameter optimization is vital for model performance, encompassing both model parameters, learned from training data, and hyperparameters, set by the model expert. Common methods for hyperparameter search include Grid Search, Random Search, and Bayesian Optimization Search [13], the latter utilizing the Bayesian theorem to estimate the target function's posterior distribution and select the subsequent hyperparameter combination.

3 The Framework

In this section, we first present the overview of LEAF and its workflow, then detail each component in LEAF.

3.1 Overview of the Framework

LEAF, as illustrated in Fig. 1, consists of three main components. It employs an adaptive query strategy in the Query procedure to select annotation-requiring samples. The Annotation Procedure manually or automatically annotates the sampled data from the Query procedure to obtain labeled data. The Training Model Procedure automates data processing, trains machine learning models for multiple tasks, and optimizes hyperparameters with AutoML's assistance.

Workflow: Given a large amount of unlabeled data, initial random sampling selects a data batch for annotation by the oracle and LLMs. The annotated data is used to train the Auto Active Learning Model (AALM). Upon completion of one iteration, another unlabeled data batch is fed into AALM. The adaptive query strategy module calculates each sample's uncertainty score, selecting confident and informative samples for annotation. Highly informative samples are manually annotated by the data expert, while confident samples are automatically annotated using soft labeling.

Following each data annotation, AALM is utilized. The initial stage involves data preparation, including data cleaning and metadata extraction. The hyperparameter optimization process then estimates the optimal hyperparameters for the dataset. These range from the model's learning rate and batch size to the number of hidden layers, ensuring optimal task performance. Finally, the model is trained for the next iteration.

3.2 Query Procedure

In the query procedure, the model fed a batch of unlabeled data, yielding predicted results and assigning an uncertainty score to each sample. This score reflects the model's confidence in its prediction. The samples with the highest and lowest uncertainty scores are selected as the most uncertain and confident samples, respectively. During the uncertainty score calculation, it is crucial to apply it at the appropriate level - sentence for classification and word for entity recognition. Given the complexity of multitasking, we have developed a solution that discerns the atomic uncertainty calculation unit per task and tailors it to diverse tasks. Further details follow.

Adaptive Query Strategy Adjustment. In multi-task scenarios, differing task outputs necessitate varied query strategies, making individual strategy designs for each task inefficient. To address this, we introduce the concept of an Atomic Target Unit (ATU), representing the smallest basic unit in a task to measure uncertainty. For instance, in classification tasks, a single sentence is the ATU, while sequence tagging tasks and sequence-to-sequence tasks use a single entity and character, respectively. We employ ATUs in uncertainty and confidence computation, adaptively adjusting the query strategy per task. Further details on uncertainty and confident sampling follow.

Uncertainty Sampling. There are many approaches to uncertainty measuring. Entropy [14] is one the popular methods to measure uncertainty. Entropy can represent the uncertainty or impurity in a probability distribution and is widely used in machine learning tasks to identify the most uncertain or unpredictable samples in a dataset [15]. LEAF computes the uncertainty by Entropy, as the sum of the negative probabilities of each possible outcome, multiplied by their logarithms (see Eq. 1).

$$UnCS(s) = \frac{\sum_i^U \operatorname{argmax} - \sum_i P_\theta\left(y_i \mid x\right) \log P_\theta\left(y_i \mid x\right)}{U}. \tag{1}$$

where x denotes an ATU in the sample, y_i ranges over all possible labelings, and x_H^* is the uncertainty of this automatic unit. A higher entropy score x_H^* indicates the unit is of more uncertainty or impurity. Since a sample may have more than one ATU, we use $UnCS(s)$ to compute the uncertainty of a sample. As shown in Eq. 2, U donates the number of ATU in a sample s.

Confident Sampling. Contrary to uncertainty sampling, a low entropy score indicates low uncertainty or impurity, and thus, these samples are considered to be more confident and certain. Thus, we use $x_{H_c}^*$ (see Eq. 2) to compute the confidence of an ATU.

$$CS(s) = \frac{\sum_i^U \text{argmin} - \sum_i P_\theta\left(y_i \mid x\right) \log P_\theta\left(y_i \mid x\right)}{U}. \tag{2}$$

$CS(s)$ represents the confidence of a sample in the current task, where U donates the number of ATU in the sample s.

3.3 Annotation Procedure

The procedure for annotation is a crucial step in the development of our model and involves a dual-pronged approach to categorize selected samples, as outlined in Sect. 3.1. Contrary to the standard practice of solely utilizing manual labeling in active learning methods, our innovative methodology employs a blend of manual and automatic annotation procedures. The selection of samples is bifurcated based on their nature: the informative samples are manually annotated by the oracle or expert human annotator, providing hard labels, At the beginning of annotation LLMs help Oracle cold start, while the confident samples are automatically assigned labels via the Soft Labeling method.

Algorithm 1. Annotation Procedure

Input: $X_{CS}, Model_{task}, Threshold, meanUnS$
Output: Soft Labeled SamplesX_{sl}
$\mathcal{M}_r \leftarrow Model_{task}(X_{CS})$
  for $i \leftarrow |\mathcal{M}_r|$ to $Numoftrials$ do
    $if\ AccuracyCheck(M_{ri}) > ThresholdandUnCS(M_{ri}) > meanUnS$
      $X_{sl} \leftarrow SoftLabeling(M_{ri})$
    $elseif\ CS(M_{ri}) < meanUnS$
      $X_{sl} \leftarrow ManualAnnotation(M_{ri})$
  end for
Return: X_{sl}

Manual Data Annotation. The efficient annotation of uncertain and informative samples can be challenging due to their complex characteristics and consequent difficulty in formulating annotation rules. To address this, Oracle, leveraging its extensive experience and domain knowledge, manually labels these samples, resulting in definitive binary membership. We have developed a user-friendly interface to expedite this process, which assists annotation and suggests potential labels based on annotation history. This solution helps overcome the initial cold-start problem faced by human annotators. To address the cold-start problem, we apply ChatGPT to associate the annotation as shown in Fig. 2. Firstly, we design data annotation prompts for each task and feed the unlabeled data to ChatGPT. Next, the annotated results generated by ChatGPT are presented at the interface to recommend to the human annotator the more possible labels. Finally, the human annotator checks and corrects the annotation results.

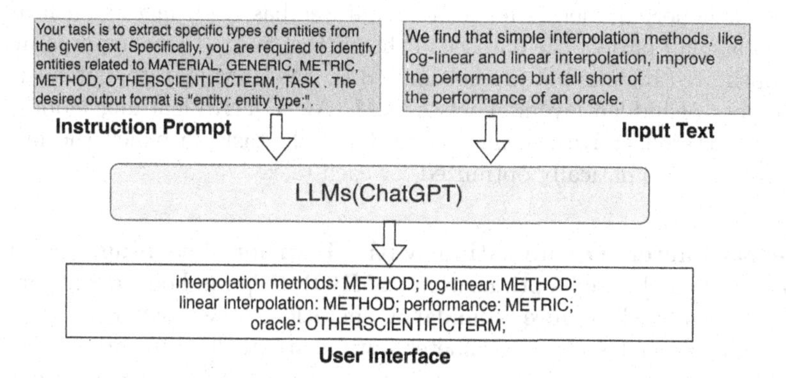

Fig. 2. The LLMs-based data annotation in the process of manual data annotation

Automatic Data Annotation. The types of samples that are clearer and yield higher confidence for the model are simpler to annotate. Therefore, within LEAF, we deploy the Validation and Soft Labeling method [16] for automated annotation of these confident samples. The Confident Samples validate by certain metrics(which precision) threshold. Soft labels assign a relative score to an element, often indicating the probability or likelihood of the element's membership to a particular class, thereby allowing for the possibility of an element belonging to multiple classes.

Active Learning seeks to enrich the training process through the integration of these intentionally unlabeled data. The procedure commences by training a classifier using a collection of hard labels obtained via supervised learning. Following this, the trained classifier is applied to the unlabeled data, assigning soft labels based on the classifier's prediction.

Upon selection of the confident samples, CS(X), by Eq. 2, we deploy soft labeling to designate labels to these samples as demonstrated in Algorithm 1. Ultimately, this results in an expanded dataset comprising both hard and soft labels that prove instrumental for further training.

Firstly, those confident samples X_{CS} are fed into the $Model_{task}$ and return the M_r prediction result. Then, the $AccuracyCheck(M_{ri})$ method checks every atomic target unit in the sample, samples that have a higher accuracy score than a certain $Threshold$ are given a soft label with $SoftLabeling(M_{ri})$ automatically.

3.4 Training Model Procedure

In the training model procedure, the annotated data feeds into AutoML. Firstly, the data preparation which is data cleaning and extracting metadata. Secondly, the transfer learning hyperparameter optimization method estimates the optimal set of hyperparameters for a given dataset based on factors such as data size and a small part of the dataset in the current task. These hyperparameters to be optimized include but are not limited to, the learning rate, batch size, and the number of hidden layers. Then AALM (Auto active learning model) is an AutoML-powered active learning model for multi-task, in which the hyperparameters are automatically optimized for each task.

Hyperparameter Optimization with Transfer Learning. We utilize Sequential model-based optimization (SMBO) [17] method to optimize our multi-task model. To reduce redundancy operation, we employ a transfer learning [18] approach for hyperparameter optimization, utilizing prior knowledge from varied datasets, models, and training processes, leading to an acceleration in the search process.

4 Experiments

In this section, we empirically evaluate the performance of LEAF using different public datasets in multiple tasks. The results show LEAF is able to significantly reduce the labor of data and model experts while maintaining high accuracy. The experiment consists of two parts: an experiment in the annotation process and a case study.

4.1 Experiment in Annotation Process

In the following, we first introduce the datasets and tasks used in the experiments in Sect. 4.1. The experimental settings are given in Sect. 4.2, including the model to be compared, the model hyperparameters, and other settings for the experiment. Then, the results of the experiment are described in Sect. 4.3.

Data. To evaluate the effectiveness of our proposed framework, we conduct experiments using three real-world datasets in two different tasks. We introduce the datasets below:

Table 1. Experiment Datasets.

Dataset	Task	Trainset	Testset
SciERC	NER	5300	800
CLUE2020	NER	8000	2000
yf_amazon	Classification	5000	1000

Table 2. The AL framework model hyperparametrs

Parameter	$CALM_p$	$AALM_p$
Learning Rate	0.0001	1e-05~1e-01
Hidden Layers	2	1~12
Hidden Size	32	16~512
Batch Size	128	16~512
Data Ratio	1.0	0.2~1.0

- SciERC [19], includes annotations for scientific entities, their relations, and coreference clusters for 500 scientific abstracts. These abstracts are taken from 12 AI conference/workshop proceedings in four AI communities from the Semantic Scholar Corpus.
- CLUENER2020 [20], a well-defined fine-grained Chinese dataset for Named Entity Recognition (NER) comes from the CLUEbenchmark[2]. CLUENER2020 contains 10 categories.
- `yf_amazon`[3] Amazon e-commerce platform purchase product review content sentiment analysis dataset, contains three sentiment types.

In the experiment, we split each of the datasets into two sets including the train and test set, as shown in Table 1, to make the results more reliable. The train data are used for train models, and the test data are used to evaluate the performance.

Experimental Design. Our experiment contrasts the Less Expert Annotation Framework (LEAF) with the conventional pool-based Active Learning Framework (CALF) in annotating a dataset for a specific model post an active learning task. The primary aim is to evaluate the annotation process's accuracy and efficiency and compare the trained models' performance.

For a standardized comparison, we also trained a model with the ground-truth dataset (GTDM) to assess the Active Learning (AL) framework-trained models' performance. The experiment employs two AL model frameworks, each with its unique hyperparameters as detailed in Table 2. Here, $CALM_p$ denotes the fixed hyperparameters of the conventional active learning model (CAML), while $AALM_p$ marks the search space for the Auto-Active Learning model's hyperparameters.

During the querying procedure, 1000 unlabeled data units are input into the model. Following output generation, the highest uncertainty samples, determined by a mean uncertainty score, are manually annotated by the oracle. Concurrently, the most confident samples, below the mean uncertainty score, are selected for soft labeling, validated based on a threshold probability set at 0.9. The ChatGPT3.5 API facilitates the initiation of the annotation process.

[2] https://www.cluebenchmarks.com/.
[3] https://www.amazon.com.

Baseline. The baseline framework is a conventional pool-based active learning framework called CALF, which has been modified by integrating the traditional single model pool-based active learning framework. The main difference between LEAF and CALF is that CALF lacks AutoML, LLMs-based annotation, and adaptive query strategy.

Evaluation Metrics. In our study, we have selected multiple evaluation metrics to thoroughly assess the performance and efficiency of the proposed framework. To evaluate the accuracy, we apply precision, recall, and F1-score, which are commonly used metrics in statistical analysis in machine learning tasks. For assessing the efficiency of the framework, we have instituted two critical factors: the number of iterations necessary and the time costs, both empowering us to make a comprehensive evaluation of the proposed LEAF framework.

4.2 Case Study: Real-World Annotation

To further evaluate the efficiency of the proposed LEAF, we apply LEAF to annotate a fine-grained Chinese Scientific Named Entity Recognition dataset (named CSciNER), which includes annotations of scientific terms into 7 entity types (as shown in Table 3).

Table 3. The Annotated Result of Each Entity Type.

Types	Number of Entities
TASK	519
PROBLEM	321
METHOD	1083
MATERIAL	276
GENERIC	435
METRIC	213
OTHER TERM	552

The 1600 Chinese research paper abstracts are from cnki[4], which are unlabeled source data. For annotation, the unlabeled data are further segmented into 28,891 sentences and procedure the annotation.

5 Results and Discussion

This section delves into the comparison of experimental outcomes from two distinct active learning frameworks. The evaluation consists of three parts: first, an

[4] www.cnki.cn.

analysis of the performance of both frameworks (LEAF and CALF) during the active learning process, with a special focus on their individual training durations (refer to Table 4), which are detailed in Sect. 5.1; second, a comparative analysis of the results produced by the final models (AALM and CALM) on the test data post-task completion (refer to Table 5), which are detailed in Sect. 5.2; finally, we present the ability of LEAF in the real-world case study, and the details are shown in Sect. 5.3.

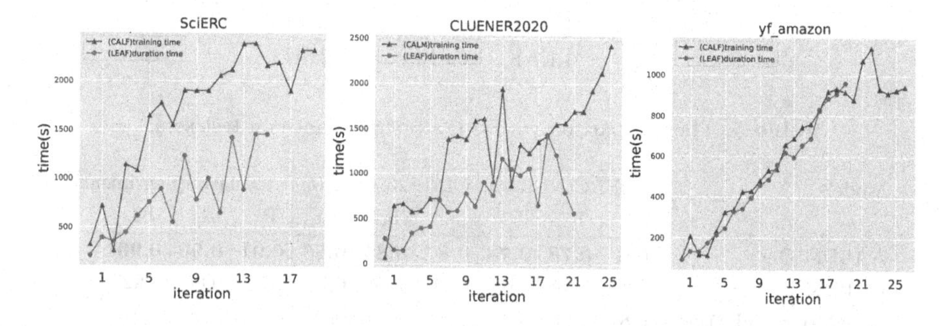

Fig. 3. The time cost in the active learning process.

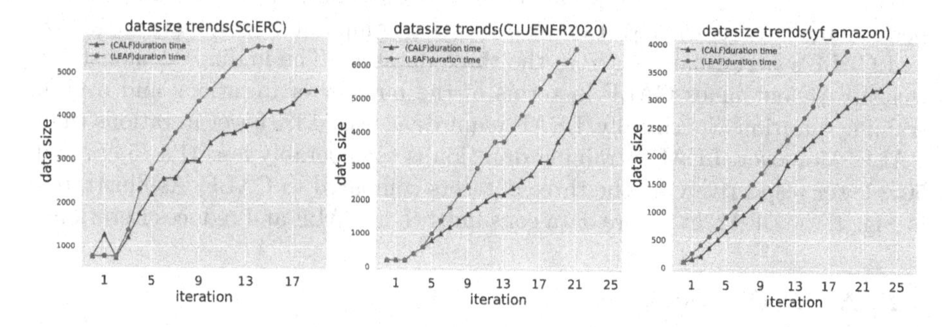

Fig. 4. The size of labeled data in the active learning process.

5.1 Efficiency Evaluation

Our appraisal of efficiency is intrinsic to understanding the performance of our framework. For this purpose, we track the evaluation for both LEAF and CALF at each iteration during the active learning process.

To deepen our insight into the model, we log the training time taken at each round. As illustrated in Fig. 3, AALM's training duration appears to be significantly less than that of the CALF. LEAF's superiority over CALF is clearly exhibited in the SciERC and CLUENER2020 datasets. Even though yf_amazon

Table 4. The total number of iterations and the elapsed time of the active learning process.

Datasets	Framework	Iteration	Time(Hour)
SciERC	CALF	19	8.2H
	LEAF	15	3.8H
CLUENER2020	CALF	24	8.8H
	LEAF	21	4.4H
yf_amazon	CALF	26	4.8H
	LEAF	19	2.8H

Table 5. The comparison with overall performance on test sets.

Model	SciERC(NER)			CLUE2020(NER)			yf_amazon(Sentiment)		
	P	R	F1	P	R	F1	P	R	F1
AALDM(Our)	**0.8**	**0.78**	**0.78**	**0.87**	**0.85**	**0.86**	**0.91**	**0.90**	**0.90**
CALDM	0.69	0.67	0.68	0.81	0.79	0.78	0.88	0.86	0.82
GTDM(Ground Truth)	0.81	0.81	0.79	0.9	0.9	0.89	0.91	0.90	0.91

shows no significant difference in the training time for each iteration, LEAF drives results with fewer iterations than CALF, confirming its overall superior performance and efficiency. Table 4 presents a comparative summary of LEAF and CALF's performances across the three datasets. The findings demonstrate that LEAF outshines CALF in terms of the number of iterations and overall training duration. Specifically, LEAF requires at least 13% fewer iterations than CALF. Moreover, LEAF's training duration is considerably less-41%, 53%, and 50% lower respectively on the three datasets-compared to CALF. As illustrated in Fig. 4, the LEAF has more data consumed than CALF and reduces iterations.

5.2 Accuracy Evaluation

To evaluate the quality of the annotated data, we train models on the labeled data generated by AALDM, CALDM frameworks, and the ground-truth data (referenced as GTDM). The evaluated models are then tested on the given test datasets (see Table 1). Table 5 outlines the evaluation metric results, showing that AALDM surpasses CALDM in terms of generalization capabilities.

The results of F1-score, precision, and recall during the process are presented in Fig. 5. As depicted in Fig. 5, AALM demonstrates a robust beginning with an initial F1 of 0.6 on the SciERC dataset, managing to maintain F1 throughout the active learning process. This contrasts CALM's entrance, which debuts with less precision and a recall, and it requires five rounds to reach the precision level of 0.6. Notably, AALM performs optimally on the CLUENER2020 dataset, where its score climbs from 0.6 and peaks at round 13, whereas CALM starts from zero and hits a precision of 0.6 only at round 17. A similar pattern of AALM outperforming CALM is experienced on the yf_amazon dataset. As a result, our framework's performance remains consistent, proving its model's accuracy throughout the task execution. The outcomes for all three datasets display similar trends: AALDM surpasses CALDM across the precision, recall, and F1 scores. Particularly, AALDM exhibits a lead of at least 3% in precision, 4% in recall, and 8% in F1 score over CALDM. As such, the accuracy of our LEAF's data annotation process proves to be superior to that of CALF, promising excellent generalization capacity and high precision.

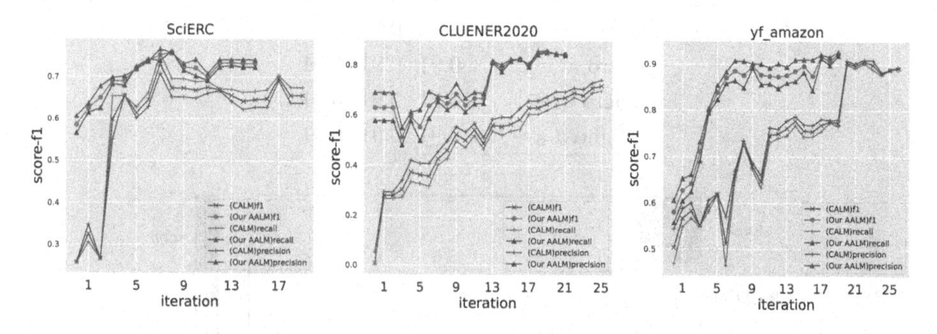

Fig. 5. The F1-score, recall, and precision in the active learning model training.

5.3 Real-World Annotation Ability

For the evaluation of the case study, we choose Two conventional named entity recognition models, LSTM+CRF [21] and Bert+LSTM+CRF [22], are used to evaluate the annotated data. The annotated data were divided into training data and testing data, then, the base models are trained on the training set and tested on the test set. The results are shown in Table 6.

Additionally, we evaluate the entity types separately using the Confusion Matrix(as done in [23]), and the results are shown in Fig. 6. The X and Y axes in the figure mark the recognition rate of each entity type. For example, the recognition rate of the entity type GENERIC, which is recognized as GENERIC, is high, which is why the figure shows a darker color. The probability of all entity types being recognized as their own is high. Therefore, the diagonal line in the figure is darker and the rest is not obvious. It is clear that the annotated data are well annotated and contain less mixed entities. The results of the experiments provide valuable insights into the performance of the proposed framework and highlight the effectiveness of the annotated data in facilitating accurate named entity recognition. As far as we know, there are few Chinese Scientific datasets for research, we hope this dataset serves as a useful reference for future research in named entity recognition and related areas.

Table 6. The Evaluation Results of Each Entity Type.

Entity type	P	R	F1
GENERIC	0.92	0.94	0.93
MATERIAL	0.89	0.89	0.89
METHOD	0.86	0.85	0.86
METRIC	0.93	0.92	0.92
OTHER TERM	0.93	0.90	0.92
PROBLEM	0.98	0.95	0.97
TASK	0.87	0.89	0.88
micro avg	0.91	0.91	0.91
macro avg	0.91	0.91	0.91
weighted avg	0.91	0.91	0.91

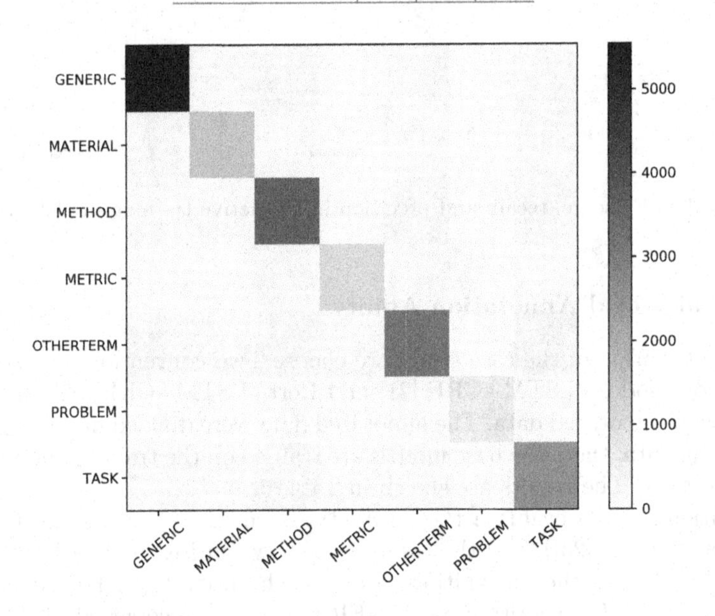

Fig. 6. Entity Prediction Confusion Matrix.

6 Conclusions and Future Work

In this paper, we present LEAF, which stands for a Less Expert Annotation Framework, specifically designed for data annotation in multiple machine learning tasks. The framework particularly aims to reduce the human labor in data annotations. The experimental results demonstrate that LEAF can effectively reduce the reliance on domain experts in both the data annotation and technical aspects of training models while maintaining a high level of accuracy in the annotated data and model training processes. As for future work, we expect to explore the potential of extending LEAF's application to encompass multi-modal data annotation tasks. This expansion aims to enhance LEAF's performance and broaden its usability in the continuously evolving field of machine learning.

Acknowledgements. This work was supported by the National Natural Science Foundation of China (NSFC) via grant 62202043,62172423,62166039,62366050.

References

1. Marcus, M.P., Santorini, B., Marcinkiewicz, M.A.: Building a large annotated corpus of English: the penn treebank. Comput. Linguis. **19**(2), 313–330 (1993)
2. Miltsakaki, E., Robaldo, L., Lee, A., Joshi, A.: Sense annotation in the penn discourse treebank. In: Gelbukh, A. (ed.) CICLing 2008. LNCS, vol. 4919, pp. 275–286. Springer, Heidelberg (2008). https://doi.org/10.1007/978-3-540-78135-6_23
3. Cohn, D.A., Ghahramani, Z., Jordan, M.I.: Active learning with statistical models. J. Artifi. Intell. Res. **4**, 129–145 (1996)
4. Lin, B.Y., Lee, D.-H., Xu, F.F., Lan, O., Ren, X.: AlpacaTag: an active learning-based crowd annotation framework for sequence tagging. In: Proceedings of the 57th Annual Meeting of the Association for Computational Linguistics: System Demonstrations, Florence, Italy, pp. 58–63. Association for Computational Linguistics (2019)
5. Ning, K.-P., Zhao, X., Li, Y., Huang, S.-J.: Active learning for open-set annotation (2022)
6. Rotman, G., Reichart, R.: Multi-task active learning for pre-trained transformer-based models. Trans. Assoc. Comput. Linguis. **10**, 1209–1228 (2022)
7. Reichart, R., Tomanek, K., Hahn, U., Rappoport, A.: Multi-task active learning for linguistic annotations (2008)
8. Gilardi, F., Alizadeh, M., Kubli, M.: ChatGPT outperforms crowd-workers for text-annotation tasks. CoRR, vol. abs/2303.15056 (2023). https://doi.org/10.48550/arXiv.2303.15056
9. Kuzman, T., Mozetic, I., Ljubesic, N.: ChatGPT: beginning of an end of manual linguistic data annotation? Use case of automatic genre identification. CoRR, vol. abs/2303.03953 (2023)
10. Reiss, M.V.: Testing the reliability of ChatGPT for text annotation and classification: a cautionary remark. CoRR, vol. abs/2304.11085 (2023). https://doi.org/10.48550/arXiv.2304.11085
11. Wei, X., et al.: Zero-shot information extraction via chatting with ChatGPT. CoRR, vol. abs/2302.10205 (2023)
12. Rotman, G., Reichart, R.: Multi-task active learning for pre-trained transformer-based models (2022)
13. Snoek, J., Larochelle, H., Adams, R.P.: Practical Bayesian optimization of machine learning algorithms, pp. 2960–2968 (2012)
14. Shannon, C.E.: A mathematical theory of communication. Bell Syst. Tech. J. **27**(3), 379–423 (1948)
15. Settles, B., Craven, M.: An analysis of active learning strategies for sequence labeling tasks, pp. 1070–1079. In: ACL (2008)
16. Galstyan, A., Cohen, P.R.: Empirical comparison of hard and soft label propagation for relational classification. In: Blockeel, H., Ramon, J., Shavlik, J., Tadepalli, P. (eds.) ILP 2007. LNCS (LNAI), vol. 4894, pp. 98–111. Springer, Heidelberg (2008). https://doi.org/10.1007/978-3-540-78469-2_13

17. Hutter, F., Hoos, H., Leyton-Brown, K.: Sequential model-based optimization for general algorithm configuration (extended version) (2010)
18. Yogatama, D., Mann, G.: Efficient transfer learning method for automatic hyperparameter tuning (2014)
19. Luan, Y., He, L., Ostendorf, M., Hajishirzi, H.: Multi-task identification of entities, relations, and coreference for scientific knowledge graph construction. In: Proceedings of the 2018 Conference on Empirical Methods in Natural Language Processing, pp. 3219–3232. Association for Computational Linguistics (2018)
20. Xu, L., et al.: Cluener2020: fine-grained named entity recognition dataset and benchmark for Chinese. CoRR, vol. abs/2001.04351 (2020)
21. Lample, G., Ballesteros, M., Subramanian, S., Kawakami, K., Dyer, C.: Neural architectures for named entity recognition. The Association for Computational Linguistics, pp. 260–270 (2016)
22. Dai, Z., Wang, X., Ni, P., Li, Y., Li, G., Bai, X.: Named entity recognition using BERT BiLSTM CRF for Chinese electronic health records, pp. 1–5. IEEE (2019)
23. Deng, Y., et al.: A Chinese conceptual semantic feature dataset (CCFD). Behav. Res. Methods **53**(4), 1697–1709 (2021)

MLT-Trans: Multi-level Token Transformer for Hierarchical Image Classification

Tanya Boone Sifuentes[1](✉), Asef Nazari[1], Mohamed Reda Bouadjenek[1], and Imran Razzak[2]

[1] Deakin University, Waurn Ponds, VIC 3216, Australia
{tboonesifuentes,asef.nazari,reda.bouadjenek}@deakin.edu.au
[2] University of New South Wales, Sydney, NSW 2052, Australia
imran.razzak@unsw.edu.au

Abstract. This paper focuses on Multi-level Hierarchical Classification (MLHC) of images, presenting a novel architecture that exploits the "[CLS]" (classification) token within transformers – often disregarded in computer vision tasks. Our primary goal lies in utilizing the information of every [CLS] token in a hierarchical manner. Toward this aim, we introduce a Multi-level Token Transformer (MLT-Trans). This model, trained with sharpness-aware minimization and a hierarchical loss function based on knowledge distillation is capable of being adapted to various transformer-based networks, with our choice being the Swin Transformer as the backbone model. Empirical results across diverse hierarchical datasets confirm the efficacy of our approach. The findings highlight the potential of combining transformers and [CLS] tokens, by demonstrating improvements in hierarchical evaluation metrics and accuracy up to 5.7% on the last level in comparison to the base network, thereby supporting the adoption of the MLT-Trans framework in MLHC.

Keywords: Hierarchical classification · Image processing · Transformer · Class tokens · Hierarchy taxonomy

1 Introduction

Traditional image classification models often employ flat classification schemes, treating each category independently and disregarding potential hierarchical relationships that may exist between classes [1]. This lack of hierarchical consideration hinders the ability of algorithms to capture and leverage the underlying semantic structure of complex visual datasets [16]. This limitation is particularly influential in domains such as fine-grained categorization, where classes often exhibit nested relationships [12]. Recent works such as HERBS [6] and Metaformer [7] display the importance of incorporating supplementary information, such as background cues or object attributes, improving classification accuracy. However, HERBS and Metaformer are designed for flat classification,

D.-N. Yang et al. (Eds.): PAKDD 2024, LNAI 14647, pp. 385–396, 2024.
https://doi.org/10.1007/978-981-97-2259-4_29

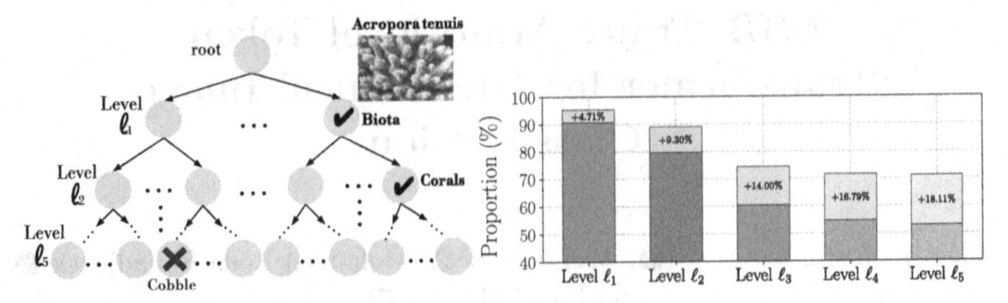

(a) Snapshot of a complex taxonomy. (b) Accuracy for each level of Marine-tree.

Fig. 1. **(a)** An image of an "Acropora tenuis" classified by five independent classifiers as a "Biota", "Corals", and on the last level, incorrectly classified as a"Cobble" for the Marine-tree dataset. **(b)** Proportion of correctly classified images (depicted in a darker color) for each level of the taxonomy for Marine-tree dataset and the proportion of images incorrectly classified (depicted in a light color) but for which the other levels in the taxonomy were correctly identified.

ignoring the potential benefits that could come from incorporating hierarchical relationships as supplementary information.

MLHC has been attracting attention over the past few years mainly because it can provide order and structure to complex real-life datasets. In contrast to flat classification, Multi-level hierarchical classification (MLHC) aims to correctly classify objects organized in a tree-based taxonomy [15]. Moreover, it significantly contributes to the development of recommendation systems, image captioning, annotation, scene graph generation, and visual question answering (VQA) leading to more accurate and interpretable data analysis [19]. To the best of our knowledge, the majority of approaches for MLHC frequently neglect the incorporation of taxonomy structures and employ backbone networks whose outputs lack mutual constraint, leading to potential inconsistency in predictions [21,25,27]. While efforts have been made within the MLHC field, there is still considerable scope for future research in developing architectures that address these limitations. To illustrate the aforementioned advantages of MLHC, Fig. 1 provides a concrete example featuring the classification of an image by five distinct independent classifiers. Figure 1b displays the proportions of accurate classifications at each hierarchical level, represented in a darker color. Furthermore, these figures include the proportion of images that were misclassified at that particular level but were correctly identified in previous levels represented in a lighter color. If we hypothesize that coarser levels of the hierarchy exhibit higher accuracy compared to finer-grained levels, it becomes evident that promoting communication between these coarser levels can enhance the classification accuracy of the fine-grained levels. This highlights our belief that MLHC warrants further exploration as an appealing solution for addressing such issues.

2 Related Work

There have been several methods proposed for MLHC, and they can be categorized according to how the hierarchical structure is explored [22]. In particular, we distinguish: (i) *flat classification approaches*, typically involve a complete disregard for the class hierarchy, wherein predictions are made exclusively at the leaf nodes, with the implicit assumption that all ancestor classes are also attributed to the given instance. This provides an indirect solution to the hierarchical classification problem that could become suboptimal when dealing with a high number of classes, images with low intra-class, low inter-class separability or a combination [5,12]; (ii) *local classification approaches*, a multi-class classifier is trained for every parent node within the class hierarchy. In the context of CNNs, we can distinguish HD-CNN [25] as one of the first attempts to embed CNNs using a taxonomy. As other predecessors, it lacks versatility and efficiency as it requires training several classifiers; and (iii) *global classification approaches* where in contrast, a single classifier is employed to manage the complete class hierarchy structure. Within this category, we can identify what is commonly referred to as "branch methods," exemplified by models such as Branch-CNN (B-CNN) [27] and Hierarchical-CNN (H-CNN) [21]. These approaches incorporate a backbone network that produces hierarchical-level outputs ("branches") at specific layers. While these methods improved performance, they are still susceptible to inconsistent predictions due to the absence of inter-level penalization.

As transformers continue to gain efficiency and importance in the field of Machine Learning, there have been some efforts to incorporate a taxonomy within these models. For instance, the Coarse-to-Fine Transformer, as proposed in [5], aims to enhance final classification accuracy. However, similar to other global approaches, it does not account for the hierarchical structure within the data. In contrast, the Nested Hierarchical Transformer (NesT) [26] and the ViT neural tree decoder (ViT-NeT) [14] introduce architectural designs that embrace the concept of taxonomy by generating decision trees. While these models demonstrate promising results, their effectiveness as MLHC methods remains challenging to evaluate because they do not present their results in hierarchical metrics.

The success of transformers can be attributed to several factors, one of them is their robust ability to model long-range dependencies. The Vision Transformer (ViT) proposed by Alexey Dosovitskiy et al. [9] introduced a special token, denominated [CLS] token or classification token, which was used to aggregate information from the entire sequence of the patch tokens from the image. This token was later replaced by average or global pooling as it would serve the same purpose. Within the domain of image segmentation, recent studies [8,24] have emphasized the significance of incorporating [CLS] tokens to exploit class-specific information, ultimately yielding state-of-the-art results. This insight has served as motivation for harnessing class-specific information in the context of MLHC. However, our approach extends beyond the singular use of [CLS] tokens; we also seek to model the interrelationships between multiple levels of [CLS] tokens. In this work, we argue that the majority of existing MLHC transformer-based models do not capture attention between different hierarchical levels and therefore, do not learn interactions between them. To tackle this issue, we propose MLT-Trans, which incorporates [CLS] token interactions between several levels of hierarchy. In summary, our main contributions are: (i) We propose MLT-Trans to exploit class-specific transformer attention for MLHC, (ii) We suggest a simple but effective transformer framework denominated Multi-level Token Transformer (MLT-Trans) which includes a multi-level token strategy to learn class-specific interactions between levels, (iii) We trained our proposed framework with sharpness-aware minimization and a hierarchical loss function based on the knowledge distillation loss, and finally, (iv) we performed extensive experimental evaluation on multiple baselines across diverse hierarchical datasets reporting our results in terms of hierarchical metrics.

3 Mathematical Notation and MLT-Trans

Multi-level Hierarchical Classification: The MLHC problem is defined by [2] as learning a mapping function $f : \mathbb{X} \rightarrow \mathcal{Y}$, which assigns to each feature vector $\mathbf{x}^{(i)}$ a prediction vector $\hat{\mathbf{y}}^{(i)} = (\hat{y}^{[\ell_1]}, \hat{y}^{[\ell_2]}, \cdots, \hat{y}^{[\ell_n]})$ s.t. $\hat{y}^{[\ell_i]}$ is the class label that f assigns for each level ℓ_i.

Taxonomy Encoding: In contrast to flat classification, where classes are perceived as unrelated entities, hierarchical classification entails the organization of classes within a taxonomic structure. In this paper, our focus is only on *tree* taxonomies, which are organized with a hierarchy structure of n levels ℓ_i, such that $\ell_i \subset \mathcal{Y}$, $\ell_1 \cup \ell_2 \cdots \cup \ell_n = \mathcal{Y}$, and $\forall y_j \in \ell_{i+1}, \exists y_k \in \ell_i$ s.t. $y_k \prec y_j$, where $\prec$ is the "subclass-of" relationship [2]. Lastly, we encode the relationship between two successive levels ℓ_i an ℓ_{i+1} in a taxonomy using an $|\ell_i| \times |\ell_{i+1}|$ matrix $M^{[\ell_i, \ell_{i+1}]}$, where the binary value $M^{[\ell_i, \ell_{i+1}]}_{y_k, y_j} \in \{0(y_k \not\prec y_j), 1(y_k \prec y_j)\}$, with $y_k \in \ell_i$ and $y_j \in \ell_{i+1}$.

MLT-Trans: Figure 2 shows an overview of our proposed MLT-Trans. Given an image $\mathbf{I}$, we first follow the conventional SwinTransformer-L [17] to obtain the patch embedding from the last transformer block $\mathbf{Z}$. According to [8], plugging cross-attention mechanisms after the last transformer block avoids large computational costs and additionally, there is no proof that performing these operations in the initial transformer blocks leads to a better performance. To calculate the attention between patch tokens $\mathbf{Z}$ and class tokens for every level ℓ_n, we first define class tokens as a matrix $\mathbf{T}_{\ell_i}$ for each level of the taxonomy. The size of this matrix is the number of classes of each ℓ_i and the embedding dimension of the transformer. To capture interactions between tokens $\mathbf{T}_{\ell_i}$, we calculate the multihead cross-attention [8] between each pair of tokens. Consider H as the number of heads where $h \in \{1, \ldots, H\}$. We have the attention between $\mathbf{T}_{\ell_i}$ and $\mathbf{Z}$ for the hth head as:

$$AT_h = \delta\left(\frac{Q_h(\mathbf{T}_{\ell_{i+1}})K_h(\mathbf{T}_{\ell_i})^T}{\sqrt{d_{K_h}}}\right)V_h(\mathbf{T}_{\ell_1}) \tag{1}$$

where δ corresponds to the Softmax function and $\sqrt{d_{K_h}}$ is the dimension of the key K_h. This attention mechanism occurs for each pair of subsequent levels. To calculate the attention between all levels, we decided to concatenate these attention outputs and use them to calculate the attention between them and the patch tokens $\mathbf{Z}$ similar to Eq. (1). To update our multi-class tokens $\mathbf{T}_{\ell_i}$, we take the concatenated attention output AT and obtain the top-k tokens with the highest activation where $k = \ell_i$. Consequently, we average it on the batch dimension to update our trainable token $\mathbf{T}_{\ell_i}$. Finally, to update our patch tokens $\mathbf{Z}$, we concatenate the norm of the concatenated attention that we calculated before and we represent this as $\tilde{\mathbf{Z}}$.

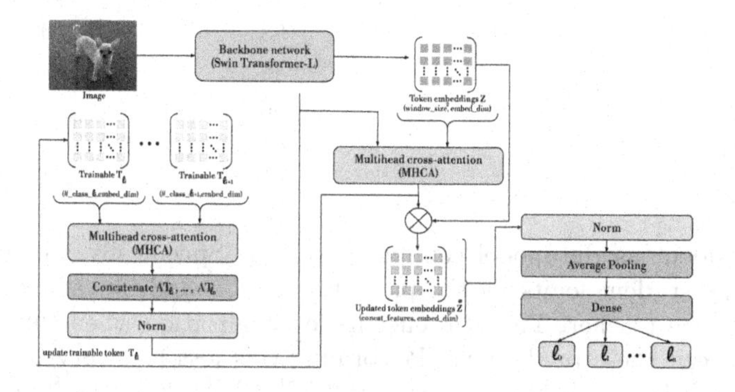

Fig. 2. MLT-Trans architecture.

Hierarchical Loss Function: In the realm of hierarchical image classification, the hierarchical loss function has become a foundational element [1]. Consequently, in this work we propose the computation of a loss, which is the summation of classification losses across all levels ℓ_i. Our strategy takes inspiration from knowledge distillation loss proposed by Hinton et al. [11]. Essentially, this loss function is a pedagogical method in Machine Learning wherein a smaller model (denominated student model) learns from a larger, more complex model (teacher model) by capturing the teacher's learned knowledge, typically in the form of softened probability distributions over classes. This process allows for model compression, enabling more efficient and compact models to retain the insights and generalization abilities of their larger counterparts, which can be especially beneficial for tasks demanding resource-efficient solutions. The knowledge distillation loss (KD_{Loss}) [11] is defined as $KD_{Loss} = CE\big(y, \delta(z_S)\big) + \lambda KL(\delta(z_T/\tau)||\delta(z_S/\tau))$, where z_S are the logits coming from the student model and z_T are the logits from the teacher model. Also, CE stands for cross-entropy, and δ stands for the softmax function; the Kullback-Leibler (KL) divergence loss is used to minimize the discrepancy between the soft outputs probabilities of the student and teacher model. These soft probabilities are obtained by dividing these logits by a temperature factor τ. By transferring KD_{Loss} to the context of MLHC, we can consider the teacher model as the coarser level ℓ_i

Table 1. Summary of the hierarchical datasets employed in our experiments.

Dataset	Stanford Cars	CUB 200-2011	Marine tree	Stanford Dogs	FGVC Aircraft	Food 101
Training set	8,144	5,944	118,260	9,600	3,334	60,600
Validation set	4,020	3,000	16,126	2,400	3,333	15,150
Test set	4,021	2,071	26,798	8,580	3,333	25,250
#classes	196	200	60	120	100	101
Taxonomy						
#classes ℓ_1	13	39	2	8	30	3
#classes ℓ_2	113	123	10	120	70	15
#classes ℓ_3	196	200	38	–	100	101
#classes ℓ_4	–	–	46	–	–	–
#classes ℓ_3	–	–	60	–	–	–
#classes ℓ_3	–	–	60	–	–	–

and the student as the model coming from ℓ_{i+1}. Specifically, we achieve this by replacing student logits by the prediction $\hat{y}^{(j)[\ell_i]}$ and its taxonomic parent $\breve{y}^{(j)[\ell_{i-1}]}$ as the teacher. This will encourage penalization when there is a mismatch between these predictions. Performing this loss for every pair of levels, we obtain HKD_{Loss} as the summation of $\frac{1}{n}\sum_{i=1}^{n}\sum_{j=1}^{m} CCE(y^{(j)[\ell_i]}, \delta(\hat{y}^{(j)[\ell_i]}))$ and $\sum_{i=2}^{n}\sum_{j=1}^{m} \lambda KL(\delta(\breve{y}^{(j)[\ell_{i-1}]}/\tau)||\delta(\hat{y}^{(j)[\ell_i]}/\tau))$, where CCE denotes the categorical cross-entropy function. λ and τ are hyperparameters that need to be tuned to calibrate the importance of HKD_{Loss} and the effect of temperature on the probabilities. Note that to obtain the taxonomic parent of prediction $\hat{y}^{(j)[\ell_i]}$ we use the taxonomy $M^{[\ell_i,\ell_{i+1}]}$ explained in this section.

4 Experimental Evaluation

Our experiments draw upon a set of six hierarchical datasets and eight models which are detailed below. Our results are expressed in terms of hierarchical metrics [15] including Hierarchical Recall (HRecall), Hierarchical Precision (HPrecision), and Hierarchical F1 Score (HF1 Score). The other metrics are Consistency and Exact Match (proportion of examples that are correct and consistent) were also reported by Boone et al. [2].

Datasets: We selected publicly accessible datasets, mainly employed for the purpose of coarse-to-fine classification, and subsequently transformed them into hierarchical datasets. This transformation involved utilizing existing semantic taxonomies wherever available, while also formulating our own taxonomic structures when none were provided. We've summarized this process in Table 1. Specifically, we performed experiments on FGVC-Aircraft [18], hierarchical versions of CUB-200-2011, Stanford Cars, and Marine-tree [3] where you can find their construction details used in [2].

- **Stanford Dogs** [13]: Following the guidelines established by the American Kennel Club (AKC), our classification scheme was aligned to include an additional hierarchical level that corresponds to the AKC's seven distinct dog groups: "hound group", "herding group", "working group", "sporting group", "non-sporting group", "toy group", and "terrier group". We added an additional super-class "wild dogs" for classes such as "dingo", "dhole", and "African hunting dog".
- **Food-101** [4]: We transformed this dataset into a non-overlapping three-tier hierarchical structure. The initial level comprises three categories: "side dish," "main dish," and "dessert". Subsequently, the second level is comprised of 15 classes: "beef dish", "pork dish", "rice dish", "egg dish," "poultry dish," "pasta dish", "seafood dish", "sandwich", "spread", "appetizer", "salad", "soup", "cake", "pastry," and "cup confections".

Implementation Details: For our augmentation and training strategy we followed [6]. The entire experimental setup was implemented using Tensorflow, and was executed on an NVIDIA Tesla V100 GPU with a batch size of 8. Equal weights were assigned to all output components for the loss function. Additionally, we experiment with three different temperature and λ configuration similar to [20] and we selected the best of the three settings **Low(L)**($\tau = 2.5, \lambda = 500$), **Medium(M)**($\tau = 5, \lambda = 1000$), **High(H)**($\tau = 10, \lambda = 2000$). Additionally, our MLT-Trans is trained using sharpness aware minimization (SAM) [10] with default settings. We trained eight state-of-the-art networks that display the highest accuracy on every datasets on their non-hierarchical representation. These baselines are: Swin Transformer, EfficientNetV2, InceptionNext, Uniformer, CoAtNet, CAFormer, TinyViT and MaxViT. You can find more details about the implementation of each baseline on the documentation from KerasCV [23].

Performance: Figure 4 shows the comparison between our proposed architecture against the baseline models. From the obtained results, we make the following key observations: (i) Our proposed MLT-Trans improved their backbone network Swin-L in all datasets for almost all metrics. Especially accuracy ℓ_i and Hierarchical F1-Score. (ii) The biggest increase in accuracy is observed at the last level while ℓ_0 is the level with the least increase in accuracy. This indicates that it is easier to improve the finer levels than the coarser levels even though we have a smaller number of classes. This problem could be related to over-fitting and top-down approaches because there is no feedback from the bottom levels to the top level. (iii) Stanford Dogs is the dataset that benefited the most from our proposed transformer, it improved 5.7% on its last level ℓ_2. (iv) Stanford Cars presents an improvement of 3%, 2.7% and 1.7% respectively compared to Swin-L. Also, there is a 2.5% in HRecall meaning that the model is able to provide examples more relevant to the class. Although there is a small increase in consistency, there is a 3% increase in Exact Match due to the increase in accuracy for every level. Of all the baseline models, MLT-Trans is the model with the best performance in all metrics. (v) Similarly, for CUB-210-2011 MLT-Trans is the model with the best performance in all metrics but with a smaller increase due to CUB-210-2011 being a dataset that is considered harder to train. The final accuracy is 92% which is only 1% less than the current state-of-the-art models HERBS [6] which also utilizes Swin-L as baseline model with the difference that we are not using extra training information such as background datasets. CAFormer presents slightly higher consistency than our method but our method outperforms their accuracy for all ℓ_i. (vi) MLT-Trans for Food-101 presents the highest accuracy against all models for the last two levels with the exception of ℓ_0 where there is a small decrease and lastly (vii) Marine-tree, the most complex dataset to train, presents a 74.58% on its last level (ℓ_5) establishing a state-of-the-art performance. The performance in all metrics is superior by a large margin with respect to the baseline models. The closest model in performance is its baseline model Swin-L. (vii) The number of parameters for Swin-L is $\approx$ 195M parameters, adding SAM minimization adds another 3M parameters and the complete MLT-trans architecture contains 270M parameters. Additionally, we want to point out that for datasets with a 3-level hierarchy, we only calculated the attention between the first two levels, and for Marine-tree we used 3 levels. Using all the levels would cause the attention calculation to slow the training process without a dramatic increase in performance. In conclusion, we observe that our proposed MLT-Trans manages to improve the results of their baseline model for most datasets but other datasets show that Swin-L might not be the best pick. For instance, for Stanford Dogs our InceptionNext implementation manages to beat the state-of-the-art performance by more than 1%, which opens the possibility of using InceptionNext as the backbone network for our proposed architecture. In terms of ablation study, Fig. 3 shows the performance of the backbone network Swin-L, the backbone network plus SAM, and finally our proposed MLT-Trans. The combination of Swin-L plus SAM increases both ℓ_1 and ℓ_2 and MLT-Trans increases them slightly more. We have observed that in some datasets, SAM causes a small decrease in Consistency compared to the backbone network. MLT-Trans is able to recover such loss and in some cases, it improves it. Nevertheless, the increase in accuracy that SAM provides is worth the small decrease in Consistency.

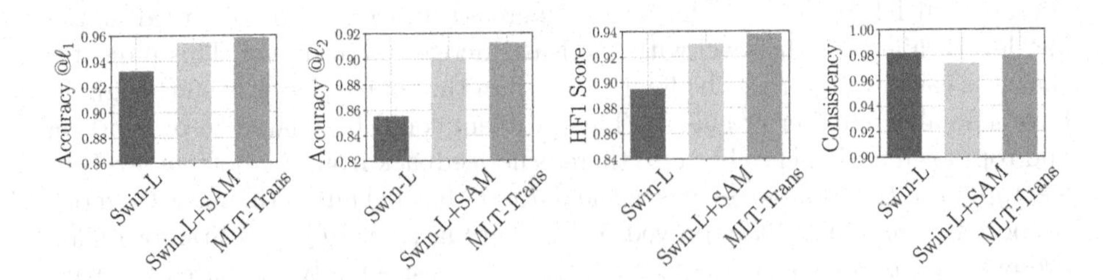

Fig. 3. Ablation study on Stanford Dogs.

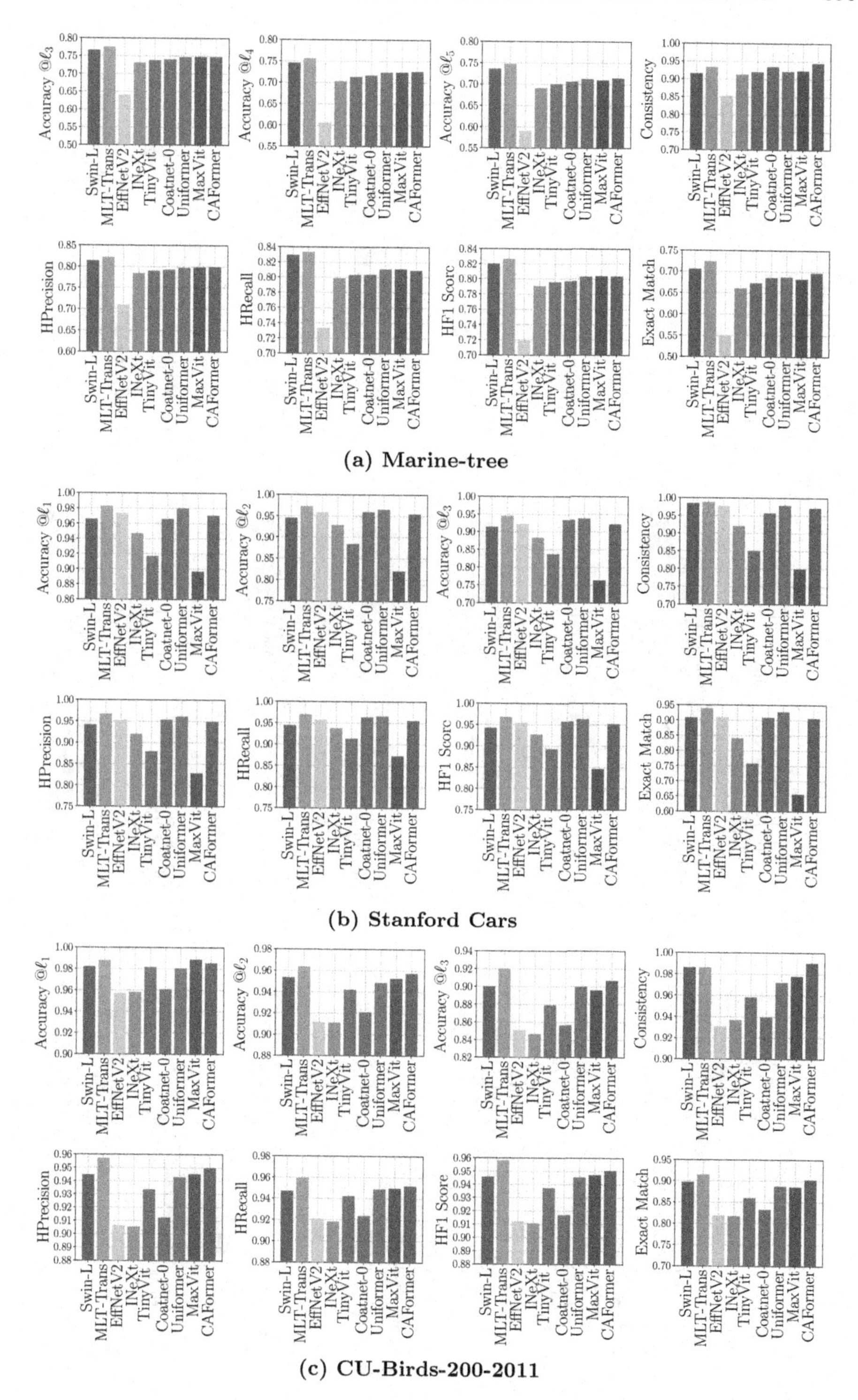

Fig. 4. Performance comparaison.

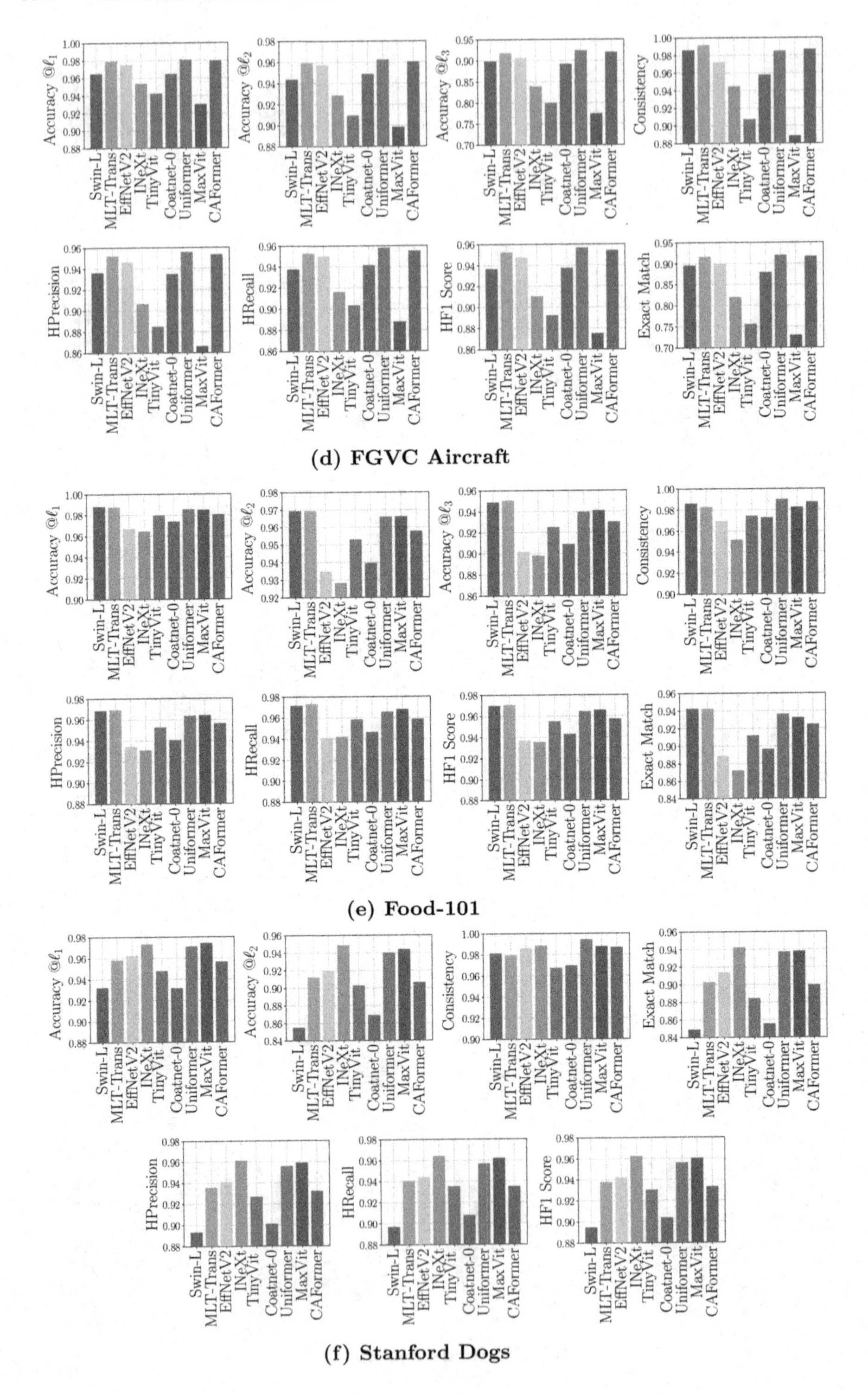

(d) **FGVC Aircraft**

(e) **Food-101**

(f) **Stanford Dogs**

Fig. 4. (*continued*)

5 Conclusions and Future Work

In this work, we introduced a new transformer for Multi-level Hierarchical Classification we denoted Multi-level Token Transformer (MLT-Trans). Our proposed transformer implements an attention mechanism between trainable classification tokens ([CLS] tokens) per level of a predefined taxonomy plus the addition of a hierarchical loss based on knowledge distillation loss. Focusing on image classification, we presented a thorough experimental evaluation of the performance of our method on six different datasets varying from two levels of hierarchy up to five. Experiments on these datasets show that our proposed architecture improves several hierarchical metrics, most importantly accuracy, demonstrating that MLHC could be a strategy to beat state-of-the-art performances without doubling or triplicating the number of parameters. Future work includes testing our proposed methodology on different backbone networks as well as different hierarchical losses. Additionally, we could improve our attention mechanism to take into account all levels of hierarchical datasets with a large number of classes without being computationally expensive.

References

1. Bertinetto, L., Mueller, R., Tertikas, K., Samangooei, S., Lord, N.A.: Making better mistakes: leveraging class hierarchies with deep networks. In: Proceedings of the IEEE/CVF Conference, pp. 12506–12515 (2020)
2. Boone-Sifuentes, T., Bouadjenek, M.R., Razzak, I., Hacid, H., Nazari, A.: A mask-based output layer for multi-level hierarchical classification. In: CIKM'22, pp. 3833–3837 (2022)
3. Boone-Sifuentes, T., et al.: Marine-tree: large-scale marine organisms dataset for hierarchical image classification. CIKM '22, New York, NY, USA (2022)
4. Bossard, L., Guillaumin, M., Van Gool, L.: Food-101 – mining discriminative components with random forests. In: Fleet, D., Pajdla, T., Schiele, B., Tuytelaars, T. (eds.) ECCV 2014. LNCS, vol. 8694, pp. 446–461. Springer, Cham (2014). https://doi.org/10.1007/978-3-319-10599-4_29
5. Chen, M., et al.: Coarse-to-fine vision transformer. arXiv preprint arXiv:2203.03821 (2022)
6. Chou, P.Y., Kao, Y.Y., Lin, C.H.: Fine-grained visual classification with high-temperature refinement and background suppression. arXiv preprint arXiv:2303.06442 (2023)
7. Diao, Q., Jiang, Y., Wen, B., Sun, J., Yuan, Z.: MetaFormer: a unified meta framework for fine-grained recognition. arXiv preprint arXiv:2203.02751 (2022)
8. Dong, B., Zhou, P., Yan, S., Zuo, W.: Towards class interpretable vision transformer with multi-class-tokens. In: Chinese Conference on Pattern Recognition and Computer Vision (PRCV), pp. 609–622. Springer (2022). https://doi.org/10.1007/978-3-031-18913-5_47
9. Dosovitskiy, A., et al.: An image is worth 16×16 words: transformers for image recognition at scale. arXiv preprint arXiv:2010.11929 (2020)
10. Foret, P., Kleiner, A., Mobahi, H., Neyshabur, B.: Sharpness-aware minimization for efficiently improving generalization. arXiv preprint arXiv:2010.01412 (2020)

11. Hinton, G., Vinyals, O., Dean, J.: Distilling the knowledge in a neural network. arXiv preprint arXiv:1503.02531 (2015)
12. Huo, Y., Lu, Y., Niu, Y., Lu, Z., Wen, J.R.: Coarse-to-fine grained classification. In: Proceedings of the ACM SIGIR Conference, pp. 1033–1036. SIGIR'19 (2019)
13. Khosla, A., Jayadevaprakash, N., Yao, B., Li, F.F.: Novel dataset for fine-grained image categorization: Stanford dogs. In: Proceedings of CVPR Workshop on Fine-Grained Visual Categorization (FGVC). vol. 2. Citeseer (2011)
14. Kim, S., Nam, J., Ko, B.C.: ViT-NeT: interpretable vision transformers with neural tree decoder. In: International Conference on Machine Learning, pp. 11162–11172. PMLR (2022)
15. Kosmopoulos, A., Partalas, I., Gaussier, E., Paliouras, G., Androutsopoulos, I.: Evaluation measures for hierarchical classification: a unified view and novel approaches. Data Min. Knowl. Disc. **29**(3), 820–865 (2015)
16. Liu, Y., Dou, Y., Jin, R., Qiao, P.: Visual tree convolutional neural network in image classification. In: 2018 24th International Conference on Pattern Recognition (ICPR), pp. 758–763. IEEE (2018)
17. Liu, Z., et al.: Swin Transformer: hierarchical vision transformer using shifted windows. In: Proceedings of the IEEE/CVF Conference, pp. 10012–10022 (2021)
18. Maji, S., Kannala, J., Rahtu, E., Blaschko, M., Vedaldi, A.: Fine-grained visual classification of aircraft. Tech. rep. (2013)
19. Parag, T., Wang, H.: Multilayer dense connections for hierarchical concept classification. arXiv preprint arXiv:2003.09015 (2020)
20. Schmid, F., Masoudian, S., Koutini, K., Widmer, G.: Knowledge distillation from transformers for low-complexity acoustic scene classification. In: Proceedings of the Detection and Classification of Acoustic Scenes and Events 2022 Workshop (2022)
21. Seo, Y., Shin, K.S.: Hierarchical convolutional neural networks for fashion image classification. Expert Syst. Appl. **116**, 328–339 (2019)
22. Silla, C.N., Freitas, A.A.: A survey of hierarchical classification across different application domains. Data Min. Knowl. Disc. **22**(1), 31–72 (2011)
23. Wood, L., Tan, Z., Stenbit, I., Bischof, J., Zhu, S., Chollet, F., et al.: Kerascv. https://github.com/keras-team/keras-cv (2022)
24. Xu, L., Ouyang, W., Bennamoun, M., Boussaid, F., Xu, D.: Multi-class token transformer for weakly supervised semantic segmentation. In: Proceedings of the IEEE/CVF Conference, pp. 4310–4319 (2022)
25. Yan, Z., et al.: HD-CNN: hierarchical deep convolutional neural networks for large scale visual recognition. In: Proceedings of the IEEE ICCV Conference (2015)
26. Zhang, Z., Zhang, H., Zhao, L., Chen, T., Arik, S.Ö., Pfister, T.: Nested hierarchical transformer: towards accurate, data-efficient and interpretable visual understanding. In: Proceedings of the AAAI Conference on Artificial Intelligence, vol. 36, pp. 3417–3425 (2022)
27. Zhu, X., Bain, M.: B-CNN: branch convolutional neural network for hierarchical classification. arXiv preprint arXiv:1709.09890 (2017)

Improving Knowledge Tracing via Considering Students' Interaction Patterns

Shilong Shu, Liting Wang, and Junhua Tian(✉)

School of Education Science, Nanjing Normal University, Nanjing, China
{shushilong,ltwang}@nnu.edu.cn, 76920466@qq.com

Abstract. Knowledge Tracing (KT), which aims to accurately identify students' evolving mastery of different concepts during their learning process, is a popular task for providing intelligent tutoring in online learning systems. Recent research has leveraged various variants of single-state recurrent neural networks to model the transition of students' knowledge states. However, students' interaction patterns implicit in learning records are overlooked which plays an important role in reflecting students' mental state and learning habits. Additionally, interaction patterns affect an individual's self-efficacy and knowledge acquisition. To fill this gap, we propose the Interaction Pattern-Aware Knowledge Tracing (IPAKT) model that uses two hidden states to model knowledge state and interaction patterns separately. Specifically, we first extract the interaction patterns from two types of interaction responses: hint and time. Subsequently, these interaction patterns are employed to regulate the update of the knowledge state. Extensive experiments on three common datasets demonstrate that our method achieves state-of-the-art performance. We also present the reasonableness of IPAKT by ablation testing. Our codes are available at https://github.com/SummerGua/IPAKT.

Keywords: Knowledge tracing · Interaction pattern · Deep learning

1 Introduction

Over the past few decades, computer-assisted education systems, especially online learning platforms, have been proven to have unique advantages over traditional classroom education, such as their wide range of educational resources and high accessibility [6]. In these applications, to better recommend personalized learning resources to students as well as to help teachers understand students' learning progress, it becomes a problem to effectively track students' mastery of each concept, or knowledge component (KC). This task is called knowledge tracing (KT). The purpose of KT is to assess students' knowledge state based on their historical records of interactions, i.e., to predict their performance on the next question based on their answer history.

Numerous methods have been proposed to improve the accuracy of prediction. Traditionally, there are two lines of research: Bayesian Knowledge Tracing and Factor Analysis Methods. They attempt to model students' skill mastery

D.-N. Yang et al. (Eds.): PAKDD 2024, LNAI 14647, pp. 397–408, 2024.
https://doi.org/10.1007/978-981-97-2259-4_30

transitions via traditional statistical methods (e.g. Hidden Markov Models and Bayesian Belief Network) combined with educational psychology theories (e.g., Learning Mastery Theory and Item Response Theory). However, these methods assume that knowledge states are binary and exhibit limitations in their functional forms.

Inspired by the success of deep learning, recent research has introduced various deep learning techniques into KT. DKT [13] uses Recurrent Neural Networks (RNNs) [9], DKVMN [17] uses memory augmented networks and AKT [4] uses attention mechanisms [16]. Initially, some researchers believe that a correct answer brings about an increase in knowledge state and vice versa. To better model actual learning activities, some studies incorporate other responses to model students' learning psychology, such as forgetting behavior. However, there are two problems with these models. First, they simply add or concatenate other responses to answers as inputs, which makes it difficult to uncover the relationship between these features. Second, they transfer models from other domains to the KT task by simply replacing the inputs and outputs, resulting in a black box. To address these problems, some studies have begun to change the internal structure of the model such as LPKT [15] and DIMKT [14]. They extend an RNN cell into multiple modules and interpret them as different learning steps. Despite the great success of their high accuracy and interpretability, they ignore the implicit information about interaction patterns in the records.

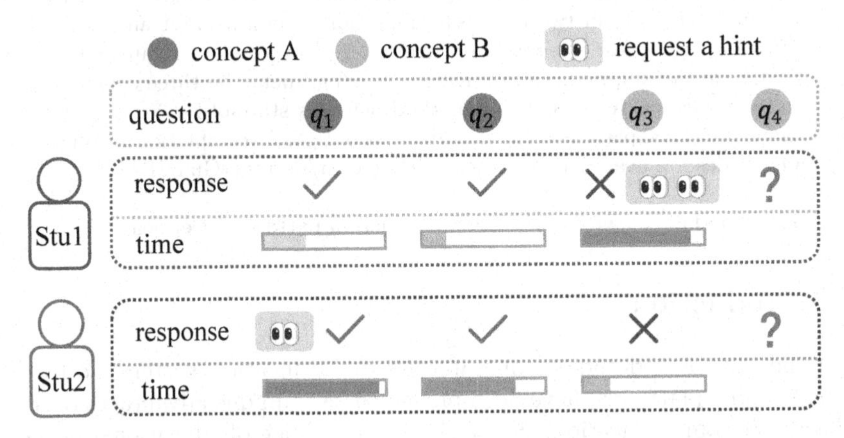

Fig. 1. Two students' interactions with the same sequence of questions. Our task is to predict the probability that they will answer q_4 correctly.

We argue that exploring the interaction patterns of online learners can help reveal their subtle learning characteristics involving cognition, emotion, and behavior. Some studies find that students who interact more with learning materials get more task value and self-efficacy [7]. We give an example in Fig. 1 where two students have done the same three questions and their answers are the same, but they show different interaction responses on each question. To be specific,

Student 1 spends much less time on questions q_1 and q_2 corresponding to concept A than Student 2, indicating that he/she is more familiar with the concept. Even though Student 2 also answers the two questions correctly, the extended time taken suggests that he/she is not proficient in concept A. Meanwhile, Student 1 answers q_3 incorrectly at the first attempt but later reviews two hints. On the contrary, Student 2 does not choose to request hints, resulting in less knowledge gained from q_3.

Considering the above problems, in this paper we propose a novel Interaction Pattern-Aware Knowledge Tracing (IPAKT) model, which enhances model performance by exploring the relationship between interaction patterns and knowledge states. We define the model as two stages: interaction pattern discovery and knowledge acquisition. In the two stages, attentive modules and fusion gates are designed to automatically measure how different interaction patterns contribute to knowledge acquisition. In summary, our main contributions are as follows:

- To the best of our knowledge, we are the first to consider multiple interaction responses at the same time and define them as interaction patterns. We analyze interaction patterns' significant impact on knowledge acquisition.
- We propose the IPAKT model to capture interaction patterns and knowledge states in two hidden states. In IPAKT, attentive modules are designed to evaluate the interaction responses and they are fused as interaction patterns, which control the knowledge acquisition.
- Extensive experiments demonstrate that we achieve state-of-the-art performance. Also, IPAKT is very interpretable.

2 Related Work

2.1 Knowledge Tracing

Knowledge tracing (KT) has been studied for nearly 30 years with the emergence of intelligent tutoring systems. Many methods have been proposed that share the common goal of improving the accuracy of predicting students' future performance based on their historical study records. We divide the KT methods into traditional methods and deep learning-based methods.

The first proposed KT model is Bayesian Knowledge Tracing (BKT) [2], which is a special case of Hidden Markov Model (HMM). BKT and its variants only consider the most recent observation and their binary assumption about students' knowledge state limits their performance. Another traditional KT model is factor analysis models (FAMs) which estimate students' performance by learning a logistic function. FAMs have difficulty capturing the relationship between different questions since it does not consider the sequence order.

DKT [13] firstly introduces deep learning techniques into KT, which utilizes RNNs to model students' knowledge state. DKT-DSC [11] uses K-means to dynamically cluster students with similar abilities into distinct groups and adds cluster information as additional input to DKT. Different from using RNNs as the KT model directly, some methods attempt to match the structure of the RNN cell to the student's learning process. LPKT [15] formalized the basic learning cell as the tuple (*exercise, answer time, answer*). Specifically, the prediction process of LPKT is defined as three steps: learning, forgetting, and predicting. IEKT [10] assesses students' individual cognition and knowledge acquisition sensitivity on each question to model the student characters. They attempt to model students' learning process at a time step by modifying the RNN cell. However, the single-state nature limits them to model the knowledge state and to be unable to capture other learning features.

2.2 Students' Interaction Patterns

Note that the interaction patterns mentioned in this paper refer specifically to those in online learning environments. Although the term interaction pattern is not strictly defined academically, it mainly represents the behavioral characteristics of students that are uncovered from the learning tracks accumulated in the logs of the learning system [7]. Analyzing these interaction sequences can provide a deeper understanding of how students solve problems.

Many studies have discussed or confirmed the existence of a specific relationship between interaction patterns and mental activity. [5] examines the relationship between interaction patterns and students' engagement in an online learning environment. [8] visualizes sequential interaction patterns to explore students' knowledge construction process. [1] indicates that participation in learning materials is positively related to learning achievement. However, existing KT methods have ignored the relationship between interaction patterns and students' performance.

3 Preliminary

In this section, we describe the knowledge tracing task. Besides, we present two interaction responses' embeddings and two hidden states' embeddings.

3.1 Knowledge Tracing Task

Knowledge tracing is formulated as a task to predict the probability of students' answers being correct to future questions given their historical learning interactions. A student's interaction log is recorded as a sequence $(q_1, a_1), ..., (q_{t-1}, a_{t-1})$, where q_i is the question at time step i and $a_i \in \{0, 1\}$ denotes whether they correctly answer the question ($a_i = 1$) or not ($a_i = 0$). Each question is related to a specific KC and has a certain number of hints available. Each student can request hints before or after answering the question.

3.2 Embedding

To better exploit the relationships between different features and mitigate data sparseness, we adopt interaction response embedding and hidden state embedding in IPAKT.

Interaction Response Embedding. To explore students' interaction patterns from limited attributes in records, two responses are used in the IPAKT model: i) the number of hints they request on the question, η; ii) the elapsed time they spend to solve the question, τ. We utilize embedding matrices $M_\tau \in \mathbb{R}^{C_\tau \times d}$ and $M_\eta \in \mathbb{R}^{C_\eta \times d}$ to get their embeddings, $\boldsymbol{\tau}$ and $\boldsymbol{\eta}$. d is the dimension, C_τ is the maximum number of hints students request, and C_η is the maximum elapsed time they spend to finish a question in seconds. C_τ is set to 10 and C_η is set to 300. Any η or τ more than the corresponding maximum value is capped off to the maximum value.

Hidden State Embedding. Similar to LSTM, there are two hidden states represented as vectors in IPAKT. One is $h_t \in \mathbb{R}^d$ representing a student's knowledge state at time step t, where d is the number of hidden state dimensions. The constantly updated h_t models a student's knowledge transition over time. The other is $c_t \in \mathbb{R}^d$ representing a student's interaction pattern the model captures at time step t. To be specific, c_0 and h_0 are initialized to $\mathbf{0}$.

4 Method

In this section, we present the IPAKT model in detail. The main structure of IPAKT is depicted in Fig. 2.

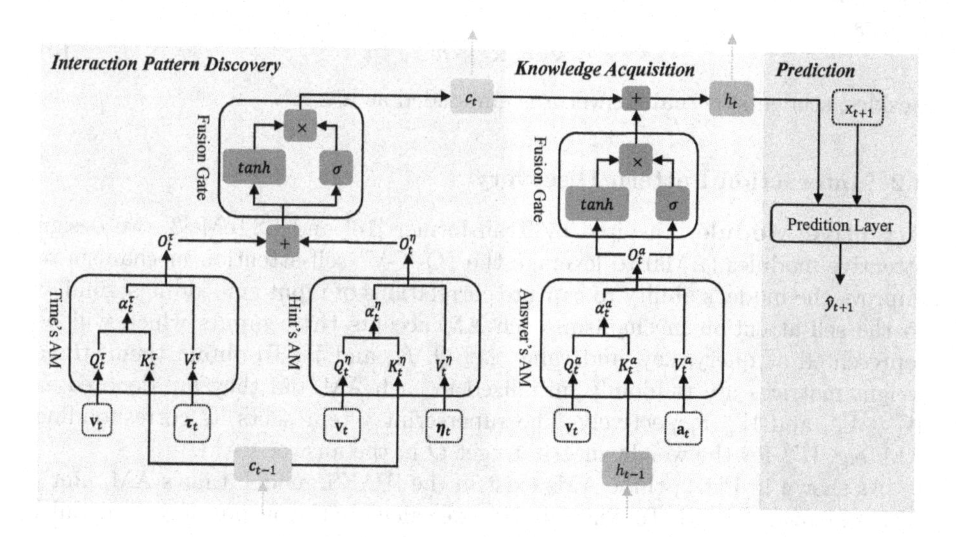

Fig. 2. The architecture of IPAKT. IPAKT has two hidden states and consists of three attentive modules and two fusion gates.

4.1 Input Embedding

To better represent a question, we do not singly use either KC or the question itself as embedding. Instead, we use an embedding method similar to the difficulty-enhanced question embedding method proposed in [14]. To be specific, We use both KC embedding and difficulty level (DL) embedding to get the final question embedding $\mathbf{x}_t$. The KC embedding matrix is denoted as $\mathbf{KC} \in \mathbb{R}^{C_{kc} \times d}$ where C_{kc} is the number of KC in a dataset and d is the dimension. The DL of each question is calculated as follows:

$$DL = \sum_{i}^{|S_i|} \frac{|\{a_{ij} = 1\}|}{|s_i|} \cdot C_{dl} \tag{1}$$

where S_i is the set of students who answer the question q_i, $a_{ij} = 1$ refers to a correct answer and C_{dl} is the predefined number of difficulty level. We represent the embedding of DL by an embedding matrix $\mathbf{DL} \in \mathbb{R}^{C_{dl} \times d}$. Then we can get the question embedding $\mathbf{x}_t$ as follows:

$$\mathbf{x}_t = W_1^T[\mathbf{KC}_t \oplus \mathbf{DL}_t] + b_1 \tag{2}$$

where $\oplus$ is the concatenation operation, $W_1 \in \mathbb{R}^{(d_{kc}+d_{dl}) \times d}$ is the weight matrix and b_1 is the bias term. To explore the mutual influence of students' knowledge state and their interaction pattern, we concatenate current question x_t and the knowledge state at the last time step h_{t-1} as the final input to our model as follows:

$$\mathbf{v}_t = \mathbf{x}_t \oplus h_{t-1} \tag{3}$$

Besides, students' actual answer a is embedded as $\mathbf{a} \in \mathbb{R}^d$.

4.2 Interaction Pattern Discovery

Attentive Module. Inspired by Transformer [16] and LSTeM [3], we design attentive modules (AMs) to leverage the "Q-K-V" self-attention mechanism to improve the model's ability to capture correlations of input embeddings. Similar to the self-attention mechanism, each AM accepts three inputs which will be represented as query, key, and value, i.e. Q, K, and V. To obtain them, three weight matrices are randomly initialized in each AM and they are denoted as W_q^*, W_k^*, and W_v^* respectively. The superscript $*$ represents its corresponding AM, e.g. W_q^a for the weight matrix to get Q in the answer's AM.

As shown in Fig. 2, three AMs exist in the IPAKT model: time's AM, hint's AM, and answer's AM. To illustrate how we calculate the output of each module, we use time's AM as an example. First, we calculate Q, K and V as follows:

$$Q_t^\tau = W_q^\tau \mathbf{v_t} \tag{4}$$

$$K_t^\tau = W_k^\tau c_{t-1} \tag{5}$$

$$V_t^\tau = W_v^\tau \tau_t \tag{6}$$

Then, we calculate the attentive score that is used to measure how much we can get from V:

$$\alpha_t^{\tau} = \sigma(\frac{Q_t^{\tau} K_t^{\tau}}{\sqrt{d}}) \tag{7}$$

where σ is the sigmoid function. Finally, we get the output of time's AM as follows:

$$O_t^{\tau} = \alpha_t^{\tau} V_t^{\tau} \tag{8}$$

In this manner, the model is able to flexibly find the connections between the current question and the student's strategy to answer it, instead of paying too much attention to the relationship between interaction responses at different time steps. In the same way, we can get the output of hint's AM, O_t^{η}.

Fusion Gate. To fuse representations of different interaction responses, fusion gates are designed. We directly add two attentive modules' output as follows:

$$SIR_t = O_t^{\eta} + O_t^{\tau} \tag{9}$$

where SIR represents the additive students' interaction response. Then the output of the fusion gate, i.e. the increment of behavior pattern is calculated as follows:

$$f_t = tanh(W_2 SIR_t + b_2)\sigma(W_3 SIR_t + b_3) \tag{10}$$

where W_2 and W_3 are weight matrix, and b_2 and b_2 are bias terms. Finally, students' behavior pattern is updated:

$$c_t = c_{t-1} + f_t \tag{11}$$

4.3 Knowledge Acquisition

After obtaining the current student's interaction pattern, it will have an impact on the student's knowledge acquisition. In a similar way, we get the output of the answer's AM, O_t^a, and send it into a fusion gate:

$$i_t = tanh(W_4 O_t^a + b_4)\sigma(W_5 O_t^a + b_5) \tag{12}$$

where i_t is the output of the fusion gate. Then the knowledge state is updated:

$$h_t = i_t + c_t \tag{13}$$

4.4 Prediction and Objective Function

Knowledge state h_t is used to calculate the probability that a student will correctly answer the next question. After getting the inner product of the embedding of the next question $\mathbf{x_{t+1}}$ and h_t, we pass it through a sigmoid function. Finally, we get a probability value between 0 and 1:

$$\hat{y}_{t+1} = \sigma(\sum(\mathbf{x_{t+1}} \cdot h_t)) \tag{14}$$

The model is optimized by minimizing the binary cross-entropy loss of all learner responses,

$$\mathcal{L} = \sum_i \sum_t -(r_t^i \log \hat{y}_t^i + (1 - r_t^i) \log(1 - \hat{y}_t^i)) \tag{15}$$

where i represents the i-th student and t represents the t-th time step.

5 Experiment

In this section, we first introduce real-world datasets used in our experiments. Then we give our choice of hyper-parameters and baselines. Finally, we conduct an ablation study to validate the effectiveness of interaction patterns for model performance improvement.

5.1 Datasets

To validate the accuracy of the IPAKT model, We train and test it on three datasets: Assist2009, Assist2012, and Assist2017, which are all collected on the ASSISTments online tutoring system. The users' information contained within the datasets has undergone randomization to ensure anonymity and thus mitigate ethical concerns. We briefly introduce them and statistics of the data used in our experiments are presented in Table 1.

- Assist2009[1] is collected in the school year 2009–2010. Massive duplicated records and repeated response sequences are observed in the raw data, so a modified version is used in experiments.
- Assist2012[2] is gathered in the school year 2012–2013. Due to the large amount of data, we randomly selected 999 interactions of 5000 students.
- Assist2017[3], also known as ASSISTments Challenge dataset, contains data in the school year 2004–2005 and 2005–2006. As a data mining competition dataset, Assist2017 contains the most descriptive information among the ASSISTments datasets.

5.2 Experimental Setup

The number of dimension d is selected from $\{50, 100, 150, 200\}$. We observed that the dimension has a minor impact on the experimental results, but an increase in the number of dimensions significantly affects the training time. Consequently, we opted for a dimension of 50. The number of difficulty level C_{dl} is set to 100. Sequences of length less than 3 do not contain sufficient historical information, so

---

[1] https://sites.google.com/site/assistmentsdata/home/2009-2010-assistment-data.
[2] https://sites.google.com/site/assistmentsdata/datasets/2012-13-school-data-with-affect.
[3] https://sites.google.com/view/assistmentsdatamining.

Table 1. Statistics of all datasets.

Statistics	Datasets		
	Assist2009	Assist2012	Assist2017
# of students	3,862	5000	1,709
# of KCs	111	259	102
# of questions	16,799	104,793	4,117
# of interactions	401,756	702,034	942,816
Avg. interactions per student	104.03	140.41	551.68

they are simply removed. We set the maximum length of a student interaction sequence to be 200, and sequences longer than 200 are split into several subsequences. For each experiment, we randomly divide each dataset into two parts, of which 80% is used as a training set and the remaining 20% as a test set. We use Adam Optimizer. The batch size is configured as 32 to align with the GPU memory, which enhances the training speed. The maximum number of epochs is set to 200. A learning rate scheduler is used to adjust the learning rate after every k iteration and $k = 50$.

5.3 Baselines

To validate the performance of IPAKT, we compare it to six classical KT models, all of which are reproduced following the hyper-parameters proposed in their respective papers. All training processes are performed on a server with an NVIDIA RTX 2060 Super GPU. The details of the baselines are as follows:

- **DKT** is the first deep learning-based KT method [13]. It uses RNNs to learn the students' knowledge state.
- **SAKT** adopts self-attention mechanism to model the whole process of student learning [12], which helps to identify the relevance between KCs.
- **AKT** uses a monotonic attention mechanism where attention weights are computed using exponential decay [4]. Two encoders are designed to encode questions and answers.
- **DKVMN** implements a key-value memory network [17], it utilizes two matrices to discover the underlying KCs of exercises and trace a student's knowledge state of all KCs over time.
- **LPKT** monitors students' knowledge state through directly modeling their learning process of learning and forgetting in an RNN-like cell [15].
- **DIMKT** designs an adaptive sequential neural network to assess a user's subjective difficulty feeling on a given question, which is represented by difficulty-enhanced question embedding [14].

5.4 Students' Performance Prediction

While the primary aim of introducing IPAKT is to achieve a more reasonable KT model, the assessment of KT methods remains significantly reliant on experimental results related to students' performance prediction. We adopt Area Under Curve (AUC) and Accuracy (ACC) as evaluation metrics. We set the threshold to 0.5 when computing ACC. Following the previously outlined experimental setup, we conduct a comparative analysis against all baselines on students' performance prediction and present the findings in Table 2. We can see that IPAKT outperforms all baselines in both AUC and ACC. We notice that compared to the DKT model on Assist2009, we achieve the largest AUC improvement of 7.98%. In particular, although IPAKT does not outperform the single-state model by much, it still shows adding a new hidden state to model students' interaction patterns is meaningful. From Table 2, we observe that the models' performance on Assist2017 is worse than the other two datasets. We speculate that this is because, in Assist2017, students' every response to each question, including requesting hints or answering, is recorded in detail, unlike other datasets where only the first response is retained.

Table 2. Results of models on three open ASSISTments datasets, where bold indicates the best performance.

Datasets	Metrics	DKT	SAKT	AKT	DKVMN	LPKT	DIMKT	IPAKT (ours)
Assist2009	AUC	0.7456	0.7136	0.7750	0.7435	0.7889	0.7934	**0.8190**
	ACC	0.7392	0.7244	0.7647	0.7562	0.7751	0.7695	**0.7797**
Assist2012	AUC	0.7304	0.7264	0.7711	0.7221	0.7714	0.7793	**0.7801**
	ACC	0.7375	0.7348	0.7509	0.7340	0.7462	0.7622	**0.7694**
Assist2017	AUC	0.7214	0.6973	0.7503	0.7107	0.7650	0.7598	**0.7680**
	ACC	0.7018	0.7043	0.7392	0.6827	0.7183	0.7371	**0.7432**

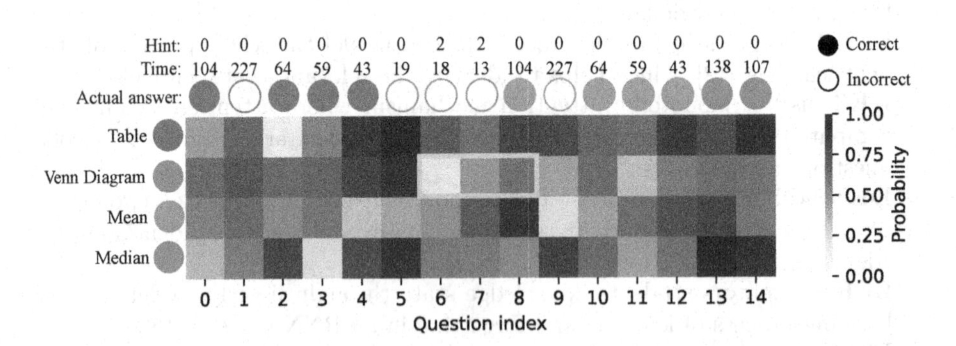

Fig. 3. The visualization of the prediction for student No. 90137 in Assist2009.

To make the prediction more intuitive, we draw a heat map of the predicted values in Fig. 3, where a student practices 15 questions related to four KCs. Note that the prediction in the first column relies on a correct answer to the KC "Stem and Leaf Plot", which is not shown in the heat map. We can see that even though the student answered q_5, q_6, and q_7 incorrectly, the predicted values labeled in the red box still show an increase, indicating that he/she acquires knowledge from the requested hints. This phenomenon indicates that our model is performing as expected and interaction patterns have an impact on knowledge states.

5.5 Ablation Study

To demonstrate the effectiveness of integrating students' interaction patterns, we conduct an ablation study by modifying the IPAKT models and then comparing their performance with that of the original IPAKT. The information of the ablation models is as follows: i) **IPAKT-hint** removes the hint's AM; ii) **IPAKT-time** removes the time's AM; iii) **IPAKT-both** removes both aforementioned AMs and hidden state c, which means no interaction patterns contribute to the model. Comparison results are presented in Table 3. IPAKT model outperforms other ablation models, which demonstrates the reasonableness of incorporating interaction patterns derived from both interaction responses.

Table 3. The AUC results (%) of our ablation study.

Ablation	Assist2009	Assist2012	Assist2017
IPAKT	**0.8190**	**0.7801**	**0.7680**
IPAKT-hint	0.7937	0.7568	0.7572
IPAKT-time	0.7940	0.7611	0.7556
IPAKT-both	0.7851	0.7503	0.7475

6 Conclusion

In this paper, we propose a novel Interaction Pattern-Aware Knowledge Tracing model, leveraging a two-state RNN-based method. Specifically, we first adopt a difficulty-enhanced question embedding to represent each question. This method not only mitigates the sparseness problem arising from a vast number of questions but also enhances the correlation among embeddings. Furthermore, we define two distinct hidden states to model students' interaction patterns and knowledge states. In the subsequent stage of interaction pattern discovery, we present a time's attentive module (AM), a hint's AM, and a fusion gate to explore how a student's specific interaction responses contribute to their interaction pattern. Finally, in the stage of knowledge acquisition, we use an answer's AM to explore the mutual influence of interaction patterns and knowledge states. The current increment of knowledge is determined by the current interaction pattern extracted from historical records. The results of experiments conducted on three ASSISTments datasets comparing the other six classical deep learning-based KT models affirm state-of-the-art performance of our IPAKT model in both AUC and ACC.

References

1. Cheng, G., Chau, J.: Exploring the relationships between learning styles, online participation, learning achievement and course satisfaction: an empirical study of a blended learning course. Br. J. Edu. Technol. **47**(2), 257–278 (2016)
2. Corbett, A.T., Anderson, J.R.: Knowledge tracing: modeling the acquisition of procedural knowledge. User Model. User-Adap. Inter. **4**, 253–278 (1994)
3. Duan, J., Zhang, P.F., Qiu, R., Huang, Z.: Long short-term enhanced memory for sequential recommendation. World Wide Web **26**(2), 561–583 (2023)
4. Ghosh, A., Heffernan, N., Lan, A.S.: Context-aware attentive knowledge tracing. In: Proceedings of the 26th ACM SIGKDD International Conference on Knowledge Discovery & Data Mining, pp. 2330–2339 (2020)
5. Goggins, S., Xing, W.: Building models explaining student participation behavior in asynchronous online discussion. Comput. Educ. **94**, 241–251 (2016)
6. Lee, Y.C.: The role of perceived resources in online learning adoption. Comput. Educ. **50**(4), 1423–1438 (2008)
7. Li, L.Y., Tsai, C.C.: Accessing online learning material: quantitative behavior patterns and their effects on motivation and learning performance. Comput. Educ. **114**, 286–297 (2017)
8. Lin, T.J., Duh, H.B.L., Li, N., Wang, H.Y., Tsai, C.C.: An investigation of learners' collaborative knowledge construction performances and behavior patterns in an augmented reality simulation system. Comput. Educ. **68**, 314–321 (2013)
9. Lipton, Z.C., Berkowitz, J., Elkan, C.: A critical review of recurrent neural networks for sequence learning. arXiv preprint arXiv:1506.00019 (2015)
10. Long, T., Liu, Y., Shen, J., Zhang, W., Yu, Y.: Tracing knowledge state with individual cognition and acquisition estimation. In: Proceedings of the 44th International ACM SIGIR Conference on Research and Development in Information Retrieval, pp. 173–182 (2021)
11. Minn, S., Yu, Y., Desmarais, M.C., Zhu, F., Vie, J.J.: Deep knowledge tracing and dynamic student classification for knowledge tracing. In: 2018 IEEE International conference on data mining (ICDM), pp. 1182–1187. IEEE (2018)
12. Pandey, S., Karypis, G.: A self-attentive model for knowledge tracing. arXiv preprint arXiv:1907.06837 (2019)
13. Piech, C., et al.: Deep knowledge tracing. In: Advances in Neural Information Processing Systems 28 (2015)
14. Shen, S., Huang, Z., Liu, Q., Su, Y., Wang, S., Chen, E.: Assessing student's dynamic knowledge state by exploring the question difficulty effect. In: Proceedings of the 45th International ACM SIGIR Conference on Research and Development in Information Retrieval, pp. 427–437 (2022)
15. Shen, S., et al.: Learning process-consistent knowledge tracing. In: Proceedings of the 27th ACM SIGKDD Conference on Knowledge Discovery & Data Mining, pp. 1452–1460 (2021)
16. Vaswani, A., et al.: Attention is all you need. In: Advances in Neural Information Processing Systems 30 (2017)
17. Zhang, J., Shi, X., King, I., Yeung, D.Y.: Dynamic key-value memory networks for knowledge tracing. In: Proceedings of the 26th International Conference on World Wide Web, pp. 765–774 (2017)

MDAN: Multi-distribution Adaptive Networks for LTV Prediction

Wenshuang Liu(✉), Guoqiang Xu, Bada Ye, Xinji Luo, Yancheng He, and Cunxiang Yin

Dayu, Tencent, Shenzhen, China
{wsliu,chybotxu,piterye,leevenluo,collinhe,jasonyin}@tencent.com

Abstract. In industry, Customer Lifetime Value (LTV) represents the entire revenue generated from a single user within an application. Accurate LTV prediction can help marketers make more informed decisions about acquiring high-quality new users and increasing revenue. However, LTV prediction is a complex and challenging task, and the LTV of most application users is prone to bias and sparsity. To address these issues, this paper proposes a Multi-Distribution Adaptive Networks (MDAN) to predict LTV. In terms of classification debiasing, we leverage multi-channel networks to simultaneously learn disparate distributions and a Channel Learning Controller (CLC) is used to advance the learning of different channels. Moreover, in the context of regression debiasing, a novel loss function called Distance Similarity Loss is introduced for the specific purpose of predicting LTV. This loss function is designed to distinguish between the feature representations associated with different LTV values, thus improving the ability to represent user characteristics within the model. The MDAN framework has been successfully deployed in multiple applications within Tencent, leading to considerable increases in revenue. Extensive experiments on three million-level datasets, QB, YYB, and WeSing, demonstrate the superiority of the proposed method compared to state-of-the-art baselines such as DNN, RankSim, ZILN and ODMN models.

Keywords: LTV prediction · Data bias · Multi-distribution

1 Introduction

Currently, Customer Life Time Value (LTV) holds a significant position as a crucial business metric. It pertains to the revenue that a singular customer or user can potentially generate for the application. LTV has found extensive utilization in the fields of Marketing [6] and Customer Relationship Management (CRM) [7]. Accurate prediction of LTV enables companies to optimize their allocation of limited resources towards various users, thereby maximizing return on investment (ROI). Such optimization encompasses the cost of advertising to acquire new users from other platforms, as well as providing in-app personalized promotional services for existing users.

D.-N. Yang et al. (Eds.): PAKDD 2024, LNAI 14647, pp. 409–420, 2024.
https://doi.org/10.1007/978-981-97-2259-4_31

The Recency, Frequency, and Monetary Value (RFM) characteristics of a customer's past behaviors are widely recognized as fundamental indicators of future consumption. To model these characteristics, researchers have attempted to describe them utilizing assumed prior distribution and proposing probability and statistics models [4,6,9] to infer the LTV of a customer in a two-stage approach. Subsequent investigations have explored the use of machine learning or deep learning models [2,3,8,11–13] for LTV prediction, which have demonstrated superior performance compared to the former, as they can incorporate more user characteristics as feature inputs to the model and directly predict LTV.

LTV prediction has been greatly improved by these methods. However, the challenges of data imbalance and sparsity in LTV datasets still remain, especially in our scenario, which includes three Tencent products–QB, YYB and WeSing, and datasets collected from the three APP will be introduced in detail in Sect. 4. Most user contribute no revenue in the first seven days of using our product, while those in the heavy tail contribute significantly, accounting for 80% to 90%, which indicates an extremely imbalanced distribution of LTV. This extreme imbalance of LTV labels presents an obstacle to models accurately capturing the inherent laws of the data without bias. To address these issue, we introduce a novel architecture called the Channel Learning Controller (CLC) that functions as a classifier and generates output that can be utilized as gate weights. The resulting gate weights are subsequently inputted into a multi-channel network, facilitating the simultaneous training of multiple channels that encode user embeddings.

Another design in this paper to mitigate bias in the model is the Distance Similarity Loss (DSL). Intuitively, embeddings produced by different channels or models may be similar to each other due to the data feature from the minority groups being overwhelmed by the majority ones, making it difficult for models to differentiate between the feature representation of valuable users and that of zero-contribution users. [5] proposed RankSim regularizer, yielded outstanding performance in imbalanced regression scenarios. This regularizer encourages the rank of the similarity in the label space to be as close as possible to the rank of the similarity in the feature space, thus constructing global connections between items in a batch. However, the difference in rank cannot fully represent the difference of label values themselves. To address this, DSL takes advantage of the difference between labels to constrain the similarity between the feature embeddings of different users, enabling the network to distinguish high-quality users more effectively.

In summary, this paper mainly contributes to the following three aspects:

- We present a novel Multi-distribution Adaptive Networks (MDAN) framework, comprising a Channel Learning Controller and Multi-channel Network. The latter generates output embeddings for users in distinct LTV buckets, while the former produces weights to combine these embeddings, thereby ameliorating the issue of imbalanced data.

- We propose a novel regularization technique called Distance Similarity Loss, which enforces a global constraint on the similarity between the labels and learned embeddings in a batch-wise manner. This serves to effectively alleviate any potential biases that may arise during the training process.
- We provide empirical evidence of the efficacy of the proposed architecture via its application to three distinct datasets sourced from Tencent products. These datasets will be made publicly available to the research community, thereby potentially promoting and fostering future research on LTV prediction.

The organization of the this paper is as follows. Section 2 provides an overview of related work in the areas of data bias mitigation and LTV prediction. Section 3 introduces the MDAN architecture and its associated Channel Learning Controller and Distance Similarity Loss. In Sect. 4, we present our experimental evaluations of MDAN and other comparative models as well as ablation experiments on datasets collected from Tencent. Finally, in Sect. 5, We make conclusion for the paper.

2 Related Work

Our objective in this paper is to predict the seven-day LTV of newly acquired users while addressing the issue of the imbalance and sparsity of LTV labels. Prior studies have applied various models to address this challenge. Early researches solve LTV prediction problem mainly by classic statistics model, based on some prior distribution assumptions. [6] proposes Basic structural model of CLV. Markov Chain Models [9] infers the probability transition matrix, considering the value of recency as node. The Hierarchical Bayes [1] model fits lognormal distributions to monetary values.

More recently, machine learning or deep learning models have been adopted to leverage more user characteristics and directly predict LTV. For example, [2,3, 11] employ two-stage tree models (Random Forest and Xgboost) to respectively predict whether the user is willing to consume and the consumption amount. [11] calculates user engagement features while [2] add user embedding obtained from skip-gram as a feature to promote the ability of trained models and [3] incorporates social features.

The imbalance and sparsity of LTV data have also recently drawn researchers' attention. [10] employs SMOTE to directly produce minority samples. ZILN [12] assumes that the monetary value follows a lognormal distribution and proposes a zero-inflated lognormal loss function to account for churn probability as well as heavy-tailedness of LTV. TSUR [13] combines the Wavelet Transform and Graph Attention Network to train temporal and structural user representation and alleviate the impact of sparsity of user history behavior sequence data. [8] introduces ODMN to catch the ordered dependencies between LTVs, and proposes MDME to group users according to their LTV, resulting in relative balanced LTV distribution within each group. However, the classification capability of MDME determines the performance of the whole network, which requires experimentation on datasets with small distribution bias. Furthermore, the complex network architecture proposed in the paper makes it challenging to achieve comparable results, thus limiting its widespread applicability.

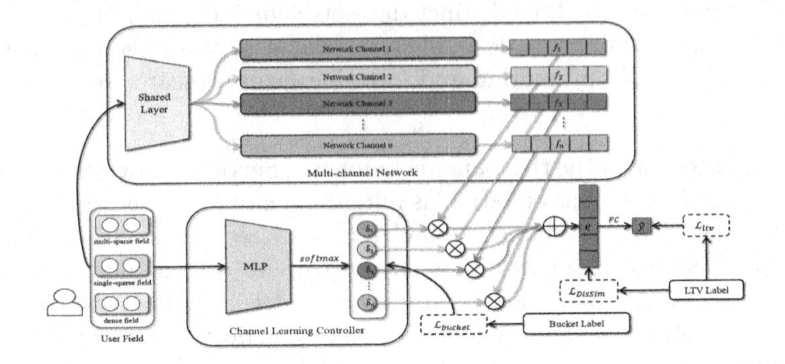

Fig. 1. The overall structure of our proposed MDAN. The upper part is the multi-channel network, which has n network channels. The lower part is Channel Learning Controller module, which is used to adaptively learn multi-distribution.

3 Method

In this section, we present a novel LTV prediction model, Multi-Distribution Adaptive Networks (MDAN), which adaptively learns different distributions using different network channels. As illustrated in Fig. 1, MDAN consists of a multi-channel network structure with a Channel Learning Controller (CLC) to control feature learning for different user LTV distributions. Moreover, it has a novel Distance Similarity Loss applied to the final feature layer, which can separate the feature representations of different LTV values and increase the feature representation capability of users. Next, we will provide a comprehensive overview of MDAN's structure and the Distance Similarity loss functions.

3.1 Multi-distribution Adaptive Networks

The present study focuses on predicting the 7-day LTV generated by users upon arrival at a particular application (app) by leveraging their features in other apps. However, the LTVs of most app users are characterized by bias and sparsity, with a majority of users having LTV values of 0 and only a select few having LTV values greater than 0. Notably, the top 20% of users with LTV values greater than 0 contribute to 80%–90% of the overall revenue. To address this issue, we employ Multi-Channel Network, as shown in Fig. 1, to learn different LTV distributions. Specifically, the orange channel network 1 is primarily responsible for learning LTV samples of 0, whereas the yellow channel network 2 is responsible for learning LTV samples within the range of $(0, 1]$. The LTV threshold values for different channel networks are set differently. The number of network channels established for different application scenarios varies, and we currently employ five network channels to learn LTVs with different distributions. The LTV thresholds for each distribution are pre-set, and the number of samples between each LTV greater than 0 distribution is made as uniform as possible. Based on the input feature x, the Multi-Channel Network yields multiple outputs, as indicated in the following formula:

$$f_i = C_i(x), i \in [1, 2, ..., n] \tag{1}$$

where i is the sequence number of the channel, n is the total number of LTV channels, C_i is each channel network respectively, and f_i is the corresponding channel network's output feature. The lower part of the figure is the Channel Learning Controller module, which is responsible for controlling the learning weights of different channels in the multi-channel network. The input feature x is passed through an MLP network to produce a feature vector k of dimension n. The formula is

$$k = MLP(x) \tag{2}$$

where MLP is the multilayer perceptron network structure with an output dimension of n in the last layer, and k is the output n-dimensional feature, the n-dimensional features cannot be used directly as the learning weights of the multi-channel network. Thus, it has to be normalized into n-dimensional features with a sum of one by softmax, which can be represented as:

$$\hat{b}_i = \frac{e^{k_i}}{\sum_{j=1}^{n} e^{k_j}} \tag{3}$$

where $\hat{b}$ is the n-dimensional feature which sums to one after normalization. To ensure that $\hat{b}$ adaptively learns the weights of different distribution samples, a multi-class loss is introduced to guide the learning of b_i and is expressed as:

$$\mathcal{L}_{bucket} = -\frac{1}{B} \sum_{x} b \log \hat{b} \tag{4}$$

where x is the real input of each sample in the batch, B is the batch size, b is the one-hot n-dimensional vector of the real bucket label, and $\hat{b}$ is the n-dimensional prediction of b. The channel learning controller and multi-channel network are then combined together such that each distribution activation value in $\hat{b}$ is multiplied with the output of each channel in f respectively, and all the outputs are added up to obtain the final embedding e. This is expressed as:

$$e = \sum_{i=1}^{n} \hat{b}_i * f_i \tag{5}$$

After obtaining the added embedding of all channels, the added embedding average is used to represent the predicted LTV value, which is expressed as:

$$\hat{y} = FC(e) \tag{6}$$

where $\hat{y}$ is the predicted LTV, FC is the function of the full connected layer without activation. Finally, the commonly used Mean Square Error loss is used to constrain the learning of the predicted LTV $\hat{y}$, with the loss function represented as:

$$\mathcal{L}_{ltv} = \frac{1}{B}\sum_{x}(\hat{y} - y)^2 \tag{7}$$

3.2 Distance Similarity Loss

The preceding section introduces the MDAN architecture, which primarily focuses on classification debiasing. In contrast, LTV prediction is a numerical regression task that emphasizes the elimination of regression bias. Recently, regression debiasing techniques have gained popularity. For example, RankSim [5] puts forth a ranking-based regression debiasing method. However, this method assumes a strong prior distribution and is primarily evaluated on datasets characterized by continuous normal distributions. On the other hand, LTV distributions often exhibit sparsity and bias. [5] emphasize that their exclusive focus on the relative arrangement of LTVs lacks comprehensiveness, highlighting the necessity for a more thorough examination of distinctions between LTVs. We propose a novel loss function for LTV regression debiasing, namely the Distance Similarity loss. This loss function employs the differences among different LTVs to regulate the similarity between distinct embeddings, so that samples with greater LTV differences exhibit more different feature embeddings. First, we compute the cosine similarity between each embedding within a batch and transform it into a corresponding matrix using the following formulas:

$$e_{norm} = [||e_1||_2, ||e_2||_2, ..., ||e_B||_2]^\top \tag{8}$$

$$T_{sim} = (e \oslash e_{norm}) * (e^\top \oslash e_{norm}^\top) \tag{9}$$

where B is the batch size, e refers to the batch embedding matrix with each row standing for the embedding of one user, e_{norm} is obtained by calculating the L2 norm of each row vector of e and $\oslash$ refers to the Hadamard division.

To incorporate variations among different LTVs in the ultimate training penalty, we employ a technique akin to converting LTV tags into a matrix. Initially, a batch of LTV tags is transformed into a matrix with identical length and width. Next, the matrix's absolute difference from its transpose is computed, denoted by:

$$T_{abs} = |tile(y) - tile(y)^\top| \tag{10}$$

where the *tile* function copies the label vector into a matrix with identical length and width. In order to encourage the embedding e to have a significant correlation with the LTV value, we propose a novel and suitable de-biasing loss - DisSim loss, which is a combination of the feature similarity matrix T_{sim} and the label difference matrix T_{abs}. The loss formula is:

$$\mathcal{L}_{DisSim} = \frac{1}{B}(norm_1(1 - T_{sim}) - norm_1(T_{abs}))^2 \tag{11}$$

where $norm_1$ represents the normalization function for features in the first dimension.

4 Experiments

In this section, we will assess the performance of our proposed model and baseline model through experiments conducted on three million-level datasets, which are openly accessible within the company. We will then discuss the datasets, evaluation metrics and baselines in subsequent subsections, and finally present and analyze the comparative experimental results.

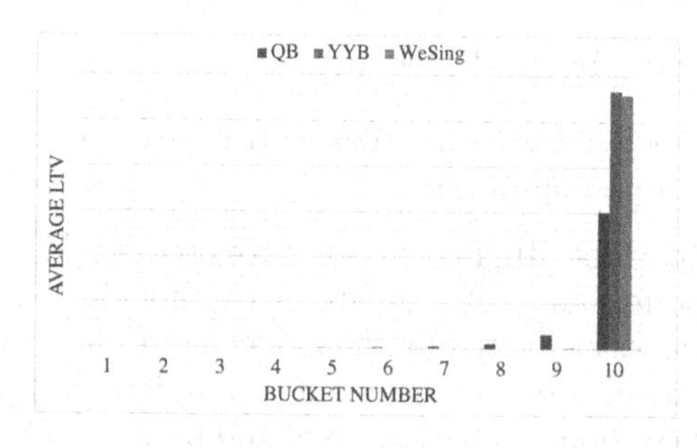

Fig. 2. The ten-bucket LTV chart of three datasets, from No. 1 bucket to bucket No. 10, and LTV values are sorted from low to high.

4.1 Dataset

Next, we will introduce the three large datasets used in the experiments, all datasets will be made available after the paper is published.

QB Dataset: The QB dataset was obtained from the QQ browser APP, where all user IDs were encrypted to protect user information. The app has more than 10 million DAU(daily active users), with multiple business scenarios such as search, novel, document, and news. To attract new novel users with higher value, we mainly obtained our data from the scenario of novel, resulting in 1,419,776 training data and 707,072 test data. The features included 76 variables such as user ID, gender, age, city, and consumption amount.

YYB Dataset: The YYB dataset was obtained from the Application Treasure APP, where all user IDs were also encrypted. This app has approximately 9 million DAU, providing services similar to the Application store of mobile phones. We randomly collected 11,576,320 new users, with 11,139,584 pieces of data in the training set and 436,736 pieces in the test set. The data included 91 features such as user ID, gender, age, city, and consumption amount.

WeSing Dataset: The WeSing dataset was obtained from the WeSing APP, with all user IDs encrypted as well. This app has over 10 million DAU, allowing users to sing songs anytime and anywhere. We randomly collected 1,650,848 new users, with 1,569,119 pieces of data in the training set and 81,729 pieces in the test set. The data included 84 features such as user ID, user gender, age, city, and consumption amount within seven days after the user first logged in. Notably, the LTV distribution of the WeSing dataset was seriously biased, with 98% of users having zero LTV.

Figure 2 illustrates the average of LTV in ten buckets for our three datasets. It can be observed from the figure that the deviation of the YYB and WeSing datasets is more typical, with more than 90% of users whose LTV equals 0, while the LTV distribution of the QB dataset is relatively even, with six buckets of which the average LTV is greater than 0.

4.2 Evaluation Metrics and Baselines

In this part, we introduce the evaluation metrics used in the experiments and the baseline models for comparison.

Evaluation Metrics: The GINI coefficient proposed by [12] is a popular metric for evaluating the performance of estimated LTV models. It reflects the proportion of top user revenues evaluated by the model and is larger when performance is better. The Spearman's rank coefficient of correlation (SC) [12] is used to evaluate the monotonic relationship between real and predict LTV. The Decile Mape (DM) [12] is used to evaluate the accuracy of different levels of LTV.

Baselines: In light of the superior performance of deep neural networks (DNNs) over traditional machine learning methods, this study does not compare with the latter. Specifically, this paper proposes the Multi-Domain Adversarial Networks (MDAN) as an enhancement of the DNN model. Therefore, we first evaluate the performance of the DNN model trained using Mean Squared Error (MSE) loss as the baseline. Additionally, we propose improvements in the loss function and compare the DNN models trained using the RankSim [5] and ZILN [12] loss functions. Finally, we replicate and compare the performance of the state-of-the-art LTV prediction framework, namely ODMN [8].

4.3 Performance Comparison

This section provides a comparative analysis of the performance of different models on three datasets. In terms of evaluation metrics, priority is given to the GINI coefficient, as our application scenario prioritizes the acquisition of high-quality users and emphasizes sorting performance over accuracy performance.

Results on QB Dataset: Table 1 presents the performance evaluation of our proposed MDAN model in comparison to other existing model structures based on various metrics. Results demonstrate that MDAN model significantly outperforms other models in terms of the GINI coefficient, exhibiting a relative improvement of 38% compared to DNN, whereby the former has increased from 0.3295 to 0.4533. Moreover, MDAN outperforms ODMN, indicating that our

Table 1. The comparative results of different models on three different dataset

DATASET	METHOD	GINI↑	SC↑	DM↓
QB	DNN	0.3295	0.1913	61%
	$DNN + \mathcal{L}_{RankSim}$	0.3983	0.2189	52%
	$DNN + \mathcal{L}_{ZILN}$	0.3813	0.2046	20%
	$ODMN$	0.3982	0.2105	100%
	$MDAN$	**0.4533**	**0.2197**	**16%**
YYB	DNN	0.2765	0.015	62%
	$DNN + \mathcal{L}_{RankSim}$	0.2907	0.0178	69%
	$DNN + \mathcal{L}_{ZILN}$	0.3097	0.0623	83%
	$ODMN$	0.0669	**0.0745**	113%
	$MDAN$	**0.4246**	0.0426	**42%**
WeSing	DNN	0.7556	0.0694	72%
	$DNN + \mathcal{L}_{RankSim}$	0.7702	0.0714	79%
	$DNN + \mathcal{L}_{ZILN}$	0.7588	0.0942	76%
	$ODMN$	0.7099	0.1015	183%
	$MDAN$	**0.8350**	**0.1098**	**47%**

network structure is the optimal choice for LTV ranking. Additionally, the GINI coefficients of $DNN + \mathcal{L}_{RankSim}$, $DNN + \mathcal{L}_{ZILN}$, and $ODMN$ are comparable, suggesting that these three models possess similar ranking capabilities on the QB dataset. Our model also demonstrates remarkable performance in predicting LTV at each level, as evidenced by the DM metric, in which ODMN achieves the best performance (16%). Lastly, MDAN exhibits a noteworthy improvement in the SC metric.

Results on YYB Dataset: Table 1 also presents empirical evidence that our proposed method performs well on the YYB dataset. Specifically, the GINI and DM evaluation metrics for MDAN are superior, with a marked improvement observed. The GINI coefficient increases from 0.2765 to 0.4246. Conversely, ODMN's GINI coefficient is only 0.0669, indicating that ODMN is unsuitable for training on the YYB dataset due to its pronounced bias, which leads to the failure of ODMN's classifier training and poor overall model ranking performance. Additionally, ODMN's DM metric performs poorly across all three datasets, suggesting that ODMN's LTV stratification ability is weak, likely due to the significant bias of the dataset, which results in the failure of the classifier network to learn. In contrast, our proposed ODMN still exhibits favorable performance on the DM metric, highlighting the generalizability of our network architecture.

Results on WeSing Dataset: The final WeSing dataset displays highly skewed characteristics, with the vast 98% of users having an LTV of 0. Nevertheless, our proposed method exhibits superior ranking performance, as demonstrated by significant improvements in both GINI and SC. The DM metric of MDAN also outperform the other four models. Furthermore, apart from MDAN, the DM metric for the WeSing dataset surpasses 70%, underscoring the increased complexity in accurately predicting LTV as dataset bias becomes more prominent.

Table 2. Ablation study of our model on YYB dataset.

METHOD	GINI↑	SC↑	DM↓
DNN	0.2765	0.015	62%
$DNN + \mathcal{L}_{DisSim}$	0.3388	0.0162	65%
MCN	0.3504	0.014	55%
$MCN + \mathcal{L}_{bucket}$	0.3963	0.0251	59%
$MDAN$	**0.4246**	**0.0426**	**42%**

4.4 Ablation Study

We evaluate the effect of each innovative component of our proposed model through ablation experiments. Our network structure is based on DNN. Thus, we mainly evaluate the optimization effect of Distance Similarity Loss, multi-channel network, and bucket classification loss. Five variants of our approach are compared, including:

(A) DNN, which trains a DNN using the MSE loss;
(B) $DNN+\mathcal{L}_{DisSim}$, which trains a DNN using the proposed Distance Similarity loss and the MSE loss;
(C) MCN, which trains the proposed multi-channel network structure using the MSE loss;
(D) $MCN + \mathcal{L}_{bucket}$, which trains the proposed multi-channel network structure using the proposed Channel Learning Controller loss and the MSE loss;
(E) $MDAN$, which trains the proposed multi-channel network structure using the proposed Channel Learning Controller loss, the proposed Distance Similarity loss and the MSE loss.

In Table 2, it can be concluded that the design of each module contributes to the performance improvement of the model, as evidenced by the comparison of performance metrics across different models. Specifically, the GINI coefficient indicates that the effect of the multi-channel network structure is greater than that of the Distance Similarity loss when comparing models (A), (B), and (C). Furthermore, when comparing models (C) and (D), it is evident that the use of a controller bucket classification loss as an auxiliary training method for the multi-channel network structure leads to a more significant improvement in performance, indicating that the controller bucket classification loss enhances the ability of the multi-channel network to learn the respective distributions. Finally, the comparison between models (D) and (E) suggests that the regression debiasing loss and the multi-channel network structure are not redundant, and their effects can be combined to optimize the overall performance of the model.

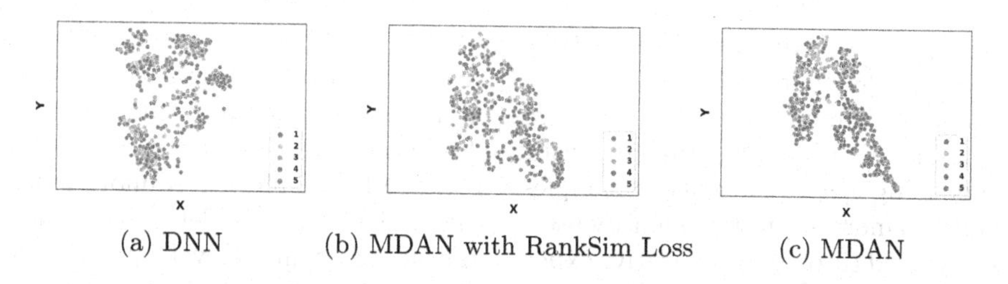

(a) DNN (b) MDAN with RankSim Loss (c) MDAN

Fig. 3. The Embeddings distribution diagram of different models. The points in different colors represent different buckets of LTV. Five buckets are mainly set here, and its numbers are set from small to large according to the value of real LTV.

4.5 Embeddings Distribution Analysis

In this part, we illustrate the mapping of high-dimensional embeddings from various models onto a two-dimensional space. To achieve this, we utilize Stochastic Neighbor Embedding (TSNE), which transforms similarity between data points into a joint probability and minimizes the KL divergence between the high-dimensional and low-dimensional embedded data. We conduct experiments on the QB dataset by randomly and uniformly sampling 500 instances, and sorting each instance based on its actual LTV value, assigning a bucket label from 1 to 5 accordingly. We subsequently apply TSNE to cluster and reduce each embedding, producing two-dimensional data that is visualized on a coordinate system. The resulting plot uses distinct color-coding to represent different LTV buckets.

As illustrated in Fig. 3(a), the two-dimensional distribution of the embedding generated by the DNN model appears to be scattered and devoid of any discernible pattern, suggesting that the simple DNN model struggles to effectively learn the embedding distribution of various LTVs. In contrast, the two-dimensional representation output by MDAN, which replaces the DisSim Loss with the RankSim Loss function (Fig. 3(b)), displays a greater degree of regularity compared to the DNN model, with blue and purple points primarily clustered in the lower right quadrant of the figure. Notably, the embedding generated by our MDAN model (Fig. 3(c)) exhibits the highest degree of regularity in two-dimensional space, with the upper portion mainly comprising red and yellow points (i.e., samples with low LTV value) and the lower portion predominantly consisting of blue and purple points (i.e., samples with high LTV value). This suggests that the proposed MDAN model possesses strong feature representation capabilities and can effectively differentiate between users with varying LTV levels.

5 Conclusion

This paper presents a novel Multi-Distribution Adaptive Networks (MDAN) model for LTV prediction, along with innovative solutions to address both classification and regression debiasing challenges. To address the classification debiasing issue, we propose a multi-channel network architecture that enables each channel to learn different LTV distributions. We also introduce a channel learning controller that allows each bucket of LTV to adaptively output in the corresponding channel. To resolve the regression debiasing issue, we introduce a new regression debiasing loss, named Distance Similarity Loss, which can reduce the embedding space similarity of samples with large LTV differences, allowing the model to more accurately identify users with high LTV. The model's effectiveness is validated through comparative experiments on three million-level datasets, and visualizations demonstrate its interpretability. Additionally, we plan to make our datasets publicly available to encourage further research in this domain.

References

1. Borle, S., Singh, S.S., Jain, D.C.: Customer lifetime value measurement. Manage. Sci. **54**(1), 100–112 (2008)
2. Chamberlain, B.P., Cardoso, A., Liu, C.B., Pagliari, R., Deisenroth, M.P.: Customer lifetime value prediction using embeddings. In: Proceedings of the 23rd ACM SIGKDD International Conference on Knowledge Discovery and Data Mining, pp. 1753–1762 (2017)
3. Drachen, A., et al.: To be or not to be... social: incorporating simple social features in mobile game customer lifetime value predictions. In: Proceedings of the Australasian Computer Science Week Multiconference, pp. 1–10 (2018)
4. Fader, P.S., Hardie, B.G., Lee, K.L.: RFM and CLV: using ISO-value curves for customer base analysis. J. Mark. Res. **42**(4), 415–430 (2005)
5. Gong, Y., Mori, G., Tung, F.: RankSim: ranking similarity regularization for deep imbalanced regression. arXiv preprint arXiv:2205.15236 (2022)
6. Jain, D., Singh, S.S.: Customer lifetime value research in marketing: a review and future directions. J. Interact. Mark. **16**(2), 34–46 (2002)
7. Kumar, V., Reinartz, W.: Customer Relationship Management. Springer, Heidelberg (2018). https://doi.org/10.1007/978-3-662-55381-7
8. Li, K., Shao, G., Yang, N., Fang, X., Song, Y.: Billion-user customer lifetime value prediction: an industrial-scale solution from Kuaishou. In: Proceedings of the 31st ACM International Conference on Information & Knowledge Management, pp. 3243–3251 (2022)
9. Pfeifer, P.E., Carraway, R.L.: Modeling customer relationships as Markov chains. J. Interact. Mark. **14**(2), 43–55 (2000)
10. Sifa, R., Runge, J., Bauckhage, C., Klapper, D.: Customer lifetime value prediction in non-contractual freemium settings: chasing high-value users using deep neural networks and SMOTE (2018)
11. Vanderveld, A., Pandey, A., Han, A., Parekh, R.: An engagement-based customer lifetime value system for e-commerce. In: Proceedings of the 22nd ACM SIGKDD International Conference on Knowledge Discovery and Data Mining, pp. 293–302 (2016)
12. Wang, X., Liu, T., Miao, J.: A deep probabilistic model for customer lifetime value prediction. arXiv preprint arXiv:1912.07753 (2019)
13. Xing, M., et al.: Learning reliable user representations from volatile and sparse data to accurately predict customer lifetime value. In: Proceedings of the 27th ACM SIGKDD Conference on Knowledge Discovery & Data Mining, pp. 3806–3816 (2021)

Author Index

D.-N. Yang et al. (Eds.): PAKDD 2024, LNAI 14647, pp. 421–422, 2024.
https://doi.org/10.1007/978-981-97-2259-4

Printed in the United States
by Baker & Taylor Publisher Services